Landslide Tsunamis: Recent Findings and Research Directions

Edited by
Jean-Pierre Bardet
Costas E. Synolakis
Hugh L. Davies
Fumihiko Imamura
Emile A. Okal

Springer Basel AG

Reprint from Pure and Applied Geophysics
(PAGEOPH), Volume 160 (2003), No. 10-11

Editor:

Jean-Pierre Bardet
University of Southern California
Department of Civil Engineering
Los Angeles, CA 90089-2531
USA
e-mail: bardet@usc.edu

Fumihiko Imamura
Tohoku University
Disaster Control Research Center
Sendai, 980-8579
JAPAN
e-mail: imamura@tsunami2.civil.tohoku.ac.jp

Costas E. Synolakis
University of Southern California
Department of Civil Engineering
Los Angeles, CA 90089-2531
USA
e-mail: costas@usc.edu

Emile A. Okal
Northwestern University
Department of Geological Sciences
Evanston, IL 60208
USA
e-mail: emile@earth.nwu.edu

Hugh L. Davies
University of Papua New Guinea
Department of Geology
POBox: 412
Port Moresby
PAPUA NEW GUINEA
e-mail: hdavies@upng.ac.pg

A CIP catalogue record for this book is available from the Library of Congress, Washington D.C., USA

Bibliographic information published by Die Deutsche Bibliothek
Die Deutsche Bibliothek lists this publication in the Deutsche Nationalbibliografie; detailed bibliographic data is available in the Internet at http://dnb.ddb.de.

ISBN 978-3-7643-6033-7 ISBN 978-3-0348-7995-8 (eBook)
DOI 10.1007/978-3-0348-7995-8

Originally published by Birkhäuser Verlag, Basel, Switzerland in 2003

9 8 7 6 5 4 3 2 1

Contents

Pure appl. geophys. 160 (2003) 1793–1809
0033 – 4553/03/111793 – 17
DOI 10.1007/s00024-003-2406-0

Pure and Applied Geophysics

Landslide Tsunamis: Recent Findings and Research Directions

J.-P. Bardet,[1] C. E. Synolakis,[1] H. L. Davies,[2]
F. Imamura,[3] and E. A. Okal[4]

Abstract—Underwater landslides can trigger local tsunamis with high runup, endangering human life and devastating coastal cities, offshore structures, communication cables, and port facilities. Unfortunately, hazards from underwater landslides are not well understood and the extents of their potential damage remain difficult to ascertain at present. There is immediate need for multidisciplinary research to improve our understanding and plan countermeasures for mitigating their hazards. Conceived in the wake of the Papua New Guinea earthquake landslide and tsunami of 1998, this volume summarizes the state-of-the-art knowledge on underwater landslides and their potential to generate tsunamis from the multidisciplinary perspectives of observational and engineering seismology, geotechnical engineering, marine geology, and hydrodynamics. These various fields of engineering and science offer new synergetic opportunities to examine landslide tsunamis. This paper makes recommendations on future research directions, and will hopefully advance scientists' and engineers' understanding of these natural hazards and assist planners in mitigating their risks.

Key words: Landslides, tsunamis, engineering seismology, marine geology, geotechnical engineering.

1. Background

Tsunamis – a Japanese word translating into "harbor waves" – have inflicted significant damage and casualties along coastlines, even after propagating vast distances across open oceans. In the last sixty years, tsunamis generated by distant earthquakes have killed more people in the United States than earthquakes within the United States. Between 1942 and 1995, tsunamis were responsible for 351 dead, whereas earthquake and tsunamis together caused a total of 627 casualties (Stover and Coffman, 1993; Lander and Lockridge, 1989). In the aftermath of the 1946 Aleutian tsunami, the Pacific Tsunami Warning Center was established with the goal of protecting coastlines from distant (far field) imminent tsunamis. Hazards from

[1] University of Southern California, Civil Engineering Department, Los Angeles, CA 90089-2531, U.S.A.
[2] Department of Geology, University of Papua New Guinea, University Post Office NCD, Port Moresby, Papua New Guinea.
[3] Disaster Control Research Center, Tohoku University, Aoba 06, Sendai 980-8579, Japan.
[4] Department of Geological Sciences, Northwestern University, Evanston, IL 60208-2150, U.S.A.

transoceanic tsunamis are now mitigated by the issuance of early warnings to evacuate coastal areas at risk; these early warnings are based on seismic and tide gage data telemetered to the warning stations, a handful of deep ocean recorders, and simulated scenarios (*e.g.*, PARARAS-CARAYANNIS, 1986). Unfortunately the existing warning systems are not in fact helpful for tsunamis triggered close to populated coastlines, as the propagation times are short from tsunami generation to runup.

Since the 1960s, and particularly after the 1964 Alaskan tsunami, considerable effort has been devoted to understanding tsunamis from generation to coastal inundation (*e.g.*, BEN-MENAHEM and ROSENMAN, 1972; WARD, 1980; OKAL, 1988; SATAKE and KANAMORI, 1991). By 1990, the problems of transoceanic propagation of tsunamis and harbor resonance effects were believed well understood. In the 1990s, computational advances led to better modeling of tectonic source mechanisms and new generations of coastal inundation models (*e.g.*, IMAMURA *et al.*, 1995; TITOV and SYNOLAKIS, 1998). These advances were also made possible due to large-scale laboratory experiments, analytical results, and especially post-tsunami field surveys (*e.g.*, YEH *et al.*, 1993; SATAKE *et al.*, 1993; SYNOLAKIS *et al.*, 1994; TSUJI *et al.*, 1995; KAWATA *et al.*, 1999). Indeed, post-tsunami reconnaissance of coastal inundation after the ten large tsunamis of the 1990s have provided hydrodynamic modelers with opportunities to understand the natural processes at work and to validate computational models. Despite such progress in modeling, some tsunamis are still defying all analyses – underwater landslide tsunamis. Landslide tsunamis present a particularly vexing problem, not only because they can be triggered sometimes unexpectedly after a strong earthquake, but also because the propagation times to the target coastlines can be as short as one minute, as in 1994 at Skagway, Alaska (KULIKOV *et al.*, 1996; SYNOLAKIS *et al.*, 2000).

The potential that major tsunamis could be generated by massive submarine mass failure (a broad geological term that includes underwater slides and slumps, SCHWAB *et al.*, 1993; PRIOR and COLEMAN, 1979) was recognized a century ago by MILNE (1898), MONTESSUS DE BALLORE (1907), and later by GUTENBERG (1939) and AMBRASEYS (1960). In more recent years, many studies have supported the scenario that a major tsunami could be generated by a large submarine mass failure, itself induced or triggered by a large earthquake in a coastal area. In addition to the well documented cases of Grand Banks in 1929 (HASEGAWA and KANAMORI, 1987), Kalapana, Hawaii in 1975 (EISSLER and KANAMORI, 1987), and the ongoing speculation about the great 1946 Aleutian tsunami (KANAMORI, 1985; OKAL, 1992), careful analyses of runup patterns along shorelines often reveal a peaked distribution, with very intense and localized maxima, generally attributed to a local submarine mass failure, against the background of a more regular wave amplitude reflecting the coseismic dislocation. Such observations were reported for localities in Prince William Sound during the great 1964 Alaska earthquake (PLAFKER *et al.*, 1969).

Surprisingly, the significance of these early reports of landslide tsunamis has not been fully appreciated in coastal planning and management. As urban and industrial development has intensified along the world's coastlines, people and facilities have become exposed to greater risks from underwater landslide tsunamis. The U.S. Geological Survey has devoted substantial effort to identify hazards from aerial and underwater landslides and to infer probable areas of historic landslides. Yet considerable portions of the U.S. coastlines remain unmapped with high-resolution imaging techniques, leading to the underappreciation and underestimation of offshore landslide hazards.

Recently, a few well-documented events have helped focus the attention on landslide tsunamis. During the 1992 Flores Island, Indonesia earthquake (YEH *et al.*, 1993), at the village of Riangkroko, far east along the coast from the most affected region, runup was measured at 26 m, the highest anywhere on Flores Island. The waves that utterly destroyed the village of Riangroko and claimed 122 lives probably originated from a nearby underwater landslide. Some researchers (*e.g.*, IMAMURA *et al.*, 1995) hypothesized that the mechanism was probably similar to those observed during the 1964 Good Friday earthquake in Alaska (*e.g.*, LANDER and LOCKRIDGE, 1989; NATIONAL ACADEMY OF ENGINEERING, 1973; PLAFKER *et al.*, 1969). The 1994 Mindoro, Philippines strike-slip earthquake triggered a tsunami that moved a 6,000 ton barge, 1.5 km inland, most probably from a submarine landslide (IMAMURA *et al.*, 1995). These events were not then recognized as the messengers they were, until the 1998 Papua New Guinea (PNG) catastrophe. Shortly after a magnitude 7 earthquake, 10-m waves swept over a sand spit, completely destroying three villages and claiming more than 2200 lives. The field observations – exceedingly large wave amplitude, delayed time of wave arrival, and limited extent of coastal devastation – all pointed to a landslide tsunami. Shortly after the PNG tsunami, the Japan Marine Science and Technology Center and the National Science Foundation funded marine surveys to investigate the offshore geology and its relation to the tsunami source. Since then, the source mechanism of the PNG tsunami has been extensively investigated (*e.g.*, GEIST, 2000; HEINRICH *et al.*, 2000; IMAMURA and HASHI, 2003; KAWATA *et al.*, 1999; KIKUCHI *et al.*, 1998; SATAKE and TANIOKA, 2002; SWEET *et al.*, 1999; SYNOLAKIS *et al.*, 2002; TANIOKA and RUFF, 1998; TANIOKA, 1999; TAPPIN *et al.*, 1999, 2001). The source mechanisms of many other tsunamis have also been retrospectively evaluated to investigate whether or not landslide tsunamis could be possible explanations for unexplained wave measurements and observations. Examples of such investigations include those during the 1992 Flores, Indonesia event (IMAMURA *et al.*, 1995), and during the recent Kocaeli, Turkey earthquake (YALÇINER *et al.*, 1999), and the 1999 Pentecost Island, Vanuatu tsunami (CAMINADE *et al.*, 2001). Landslides can also explain minor tsunamis during strike-slip earthquakes on nearby on-land faults, for example, following the 1989 Loma Prieta earthquake (MA *et al.*, 1991). It is clear that landslide tsunamis could be delayed for a few minutes to a few tens of minutes after the earthquake main events due to the delayed failure of metastable sediment, or to a mild secondary trigger (aftershock) tipping a precarious balance (BJERRUM, 1971; MURTY, 1979; TURNER and SCHUSTER, 1996).

Characteristics of tsunamis generated by coseismic dislocations and mass failures (*e.g.*, landslides) can be compared in general terms by considering the vertical deformation of the seafloor associated with each type of event (SYNOLAKIS *et al.*, 2002). In the case of underwater mass failure, the vertical deformation of the seafloor is controlled by a combination of the dimension of the sliding mass and the motion of its center of mass, both easily reaching hundreds of meters. In the case of coseismic dislocations, a practical upper bound on seafloor deformation during even the largest earthquakes is usually several meters (PLAFKER, 1965; KANAMORI, 1970), exceptionally reaching 20–30 m (PLAFKER and SAVAGE, 1970; KANAMORI and CIPAR, 1974). On the other hand, the linear dimension of an underwater landslide will rarely exceed 100 km (although the resulting turbidity current could extend over a greater distance, *e.g.*, PIPER and AKSU, 1987), while catastrophic earthquakes can rupture faults along close to 1000 km (KANAMORI, 1970). These order-of-magnitude arguments then predict that seafloor dislocations will result in tsunamis featuring greater wavelength and longer periods, and in turn, in a potential for transoceanic devastation, whereas those caused by mass failures are more geographically contained, even though they may give rise to higher amplitudes in the near field (PLAFKER *et al.*, 1969; SCHWAB *et al.*, 1993).

Current research has demonstrated that modeling of landslide tsunami hazards requires information and data from seismology, marine geology, geotechnical engineering and hydrodynamics. The outcomes of hydrodynamic simulations were found to depend largely on the assumptions made on the geological and geotechnical processes governing mass failures. These discoveries raised fundamental issues in the modeling of tsunamis, especially about the prediction of future mass failure events.

2. Objective of Present Volume

In the wake of the devastating 1998 Papua New Guinea tsunami, the editors of the present monograph concerted to organize a comprehensive volume hosted by the journal *Pure and Applied Geophysics*. Some of the papers in this volume originated from a workshop on landslide tsunamis, which was sponsored by the National Science Foundation at the University of Southern California on March 10–11, 2000. The main objective of the workshop was the prediction of the occurrence, location, and dimensions of underwater landslides as well as their tsunamigenic capabilities for continental margins located in earthquake areas. Other specific objectives were (1) to summarize the state-of-the-art in various disciplines dealing with underwater landslides and tsunamis; (2) to document existing case studies on landslide tsunamis; (3) to review the capabilities of existing models; and (4) to provide recommendations to the U.S. National Science Foundation for future research leading eventually to a better assessment of landslide tsunami hazards. The workshop, which was attended by 67 participants, presented interdisciplinary perspectives of engineering seismology, geotechnical engineering, field investigation in marine geology, and hydrodynamic modeling. Underwater landslides were examined from different perspectives, including physical mechanisms, location, size, motion, and implications on wave generation.

3. *Main Findings*

The 19 papers of this volume document the findings on landslide tsunamis in the wake of the devastating 1998 Papua New Guinea tsunami. The papers were edited to remove technical jargon and to facilitate exchanges of information across the multidisciplinary fields that study landslide tsunamis. The papers, which complement each other and present multidimensional facets of the ongoing research on landslide tsunamis, have been regrouped into the following five categories.

3.1 Modeling Earthquake Ground Motions and Source Characteristics

The characterization of earthquake ground motion is critical to predicting landslide-induced tsunamis. Earthquake shakings are regarded as one of the main triggering mechanisms of mass failures along the U.S. West Coast. Strong ground motions may destabilize slopes by overtopping static gravity forces with transient inertial forces, building up pore water pressure within slope materials, and therefore weakening their shear resistance.

SOMERVILLE and GRAVES (2003) review the current concepts in engineering seismology and the recent advances in modeling earthquake strong ground motions. They provide a useful list of state-of-the-art references on the modeling of strong ground motion. They document that earthquake ground motion can be modeled even near fault ruptures and within basins, in areas where submarine landslides are likely to occur. They report on numerical ground motion models based on seismological theory, which include the effects of near-fault rupture directivity effects, crustal waveguide effects, and basin response effects. These models have been validated against recorded ground motions, and used to estimate the ground motions of past earthquakes and predict the ground motions of future scenario earthquakes.

Engineering seismology is essential in modeling landslide tsunamis as it helps to define the seismic conditions (*i.e.*, intensity and duration of earthquake ground motions) that may induce the failure of marine sediments in study areas. Most engineering seismology studies have mainly been focused on the modeling of transient ground motions relevant to structural responses on land; unfortunately, to our knowledge, they have not been calibrated as extensively from strong motion recording data on the seafloor and at the probable depth of tsunami sources. In view of the recent progress in force-balanced strong motion instrumentation, there is hope that new strong motion instruments will be deployed and data collected on seafloors of coastal areas and will in the future yield a better definition of seafloor transient motion.

The modeling of landslide tsunamis requires, in addition to a better characterization of earthquake ground motions, an understanding of earthquake source mechanisms and of the waves originating from mass failures.

HURUKAWA *et al.*, (2003) present a detailed investigation of the three-dimensional distribution of the aftershocks of the 1998 Papua New Guinea earthquake, with the help of data recorded by a portable network of stations deployed in the epicentral area for a period of two months following the mainshock. They conclude that the shallow-dipping plane is indeed the fault plane, thus supporting the interpretation of the event as underthrusting of the Australian plate by the Bismarck Sea platelet, and constraining the geometry of the field of seafloor deformation in any dislocative model of the source of the tsunami.

OKAL (2003a) addresses the question of the hydroacoustic signature of landslides by examining *T*-phase wavetrains recorded across the Pacific Ocean following the 1998 Papua New Guinea earthquake. He identifies one such wavetrain generated 13 minutes after the mainshock and featuring an enhanced duration, as compared to events of similar seismic magnitudes; the signal is also richer at high frequencies (7 Hz) than typical of small aftershocks. These characteristics, interpreted as expressing a different kind of hydroacoustic source, are generally compatible with a sliding or slumping process. This provides a scenario for the generation of the local tsunami by an underwater mass movement, complete with a precise timing, the latter resolving the reported time discrepancy between the mainshock and the onslaught of the tsunami on the local beach. If confirmed in future studies of underwater events, spectral properties of hydroacoustic signals could open – at least in principle – the possibility of identifying major, possibly tsunamigenic, landslides in the hydroacoustic record, along the lines suggested by CAPLAN-AUERBACH *et al.* (2001) for underwater volcanic events in Hawaii.

3.2 Modeling Landslides in Geotechnical Engineering

Geotechnical engineering is a discipline that complements seismology as it explains some of the physical mechanisms causing the failure of submarine masses. Geotechnical engineering has had some prior experience with the relationship between landslide and tsunami (*e.g.*, SEED *et al.*, 1988). This volume contains two papers in geotechnical engineering useful to ascertain the parameters controlling sediment failure, and to identify the geotechnical methodologies available for predicting the instability of inclined seafloors.

WRIGHT and RATHJE (2003) review earthquake-related triggering mechanisms for submarine and shoreline slope failures. Triggering mechanisms range from direct effects, such as inertial forces from earthquake shaking, to indirect effects, such as rapid drawdown that occurs when an earthquake-generated tsunami first approaches a shoreline. Soil shear strength is an important factor in earthquake-related slope failures. Earthquakes change the shear strength of the soil by inducing excess pore water pressures. These excess pore water pressures change with time after the earthquake, resulting in changes in shear strength and slope stability with time. WRIGHT and RATHJE (2003) show that delayed slope failures after an earthquake can occur as a result of changes in earthquake-induced excess pore water pressures and shear strength with time.

FINN (2002) emphasizes the effects of liquefaction on the stability of sediment deposits. With few exceptions, large landslides have occurred in material susceptible to liquefaction, particularly during earthquakes. FINN (2003) examines the case history of the Seward landslide during the 1964 Alaska earthquake, and reviews the analysis methods used for determining the triggering of liquefaction, and the residual strength of liquefied materials. He also reviews numerical methods available for estimating the displacement of liquefied soils, with specific applications to liquefaction-induced ground deformation in gently sloping grounds and earth dams.

3.3 Post-tsunami Field Surveys

Post-tsunamis field surveys are critical to the studies of underwater landslide tsunamis, as they document what actually occurred in the field. Based on the lessons learned from field observations of past landslides, one may hope to infer the physical and mechanical conditions at the onset of failure, and to distinguish the morphologies of stable and unstable seafloors. Marine geological observations on sediment slopes, including boring, failure scars and slumps, can assist in assessing the probability of submarine mass failure. This volume contains five papers related to the field surveys of the 1998 Papua New Guinea (PNG) tsunami.

DAVIES *et al.* (2003) reconstruct the sequence of events during the tsunami attack, based on a comprehensive program of interviews of survivors and field mapping, started in the immediate aftermath of the disaster, and continuing to this date. Their most important results concern the extent of the zone of extreme devastation along the coast (12 km), the quantification of the runup, their interpretation of the largest amplitudes as the result of interaction with the shore, and the constraints put on the timing of the arrival of the wave at the various locations along the shoreline. The compilation of additional information on subjects as diverse as the possible escape of gases before and during the earthquake and tsunami, the nature of the vegetation species most resistant to uprooting by the wave, and the sighting of unusual lights immediately following the main earthquake, raises challenging problems of interpretation, but may eventually shed light on the processes accompanying the destabilization of the ocean floor during a major earthquake and/or landslide.

MATSUMOTO and TAPPIN (2003) report the result of three cruises carried out in the immediate aftermath of the PNG disaster, and aimed at documenting underwater topographic features which may have played a role in the generation of the tsunami. The first one, using *R/V Kairei*, performed a detailed bathymetric survey and identified the major faults in the area, which were found to be too old to have been involved in the 1998 earthquake. The second and third surveys used the Remotely Operated Vehicle *Dolphin-3K* and the submersible *Shinkai 2000* to conduct visual exploration of the ocean floor. These surveys identified a number of slumps, including a very fresh one in the amphitheater, which can be interpreted as the source of the tsunami. Piston core analyses suggest that, at that location, the slumping involved the motion of a cohesive block.

SWEET and SILVER (2003) report the result of their high-resolution two-dimensional seismic reflection survey along the northern continental margin of Papua New Guinea, in the source region of the 1998 tsunami. Their two-dimensional seismic reflection profiles image several faults, bottom simulating reflectors indicative of gas hydrates, and importantly a large rotational slump. One of their profiles displays the internal structure of a slump within a large amphitheater previously recognized in the bathymetry. They conclude that the central part of the slump is a coherent block with parallel reflections that are tilted back toward the head of the slump, indicating rotation. They also infer a basal failure plane at a maximum depth of 760 m below the seafloor, which outcrops at a steep escarpment, about 100 meters high, located 4.5 km north of the head scarp. The cross-sectional area of the displaced mass is 2.3 km^2. From the bathymetry, the width is approximately 2.5–3 km, yielding a total volume of 3.8–4.6 km^3. Based on these interpretations, they reconstitute the undeformed position of the slump, and found that the center of mass dropped 380 m vertically, moved 840 m horizontally and slipped 980 m along the slide plane.

GELFENBAUN and JAFFE (2003) describe the results of a survey of erosion and sedimentation during the 1998 Papua New Guinea tsunami. Based on four transects spread over a 20-km section of coastline, they document erosion of the beach and berm up to 150 m inland, followed by deposition up to 750 m inland, the average thickness of the deposit being 8 cm, with a maximum of 26 cm. They estimate that as much as 2/3 of the deposits were carried from offshore, rather than moved from the beach and berm, a significantly higher fraction than observed for other recent tsunamis (SHI *et al.* 1995; SATO *et al.* 1995), suggestive of a different mode of interaction between the wave and the deep ocean floor.

Finally, DENGLER and PREUSS (2003) report on the social aspects of the 1998 Papua New Guinea disaster, in the general context of the mitigation of tsunami hazard along comparable segments of coastlines. The compilation of the destruction and loss of life in the various villages along the coast is an important step in the reconstruction of the physical agents and mechanisms directly responsible for the destruction. The authors conclude by listing a number of lessons learnt in Papua New Guinea for tsunami mitigation. In particular, and beyond the obvious danger of the narrow sand spits and the value of education and preparedness, they point out the mixed effect of vegetation whose debris can become battering rams when floated by successive waves.

3.4 Studies on Past, Present, and Future Potential Tsunamis

Besides field surveys, there is a great need for studies to reinvestigate past tsunamis (*e.g.*, TANIOKA and SATAKE, 1996; GUIBOURG *et al.* 1997). There is also a need for marine geology studies and pre-event field surveys (LEGG and KAMERLING, 2003) to detect seafloor morphology that may generate tsunamis; a need for computer simulation to understand the formation of seafloor deposits and time-dependent stability (SYVITSKI and HUTTON, 2003); and a need to understand parameters controlling the generation and propagation of the resulting waves (IMAMURA and HASHI, 2003; SATAKE and TANIOKA, 2003) and in particular the influence of frequency dispersion on near-field propagation (LYNETT *et al.* 2003).

LEGG and KAMERLING (2003) remark that there are major potential submarine slides in the California Continental Borderland, which are capable of generating local tsunamis, invoking large seafloor relief, shallow metamorphic basement, and seismic activity. They report on the Catalina Schist basement complex of the Borderland, which contains clear evidence of a history of deep subduction underthrusting and subsequent tectonic exhumation, and displays steep escarpments which provide slip surfaces with free faces susceptible to failure. They describe two slope failure examples west of San Diego, including large block-glides and progressive rotational slumps along the steep northeast-facing Thirtymile Bank escarpment and along the southwest flank of Fortymile Bank.

SYVITSKI and HUTTON (2003) recommend the use of computer simulation to model the fill of sedimentary basins, in order to examine the location and attributes of sediment failure on continental margins and to infer the runout of their associated sediment gravity flows. Based on numerical experiments, they show how the evolving boundary conditions of sea-level fluctuations, floods, storms, and tectonics control the rate and size of slope instabilities. They characterize the failure potential of finite-slope of marine deposits using a factor-of-safety analysis and tracking deposit properties (*i.e.*, pore pressures, grain size, bulk density, and porosity). They are able to determine whether the failed material will travel down slope as turbidity current or a debris flow. Their computational methodology is illustrated with examples explaining the presence of debris flow deposits in fjords dominated by turbidity current deposition, and the seaward progression of glaciated margins. SYVITSKI and HUTTON (2003) synthesize many of the physical processes that shape seafloor morphology, and simulate the formation of observed features of diverse margin evolutions that took place over thousands of years.

IMAMURA and HASHI (2003) present a number of hydrodynamic simulations of the generation and propagation of the 1998 Papua New Guinea tsunami in the near field, based on precise bathymetry even in the shallowest waters. They conclude that the three key observations revealed by the field surveys, *i.e.*, the maximum runup, its distribution along the coast, and the timing of the waves, require supplementing the earthquake with a small, concentrated landslide of 4 to 8 km^3, delayed on the order of 10 minutes after the mainshock. They also conclude that the tsunami generated by the seismic dislocation was locally too small to be observed or reported by the population.

SATAKE and TANIOKA (2003) use hydrodynamic simulations both in the near and far fields to examine several models of tsunami generation following the 1998 Papua New Guinea earthquake. In particular, they model the effects of a splay fault, branching off the fault ruptured during the mainshock, in a geometry reminiscent of FUKAO's (1979) model of the 1975 tsunami earthquake in the Kurile Islands. SATAKE and TANIOKA (2003) conclude that far-field observations cannot resolve a landslide or a splay fault, and can generally be modeled using the earthquake as the source of the tsunami. On the other hand, the near-field runup requires an additional source, which they suggest could be either a landslide or a splay fault. They also estimate the volume of displaced water and its potential energy from the order of magnitude of tsunami amplitudes in the far and near fields, respectively.

LYNETT *et al.*, (2003) examine the influence of both initial conditions and analytical approximations on the results of hydrodynamic simulations. They examine critically the database on the runup and inundation surveyed in Papua New Guinea, and discuss the correct interpretation of vertical measurements as depth flow, rather than runup, under the particular situation of a sand spit. They carry out numerical simulations using both the Non-Linear Shallow Water (NSLW) approximation, and the Boussinesq equations, which assume a less stringent condition on the ratio of water depth to wavelength, and take into account the frequency dispersion of the wave. They conclude that while the amplitude of the free surface disturbance is largely insensitive to the choice of approximation, both its timing and the resulting wave shapes are strongly affected by dispersion. They similarly test several sets of initial conditions, corresponding to different shapes of initial deformation of the sea surface, and conclude to the general robustness of free surface elevation in front of the beach (and eventually of runup).

3.5 Mechanism of Tsunami Generation by Landslides

All submarine landslides do not have identical capabilities of generating tsunamis (WATTS, 2000; WARD, 2001). A critical aspect of landslide tsunami research is to define the circumstances under which landslides may efficiently generate tsunamis. This volume contains four papers that deal with the efficiency of landslides to generate tsunamis.

MURTY (2003) invokes basic physics to scrutinize the validity of hydrodynamic simulations as applied to landslide tsunamis. He reminds us that tsunami generation from submarine landslides depends mainly on the volume of the slide material and to a lesser degree on other factors including: angle of the slide, water depth, density and speed of the slide material, duration of the slide, etc. Based on a data set of volume V of slide and maximum amplitude H of the resulting tsunami waves, he derives a simple linear regression relationship between H and V. He repeats the exercise with another data set from published literature but using numerical simulation results, instead of observations. Surprisingly, he finds a rather poor agreement between the results of numerical simulations and observations. He concludes that the origins of this discrepancy are not clear, but are large enough to justify some cautions when applying hydrodynamic simulation codes to landslide tsunamis.

RUFF (2003) examines energy balances for tsunamis originating from earthquakes and landslides. He notes that submarine landslides have generated tsunami hazards to near-source coasts, and that ongoing exploration of the oceans has revealed large paleo-landslides in many places. He compares the relative contribution of faulting and landslides to tsunami generation. RUFF (2003) points out that only a small fraction of earthquake energy (less than 1%) is converted into tsunami energy, which is anyway enough energy to generate large tsunamis. For simulating landslide tsunamis, he uses a physical model based on a 2-D block sliding on a constant-angle slope, and solves the problem of wave transients using a Green's function approach. His main finding is that landslides generate the largest waves when the ratio of initial water depth (H_0) above the block to downslope vertical drop ($W \sin \theta$) of the block is less than 1. His model predicts that the conversion factor of gravitational energy into tsunami energy varies from 0% (slow-velocity slide in deep water) to about 50% (fast-velocity slide in shallow water). Great earthquakes produce wave amplitudes of a few meters at a wavelength fixed by the fault width in the range of 100 km. Submarine landslides produce wave heights related to block height b, and have wavelengths that scale with block width. In most case scenarios, landslide tsunamis have small wave amplitudes. However, for small friction and $H_0/W \sin \theta < 1$, wave heights can reach b and $b/4$ in the downslope and upslope directions, respectively. For submarine slides, b can have values ranging from a few meters to several kilometers. RUFF (2003) concludes that under some circumstances submarine landslides can generate large tsunamis.

OKAL and SYNOLAKIS (2003) propose a simpler physical model than RUFF (2003). They evaluate and compare the orders of magnitude of the energy generated into a tsunami wave using two different source mechanisms: (1) seismic dislocations and (2) underwater slumps. Based on simple physical models, they conclude that these two sources can generate tsunamis of comparable energy. They remark that the slumping source is dipolar in nature, and has low frequency deficiency in the far field. Their conclusions corroborate the interpretation that the 1998 Papua New Guinea tsunami was indeed generated by an underwater slump, as per SYNOLAKIS *et al.* (2003).

Finally, OKAL (2003b) explores theoretically the differences in far-field tsunami excitation by dislocation and landslide sources, using the normal mode formalism introduced by WARD (1980). He shows that the spectrum of landslide-generated tsunamis is expected to be shifted to relatively high frequencies, where dispersion effects act to further reduce significantly far-field amplitudes. He proposes order-of-magnitude estimates scaling the energy channeled into the tsunami by dislocations and landslides of variable sizes, and suggests that far-field directivity can be a robust discriminant of the nature of the tsunami source, since highly directive radiation patterns cannot be generated by landslides with physically acceptable values of the sliding velocity.

4. Main Observations and Recommendations

Four main consensual points emerge from the papers in this volume:

- Underwater landslide tsunamis are definite hazards to the world coastlines, especially in areas vulnerable to strong groundshaking such as the Pacific Coast of the United States.
- The observations and modeling made in the aftermath of the Papua New Guinea event demonstrate that no comprehensive model covers all aspects of landslide-induced tsunamis from source mechanism to coastal inundation.
- Research must be carried out urgently so as to reduce the risks on the population and industrial developments along the West Coast of the United States.
- Landslide tsunamis require multidisciplinary studies that build upon experiences in engineering seismology, geotechnical engineering, marine geology, modeling of sediment deposition and runoff, and hydrodynamics.

The following recommendations are made to the U.S. National Science Foundation to improve our understanding and predictive capabilities of landslide tsunamis, and to decrease the hazards they pose to population. These recommendations are listed below without having been prioritized.

- Data remain a critical element in evaluating the risks associated with landslide tsunamis. A large amount of data presently collected is relevant to research on underwater landslide hazards. Unfortunately those data are handled by separate and uncoordinated entities, *e.g.*, Navy, Mineral Management Service, US Geological Survey. Coordination and collaboration procedures need to be established to process and archive existing data sets, and make them available to researchers, if possible through the means of the Internet.
- Important lessons have been learned from the post-tsunami field surveys. There is great need to continue the support of International Tsunami Survey Teams in future events.
- There is much to learn from areas prone to underwater landslides before failures actually occur. It is recommended to identify specific sites and to document and instrument those sites. Possible sites could include offshore Santa Barbara, California. Documentation includes bathymetry, seafloor morphology, coring and boring activities, and evaluation of gas hydrate stability. Instrumentation includes sensors measuring water pressures within the sediments, and strong ground motion on seafloor. In those specific sites, hydrodynamic modeling could be used to assess the tsunamigenic potential of these underwater landslides before they actually occur.
- Workshops should be planned in the future, gathering participants from the fields of hydrodynamics, marine geology, earthquake and engineering seismology, and

geotechnical engineering. They should also call upon local, federal and state agencies, oil and gas producers, insurance companies, and operators of port facilities, to present their views and to help formulate global plans of actions to mitigate landslide tsunami hazards.

- There is a need for producing hazard maps that delineate coastal areas hosting potential underwater landslides, and the shorelines that may be at risk from tsunamis generated by these potentially unstable areas. These preliminary, and possibly conservative, assessments of risks associated with underwater landslide tsunamis should assist in planning marine surveys of coastal areas exhibiting tsunamigenic potentials. These hazards maps could be improved based on probability distributions of landslides and tsunami events, and observed landslide and tsunami recurrence rates. Such information will be useful for risk analyses of landslide tsunamis, hazard mitigation and preparedness countermeasures, and port survivability studies.
- Research on underwater landslide hazards is a new interdisciplinary discipline, which involves several existing programs at NSF. An effective research program on the mitigation of underwater landslide hazards requires a concerted contribution from concerned NSF programs and other federal and states agencies.

5. Conclusions

Landslide tsunamis pose definite hazards to populations and industrial developments, notably along the West Coast of the United States. These hazards are not hypothetical but real as demonstrated by the 1998 Papua New Guinea catastrophe. Unfortunately, they are not fully understood and their potential cannot be as yet adequately ascertained. There is an immediate need for research to be coordinated and performed in order to improve our understanding of these phenomena and to plan countermeasures for mitigating their effects.

Intended to serve as a platform to build upon, this volume summarizes the state-of-the-art knowledge on underwater landslide tsunami hazards from the perspectives of earthquake and engineering seismology, geotechnical engineering, marine geology, and hydrodynamics. These various fields of engineering and science offer new synergetic opportunities for improving our understanding of landslide tsunamis. We hope that this volume will help to assemble interdisciplinary interactions and collaborations and will play a role in advancing our understanding of landslide tsunamis, and thus contribute to building a safer coastal environment.

Acknowledgements

The landslide tsunami workshop was funded by a grant from the U.S. National Science Foundation (CMS-9981789) to J.-P. Bardet (PI) and C.E. Synolakis (co-PI).

The Editors wish to thank Dr. Clifford Astill of the National Science Foundation for his guidance, Dr. Philip Watts for his help in organizing the workshop, Prof. T. Tobita for his assistance in preparing the workshop website, and finally all the workshop participants for their enthusiastic contribution during the workshop.

References

AMBRASEYS, N. N. (1960), *The Seismic Sea Wave of July 9, 1956 in the Greek Archipelago*, J. Geophys. Res. *65*, 1257–1265.

BEN-MENAHEM, A. and ROSENMAN, M. (1972), *Amplitude Patterns of Tsunami Waves from Submarine Earthquakes*, J. Geophys. Res. *77*, 3097–3128.

BJERRUM, L. (1971), *Subaqueous Slope Failures in Norwegian Fjords*, Nor. Geotech. Inst. Bull. *88*, 1–8.

BUGGE, T., BELDERSON, R. H., and KENYON, N. H. (1987), *The Storegga Slide*, Phil. Trans. R. Soc. London A *325*, 357–388.

CAMINADE, J. P., CHARLIE, D., KANOĞLU, U., KOSHIMURA, S., MATSUTOMI, H., MOORE, A., RUSCHER, C., SYNOLAKIS, C. E., and TAKAHASHI, T. (2001), *Vanuatu Earthquake and Tsunami Cause Much Damage, Few Casualties*, EOS, Trans. Am. Geophys. Union *81*(52), 641, 646–647.

CAPLAN-AUERBACH, J., FOX, C. G., and DUENNEBIER, F. K. (2001), *Hydroacoustic Detection of Submarine Landslides on Kilauea Volcano*, Geophys. Res. Lett. *28*, 1811–1813.

DAVIES, H. L. (1998), *The Sissano Tsunami*, Port Moresby: University of Papua New Guinea.

DAVIES, H. L., DAVIES J. M., PEREMBO, R. C. B., and LUS, W. Y. (2003), *The Atape 1998 Tsunami: Reconstructing the Event from Interviews and Field Mapping*, Pure Appl. Geophys. this volume.

DAWSON, A. G., LONG, D., and SMITH, D. E. (1988), *The Storegga Slides: Evidence from Eastern Scotland for a Possible Tsunami*, Marine Geology *82*, 271–276.

DENGLER, L. and PREUSS, J. (2003), *Mitigation Lessons from the July 17, 1998 Papua New Guinea Tsunami*, Pure Appl. Geophys. *160*, 2001–2031.

EISSLER, H. K. and KANAMORI, H. (1987), *A Single-force Model for the 1975 Kalapana, Hawaii Earthquake*, J. Geophys. Res. *92*, 4827–4836.

FINN, W. D. L. (2003), *Landslide Generated Tsunamis: Geotechnical Considerations*, Pure Appl. Geophys. *160*, 1879–1894

FUKAO, Y. (1979), *Tsunami Earthquake and Subduction Processes near Deep Sea Trenches*, J. Geophys. Res. *84*, 2303–2314.

GEIST, E. L. (2000), *Origin of the 17 July 1998 Papua New Guinea Tsunami: Earthquake or Landslide*? Seismol. Res. Lett. *71*, 344–351.

GELFENBAUM, G. and JAFFE, B. (2003), *Erosion and Sedimentation from the 17 July 1998 Papua New Guinea Tsunami*, Pure Appl. Geophys. *160*, 1969–1999.

GUIBOURG, S., HEINRICH, P., and ROCHE, R. (1997), *Numerical Modeling of the 1995 Chilean Tsunami; Impact on French Polynesia*, Geophys. Res. Lett. *24*, 775–778.

GUTENBERG, B. (1939), *Tsunamis and Earthquakes*, Bull. Seismol. Soc. Am. *29*, 517–526.

HEINRICH, P. (1992), *Non-Linear Water Waves Generated by Submarine and Aerial Landslide*, J. Wtrwy Port Coast. Ocean Engng. *118*, 249–266.

HEINRICH, P., PIATANESI, A., OKAL, E. A., and HÉBERT, H. (2000), *Near-Field Modeling of the July 17, 1998 Tsunami in Papua New Guinea*, Geophys. Res. Lett. *27*, 3037–3040.

HURUKAWA, N., TSUJI, Y., and WALUYO, B. (2003), *The 1998 Papua New Guinea Earthquake and its Fault Plane Estimated from Relocated Aftershocks*, Pure Appl. Geophys. *160*, 1829–1841.

IMAMURA, F. and HASHI, K. (2002), *Re-examination of the Tsunami Source of the 1998 Papua New Guinea Earthquake Tsunami*, Pure Appl. Geophys. *160*, 2071–2086.

IMAMURA, F., GICA, E., TAKAHASHI, T., and SHUTO, N. (1995), *Numerical Simulation of the 1992 Flores Tsunami: Interpretation of Tsunami Phenomena in Northeastern Flores Island and Damage at Babi Island*, Pure Appl. Geophys. *144*, 555–568.

IMAMURA, F., SYNOLAKIS, C. E., TITOV, V., and LEE, S. (1995), *Field Survey of the 1994 Mindoro Island, Philippines Tsunami*, Pure Appl. Geophys. *144*, 875–890.

JIANG, L. and LEBLOND, P. H. (1992), *The Coupling of a Submarine Slide and the Surface Wave which it Generates*, J. Geophys. Res. *97*, 12,731–12,744.

KANAMORI, H. and CIPAR, J. J. (1974), *Focal Process of the Great Chilean Earthquake, May 22, 1960*, Phys. Earth Planet. Inter. *9*, 128–136.

KANAMORI, H. (1970), *The Alaska Earthquake of 1964: Radiation of Long-Period Surface Waves and Source Mechanism*, J. Geophys. Res. *75*, 5029–5040.

KANAMORI, H. (1972), *Mechanisms of Tsunami Earthquakes*, Phys. Earth Planet. Inter. *6*, 346–359.

KANAMORI, H. (1985), *Non-double-couple seismic source*. In *Proc. 23rd Gen. Assemb. Int. Ass. Seism. Phys. Earth Inter.* Tokyo, p. 425 (abstract).

KAWATA, Y., BENSON, B. C., BORRERO, J. C., BORRERO, J. L., DAVIES, H. L., DE LANGE, W. P., IMAMURA, F., LETZ, H., NOTT, J., and SYNOLAKIS, C. E. (1999), *Tsunami in Papua New Guinea was as Intense as First Thought*, EOS. Trans. Am. Geophys. Union *80*, 101, 104–105.

KIKUCHI, M., YAMANAKA, Y., ABE, K., MORITA, Y., and WATADA, S. (1998), *Source Rupture Process of the Papua New Guinea Earthquake of July 17, 1998 Inferred from Teleseismic Body Waves*. EOS, Trans. Am. Geophys. Union *79, (45)* F573. (abstract).

KULIKOV, E. A., RABINOVICH, A. B., THOMSON, R. E., and BORNHOLD, B. D. (1996), *The Landslide Tsunami of November 3, 1994, Skagway Harbor, Alaska*, J. Geophys. Res. *101*, 6609–6615.

LANDER, J. F. and LOCKRIDGE, P. A. (1989), *United States Tsunamis*. Publication 41-2. U.S. Department of Commerce.

LEGG, M. R. and KAMERLING, M. J. (2003), *Large-scale Basement-involved Landslides, California Continental Borderland*, Pure Appl. Geophys. *160*, 2033–2051.

LYNETT, P. J., BORRERO, J. C., LIU, P. L.-F., and SYNOLAKIS, C. E. (2003), *Field Survey and Numerical Simulations: A Review of the 1998 Papua New Guinea Tsunami*, Pure Appl. Geophys. *160*, 2119–2146.

MA, K.-F., SATAKE, K., and KANAMORI, H. (1991), *The Origin of the Tsunami Excited by the 1989 Loma Prieta Earthquake: Faulting or Slumping*? Geophys. Res. Lett. *18*, 637–640.

MATSUMOTO, T. and TAPPIN, D. R. (2003), *Possible Coseismic Large-scale Landslide off the Northern Coast of Papua New Guinea in July 1998. Geophysical and Geological Results from SOS Cruises*, Pure Appl. Geophys. *160*, 1923–1943.

MATSUYAMA, M., WALSH, J. P., and YEH, H. (1999), *The Effect of Bathymetry on Tsunami Characteristics at Sissano Lagoon, Papua New Guinea*, Geophys. Res. Lett. *26*, 3513–3516.

MILNE, J., *Earthquakes and Other Earth Movements* (Paul, Trench, Trübner & Co., London 1898).

MONTESSUS DE BALLORE, F., *La Science Séismologique* (A. Colin, Paris 1907).

MURTY, T. S. (1979), *Submarine Slide-generated Water Waves in Kitimat Inlet, British Columbia*, J. Geophys. Res. *84*, 7777–7779.

MURTY, T. S. (2002), *Tsunami Wave Height Dependence on Landslide volume*, Pure Appl. Geophys. *160*, 2147–2153.

NATIONAL ACADEMY OF SCIENCES, *The Great Alaska earthquake of 1964*. In Eight Volumes, *Committee on the Alaska earthquake of the Div. of Earth Sciences, National Research Council* (National Acad. of Sciences, Washington, D.C. 1973), 291 pp.

OKAL, E. A. (1988), *Seismic Parameters Controlling Far-field Tsunami Amplitudes: A Review*, Natural Hazards *1*, 67–96.

OKAL, E. A. (1992), *Use of the Mantle Magnitude M_m for the Reassessment of the Seismic Moment of Historical Earthquakes. I: Shallow Events*, Pure Appl. Geophys. *139*, 17–57.

OKAL, E. A. (2003a), *T Waves from the 1998 Papua New Guinea Earthquake and its Aftershocks: Timing the Tsunamigenic Slump*, Pure Appl. Geophys. *160*, 1843–1863.

OKAL, E. A. (2003b), *Normal Modes Energetics for Far-field Tsunamis Generated by Dislocations and Landslides*, Pure Appl. Geophys. *160*, 2189–2221.

OKAL, E. A. and SYNOLAKIS, C. E. (2003), *Theoretical Comparison of Tsunamis from Dislocations and Landslides*, Pure Appl. Geophys. *160*, 2177–2188.

PARARAS-CARAYANNIS, G. (1986), *The Pacific Tsunami Warning System*, Earthquakes and Volcanoes *18*, 122–130.

PELINOVSKY, E. and POPLAVSKY, A. (1996), *Simplified Model of Tsunami Generation by Submarine Landslides*, Phys. Chem. Earth *21*, 13–17.

PIPER, D. J. W. and AKSU, A. E. (1987), *The Source and Origin of the 1929 Grand Banks Turbidity Current Inferred from Sediment Budgets*, Geol. Mar. Lett. *7*, 177–182.

PLAFKER, G. (1965), *Tectonic Deformation Associated with the 1964 Alaskan Earthquake*, Science *148*, 1675–1687.

PLAFKER, G. (1972), *Alaska Earthquake of 1964 and Chilean Earthquake of 1960: Implications for Arc Tectonics*, J. Geophys. Res. *77*, 901–925.

PLAFKER, G., KACHADOORIAN, R., ECKEL, E. B., and MAYO, L. R. (1969), *Effects of the Earthquake of March 27, 1964 on Various Communities*, US Geol. Surv. Prof. Paper 542-G, US Geological Survey, Washington, DC.

PLAFKER, G. and SAVAGE, J. C. (1970), *Mechanism of the Chilean Earthquakes of May 21 and 22, 1960*, Geol. Soc. Am. Bull. *81*, 1001–1030.

PRIOR, D. B. and COLEMAN, J. M. (1979), *Submarine Landslides: Geometry and Nomenclature*, Z. Geomorphol. *23*, 415–426.

RUFF, L. J. (2003), *Some Aspects of Energy Balance and Tsunami Generation by Earthquakes and Landslides*, Pure Appl. Geophys. *160*, 2155–2176.

SATAKE, K. and KANAMORI, H. (1991), *Use of Tsunami Waveforms for Earthquake Source Study*, Natural Hazards *4*, 193–208.

SATAKE, K. and TANIOKA, Y. (2003), *The July 1998 Papua New Guinea Earthquake: Mechanism and Quantification of Unusual Tsunami Generation*, Pure Appl. Geophys. *160*, 2087–2118.

SATAKE, K., BOURGEOIS, J., ABE, KU., ABE, KA., TSUJI, Y., IMAMURA, F., IIO, Y., KATAO, H., NOGUERA, E., and ESTRADA, F. (1993), *Tsunami Field Survey of the 1992 Nicaragua Earthquake*, EOS, Trans. Am. Geophys. Union, *74*, 145, 156–157.

SATO, H., SHIMAMOTO, T., TSUTSUMI, A., and KAWAMOTO, E. (1995), *Onshore Tsunami Deposits Caused by the 1993 Southwest Hokkaido and 1983 Japan Sea Earthquakes*, Pure Appl. Geophys. *144*, 693–717.

SCHWAB, W. C., LEE, H. J., and TWICHELL, D. C. (eds), *Submarine Landslides: Selected Studies in the US Exclusive Economic Zone*, US Geol. Surv. Bull. B-2002 (Washington, DC: US Geological Survey. 1993).

SEED, H. B., SEED, R. B., SCHLOSSER, F., BLONDEAU, F., and JURAN, I. (1988), Report no. UCB/EERC. 88/110. *Earthquake Engineering Research Center*, University of California, Berkeley, CA.

SHI, S., DAWSON, A. G., and SMITH, D. E. (1995), *Coastal Sedimentation Associated with the December 12th, 1992 Tsunami in Flores, Indonesia*, Pure Appl. Geophys. *144*, 525–536.

SOMERVILLE, P. G. and GRAVES, R. W. (2003), *Characterization of Earthquake Strong Ground Motion*, Pure Appl. Geophys. *160*, 1811–1828.

STOVER, C. W. and COFFMAN, J. L. (1993), *Seismicity of the United States, 1568–1989*, U.S. Geol. Survey Prof. Paper, 1527, US Govt. Printing Office, Washington, DC, 1993 (and updates for 1994–1995).

SWEET, S. and SILVER, E. A. (2003), *Seismic Reflection Images of the Source Region of the 1998 Papua New Guinea Tsunami*, Pure Appl. Geophys. *160*, 1945–1968.

SWEET, S., SILVER, E. A., DAVIES, H., MATSUMOTO, T., WATTS, P., and SYNOLAKIS, C. E. (1999), *Seismic Reflection Images of the Source Region of the Papua New Guinea Tsunami of July 17, 1998*, EOS, Trans. Am. Geophys. Union *80*, F750 (abstract).

SYNOLAKIS, C. E., BORRERO, J. C., PLAFKER, G., YALÇINER, A., GREENE, G., and WATTS, P. (2000), *Modeling the 1994 Skagway, Alaska Tsunami*, EOS, Trans. Am. Geophys. Union *81*, F748 (abstract).

SYNOLAKIS, C. E., IMAMURA, F., TSUJI, Y., MATSUTOMI, S., TINTI, B., COOK, B., and USHMAN, M. (1995), *Damage, Conditions of East Java Tsunami of 1994 Analyzed*, EOS, Trans. Am. Geophys. Union *76* (26), 257, 261–262.

SYNOLAKIS, C. E., BARDET, J.-P., BORRERO, J., DAVIES, H., OKAL, E., SILVER, E., SWEET, S. and TAPPIN, D. (2002), *Slump origin of the 1998 Papua New Guinea tsunami*. Proc. Roy. Soc. London, Ser. A *458*, 763–789.

SYVITSKI, J. P. M. and HUTTON, E. W. H. (2003), *Failure of Marine Deposits and their Redistribution by Sediment Gravity Flows*, Pure Appl. Geophys. *160*, 2053–2069.

TANIOKA, Y. (1999), *Analysis of the Far-field Tsunamis Generated by the 1998 Papua New Guinea Earthquake*, Geophys. Res. Lett. *26*, 3393–3396.

TANIOKA, Y. and RUFF, L. J. (1998), *The 1998 Papua New Guinea Earthquake. An Outer Rise Event?* EOS, Trans. Am. Geophys. Union, *79*, F572 (abstract).

TANIOKA, Y. and SATAKE, K. (1996), *Fault Parameters of the 1896 Sanriku Tsunami an Outer Rise Event?* Geophys. Res. Lett. *23*, 1549–1552.

TAPPIN, D. R. and 18 others (1999), *Sediment Slump Likely Caused 1998 Papua New Guinea Tsunami*, EOS, Trans. Am. Geophys. Union *80*, 329, 334, 340.

TAPPIN, D. R., WATTS, P., MCMURTRY, G. M., LAFOY, Y., and MATSUMOTO, T. (2001), *The Sissano, Papua New Guinea Tsunami of July 1998 – Offshore Evidence on the Source Mechanism*, Mar. Geol. *175*, 1–23.

TITOV, V. V. and SYNOLAKIS, C. E. (1998), *Numerical Modeling of Tidal Wave Runup*, J. Wtrwy Port Coast. Ocean Engng *124*, 157–171.

TSUJI, Y., IMAMURA, F., MATSUTOMI, H., SYNOLAKIS, C. E., NANANG, P. T., JUMADI, HARADA, S., HAN, S. S., ARAI, K., and COOK, B. (1995), *Field Survey of the East Java Earthquake and Tsunami of June 3, 1994*, Pure Appl. Geophys. *144*, 839–854.

TURNER, A. K. and SCHUSTER, R. L. (eds.) (1996), Special Report 247, Transportation Research Board, Washington, DC.

U.S. GEOLOGICAL SURVEY (1966–1970), *The Alaska Earthquake March 27, 1964 Series*, Geological Survey Professional Papers 542, 543, 544, 545, and 546, U.S. Dept. of Interior, USGS.

WARD, S. N. (1980), *Relationships of Tsunami Generation and an Earthquake Source*, J. Phys. Earth *28*, 441–474.

WARD, S. N. (2001), *Landslide Tsunami*, J. Geophys. Res. *106*, 11,201–11,215.

WATTS, P. (2000), *Tsunami Features of Solid Block Underwater Landslides*, J. Wtrwy Port Coast. Ocean Engng. *126*, 144–152.

WRIGHT, S. G. and RATHJE, E. M. (2002), *Triggering Mechanisms of Slope Instability and their Relationship to Earthquakes and Tsunamis*, Pure Appl. Geophys. *160*, 1865–1878.

YALÇINER, A. C., BORRERO, J. C., KANOĞLU, U., WATTS, P., SYNOLAKIS, C. E., and IMAMURA, F. (1999), *Field Survey of the 1999 Izmit Tsunami and Modeling Effort of New Tsunami Generation Mechanism*. EOS, Trans. Am. Geophys. Union *80*, F751 (abstract).

YEH, H., IMAMURA, F., SYNOLAKIS, C. E., TSUJI, Y., LIU, P., and SHI, S. (1993), *The Flores Island Tsunami*, EOS, Trans. Am. Geophys. Union *74*(33), 369, 371–373.

Pure appl. geophys. 160 (2003) 1811–1828
0033–4553/03/111811–18
DOI 10.1007/s00024-003-2407-z

Pure and Applied Geophysics

Characterization of Earthquake Strong Ground Motion

P. G. Somerville[1] and R. W. Graves[1]

Abstract—Some underwater landslides are triggered by strong ground motions caused by earthquakes. This paper reviews current concepts and trends in the characterization of strong ground motion. Improved empirical ground motion models have been derived from a strong motion data set that has grown markedly over the past decade. However, these empirical models have a large degree of uncertainty because the magnitude-distance-soil category parameterization of these models often oversimplifies reality. This reflects the fact that other conditions that are known to have an important influence on strong ground motions, such as near-fault rupture directivity effects, crustal waveguide effects, and basin response effects, are not treated as parameters of these simple models. Numerical ground motion models based on seismological theory that include these additional effects have been developed and extensively validated against recorded ground motions, and used to estimate the ground motions of past earthquakes and predict the ground motions of future scenario earthquakes.

Key words: Earthquake source, wave propagation, rupture directivity, basin response, ground motion prediction, numerical simulation.

1. Introduction

This paper reviews current concepts and trends in the characterization of earthquake ground motion, which can trigger underwater landslides. The focus of this paper is on the prediction of ground motions on rock and firm soil. The response to these ground motions of soft soils that overly rock and firm soil is beyond the scope of this paper, however it is addressed by other papers in this volume.

The tectonic environment of the offshore region of Southern California is described by the Working Group on California Earthquake Probabilities (1995). The offshore region is characterized by strike-slip and reverse faulting. Faulting gives rise to volume strains that may influence normal stresses and the hydrological environment (Muir-Wood and King, 1993). Generally, thrust faults give rise to reduction in normal stresses and strike-slip faults give rise to small changes in normal stress. Normal faulting, which generally gives rise to increases in normal stresses that could potentially induce slope instability, is not prevalent in the offshore region.

Earth scientists have a partial understanding of the long-term seismic behavior of some of the more active faults in the southern California region. However, with few

[1] URS Corporation, Pasadena, CA, U.S.A.

exceptions, large earthquakes do not appear to occur at uniform time intervals, or to rupture exactly the same segment of a fault from one earthquake to the next. Consequently, even for sites near the best understood faults, earth scientists are not able to predict where and when the next large earthquake is going to occur, and so a probabilistic approach to seismic hazard evaluation is most compatible with our state of knowledge of seismic potential.

Conventional ground motion models predict ground motion parameters using a simplified model in which the effects of the earthquake source are represented by earthquake magnitude, the effects of wave propagation from the earthquake source to the site region are specified by a distance, and the effects of the site are specified by a site category. These ground motion models have a large degree of uncertainty because other conditions that are known to importantly influence strong ground motions, such as near-fault rupture directivity effects and the response of sedimentary basins, are not treated as parameters of these simple models. In order to reduce the uncertainty in ground motion prediction at a given site, the parameterization of ground motion models is being augmented to include more realistic representations of source, path and site effects.

2. *Earthquake Source Characterization*

In a seismic hazard evaluation for a given site, it is necessary to identify the seismic sources on which future earthquakes are likely to occur, to estimate the magnitudes and frequency of occurrence of earthquakes on each seismic source, and to identify the distance and orientation of each seismic source in relation to the site. If a deterministic approach is used to characterize the ground motions, then a single scenario earthquake is usually used to represent the seismic hazard, and its frequency of occurrence does not directly influence the level of the hazard. If a probabilistic approach is used, then the ground motions from numerous possible earthquakes are considered, and their frequencies of occurrence are key parameters in the analysis.

Earthquake sources may be characterized as discrete faults, or as zones having uniform seismic potential. In tectonically active regions, it is generally possible to identify discrete earthquake faults. The regions surrounding these faults typically have some level of seismic activity, and are characterized by distributed source zones with uniform seismic potential.

The size of the earthquake is characterized by its magnitude or seismic moment. Earthquake size is most conveniently characterized by seismic moment (or moment magnitude), because the seismic moment is directly related to the product of the fault rupture area and average fault displacement. Thus the seismic moment of an earthquake on a given fault can be directly estimated from fault-specific estimates of the fault length, fault width and average displacement. Alternatively, the seismic moment can be estimated from empirical relationships between seismic moment and fault parameters such as fault area, fault length, and average displacement that are based on the average properties of a large number of earthquakes (WELLS and COPPERSMITH, 1994; SOMERVILLE *et al.*, 1999). The most reliable method of estimating seismic moment is from fault area.

In order to identify the distance between the source and the site, the dimensions of the fault for a given magnitude must be specified. Line source representation of distant sources may be adequate, nonetheless the fault dimensions both along strike and downdip may need to be specified for measuring the distance to the site when the source is close to the site. In some ground motion models, additional source parameters specifying the dimensions and geometry of the source, the location of the site in relation to the source, and the location of the hypocenter, are required for represent near-fault effects such as rupture directivity effects and hanging wall effects, which are described later.

If numerical simulation methods are used to estimate the ground motions, then the spatial distribution of slip on the fault and the time function of slip on the fault also needs to be characterized. Detailed studies of the spatial distribution of slip on the fault plane for earthquakes in tectonically active regions, derived from strong motion recordings and other data, have shown that the slip distribution is highly variable, characterized by asperities (regions of large slip) surrounded by regions of low slip. These slip models have been used to develop relationships between seismic moment and a set of fault parameters that are needed for predicting strong ground motions (SOMERVILLE *et al.*, 1999). These parameters include fault length, fault width, rise time (duration of slip at a point on the fault), and the size, slip contrast and location of asperities.

3. Ground Motion Prediction Models

3.1 Empirical Ground Motion Models

The simplest ground motion attenuation relations predict ground motion parameters using a simplified model in which the effects of the earthquake source are represented by earthquake magnitude; the effects of wave propagation from the earthquake source to the site region are specified by a distance; and the effects of the site are specified by a site category.

Moment magnitude is the most appropriate measure of earthquake size, because it is directly related to the seismic moment of the earthquake, and thus to the fault area and the average displacement on the fault. When earthquake occurrence rates are instead based on historical seismicity in seismic hazard analyses, other magnitude measures are sometimes used because the seismic moments of the historical earthquakes may be poorly known.

The distance between the source to the site is measured in a variety of different ways, including closest distance to the rupture surface, closest distance to the seismogenic rupture surface, closest distance to the vertical projection of the rupture to the surface, and hypocentral distance, as summarized by ABRAHAMSON and SHEDLOCK (1997). The lack of a standard definition of distance hinders the comparison and use of multiple ground motion attenuation relations, which is necessary to fully account for uncertainty in estimated ground motions.

Site categories are usually based either on geological criteria or on shear wave velocity of the surficial materials. The use of shear wave velocity has the advantage of being based on an objective measure which affects ground motions in a way that can be modeled. However, it cannot be directly applied to sites that lack shear wave velocity measurements. Also, deeper geological structure such as sedimentary basins and laterally varying structure may have an equally strong or even stronger effect on site response.

A major shortcoming of existing attenuation relations is the lack of a standard definition of site categories. Consequently, a large source of difference in the ground motion models of different investigators lies in the different criteria that are used for site conditions, and in the assignment of site categories to individual recording sites. The use of a standard site classification system, and the assignment of these categories to individual recording stations, have the potential to reduce these unnecessary discrepancies between different model predictions. The important effect of nonlinear soil behavior on site response is recognized in the site response factors that are embodied in current building codes and provisions in the United States such as the 1997 UBC (SEAOC, 1996) and 1997 NEHRP (FEMA, 1998) provisions.

During the past several decades, large sets of strong motion recordings were obtained from numerous earthquakes, significantly expanding the database of strong motion recordings available for the derivation of empirical ground motion models. A collection of recent models based on recorded data and in some cases on seismological models, accompanied by an overview (ABRAHAMSON and SHEDLOCK, 1997), was published in Seismological Research Letters. These ground motion models are for distinct tectonic categories of earthquakes: shallow crustal earthquakes in tectonically active regions; shallow crustal earthquakes in tectonically stable regions; and subduction zone earthquakes. Mixing data from different earthquake categories gives rise to attenuation relations that overpredict the ground motions from some tectonic categories of earthquakes and underpredict those from others.

3.2 Variability in Ground Motions

There is substantial variability in ground motion characteristics due to effects that are more complex than the simple parameterization described above based on magnitude, distance and site category. New methods for analyzing the origins of this variability have provided important insight into the nature of strong ground motions, and indicate the directions in which further research may be able to reduce the uncertainty in the estimation of ground motions for engineering application. Specifically, the random effects approach (ABRAHAMSON and YOUNGS, 1992) has been applied to the strong motion database to separately quantify two sources of variability: the variability in the average ground motions from one earthquake to the next, and the variability in ground motions from one site to another at the same closest distance from a given earthquake. For earthquakes of a given tectonic category larger than about magnitude 6, the event-to-event variability is found to be insignificant compared with the intra-event variability (YOUNGS *et al.*, 1995). The overall variability thus decreases significantly for the larger magnitudes. The decrease in the variability of ground motion amplitudes with increasing magnitude can significantly effect the estimation of ground motions for engineering analysis and design.

This finding indicates that while the average ground motions from one large earthquake are very similar to those of another, there are conditions that cause the ground motions to vary significantly from one location to another at the same distance from a given event. The factors that cause this variability are related to aspects of the earthquake source process, the propagation of seismic waves from the source region to the site region, and the site response that are not contained in the simple magnitude-distance-site category parameterization of standard attenuation relations. In these simple models, these other sources of variability are treated as randomness, whereas they potentially could be treated as resulting from specific effects which may be predictable.

3.3 Enhanced Ground Motion Models

The simple ground motion models described above have a large degree of uncertainty because other conditions that are known to have an important influence on strong ground motions, such as near-fault rupture directivity effects and basin response effects are not treated as parameters of these simple models. In order to reduce the uncertainty in predicting the ground motions at a given site, we need to augment these ground motion prediction models to include more realistic representations of source, path and site effects.

One approach is to use seismological-based ground motion models that take account of the specific source, path and site conditions. These methods can be applied to augment the recorded data used to generate empirical models, or to generate suites of ground motion estimates that can be used to develop independent ground motion models. Also, considerable progress has been made in understanding and predicting ground motion variability, especially at periods longer than about 1 second, through the modeling of specific effects such as rupture directivity and basin response. By incorporating these effects into ground motion models, in addition to the standard parameters of magnitude, distance and site category, we will be able to reduce the uncertainty in the ground motion estimates for a specified site.

3.4 Numerical Ground Motion Models

The broadband Green's function method (*e.g.*, SOMERVILLE *et al.*, 1996) has a rigorous basis in theoretical and computational seismology (HELMBERGER, 1983). The earthquake source is represented as a shear dislocation on an extended fault plane, whose radiation pattern, and its tendency to become subdued at periods shorter than about 0.5 sec., are accurately represented. The spatial and temporal variation of slip of the fault surface is represented using methods described by SOMERVILLE *et al.* (1999). Wave propagation is represented by Green's functions computed for the seismic velocity structure which contains the fault and the site, or by empirical Green's functions derived from strong motion recordings of small earthquakes. These Green's functions contain both body waves and surface waves. The ground motion time history is calculated in the time domain using the elastodynamic representation theorem. This involves integration over the fault surface of the convolution of the slip time function on the fault with the Green's function for the appropriate depth and distance.

An example of broadband simulation of strong ground motions is shown in Figure 1, which compares the recorded and simulated ground motions at Arleta from the 1994 Northridge earthquake. The broadband nature of the simulation is demonstrated by comparing the recorded and simulated acceleration, velocity and displacement time histories. When compared to observed records of actual earthquakes, such ground motion simulation procedures can reproduce the duration, peak accelerations, and short-period (less than 1 sec.) response spectral accelerations within a factor of about 1.5, which is comparable to the uncertainty in empirical ground motion prediction models. For peak velocities and response spectral accelerations at periods longer than 1 sec, the error in prediction by these simulation methods may be lower than that of empirical ground motion models, especially if rupture directivity and basin response affect the data and are included in the predictions. This kind of validation against recorded data constitutes an important criterion for the selection of ground motion simulation procedures for use in earthquake engineering.

New computational procedures have the potential to greatly enhance the reliability of ground motion simulations. It is expected that wave propagation computations using 3-D finite difference and other methods will largely supplant the use of simple 1-D models. The rapid increase in the speed and memory of computers, and the development of efficient computational methods for modeling seismic wave propagation in laterally varying geological structure, enable us to model the effects of sedimentary basins on ground motions generated by scenario earthquakes. By using finite-difference techniques (GRAVES, 1996), we are able to compute the complete wave field up to a given frequency threshold on a dense grid of points. Thus, instead of obtaining ground motion estimates at a set of discrete locations using local flat layered geological models, we are able to calculate the

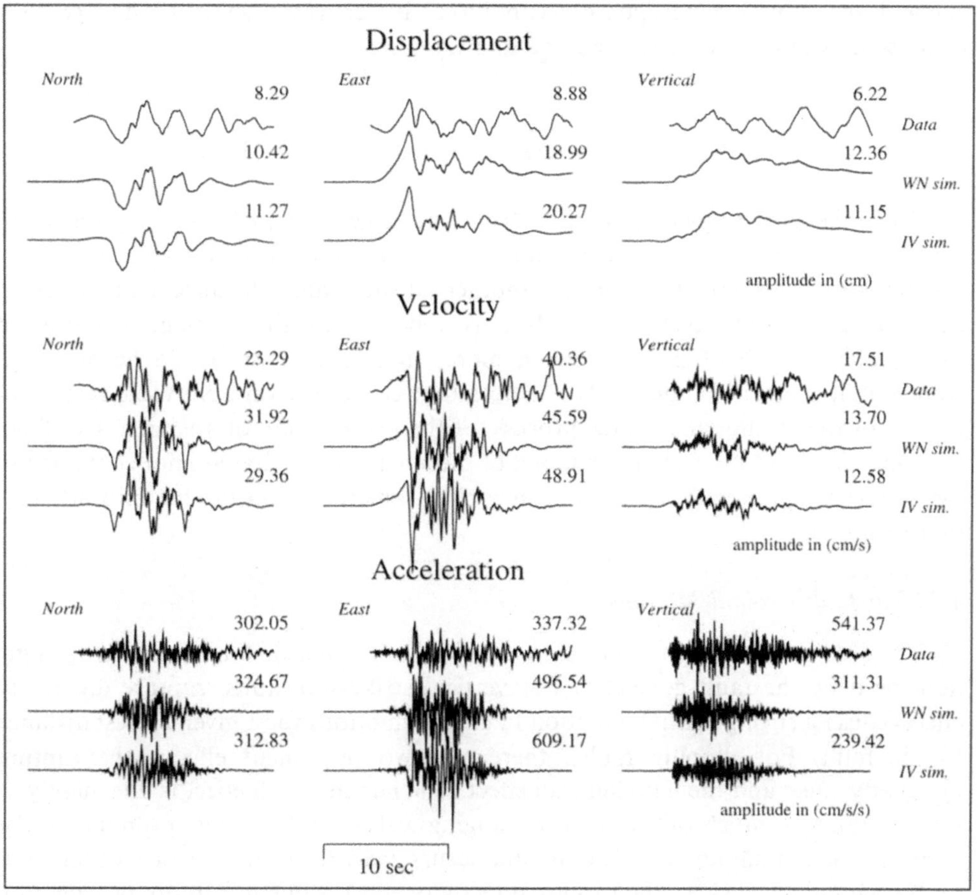

Figure 1
Comparison of recorded (top row) and simulated (middle and bottom rows) displacement, velocity and acceleration time histories at Arleta from the 1994 Northridge earthquake. Traces are plotted on a common scale, with peak value given in the top left.

seismic wave field throughout the region of interest, with the ground motion characteristics at each location reflecting the effects of the laterally varying geologic structure in its vicinity.

Through the use of these seismological models of strong ground motion, the number of parameters used in the estimation of strong ground motion has grown beyond the simple set of magnitude, distance and site category. For example, in recent ground motion models, the source and path parameters allow for differences in the ground motions between the hanging wall and foot wall of dipping faults, and for the spatial variations in ground motions around faults due to rupture directivity effects. Additionally, in place of simple geological descriptors of site conditions or ranges of seismic velocity, the complexity of the interaction of waves arriving at the site with the strongly heterogeneous conditions that characterize the shallow geology of most sites has been recognized and analyzed.

4. Additional Conditions that Influence Ground Motions

While the average ground motions from one large earthquake are very similar to those of another, there are conditions that cause the ground motions to vary significantly from one location to another at the same distance from a given earthquake, as described above. This is reflected in the damage patterns of earthquakes, which often show much more irregularity than would be predicted from simple empirical models. The factors that cause this variability are related to aspects of the earthquake source process, of the propagation of seismic waves from the source region to the site region, and of the site response that are not contained in the simple magnitude-distance-site category parameterization of standard attenuation relations.

4.1 Near-fault Ground Motions

The distribution of ground motion amplitudes in the near-fault region is strongly influenced by the fault geometry. For vertical strike-slip faults, rupture directivity effects cause a strong spatial variation in ground motions for a given closest distance to the fault. For dipping faults, there are two prominent effects: the rupture directivity effect and the hanging wall effect. The hanging wall effect is due mainly to the proximity of much of the fault to hanging wall sites. It is most pronounced for periods shorter than about 1 second, and occurs away from the top edge of the fault on the hanging wall side. The rupture directivity effect is due to rupture propagation and radiation pattern effects. It is most pronounced at periods longer than 1 second, and is concentrated over the top edge of the fault. The relationship between the rupture directivity effect and the hanging wall effect is thus complementary both spatially and in period range, enhancing the degree of spatial variation of strong ground motion around dipping faults.

4.2 Hanging Wall Effects

Sites on the hanging wall of a dipping fault have closer proximity to the fault as a whole than do sites at the same closest distance on the foot wall. This causes larger short-period ground motions on the hanging wall than on the foot wall at the same closest distance. The hanging wall adds to the style of faulting factor for short-period ground motions from reverse and thrust earthquakes, which are typically 1.3 to 1.4 times larger than those from strike-slip earthquakes. The empirical model of hanging wall effects developed by ABRAHAMSON and SOMERVILLE (1996) distinguishes the ground motions for sites on the hanging wall from those on the foot wall and off the end of the fault rupture. The effect is greatest (a factor of 1.45) in the closest distance range of 8 to 18 km in the period range of 0 to 0.6 seconds, and decreases to unity at 5 seconds. This model of hanging wall effects is incorporated into the ground motion model of ABRAHAMSON and SILVA (1997).

4.3 Near-fault Rupture Directivity Effects

The propagation of fault rupture toward a site at a velocity that is almost as large as the shear wave velocity causes most of the seismic energy from the rupture to arrive coherently in a single large long-period pulse of motion which occurs at the beginning of the record. This pulse of motion represents the cumulative effect of most of the seismic radiation from the fault. The radiation pattern of the shear dislocation on the fault causes this large pulse of motion to be oriented in the direction perpendicular to the fault, causing the strike-normal peak velocity to be larger than the strike-parallel peak velocity. The enormous destructive potential of near-fault ground motions was manifested in the 1994 Northridge and 1995 Kobe earthquakes. In each of these earthquakes, peak ground velocities as high as 175 cm/sec. were recorded (Fig. 2). The period of the near-fault pulses recorded in both of these earthquakes lie in the range of 1 to 2 sec, comparable to the natural periods of structures such as bridges and mid-rise buildings, many of which were severely damaged.

Forward rupture directivity effects occur when two conditions are met: the rupture front propagates toward the site, and the direction of slip on the fault is aligned with the site. The conditions for generating forward rupture directivity effects are readily met in strike-slip faulting, where the rupture propagates horizontally along the strike either unilaterally or bilaterally, and the fault slip direction is oriented horizontally in the direction along the strike of the fault. However, not all near-fault locations experience forward rupture directivity effects in a given event. Backward directivity effects, which occur when the rupture propagates away from the site, give rise to the opposite effect: long duration motions having low amplitudes at long periods (Fig. 3).

The conditions required for forward directivity are also met in dip-slip faulting, including both reverse and normal faults. The alignment of both the rupture direction and the slip direction updip on the fault plane produces rupture directivity effects at sites located around the surface exposure of the fault (or its updip projection if it does not break the surface). Consequently, it is generally the case that all sites located near the surface exposure of a dip-slip-fault experience forward rupture directivity when an earthquake occurs on that fault. Unlike the case for strike-slip faulting, where forward rupture directivity effects are expected to be most concentrated away from the hypocenter, dip-slip faulting produces directivity effects on the ground surface that are most concentrated updip from the hypocenter.

The effect of forward rupture directivity on the response spectrum is to increase the level of the response spectrum of the horizontal component normal to the fault

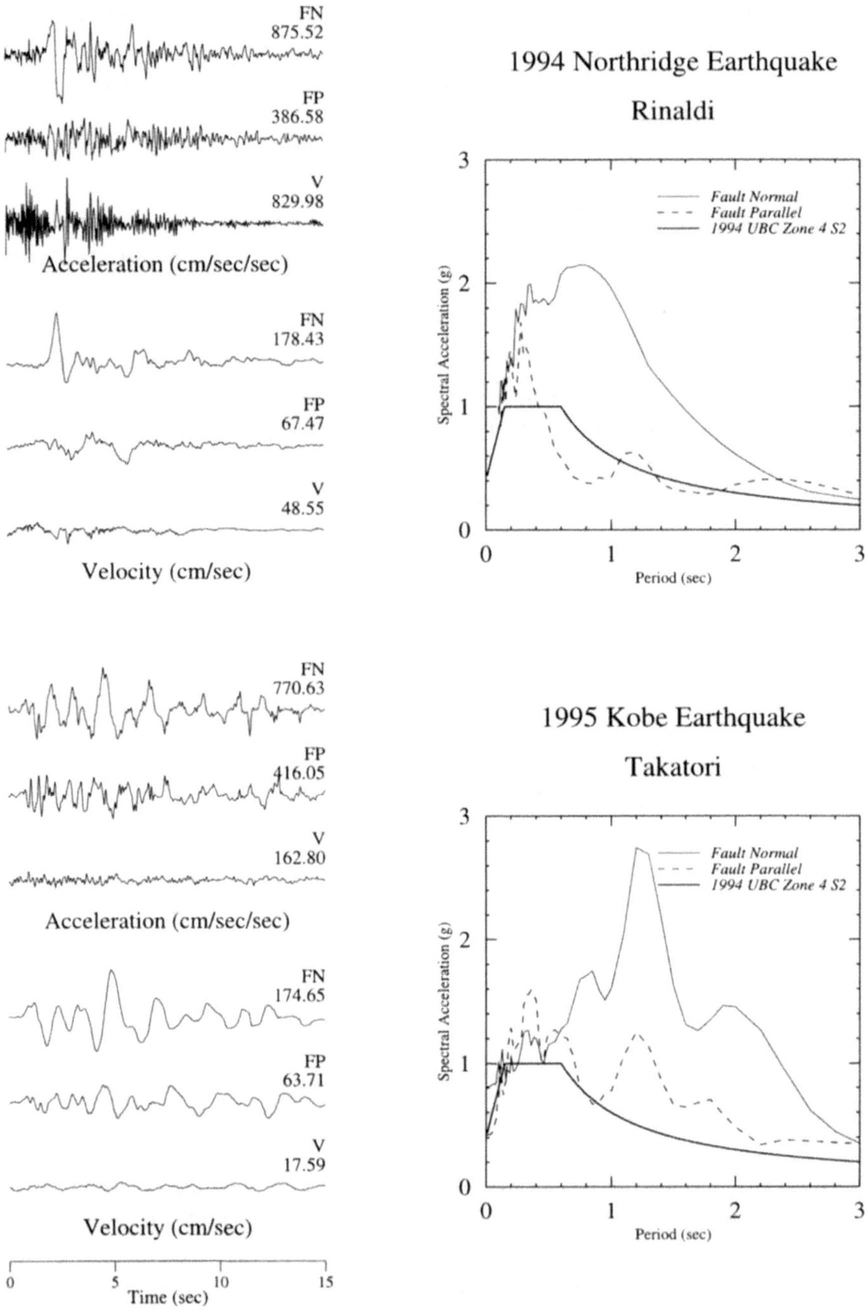

Figure 2
Near-fault acceleration and velocity time histories (left) and response spectra (right) recorded during the 1994 Northridge and 1995 Kobe earthquakes.

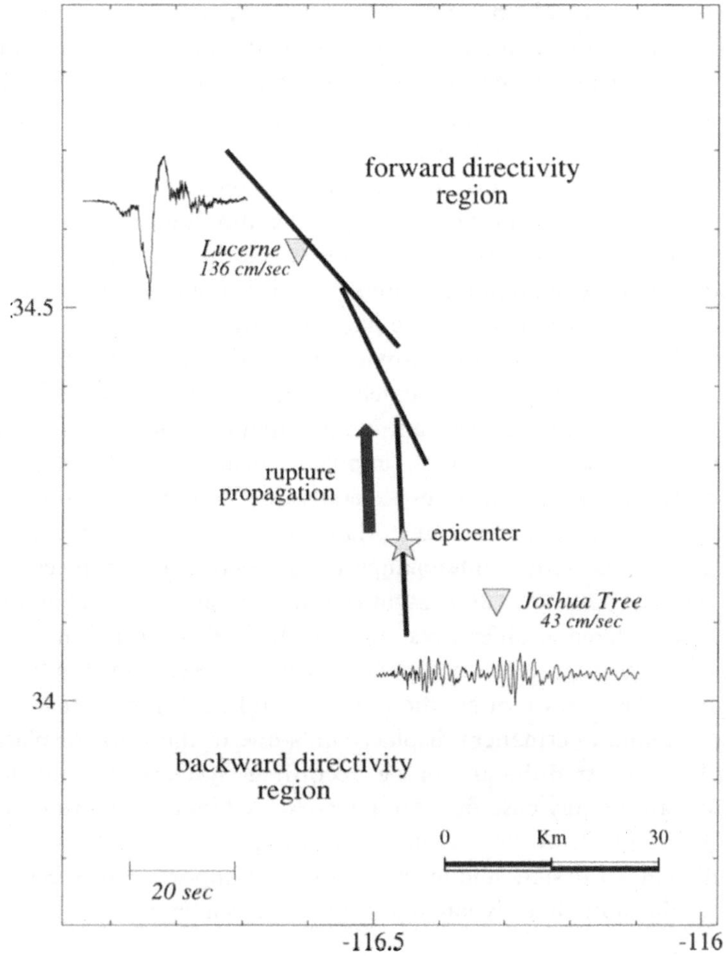

Figure 3
Map of the Landers region showing the location of the rupture of the 1992 Landers earthquake (which occurred on three fault segments), the epicenter, and the recording stations at Lucerne and Joshua Tree. The fault normal velocity time histories at Lucerne and Joshua Tree exhibit forward and backward rupture directivity effects, respectively.

strike at periods longer than 0.5 seconds. This causes the peak response spectral acceleration of the strike-normal component to shift to longer periods. Near-fault effects cannot be adequately described by uniform scaling of a fixed response spectral shape; the shape of the spectrum becomes richer in long periods as the level of the spectrum increases. Near-fault rupture directivity effects can be significant at periods longer than 1 second and at distances less than about 50 km, with the size of the effect dependent on the earthquake magnitude and on the geometry of the site in relation to the fault.

Based on an empirical analysis of near-fault data, SOMERVILLE *et al.* (1997) developed modifications to empirical strong ground motion attenuation relations to account for the effects of rupture directivity on strong motion amplitudes and durations. In the near-fault rupture directivity model, amplitude variations due to rupture directivity depend on two geometrical parameters. First, the smaller the angle between the direction of rupture propagation and the direction of waves travelling from the fault to the site, the larger the amplitude. Second, the larger the fraction of the fault rupture surface that lies between the hypocenter and the site, the larger the amplitude. Attenuation relations modified to include directivity effects can be used directly in probabilistic seismic hazard analysis (PSHA).

Although the response spectrum provides the basis for the specification of design ground motions in all current design guidelines and code provisions, there is a growing recognition that the response spectrum is not capable of adequately describing the seismic demands presented by brief, impulsive near-fault ground motions. This indicates the need to use time histories to represent near-fault ground motions.

Many analysis procedures assume that the two horizontal components are uncorrelated, and prescribe interchanging the two horizontal components in structural analyses. This is clearly inappropriate for near-fault ground motions, in which there are systematic differences between the strike-normal and strike-parallel components. Interchanging the components can represent physically unrealizable scenarios given the known orientation of the fault. Although near-fault ground displacements contain permanent displacements due to the static displacement field of the earthquake, traditional analog recording systems do not retain these displacements, and in any case they are removed by highpass filtering in traditional processing methods. However, in some cases the permanent ground displacements may be significant for design, and in these cases it is important to specify the correct orientation of the static and dynamic ground displacements.

4.4 Effects of Sedimentary Basins on Strong Ground Motions

Many regions are situated on deep sediment-filled basins. A basin consists of alluvial deposits and sedimentary rocks that are geologically younger and have lower seismic wave velocities than the underlying rocks upon which they have been deposited. Basins have thicknesses ranging from a hundred meters to over ten kilometers. Waves that become trapped in deep sedimentary basins can potentially be very damaging. Conventional criteria for characterizing site response are typically based on the distribution of shear wave velocity with depth in the upper 30 meters as determined from field measurements. The response of this soil layer is usually modeled assuming horizontal layering, following the method illustrated on the left side of Figure 4. The wave that enters the layer may resonate in the layer but cannot become trapped within the layer.

However, at periods longer than about one second, seismic waves have wavelengths that are considerably longer than 30 meters, and their amplitudes are

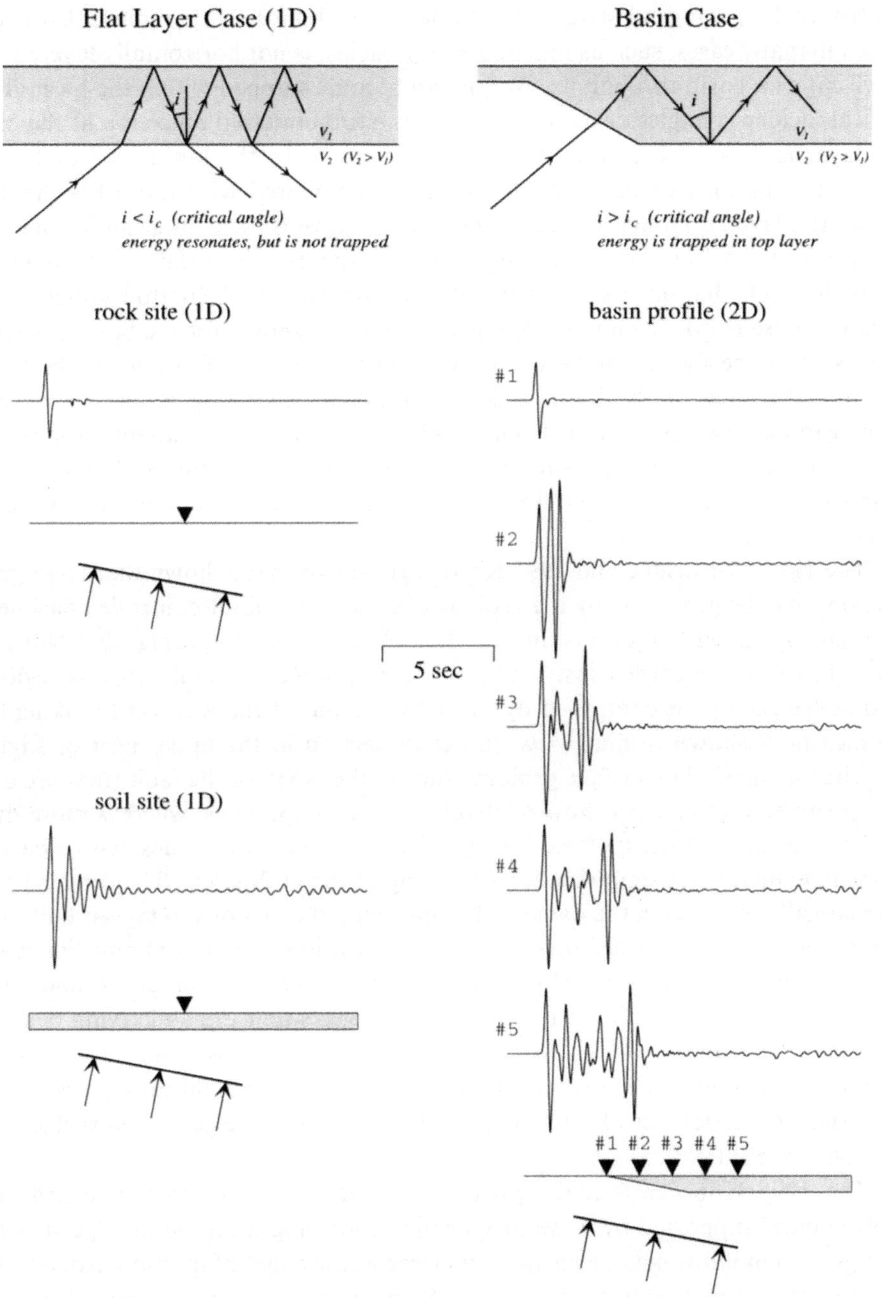

Figure 4
Schematic diagram showing that seismic waves entering a sedimentary layer from below can escape if the layer is flat (upper left), but can become trapped in the layer if it has varying thickness (as in the case of a basin) and the waves enter through its edge (upper right).

controlled by geological structure with depths of hundreds or thousands of meters which in many cases, such as in sedimentary basins, is not horizontally layered. If the wave enters a basin through its edge, it can become trapped within the basin if post-critical incidence angles develop. The resulting total internal reflection at the base of the layer is illustrated at the top right of Figure 4. In the lower part of Figure 4, simple calculations of the basin response are compared with those for the simple horizontal layered model. In each case, a plane wave is incident at an inclined angle from below. The left side of the figure shows the amplification due to impedance contrast effects that occurs on a flat soil layer overlying rock (bottom) relative to the rock response (top). A similar amplification effect is shown for the basin case on the right side of the figure. However, in addition to this amplification, the body wave entering the edge of the basin becomes trapped, generating a surface wave that propagates across the basin. Current empirical ground motion attenuation relations do not distinguish between sites located on shallow alluvium and those in deep sedimentary basins, and tend to underestimate the ground motions recorded in basins.

The 1994 Northridge and 1995 Kobe earthquakes have shown that large ground motions may be generated by the geological structure of fault-controlled basin edges. The largest ground motions in the Los Angeles basin during the Northridge earthquake were recorded just south of the Santa Monica fault. In this region the basin-edge geology is controlled by the active strand of the westward striking Santa Monica fault shown in map view and cross section in the upper part of Figure 5. Despite having similar surface geology, sites to the north of the fault (that are closest to the earthquake source) show relatively modest amplitudes, whereas more distant sites to the south of the fault exhibit significantly larger amplitudes, with a clear and immediate increase in amplification occurring at the fault scarp. The same pattern is dramatically reflected in the damage distribution indicated by red tagged buildings at the top of Figure 5, which shows a large concentration of damage immediately south of the fault scarp in Santa Monica. The strong correlation of ground motion amplification pattern with the fault location suggests that the underlying basin-edge geology is controlling the ground motion response. This is confirmed in the 2-D synthetic seismograms shown in Figure 5. The large amplification results from constructive interference of direct waves with the basin-edge generated surface waves (GRAVES *et al.*, 1998).

The 1995 Kobe earthquake provided further evidence from recorded strong motion data, supported by wave propagation modeling using basin edge structures, that ground motions may be particularly large at the edges of fault-controlled basins. Severe damage to buildings due to the Kobe earthquake was observed in a zone about 30-km long and 1-km wide, and offset about 1-km southeast of the fault on which the earthquake occurred (Fig. 6). The near-fault ground motions generated by rupture directivity effects in the Kobe earthquake were further amplified by the basin edge effect. This effect was caused by the constructive interference between direct

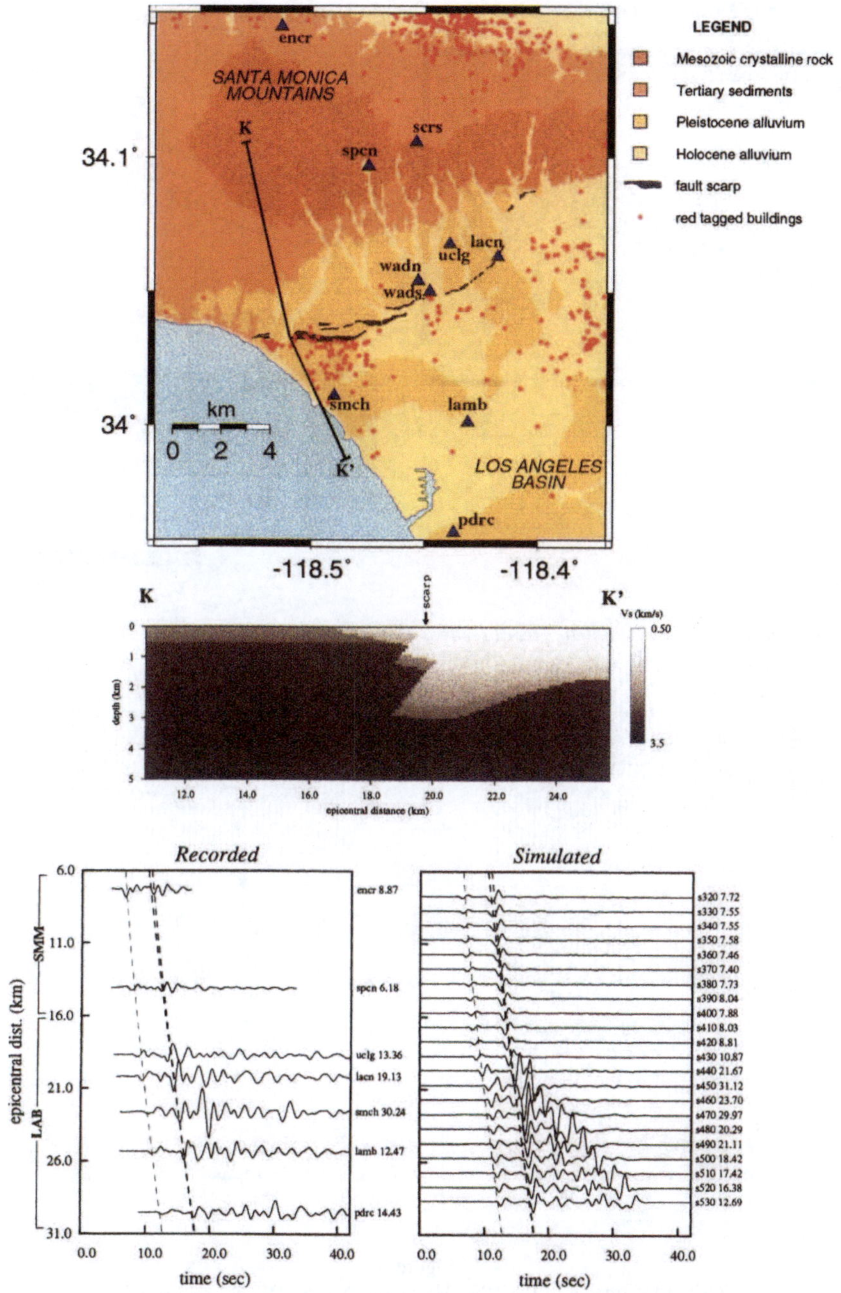

Figure 5

Basin edge effects in Santa Monica during the 1994 Northridge earthquake. Large ground motion and associated damage occurred immediately south of the active strand of the Santa Monica fault.

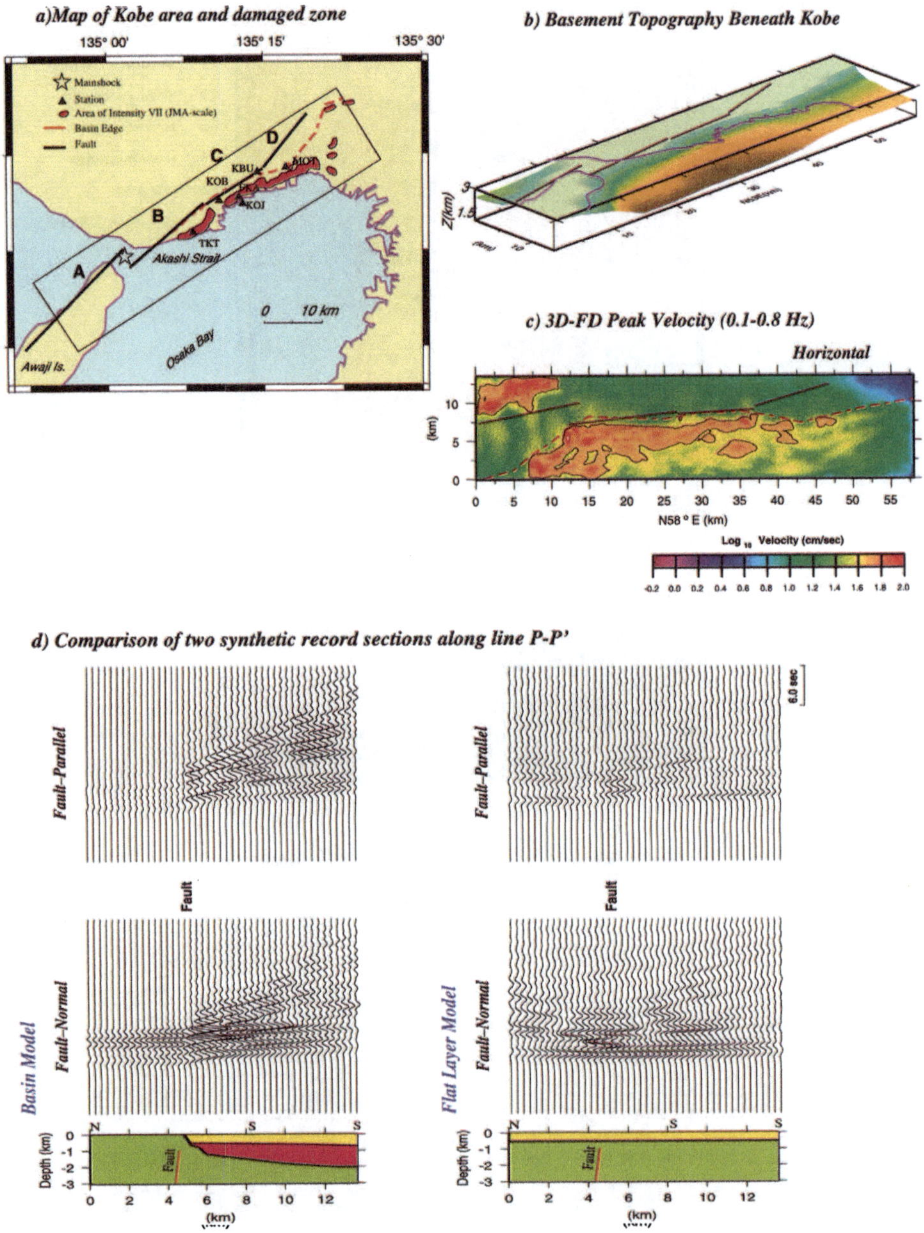

Figure 6
Basin edge effects in the 1995 Hygo-ken Nanbu (Kobe) earthquake.

seismic waves that propagated vertically upward through the basin sediments from below, and seismic waves that diffracted at the basin edge and proceeded laterally into the basin (KAWASE, 1996; PITARKA *et al.*, 1998). The basin edge effect caused a concentration of damage in a narrow zone running parallel to the faults through Kobe and adjacent cities (Fig. 6). The observed band of damage, represented by the zone of JMA Intensity VII shown in panel a, is reproduced in the map of simulated peak velocity (panel c), generated using a 3-D model of the basin edge (panel b). The strong influence of the basin edge is demonstrated in the lower part of the panel d, which shows vertically aligned time histories from 2-D simulations for the actual basin geometry (left side) and a flat-layered model with no basin edge (right side). The basin structure produces a strongly asymmetrical zone of large amplitudes along the edge of the basin.

5. Summary

This paper reviews current concepts and trends in the characterization of earthquake ground motion, which can trigger underwater landslides. The focus of this paper is on the prediction of ground motions on rock and firm soil. The response to these ground motions of soft soils that overly rock and firm soil is beyond the scope of this paper, but is addressed by other papers in this volume. New methods have been developed for characterizing potential earthquake sources that use geological and geodetic data in conjunction with historical seismicity data. Scaling relationships among earthquake source parameters have been developed to provide a more detailed representation of the earthquake source for ground motion prediction. Improved empirical ground motion models have been derived from a strong motion data set that has grown markedly over the past decade. However, these empirical models have a large degree of uncertainty because the magnitude-distance-soil category parameterization of these models often oversimplifies reality. This reflects the fact that other conditions that are known to have an important influence on strong ground motions, such as near-fault rupture directivity effects, crustal waveguide effects, and basin response effects, are not treated as parameters of these simple models. In order to reduce the uncertainty in ground motion prediction at a given site, ground motion models are being augmented to include more realistic representations of source, path and site effects through the use of seismologically based strong motion simulation procedures that have been extensively validated against recorded data.

References

Abrahamson, N. A., and Shedlock, K. (1997), *Overview, Ground Motion Attenuation Relations*, Seism. Res. Lett. *68*, 9–23.

Abrahamson, N. A., and Silva, W. J. (1997), *Empirical Response Spectral Attenuation Relations for Shallow Crustal Earthquakes*, Seism. Res. Lett. *68*, 94–127.

Abrahamson, N. A., and Somerville, P. G. (1996), *Effects of the Hanging Wall and Foot Wall on Ground Motions Recorded during the Northridge Earthquake*, Bull. Seismol. Soc. Am. *86*, S93–S99.

ABRAHAMSON, N. A., and YOUNGS, R. R. (1992), *A Stable Algorithm for Regression Analysis Using the Random Effects Model*, Bull. Seismol. Soc. Am. *82*, 505–510.

FEDERAL EMERGENCY MANAGEMENT AGENCY (1998), *1997 Recommended Provisions for Seismic Regulations for New Buildings and Other Structures*, FEMA *302*, 337 pp.

GRAVES, R. W., PITARKA, A., and SOMERVILLE, P. G. (1998), *Ground Motion Amplification in the Santa Monica Area: Effects of Shallow Basin Edge Structure*, Bull. Seismol. Soc. Am. *88*, 1224–1242.

GRAVES, R. W. (1996), *Simulating Seismic Wave Propagation in 3-D Elastic Media Using Staggered-grid Finite-differences*, Bull. Seismol. Soc. Am. *86*, 1091–1106.

HELMBERGER, D. V. (1983), *Theory and Application Of Synthetic Seismograms*. In *Earthquakes: Observation, Theory and Interpretation* (Kanamori, H. and E. Boschi eds), *Proc. Int. Sch. Phys. Enrico Fermi Course LXXXV*:174-221 (North-Holland, Amsterdam, 1983).

KAWASE, H. (1996), *The Cause of the Damage Belt in Kobe: "The Basin-Edge Effect," Constructive Interference of the Direct S-Wave with the Basin-Induced Diffracted/Rayleigh Waves*, Seism. Res. Lett. *67*, 25–34.

MUIR-WOOD, R. and KING, G. P. C. (1993), *Hydrological Signatures of Earthquake Strain*, J. Geophys. Res. *98*, 22,035–22,068.

PITARKA, A., IRIKURA, K., IWATA T., and SEKIGUCHI, H. (1998), *Three-dimensional Simulation of the Near-fault Ground Motion for the 1995 Hyogo-Ken Nanbu (Kobe), Japan, Earthquake*, Bull. Seismol. Soc. Am. *88*, 428–440.

SOMERVILLE, P. G., IRIKURA, K., GRAVES, R. W., SAWADA, S., WALD, D., ABRAHAMSON, N. A., IWASAKI, Y., KAGAWA, T., SMITH, N., and KOWADA, A. (1999), *Characterizing Earthquake Slip Models for the Prediction of Strong Ground Motion*, Seism. Res. Lett. *70*, 59–80.

SOMERVILLE, P. G., SMITH, N. F., GRAVES, R. W., and ABRAHAMSON, N. A. (1997), *Modification of Empirical Strong Ground Motion Attenuation Relations to Include the Amplitude and Duration Effects of Rupture Directivity*, Seism. Res. Lett. *68*, 199–222.

SOMERVILLE, P. G., SAIKIA, C. K., WALD, D., and GRAVES, R. W. (1996), *Implications of the Northridge Earthquake for Strong Ground Motions from Thrust Faults*, Bull. Seismol. Soc. Am. *86*, S115–S125.

STRUCTURAL ENGINEERS' ASSOCIATION OF CALIFORNIA (1996), *Recommended Lateral Force Requirements And Commentary, 1996*. Sixth Edition.

WELLS, D. L. and COPPERSMITH, K. J. (1994), *Analysis of Empirical Relationships among Magnitude, Rupture Length, Rupture Area and Surface Displacement*, Bull. Seismol. Soc. Am. *84*, 974–1002.

WORKING GROUP ON CALIFORNIA EARTHQUAKE PROBABILITIES (1995), *Seismic Hazards in Southern California: Probable Earthquakes, 1994 To 2024*, Bull. Seismol. Soc. Am. *85*, 379–439.

YOUNGS, R. R., ABRAHAMSON, N. A., MAKDISI, F. I., and SADIGH, K. (1995), *Magnitude-dependent Variance of Peak Ground Acceleration*, Bull. Seismol. Soc. Am. *85*, 1,161–1,176.

Pure appl. geophys. 160 (2003) 1829–1841
0033–4553/03/111829–13
DOI 10.1007/s00024-003-2408-y

Pure and Applied Geophysics

The 1998 Papua New Guinea Earthquake and its Fault Plane Estimated from Relocated Aftershocks

NOBUO HURUKAWA,[1] YOSHINOBU TSUJI,[2] and BUDI WALUYO[3]

Abstract—The 1998 Papua New Guinea earthquake of M_w 7.0 occurred near the Wewak trench where the North Bismarck plate is subducting beneath the Australian plate. Its mechanism is thrust-type, and one of the nodal planes is almost parallel to the plate interface. To determine which of the two nodal planes of the main shock is the fault plane, we relocated the main shock and aftershocks using a method of modified joint hypocenter determination. We combined and employed two types of data in this study. Firstly, we used data reported by the National Earthquake Information Center (NEIC) of the U.S. Geological Survey (USGS), which includes three stations at the northeastern edge of Irian Jaya and one station in northern Papua New Guinea, from which the epicentral distances are less than 2 degrees. Secondly, in addition to the above permanent-station data, we used data from temporary aftershock observations near the epicentral area around the Sissano Lagoon carried out by TSUJI *et al.* (1998). Using three-component seismometers, they carried out observations from August 2 to October 2, 1998 at three sites. Although the network did not record the main shock and immediate aftershocks, the data obtained by temporary observation sites can clearly assist in identifying their absolute locations, since it is possible to apply the joint hypocenter determination (JHD) method. Hypocenters were relocated between the coastline and the Wewak trench, distributed along a nodal plane dipping shallowly to the southwest. Therefore, we can conclude that this nodal plane is the main shock fault and that the 1998 Papua New Guinea earthquake was an interplate earthquake between the North Bismarck and Australian plates.

Key words: Joint hypocenter determination, aftershocks, fault plane, Papua New Guinea earthquake.

Introduction

The 1998 Papua New Guinea earthquake (M_w 7.0, M_s 7.1) occurred on July 17 near the Wewak trench (Fig. 1) where the North Bismarck plate is subducting beneath the Australian plate (e.g., TREGONING *et al.*, 1998). This earthquake caused more severe damage to Papua New Guinea than its magnitude would suggest (KAWATA *et al.*, 1999). The maximum height of the tsunami was 15 m near the top of the eastern spit of the Sissano Lagoon. In general, only earthquakes of M 7.5 or

[1] International Institute of Seismology and Earthquake Engineering, Building Research Institute, 1 Tatehara, Tsukuba, Ibaraki 305-0802, Japan. E-mail: Hurukawa@kenken.go.jp
[2] Earthquake Research Institute, University of Tokyo, Tokyo, Japan.
[3] Meteorological and Geophysical Agency, Jakarta, Indonesia.

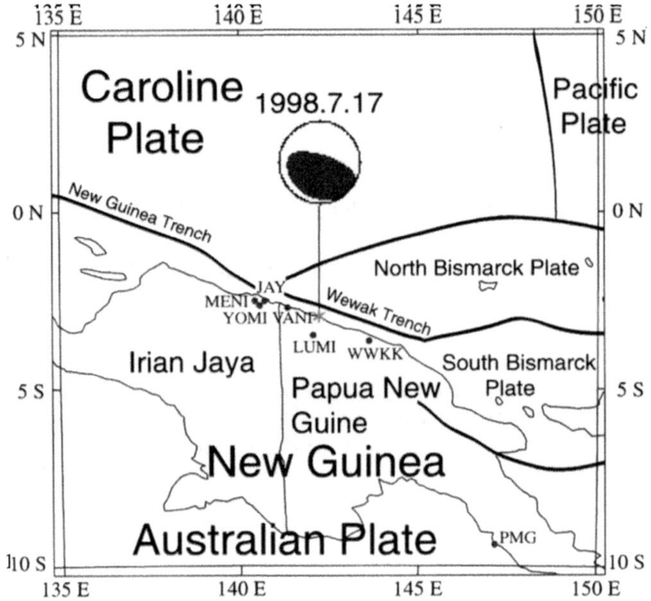

Figure 1
Tectonic map near the source region of the 1998 Papua New Guinea earthquake (M_w 7.0). The asterisk represents the epicenter of the 1998 Papua New Guinea earthquake. The Harvard CMT solution of the earthquake is also shown. Solid circles represent the nearest 7 stations. The *P*- and *S*-wave arrival times measured at 6 of these stations closest to the epicenter were used in our analysis. The central 2 stations (VANI, LUMI) are temporary observation stations set up by Tsuji *et al.* (1999).

greater cause dangerous tsunami; it is not known why the M_w 7.0 Papua New Guinea earthquake caused such extensive damage. Kawata *et al.* (1999), Tappin *et al.* (1999), and Okal and Synolakis (2001) proposed that large submarine landslides or a sediment slump accompanied the event, causing the tsunami.

According to the CMT solution by Dziewonski *et al.* (1999), this is a thrust-type event with two nodal planes of (strike = 146°, dip = 19°, slip = 127°) and (strike = 287°, dip = 75°, slip = 78°) as shown in Figure 1. Although analyses of teleseismic body waves and far-field tsunami were attempted to identify the fault plane, concluding that the steeply dipping nodal plane is the fault plane (Kikuchi *et al.*, 1999; Tanioka, 1999), there are no significant differences between the steeply and shallowly dipping fault models.

To identify which nodal plane is the fault plane of the earthquake, it is necessary to obtain an accurate aftershock distribution map. However, since the event occurred in a remote area, no global or regional network was able to locate it or its aftershocks accurately. Hurukawa (1999) relocated the main shock and aftershocks using information from the Comprehensive Test Ban Treaty's International Data Center (IDC) and employed modified joint hypocenter determination (MJHD) to determine the fault plane of the main shock. He stated that the aftershock distribution suggested that the nodal plane steeply dipping northeastward was the fault plane of the main shock. However, the hypocenter errors of his location are too large to enable reliable identification of the fault.

In this study, we relocate the main shock and its aftershocks by means of MJHD, employing data from more stations than HURUKAWA (1999) used, including NEIC stations and temporary stations deployed by TSUJI *et al.* (1998) to improve the location accuracy.

Data and Relocation

We used modified joint hypocenter determination (MJHD) as introduced by HURUKAWA and IMOTO (1992) and HURUKAWA (1995) to make an accurate identification of the hypocenters. This approach enables us to obtain accurate relative locations of earthquakes. We simultaneously relocated the main shock and the aftershocks that occurred between July 17 and August 31, 1998 to within the area 1°S to 5°S and 140°E to 144°E, with focal depths of less than 100 km.

We combined and used two types of data in our analysis. Firstly, we employed *P*-wave arrival times from worldwide stations that were reported by the National Earthquake Information Center (NEIC) of the U.S. Geological Survey (USGS). Since *S*-wave arrival times at nearer stations help to obtain good locations, we used both the *P*- and *S*-wave arrival times at the nearest four stations on New Guinea Island shown in Figure 1. The three stations of JAY, YOMI and MENI are on the northeastern edge of Irian Jaya, operated by the Meteorological and Geophysical Agency (BMG) of Indonesia. The fourth station is WWKK in Papua New Guinea. The epicentral distance between the main shock and each station is less than 2 degrees. Secondly, in addition to the above permanent network data, we used data from temporary aftershock observation stations (TEMP) near the epicentral area around the Sissano Lagoon carried out by TSUJI *et al.* (1998) (Figs. 1 and 3). Using three-component seismometers, they carried out observations from August 2 to October 2, 1998 at the three sites of VANI, LUMI and WEWA. We used both the *P*- and *S*-wave arrival times at the first two stations, since the last station did not successfully record events.

By combining these two sets of data, we were able to relocate the 1998 Papua New Guinea earthquake sequence. Excluding bad readings for which traveltime residuals were greater than 3 s, we finally located 80 events using 59 stations, including the NEIC and TEMP stations shown in Figure 2. The minimum number of stations for each event and the minimum number of events observed at each station was 10 and 10, respectively, except for the nearest 6 stations. The minimum number of observed events at these stations was 3.

Figure 2
The fifty-nine stations used in this study for relocation of the 1998 Papua New Guinea earthquake sequence. The asterisk at the center of the left-hand map represents the epicenter of the main shock.

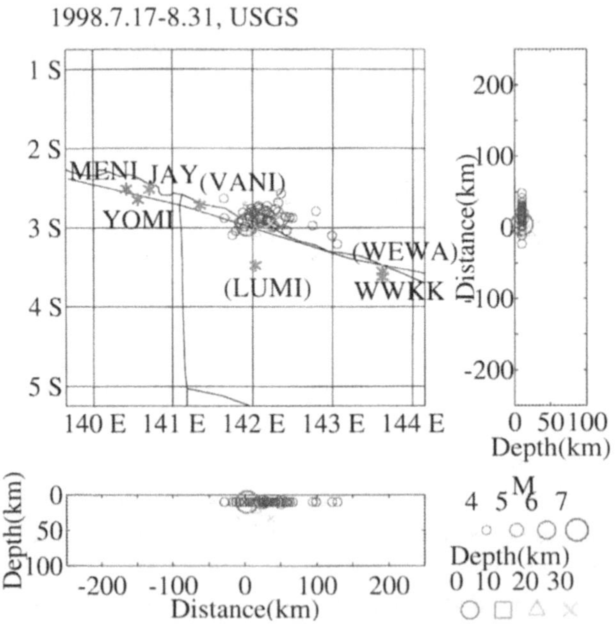

Figure 3
Hypocenter distribution of the 1998 Papua New Guinea earthquake sequence during the period from July 17 through August 31, 1998. The epicentral distribution and two vertical cross sections along NS and EW are shown. Hypocenters calculated by the NEIC of the USGS are shown in this Figure. The three stations with parentheses are temporary stations set up by Tsuji *et al.* (1998).

Hypocenters

Figures 3 and 4 show the hypocenter distribution of the 1998 Papua New Guinea earthquake sequence during the period of July 17 to August 31, 1998. Figure 3 shows hypocenters located by the NEIC. The number of events is 85. Since most events were fixed at a depth of 10 km, it is impossible to identify the fault plane of the main shock. On the other hand, Figure 4 shows the hypocenters of relocated events revealed by MJHD using NEIC and temporary observation (TEMP) data in this study. Since they show a high degree of scattering, and the hypocenters of some events have large standard errors, it is still impossible to identify the main shock fault plane. However, there is evidence that south and west of the epicentral area deeper events outnumbered shallower events. Although the location of the main shock is inaccurate because there are initial phases which exist about 20 s before the main phases, the main shock is likely to have occurred at the southwestern edge of the source area. Most of the source area is offshore.

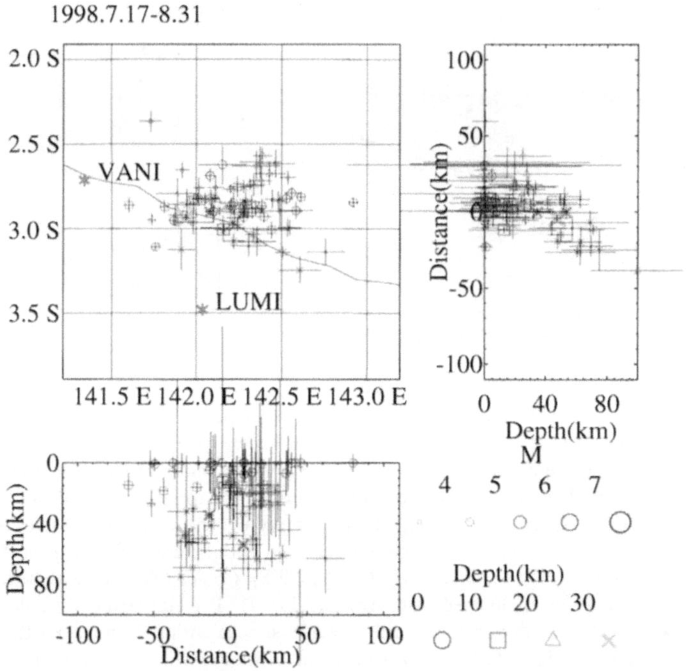

Figure 4
Hypocenter distribution of the relocated main shock and all the aftershocks of the 1998 Papua New Guinea earthquake examined in this study. The epicentral distribution and two vertical cross sections along the NS and EW axes are shown. Bars represent the standard errors of the hypocenters.

Fault Plane

To identify the main-shock fault plane, we plotted in Figure 5 only well-relocated events of which standard errors of latitude, longitude and focal depth are within 0.1°, 0.1° and 20 km, respectively. The epicentral distribution and two vertical cross sections along A-B (S56° W–N56° E) and C-D (S17° W–N17° E) are shown. These strikes are perpendicular to the two nodal planes of the Harvard CMT solution and dip shallowly to the southwest but steeply toward the north-northeast. The two thick solid lines in each Figure represent the two nodal planes of the Harvard CMT solution. Figure 5a shows all well-relocated events during

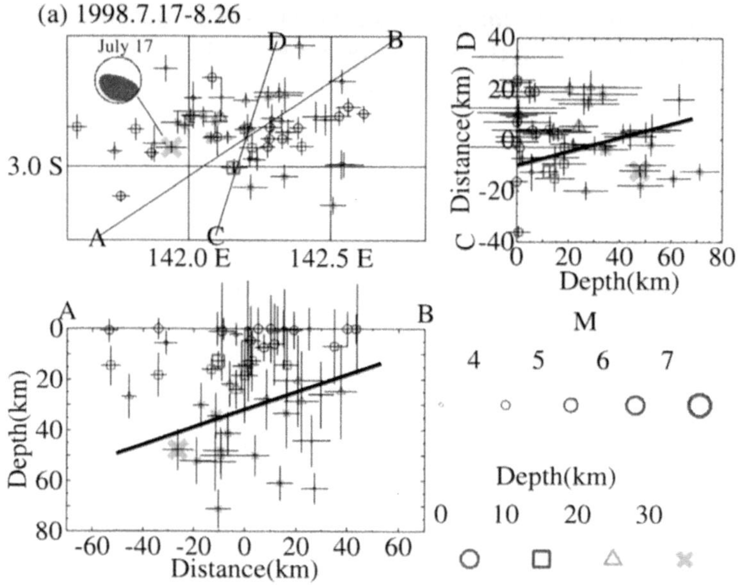

Figure 5
Hypocenter distribution of the accurately relocated main shock and aftershocks of the 1998 Papua New Guinea earthquake. The epicentral distribution and two vertical cross sections along A–B (S56°W–N56°E) and C–D (S17°W–N17°E) are shown. These strikes are perpendicular to the two nodal planes of the Harvard CMT solution. Only earthquakes with standard errors of latitude, longitude and focal depth less than 0.1°, 0.1° and 20 km, respectively are included. Bars represent the standard errors of the hypocenters. The two thick unbroken lines in each Figure represent the two nodal planes of the Harvard CMT solution. Note that the origin of the (x, y) coordinate is (142.2°E, 2.9°S). (a) All relocated events during July 17 to August 26. Note that no events are located from August 27 to 31. The Harvard CMT solution of the main shock is also shown. (b) The main shock and one-day aftershocks within 24 hours of the main shock. A solid ellipse indicates the one-day aftershock area of the main shock, which includes all aftershocks except the two that are distant from the rest. (c) Earthquakes during July 17 and 21. (d) Earthquakes during July 22 and August 26. In this period, three or more readings at the nearest six stations are used. The Harvard CMT solutions of two larger earthquakes are also shown. (e) Earthquakes during July 23 and August 26. Since the earthquakes on July 22 occurred mainly inland, we eliminated them from Figure 5d. (f) Earthquakes during August 9 and 26. All events are recorded at the temporary stations.

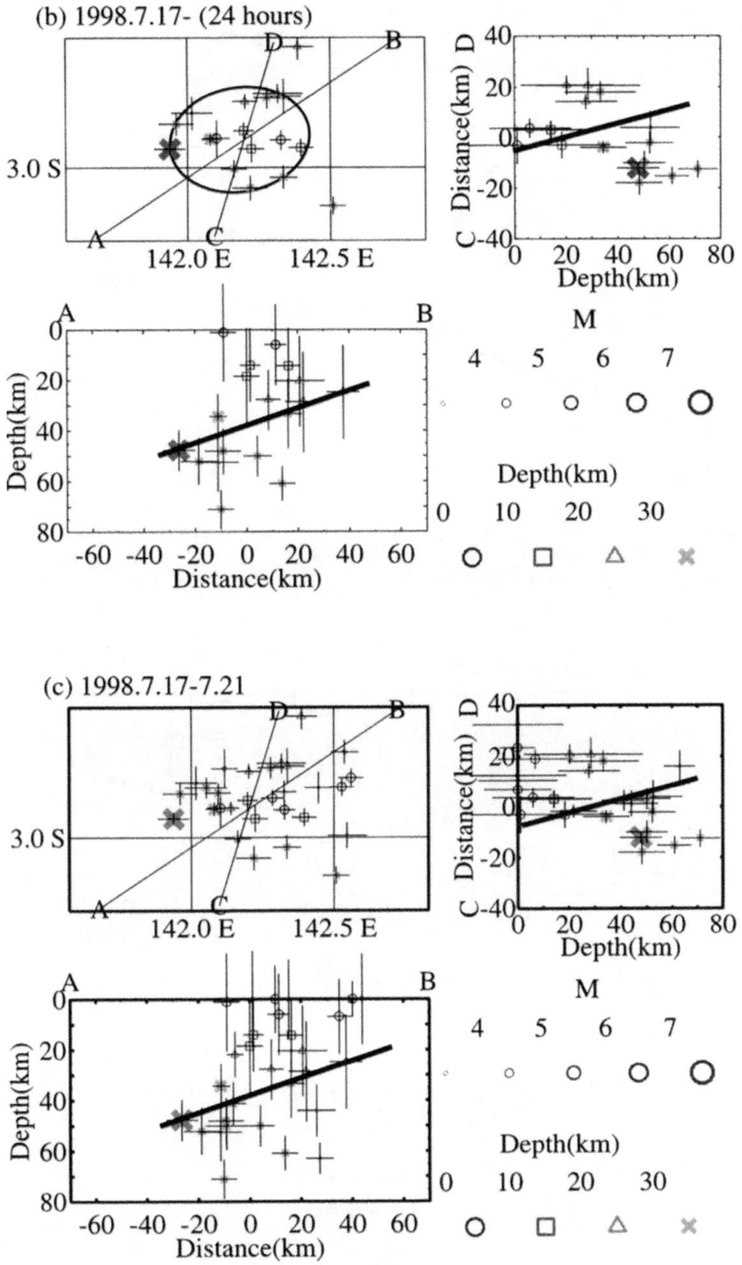

Figure 5
Continued

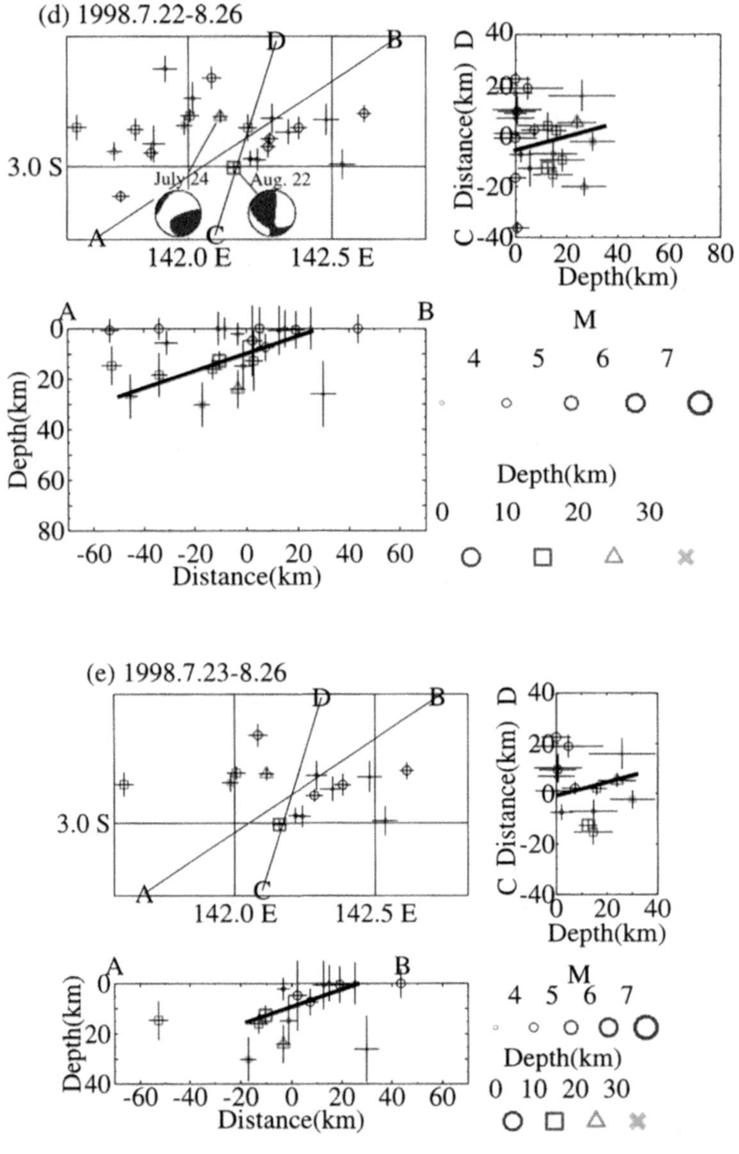

Figure 5
Continued

July 17 to August 26. Note that no events are located from August 27 to 31. The Harvard CMT solution of the main shock is also shown in this figure. Although errors in focal depths are large, it seems that aftershocks are distributed not along the plane dipping north-northeastward but along the plane dipping southwestward. To confirm this result, we plotted the events with different periods. Figure 5b

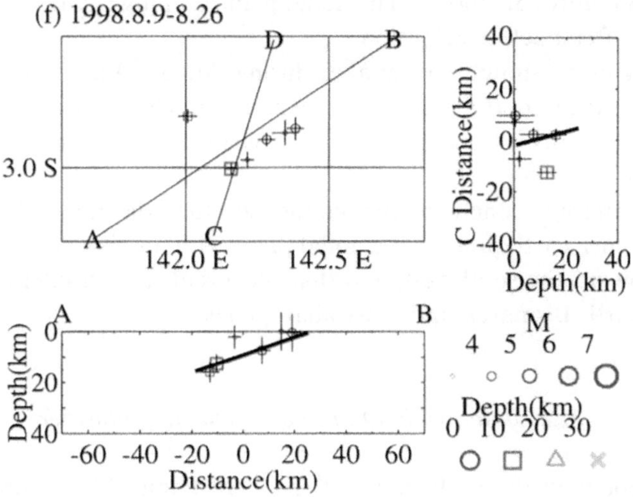

Figure 5
Continued

shows the main shock and one-day aftershocks within 24 hours of the occurrence of the main shock. Although errors in focal depths are large, it is clear that the aftershock distribution is consistent with the plane dipping southwestward and not consistent with the other nodal plane. Since all events except two are concentrated in a small area, we define the aftershock area by an ellipse that includes all events except these two. The size and shape of the area is ($r1 = 28$ km, $r2 = 21$ km, $\theta = 80°$).

As explained above, there are 4 nearer NEIC stations. The number of readings at these stations that recorded events was fewer than or equal to 2 until July 21 and was more than or equal to 3 after July 22 except for one event. Therefore, real errors in hypocenters may vary in these two periods, which could not be estimated using the calculated standard errors. Figures 5c and d show hypocenters during July 17 and 21, and during July 22 and August 26, respectively. It is clear that focal depths of events after July 22 are shallower than or equal to 30 km. Therefore, we assume that the real focal depths of events before July 21 are also shallower than 30 km. The Harvard CMT solutions of two larger aftershocks, which are the only events for which CMT solutions are reported, are also shown in Figure 5d. Their M_w values are 5.2 and 5.6. One of the nodal planes of each event dips southwestward with a gentle dip that is almost parallel to one nodal plane of the main shock. Therefore, we can consider that these two aftershocks occurred on the main shock fault plane.

Since the earthquakes on July 22 occurred mainly inland and outside the main-shock's aftershock area as defined in Figure 5b, we plotted earthquakes during July 23 and August 26 in Figure 5e in order to gain a clearer picture of aftershock distribution. It is evident that the aftershock area in this period is almost same as those shown in Figures 5b and 5c. The nodal plane dipping shallowly southwestward closely fits the aftershock distribution.

Finally, Figure 5f shows earthquakes during August 9 and 26. Since all events were recorded at the temporary stations, the real errors in hypocenters are smallest in this period. All events except one are distributed along the nodal plane dipping southwestward.

We can, therefore, conclude with confidence that the fault plane of the 1998 Papua New Guinea earthquake is the nodal plane with a strike of N34°W–S34°E and a dip of 19° toward the southwest; and that this event was an interplate earthquake between the North Bismarck and Australian plates.

Association with 40 km Fault and Amphitheater

We have obtained the fault plane of the 1998 Papua New Guinea earthquake from the aftershock distribution, in which the relative locations of hypocenters are quite accurate. In order to compare the obtained fault plane with other tectonic phenomena, absolute locations of hypocenters are essential. Station corrections at the two temporary and nearest four permanent stations are shown in Table 1. *P*- station corrections range from −2.0 to 0.1 s. Assuming that the average *P*-wave velocity along the ray paths between earthquakes and stations is 7 km/s, a station correction of ± 1 s causes a variation of ± 7 km with respect to hypocentral distance. Since station corrections at the nearest two stations, LUMI and VANI, are less than 0.1 ± 0.6 s, we can conclude that absolute errors of relocated epicenters are less than about 5 km, although errors in focal depths and origin times influence hypocentral distances.

Figure 6 shows an aftershock area and epicenters of the main shock and immediate aftershocks until 09 h 10 m. The so-called 40 km fault mapped by TAPPIN *et al.* (1999) is located at the northern or north-northeastern edge of the aftershock area; the main part of the aftershock area is thus located between the coastline and the 40 km fault. Since the dipping direction of the fault plane is S56°W, as shown by

Table 1

Station corrections at nearer stations

Type	Δ, degree	Station Code	Station Correction	
			P	S
TSUJI *et al.*	0.6	LUMI	0.1+−0.5	1.0+−0.5
	0.7	VANI	0.0+−0.6	4.1+−0.7
Irian Jaya, Indonesia	1.3	JAY	−2.0+−0.2	0.3+−0.4
	1.4	YOMI	−1.8+−0.2	0.1+−0.3
	1.6	MENI	−1.9+−0.2	−0.7+−0.4
Papua New Guinea	1.8	WWKK	−1.4+−0.2	–

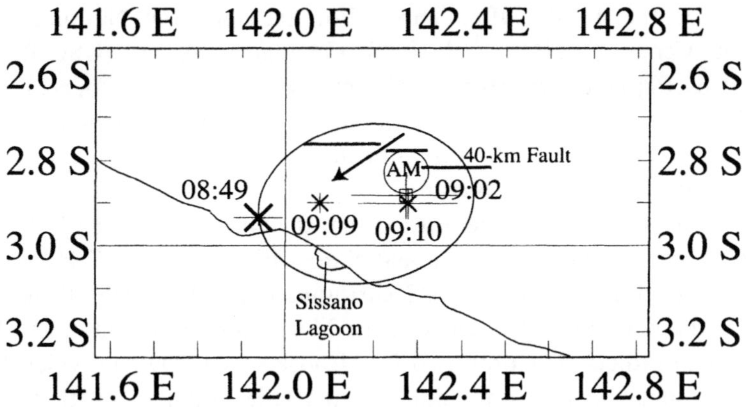

Figure 6
Epicenters of the main shock and immediate aftershocks up to 09h10m. The larger and smaller crosses are epicenters of the relocated main shock at 08:49 and the doublet at 09:09 and 09:10 GMT, respectively. The open square is the relocated epicenter of the 09:02 seismic event. Bars represent standard errors of the epicenters. The solid ellipse represents the aftershock area of the main shock obtained in Figure 5b. An arrow indicates the dipping direction of the fault plane. The approximate position of the 40-km fault (TAPPIN *et al.*, 1999) is also shown. Note that Figure 1 for TAPPIN *et al.* (1999) mislabeled its latitude and GEIST (2000, 2001) referred to the wrong location of the 40-km fault. The superimposed disk labeled "AM" schematizes the location of the amphitheater where the slump was identified by the survey (after OKAL, 2001).

the arrow, and the shallowest part of the main shock fault is close to the 40 km fault, there is a possibility that the fault relates to the event.

OKAL (1999, 2001) proposed that the seismic event at 09:02 GMT is a slump event that occurred in an amphitheater identified by OKAL (2001) and then caused the severe tsunami. Although the standard error for the focal depth is large, the 09:02 event was located at a shallower depth than the other three events shown in Figure 6. Our precise relocation indicates that the actual epicenter of the event is much closer to the amphitheater than his proposed location and coincides closely with the location of the amphitheater as shown in Figure 6. Therefore, we cannot reject his proposal in view of the aftershock location.

Discussions and Conclusion

Since digital waveform data at many stations worldwide are available, it is a simple matter to obtain a centroid moment tensor (CMT) solution and the rupture process of a large earthquake wherever it occurs (e.g., DZIEWONSKI *et al.*, 1994; SIPKIN, 1994; KAWAKATSU *et al.*, 1994; KIKUCHI *et al.*, 1999). However, it is difficult to determine the fault plane of earthquakes that occur in remote areas because of no nearer seismic stations. One method for obtaining the fault plane is via the distribution of the event's aftershocks. For example, the National Earthquake Information Center (NEIC) of the U.S. Geological Survey (USGS) reports the location of earthquakes wherever they occur. However, the hypocenter errors for aftershocks are too large to determine the fault plane of the main shock from their distribution. Therefore, relocation of aftershocks is essential to obtain the fault plane of a large earthquake in a remote area.

We can conclude by precise relocation that the fault plane of the 1998 Papua New Guinea (PNG) earthquake was the nodal plane with a strike of N34°W–S34°E and a dip of 19°. Therefore, the 1998 PNG earthquake is a low-angle thrust event that occurred at the plate interface between the North Bismarck and Australian plates. This result contradicts that proposed by HURUKAWA (1999), in which he suggests that the aftershock distribution indicates that the nodal plane that dips steeply northeastward is most likely to be the fault plane of the main shock. This implies that it is difficult to determine the fault plane using IDC data only because of lack of nearer stations and that both the *P* and *S* data of a local network are useful to determine the fault plane in addition to the global data. Furthermore, the temporary observation data near the source area can improve the absolute location significantly even if the main shock and immediate aftershocks are not recorded.

Since, in general, a high-angle thrust event can generate a larger tsunami than a low-angle type, initial examination of seismic wave analyses led to the selection of a high-angle thrust model (TANIOKA, 1999; MATSUYAMA *et al.*, 1999). MATSUYAMA *et al.* (1999), in a numerical simulation employing the newly obtained bathymetric data, concluded that only a low-angle reverse fault model accompanied by a large underwater landslide could explain the 1998 PNG tsunami. HEINRICH *et al.* (2000, 2001) showed that a submarine landslide could closely simulate the PNG tsunami. GEIST (2000, 2001), OKAL and SYNOLAKIS (2001) have discussed whether earthquakes or landslides cause tsunami. Combining these analyses and our relocation, we conclude that a combination of a low-angle reverse fault and a large underwater landslide caused the severe PNG tsunami.

REFERENCES

DZIEWONSKI, A. M., EKSTRÖM, G., and SALGANIK, M. P. (1994), *Global Seismicity of 1993, EOS, Fall Mtg. Suppl. 75*, 234.

DZIEWONSKI, A. M., EKSTRÖM, G., and MATERNOVSKAA, N. N. (1999), *Centroid-Moment Tensor Solutions for July–September, 1998*, Physics Earth Planet. Inter. *114*, 99–107.

GEIST, E. L. (2000), *Origin of the 17 July 1998 Papua New Guinea Tsunami: Earthquake or Landslide?* Seismol. Res. Letters *71*, 344–351.

GEIST, E. L. (2001), Reply to Comment by E. A. OKAL and C. E. SYNOLAKIS on *"Origin of the 17 July 1998 Papua New Guinea Tsunami: Earthquake or Landslide?" by E. L. Geist.* Seismol. Res. Lett. *72*, 367–372.

HEINRICH, P., PIATANESI, A., OKAL, E., and HEBERT, H. (2000), *Near-field Modeling of the July 17, 1998 Tsunami in Papua New Guinea*, Geophys. Res. Lett. *27*, 3037–3040.

HEINRICH, P., PIATANESI, A., and HEBERT, H. (2001), *Numerical Modeling of Tsunami Generation and Propagation from Submarine Slumps: The 1998 Papua New Guinea Event*, Geophys. J. Int. *145*, 97–111.

HURUKAWA, N. (1995), *Quick Aftershock Relocation of the 1994 Shikotan Earthquake and its Fault Planes*, Geophys. Res. Lett. *22*, 3159–3162.

HURUKAWA, N. (1999), *A Fault Plane of the 1998 Papua New Guinea Earthquake Estimated from Relocated Aftershocks Using Data of the Intentional Data Center of CTBT*, Zisin (J. Seismol. Soc. Jpn) *52*, 95–99 (in Japanese).

HURUKAWA, N. and IMOTO, M. (1992), *Subduction of Oceanic Crusts of the Philippine Sea and Pacific Plates and Weak-Zone-Normal Compression in the Kanto District, Japan*, Geophys. J. Int. *109*, 639–652.

KAWAKATSU, H., TAKANO, K., TSUBOI, S., and YAMANAKA, Y. (1994), *Automated Pseudo-realtime CMT Inversion, EOS, Fall Mtg. Suppl. 75*, 234.

KAWATA, Y., BENSON, B. C., BORRERO, J. C., BORRERO, J., DAVIES, H. L., DE LANGE, W. P., IMAMURA, F., LETZ, H., NOTT, J., and SYNOLAKIS, C. E. (1999), *Tsunami in Papua New Guinea was as Intense as First Thought*, EOS, Trans. Am. Geophys. Union *80*, 101–105.

KIKUCHI, M., YAMANAKA, Y., ABE, K., and MORITA, Y. (1999), *Source Rupture Process of the Papua New Guinea Earthquake of July 17, 1998 Inferred from Teleseismic Body Waves*, Earth Planets Space *51*, 1319–1324.

MATSUYAMA, M., WALSH, J. P., and YEH, H. (1999), *The Effect of Bathymetry on Tsunami Characteristics at Sisano Lagoon, Papua New Guinea*, Geophys. Res. Lett. *26*, 3513–3516.

OKAL, E. A. (1999), *The Probable Source of the 1998 Papua New Guinea Tsunami as Expressed in Oceanic T Waves*, EOS, Transaction, Am. Geophys. Union, 80, F750.

OKAL, E. A. (2001), *T Waves from the Papua New Guinea Sequence: Timing the Slump*, Pure Appl. Geophys *160*, 1843–1863.

OKAL, E. A. and SYNOLAKIS, C. E. (2001), *Comment on "Origin of the 17 July 1998 Papua New Guinea Tsunami: Earthquake or Landslide?" by E. L. Geist*. Seismol. Res. Lett. *72*, 362–366.

SIPKIN, S. A. (1994), *Rapid Determination of Global Moment-Tensor Solutions* Geophys. Res. Lett. *21*, 1667–1670.

TANIOKA, Y. (1999), *Analysis of the Far-Field Tsunami Generated by the 1998 Papua New Guinea Earthquake*, Geophys. Res. Lett. *26*, 3393–3396.

TAPPIN, D. R., MATSUMOTO, T., WATTS, P., and SATAKE, K. (1999), *Sediment Slump Likely Caused 1998 Papua New Guinea Tsunami*, EOS Trans. Am. Geophys. Union *80*, 329–340.

TREGONING, P., LAMBERG, K., STOLZ, A., MORGAN, P., MCCLUSKY, S. C., BECK, P., MCQUEEN, H., JACKSON, R. J., LITTLE, R. P., LAING, A., and MURPHY, B. (1998), *Estimation of Current Plate Motions in Papua New Guinea from Global Positioning System Observation*, J. Geophys. Res. *103*, 12,181–12,203.

TSUJI, Y., KAWATA, Y., IMAMURA, F., MATUTOMI, H., FUJIMA, K., and MATSUYAMA, M. (1998), *Survey of the 17 July 1998 Papua New Guinea Aitape Tsunami Earthquake*, Abstract, Fall Meeting of Seismol. Soc. Japan, C60.

Pure appl. geophys. 160 (2003) 1843–1863
0033–4553/03/111843–21
DOI 10.1007/s00024-003-2409-x

Pure and Applied Geophysics

T Waves from the 1998 Papua New Guinea Earthquake and its Aftershocks: Timing the Tsunamigenic Slump

EMILE A. OKAL[1]

Abstract—*T* waves recorded at hydrophone and seismic stations following the Papua New Guinea earthquake of 17 July 1998 and its aftershocks show that a small event at 09:02 GMT featured source properties incompatible with an elastic dislocation of appropriate body-wave magnitude ($m_b = 4.4$). These include an exceptional duration (47 s at the Wake Island hydrophone station WK31), a spectrum rich in high frequencies (7 to 12 Hz), and a generally low spectral amplitude. These characteristics can be explained by the model of an underwater slump, accelerating from a standstill and eventually slowing down. The relocation of the 09:02 event is compatible with its location within an amphitheater inside which shipboard cruises in 1998 and 1999 documented the presence of a 4 – km^3 geologically fresh slump. We propose that the slump took place at 09:02 on 17 July 1998, *i.e.*, 13 minutes after the mainshock, and that it generated the locally catastrophic tsunami, whose properties (amplitude and distribution of runup; timing) could not be explained by a dislocation model.

Key words: *T* waves, landslides, Tsunamis.

1. Introduction

We present in this paper a study of a small aftershock ($m_b = 4.4$; origin time: 09:02 GMT) of the 1998 Papua New Guinea (PNG) earthquake, which we interpret as an underwater slump that generated the locally catastrophic tsunami which followed the earthquake, killing upwards of 2100 people.

We refer to SYNOLAKIS *et al.* (2002) for a discussion of the failure of the main-shock dislocation to properly account for the characteristics of the local tsunami. Briefly summarized, the arguments are of three kinds: First, the runup amplitude observed along the PNG coast ($\zeta = 8$ to 15 m) would require a seismic slip of at least 8 m, in turn incompatible with the seismic moment of the earthquake, which DZIEWONSKI *et al.* (1999) inverted at only $M_0 = 3.7 \times 10^{26}$ dyn-cm. Second, these large runup values are concentrated along a stretch $d = 25$ km of coastline, the aspect ratio ζ/d being incompatible with dislocation models obtained from seismic scaling laws (HOFFMAN *et al.*, 2002). Finally, the arrival time of the tsunami on shore, reconstructed from survivor interviews by DAVIES (1998) to be about 09:10 to 09:12 GMT in the Arop–Malol

[1] Department of Geological Sciences, Northwestern University, Evanston, IL 60208, U.S.A.

area, is too late by at least ten minutes for the mainshock (origin time 08:49) to be its source. By the same token, the main aftershock doublet, at 09:09:32 ($m_b = 5.6$) and 09:10:02 ($m_b = 5.9$) is too late to be an appropriate source of the tsunami.

This failure to explain the local tsunami by a simple elastic dislocation, and in particular the extreme concentration of the high runup amplitudes, suggested, as early as the hours following the disaster, that an underwater landslide or slump may have generated the tsunami. This scenario was supported by shipboard explorations of the area offshore of the Sandaun coast in late 1998 and 1999, which identified and mapped a fresh slump of $\sim$4 km^3 centered in an amphitheater located at 2.83°S and 142.25°E, although obviously no precise time stamp could be put on its occurrence (TAPPIN *et al.*, 1999; SWEET and SILVER, 2003).

In this framework, we investigate in the present paper the seismic events which took place between the mainshock at 08:49 GMT and the main aftershock doublet at 09:09–09:10, and present evidence that an event detected at 09:02 GMT involved a major slump, which we propose as the source of the tsunami. Our evidence is largely based on the T waves recorded at various sites from the 09:02 event.

2. *Relocation*

The National Earthquake Information Center (NEIC) lists two earthquakes between the main shock and the main aftershock doublet: one at 09:02:06 ($m_b = 4.4$) and one at 09:06:01 (no magnitude reported). We use the technique of WYSESSION *et al.*(1991) to relocate these events, including an assessment of their 95%–confidence ellipses through a Monte Carlo method injecting Gaussian noise (with standard deviation σ_G) into the data set. Note that, while the two events considered are legitimate contenders for the generation of the tsunami in terms of their timing, their size clearly rules out explaining the amplitude of the observed runup as due to any model of those earthquakes as standard dislocations in an elastic medium.

Results, shown on Figure 1, show that the error ellipse of the 09:02 event (computed for $\sigma_G = 1$ s) does include the amphitheater where the slump was mapped during the shipboard surveys (TAPPIN *et al.*, 1999). It is then legitimate to assume that the 09:02 event took place at the location of the observed slump, inside the amphitheater. On the other hand, the 09:06 event could not be meaningfully relocated (Fig. 1), its error ellipse extending more than 150 km in the EW direction.

We also relocated all 57 further aftershocks of the PNG earthquake initially listed in the weekly Preliminary Determination of Epicenter Bulletins for the remainder of the year 1998 (a 166–day interval), in order to assess their error ellipses. Results, listed in Table 1 and shown on Figure 2, indicate a large scatter of the solutions, including in the direction perpendicular to the shoreline, which would be generally compatible with interpreting the shallow-dipping fault plane of the CMT mechanism (DZIEWONSKI

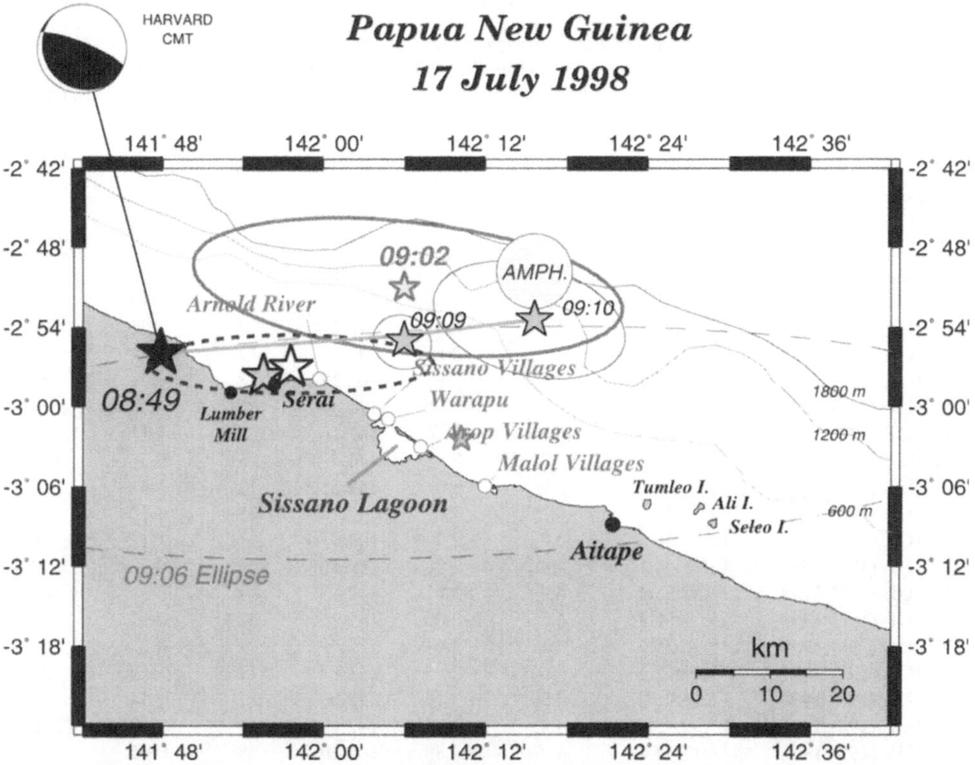

Figure 1

Map of the Sandaun coast of Northwestern Papua New Guinea. The open circles on the coastline identify villages devastated by the tsunami; the solid circles are other, mostly spared, communities. The large stars are epicenters of the mainshock (black: initial NEIC; grey: final NEIC; white: as relocated in this study, with error ellipse (dotted line)). The smaller, grey stars are epicenters of the relocated doublet at 09:09 and 09:10 GMT (with Monte Carlo ellipses). The line joining them is the extent of the seismic rupture, as inferred from the seismological modeling of KIKUCHI *et al.* (1998). The open star is the relocated epicenter of the 09:02 seismic event (with solid-line Monte Carlo ellipse). The superimposed disk labeled "*AMPH.*" schematizes the location of the amphitheater where the slump was identified by the surveys. The smaller gray star if the tentative relocation of the 09:06 event, with its large error ellipse (dashed line).

et al., 1999) as the plane of rupture. These results are essentially identical to those of MCCUE (1998) who combined preliminary NEIC epicenters with data from portable stations operated after 03 August 1998 by the Australian Geological Survey Organization, and of the more recent ones by HURUKAWA *et al.* (2003) who included data from a deployment of three portable stations in August and September 1998.

In addition to the 09:02 and 09:06 events, the records of the broadband Poseidon station at Jayapura, Indonesia (JAY; 2.52°S, 140.70°E; 155 km WNW of Sissano along the coast) feature a strongly impulsive event at 08:58. This event is not documented by the NEIC. Figure 3 shows that its duration at JAY (35 s) is significantly shorter than that of typical aftershocks of small magnitude, such as the

Table 1

Relocation of the PNG earthquake and its principal aftershocks

Date	Origin Time	Magnitude	Epicenter		Number of stations	σ	Length of major axis
D M (J) Y	GMT	m_b	°N	°E		(s)	(km)
17 JUL (198) 1998	08:49:12.7	6.8†	−2.95	141.96	54	1.30	38
17 JUL (198) 1998	09:02:05.7	4.4	−2.85	142.10	13	1.13	63
17 JUL (198) 1998	09:06:00.9		−3.04	142.17	4	0.14	150
17 JUL (198) 1998	09:09:31.2	5.6	−2.91	142.10	99	1.25	8
17 JUL (198) 1998	09:10:01.8	5.9	−2.89	142.26	38	1.13	27
17 JUL (198) 1998	09:19:12.7	4.5	−2.59	142.27	14	1.31	24
17 JUL (198) 1998	09:29:13.5	4.1	−2.82	141.82	17	1.35	23
17 JUL (198) 1998	09:40:07.6	4.5	−2.94	142.12	28	1.21	21
17 JUL (198) 1998	10:21:35.6	3.6	−3.06	142.44	5	0.37	177
17 JUL (198) 1998	11:53:36.9	4.1	−2.70	142.16	21	1.11	24
17 JUL (198) 1998	12:15:09.0	3.2	−2.99	142.47	6	0.34	156
17 JUL (198) 1998	12:59:59.5	3.9	−3.04	142.55	12	1.61	55
17 JUL (198) 1998	13:02:24.2	4.2	−2.69	142.27	15	1.00	39
17 JUL (198) 1998	13:12:56.5	4.2	−2.90	141.97	11	1.12	48
17 JUL (198) 1998	13:16:44.1	4.4	−2.75	142.16	25	0.72	22
17 JUL (198) 1998	13:32:55.4	4.4	−3.02	142.43	7	0.51	146
17 JUL (198) 1998	13:52:55.9	4.5	−3.00	142.26	32	2.11	16
17 JUL (198) 1998	15:46:35.9	3.8	−2.67	142.40	9	0.28	62
17 JUL (198) 1998	17:17:57.1	3.6	−2.76	141.87	11	1.64	27
17 JUL (198) 1998	17:19:23.8	4.3	−3.16	142.83	12	1.44	46
17 JUL (198) 1998	17:39:45.6	4.4	−2.99	142.13	21	1.29	18
17 JUL (198) 1998	18:17:14.4	4.7	−2.89	142.10	38	0.95	15
17 JUL (198) 1998	18:36:36.1	4.5	−2.89	142.18	28	1.14	16
17 JUL (198) 1998	18:50:42.0	4.7	−2.92	142.28	32	1.10	19
17 JUL (198) 1998	19:03:52.2		−2.76	142.37	5	0.34	74
17 JUL (198) 1998	20:19:56.4	4.1	−2.73	142.05	9	0.57	47
18 JUL (199) 1998	01:32:37.7	4.5	−2.91	142.16	34	1.07	25
18 JUL (199) 1998	01:41:11.5	4.9	−2.76	142.06	41	1.10	16
18 JUL (199) 1998	08:19:00.8	4.4	−2.92	141.87	20	1.66	34
18 JUL (199) 1998	09:23:34.7	4.4	−2.77	141.96	31	1.13	21
18 JUL (199) 1998	19:52:02.9	4.9	−2.61	142.14	6	0.79	86
19 JUL (200) 1998	08:47:05.7	4.8	−2.52	142.04	62	1.29	9
19 JUL (200) 1998	16:57:40.2	4.6	−2.92	142.16	35	1.31	13
19 JUL (200) 1998	19:44:53.8	4.4	−2.97	141.98	7	0.69	26
21 JUL (202) 1998	16:29:51.1	4.4	−2.89	142.42	28	0.94	17
22 JUL (203) 1998	06:30:50.2	4.4	−2.92	142.19	21	1.33	21
22 JUL (203) 1998	09:21:18.5	4.5	−3.05	141.79	31	0.90	16
22 JUL (203) 1998	10:09:51.5	4.8	−3.10	141.76	44	0.90	12
22 JUL (203) 1998	22:42:48.4	4.3	−3.10	142.21	23	1.33	17
24 JUL (205) 1998	17:11:54.3	5.0	−2.84	142.13	70	1.15	10
24 JUL (205) 1998	23:58:27.8	4.9	−2.82	142.38	5	0.56	40
25 JUL (206) 1998	17:36:49.9	4.6	−2.76	142.06	19	1.04	21
26 JUL (207) 1998	02:09:57.7	4.0	−2.92	141.66	10	1.15	35
26 JUL (207) 1998	02:30:25.5	3.4	−2.76	142.00	7	1.07	40
26 JUL (207) 1998	08:49:22.5		−2.95	141.92	5	0.32	36
27 JUL (208) 1998	01:56:03.3	3.7	−2.80	142.32	8	1.00	28

Table 1

Contd

Date	Origin Time	Magnitude	Epicenter		Number of stations	σ	Length of major axis
D M (J) Y	GMT	m_b	°N	°E		(s)	(km)
27 JUL (208) 1998	05:25:53.8	4.5	−2.67	142.75	6	0.60	38
02 AUG (214) 1998	07:35:12.7	4.5	−2.79	142.79	23	1.32	20
09 AUG (221) 1998	18:40:10.3	4.7	−2.84	141.95	26	1.08	23
22 AUG (234) 1998	11:09:45.1	5.4	−3.06	142.03	100	1.20	9
25 AUG (237) 1998	12:37:12.4	4.4	−2.99	142.26	11	0.78	22
26 AUG (238) 1998	02:04:44.9	4.8	−2.93	142.19	43	1.15	13
26 SEP (269) 1998	22:12:31.1	4.4	−2.91	141.51	16	0.94	29
28 SEP (271) 1998	23:41:37.0	4.1	−2.97	141.87	13	0.89	24
11 OCT (284) 1998	10:00:10.9	4.6	−2.93	142.16	22	1.12	17
11 OCT (284) 1998	14:35:01.3	4.7	−2.94	142.01	46	1.20	12
21 OCT (294) 1998	05:30:52.4	3.7	−3.26	142.77	21	0.85	18
25 NOV (329) 1998	05:19:15.2	4.3	−2.90	141.64	19	0.86	25
29 DEC (363) 1998	01:22:11.6	4.7	−3.04	141.91	8	0.93	28

† Magnitude M_m (OKAL and TALANDIER, 1989).

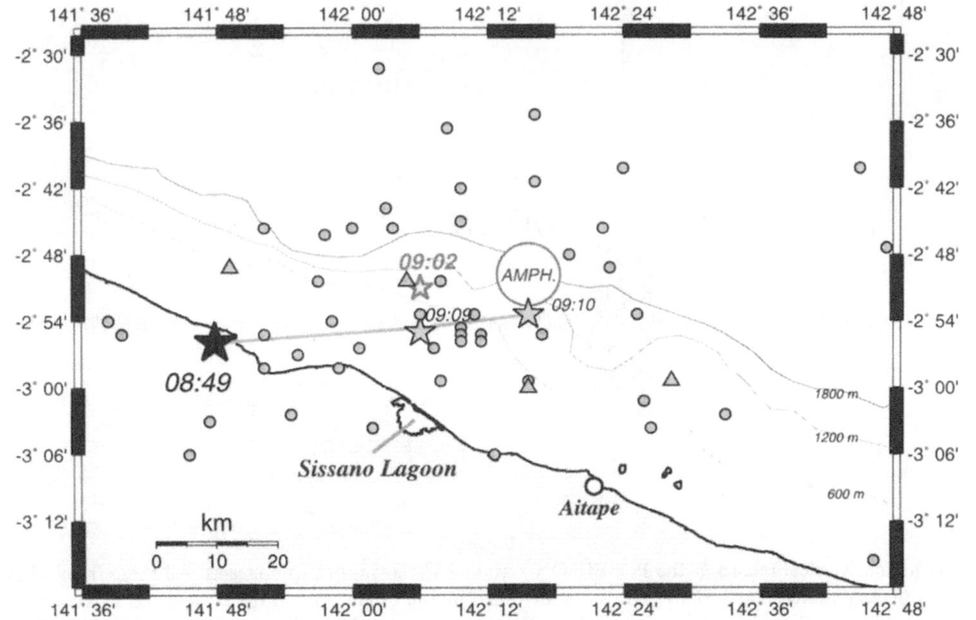

Figure 2

Map of aftershocks of the PNG sequence. The solid dots are the relocated epicenters listed in Table 1. Stars as in Figure 1. The four triangles identify those events which featured *T* wavetrains of longer duration at hydrophone station WK31.

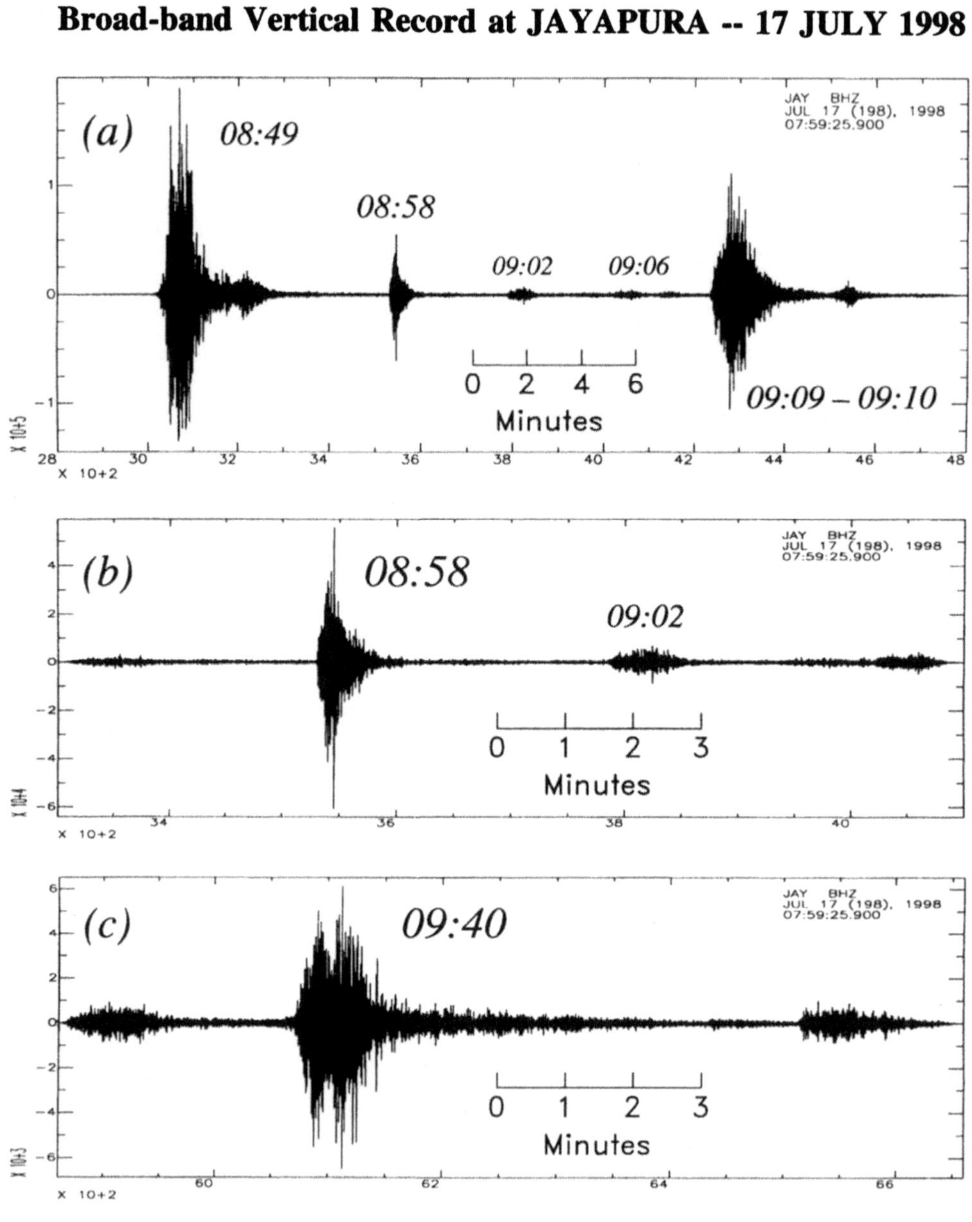

Figure 3
Broadband vertical record at the POSEIDON station JAY (Jayapura, Indonesia). All three frames have been highpass filtered at 1 Hz. Arrivals are identified by the origin time of the corresponding events. (a): 2000-second window showing P_n arrivals from the mainshock (08:49) and main aftershock doublet (09:09–09:10). Note that, in addition to the 09:02 and 09:06 events, this record shows a prominent arrival at 08:58. (b): 800-second window showing a close-up of the 08:58 arrival. Note impulsive nature and short duration of the wavetrain, as compared to the case of the 09:40 aftershock shown on Frame (c).

09:40 aftershock (70 s; $m_b = 4.5$), which we relocated at 2.94°S, 142.12°E, 165 km from JAY. In addition, the 08:58 event has a sharp impulsive arrival at JAY, characteristic of crustal propagation, as compared to the classically emergent nature of P_n wavetrains observed for the regular aftershocks. Finally, the spectrum of the 08:58 wavetrain is richer in intermediate-to-high frequencies (f = 2–4 Hz) than that of the 09:40 shock, advocating lesser attenuation, and hence a shorter path to JAY. While it was not possible to formerly relocate the 08:58 event, these characteristics of the JAY record suggest an epicentral distance no greater than 100 km and thus rule it out as a direct aftershock of the PNG earthquake located in the rupture area of the mainshock, and *a fortiori* as a source of the tsunami.

3. T-wave Records of the 09:02 Event

The small shocks at 09:02 and 09:06 GMT were recorded by very few stations (13 and 4, respectively), and their conventional seismic records are of relatively poor quality, especially in the near-field. For this reason, and in order to investigate their sources, we now turn to the *T* waves generated by the PNG aftershocks. We recall that *T* waves are acoustic (sound) waves generated in the water column of the world's oceans by a variety of sources including oceanic earthquakes, and efficiently propagated to extreme distances on account of the channeling properties of the SOFAR waveguide (EWING *et al.*, 1946). They can be recorded either at sea on hydrophones or by seismometers deployed close to shorelines, through their conversion to seismic waves when they hit the shore (*e.g.*, OKAL, 2001a).

- *The Wk31 record*

Figure 4 is a plot of the time series recorded at Hydrophone WK31, located 175 km SSE of Wake Island, at a distance of 3600 km from the amphitheater. The record has been filtered in the frequency band $1 < f < 30$ Hz. Apart from the prominent wavetrains from the mainshock and main aftershock doublet, *T* waves from the 09:02 and 09:06 events (arriving at 09:42 and 09:46, respectively) are readily identified. Even on this unprocessed record, there is a hint of a longer duration of the 09:02 *T* wavetrain. This is confirmed by the spectrograms shown on Figure 5, which reveal some exceptional characteristics of this wavetrain.

* *Duration*

First, the duration of the *T* wave at WK31 is approximately 47 seconds. As documented on Figure 5b, this is comparable to the duration of *T* waves from the mainshock ($m_b = 5.9; M_0 = 3.7 \times 10^{26}$ dyn-cm) and three times that of aftershocks of comparable magnitude (*e.g.*, 09:40 GMT; $m_b = 4.5$; Figure 5c). Note that the duration which can be measured on self-scaling plots such as the spectrograms on Figure 5, represents the duration of sustained energy at a given level of attenuation

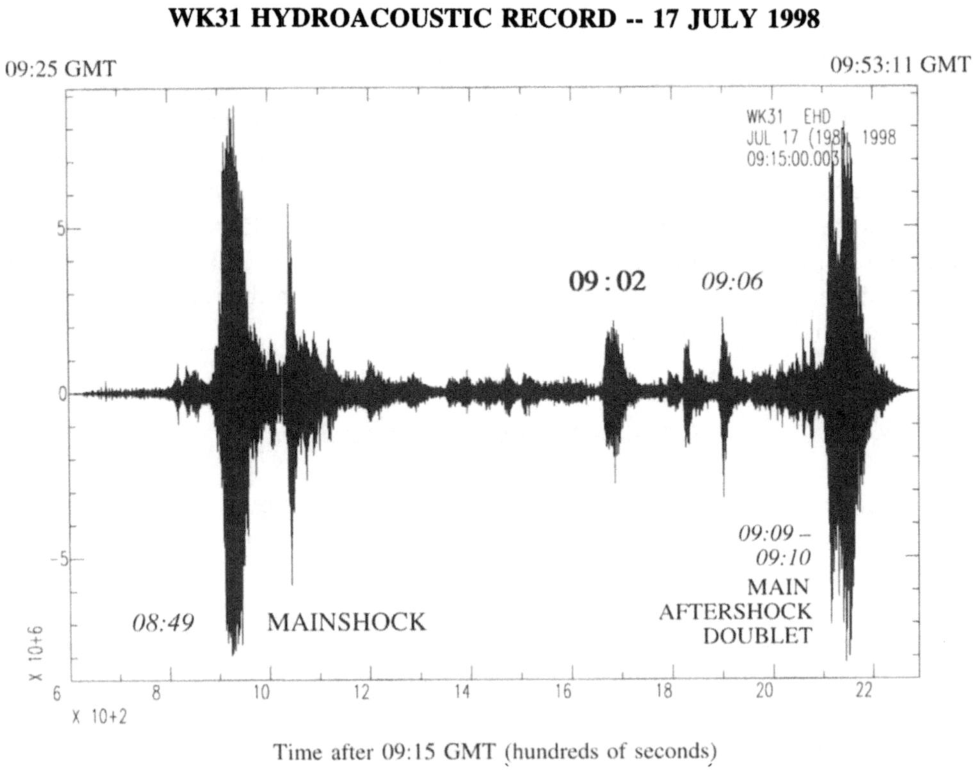

Figure 4
T-wave record of the PNG sequence at hydroacoustic station WK31. This record has been bandpass filtered between 1 and 30 Hz, in order to eliminate electrical noise contaminating the original time series at 60 Hz. The various events are identified on the record by their origin time. Note that the longer duration of the 09:02 event can be detected on this record (as compared for example with the 09:06 shock). Note also the absence of any T-wave signal for the 08:58 event recorded at JAY (see Fig. 3), suggesting the latter has a different epicenter, outside the rupture area of the mainshock.

from its peak value (in practice $\approx$ 40 dB). In terms of characterizing the duration of the source process, this concept is more robust than the use of an absolute threshold; for example, a similar approach was used for discrimination purposes on T–phase signals by TALANDIER and OKAL (2001).

▶

Figure 5
Time series and spectrograms of T waves received from the PNG sequence at the hydrophone station WK31 of the PIDC. All time series are 120-s long. The 09:02 event ($m_b = 4.4$) interpreted as the tsunami-generating slump is shown at the top (a), and compared to the mainshock (b) and an aftershock at 09:40, with comparable magnitude and location (c; $m_b = 4.5$). The spectrograms contour the energy present in the signal as a function of time (abscissa) and frequency (ordinate). Note the exceptional duration (47 s) of the 09:02 signal. The duration of the mainshock T wave is not affected by the obvious amplitude saturation of the signal.

The duration of a *T* wave recorded by a hydrophone can be expected to be controlled by (i) the duration of the seismic source itself, which is a combination of rise time and rupture time; (ii) different modes of seismic propagation before

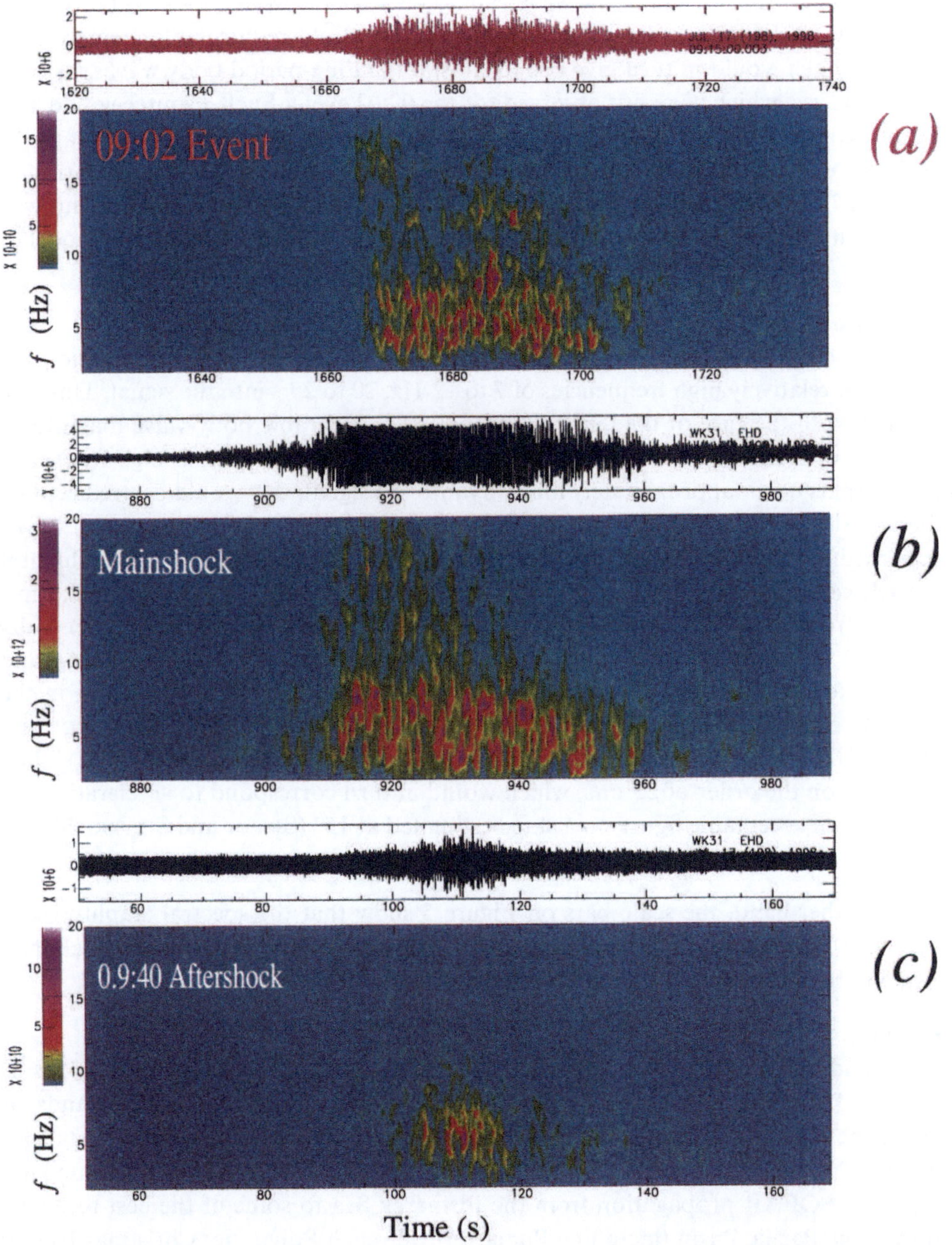

conversion to an acoustic wave in the water, such as a difference in travel time for P or S segments in the solid earth, as well as multipathing along different seismic paths; and (iii) the intrinsic dispersion of the acoustic wave in the water. Parameters (ii) and (iii) would not be expected to vary significantly between aftershocks, and certainly not to extend the duration of the wavetrain to 45 seconds. Matching a 1-s body-wave magnitude of 4.4 with a duration of 45 s or more would require an extremely slow rupture which would in turn give rise to substantial long-period body waves, or even surface waves, which were not observed for the 09:02 event. Such a source would also be an extremely poor generator of the high frequencies necessary to penetrate the SOFAR channel; indeed the so-called "tsunami earthquakes" are notoriously deficient T-wave generators (OKAL *et al.*, 2003). In summary, there is simply no way to explain the duration of the 09:02 T-wavetrain at WK31 with the simple model of an elastic dislocation compatible with earthquake scaling laws as its source.

* *Spectrum*

The frequency content of the spectrogram on Figure 5a is also remarkable: energy is present at relatively high frequencies of 7 to 12 Hz, 20 to 25 s into the signal. This is in contrast with the case of the 09:40 aftershock of comparable body-wave magnitude, whose spectrum is peaked at 5 Hz. Note in particular that these high-frequency components occur approximately mid-way into the signal; this would be predicted by the model of a block sliding on an inclined plane (HASEGAWA and KANAMORI, 1987), accelerating from a standstill and then slowing down to a stop, in which the highest particle velocities are reached at the center of the source time series. In contrast, wave packets from small-magnitude dislocations reflect very short rise times and generally feature a decay of frequency with time, illustrative of near-source scattering and accentuated in the case of T waves by their dispersion during propagation in the SOFAR channel. Finally, when combined with a total slip of $\approx$ 1 km (SWEET and SILVER, 2003), the total duration of the T wavetrain (47 s) would suggest average sliding velocities on the order of 25 m/s, which would in turn correspond to accelerations of $\approx 2\text{m/s}^2$, an acceptable figure on a slope estimated at 15° (SWEET and SILVER, 2003).

* *Amplitude*

On the other hand, the scale bars on Figure 5 show that the spectral amplitudes of the 09:02 T waves are only marginally larger than those of the 09:40 aftershock, and remain about 50 times smaller than those of the mainshock.

• Other records

In addition to the WK31 hydrophone records, we gathered as many records as possible from seismic stations located close to shore, either on Pacific Islands or along the rim of the ocean, in the configuration of the so-called T-phase stations (OKAL, 2001a). These are listed in Table 2, and spectrograms are presented on Figure 6. SOFAR propagation from the Bismarck Sea to some of the best receiving sites in the Pacific Basin (including Pomariorio, French Polynesia; Christmas Island;

Table 2

T-wave records at hydroacoustic and seismic stations

Location	Code	Nature of sensor	Network	Epicentral distance (km)	Estimated land path (km)	Record plotted on
Stations recording the 09:02 event						
Wake Hydrophone	WK31	Hydrophone	PIDC	3635		Fig. 5
Wake Island	WAKE	Broadband seismic	IRIS	3658	3	Fig. 6a
Guam	GUMO	Broadband seismic	IRIS	1863	43	Fig. 6b
Erimo, Japan	ERM	Broadband seismic	IRIS	4990	211	Fig. 6c
Petropavlovsk-Kamchatskiy, Russia	PET	Broadband seismic	IRIS	6407	280	Fig. 6d
Mount Rainier, Oregon	RAIO	Broadband seismic	IRIS	10645	164	Fig. 6e
Van Inlet	VIB	High-frequency seismic	PIDC	9973	41	Fig. 6f
Stations recording the mainshock (08:49) but not the 09:02 event						
Monterey OBS	MCGR	Ocean-bottom seismic	MBARI	10720	0	
Pin-lang, Taiwan	TWGB	Broadband seismic	Taiwan	3645	29	
Stations recording T waves after reflection						
Christmas Island	XMAS	Broadband seismic	IRIS	6762	8	
Haleakala, Maui	HLK	Broadband seismic	PeleNet	7225	27	

and most Hawaiian sites) is blocked at the source by the Admiralty Islands and adjoining structures (see Fig. 7), although reflected *T* phases are detectable at Christmas and Maui.

As shown on Figure 6, the frequency content of the 09:02 *T* phases recorded at the various stations shows a high degree of variability, reflecting the overland propagation following the acoustic-to-seismic conversion near the receiver. Table 2 documents that the land path traveled by the converted wave can vary substantially, reaching lengths of more than 150 km at RAIO, PET, and ERM (these estimates were obtained from a combined interpretation of arrival times of *T* phases and nearshore bathymetry). As discussed for example by TALANDIER and OKAL (1998) and OKAL (2001b), the nature of the converted wave and the characteristics of its on-land propagation are controlled by local structures on a scale comparable to the acoustic wavelengths involved (200 m at

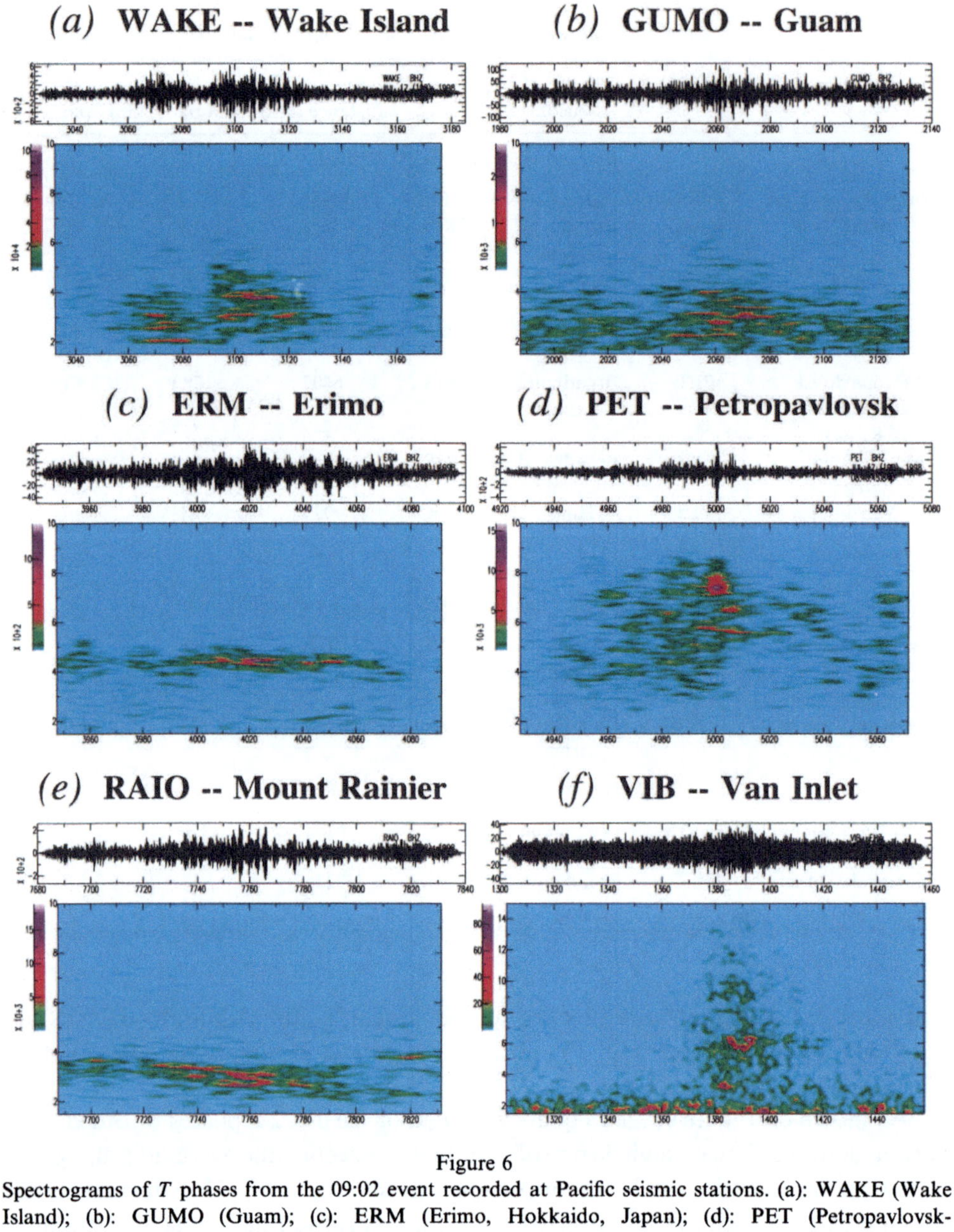

Figure 6

Spectrograms of *T* phases from the 09:02 event recorded at Pacific seismic stations. (a): WAKE (Wake Island); (b): GUMO (Guam); (c): ERM (Erimo, Hokkaido, Japan); (d): PET (Petropavlovsk-Kamchatskiy, Russia); (e): RAIO (Mount Rainier, Oregon); (f): VIB (Van Inlet, B.C., Canada). In all cases, the processed window lasts 160 seconds; the frequency band is 1.5–10 Hz, except at VIB (1.5–15 Hz) which records on a higher frequency channel.

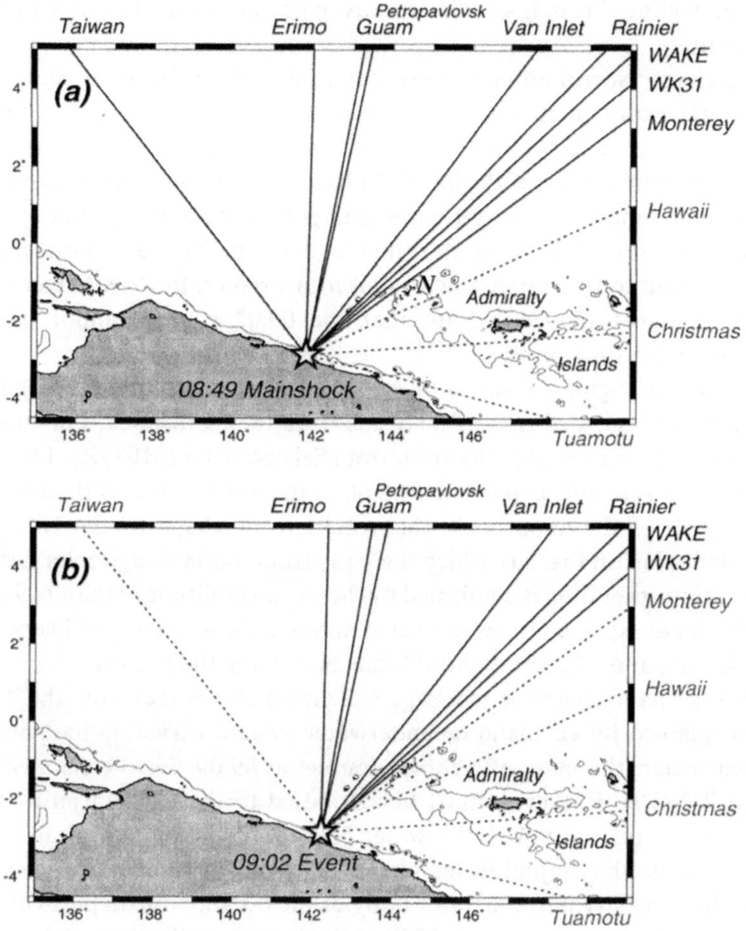

Figure 7

(a): Source-side blockage of *T* waves from the mainshock epicenter. The thin continuous line is the 1200-m isobath, representative of the axis of the SOFAR channel. Solid lines show great circle paths from the 08:49 NEIC epicenter to those stations where its *T* waves are recorded; the grey dashed lines show the paths to stations failing to record, or where the *T* wave involves a significant off-great circle reflection (Christmas; Maui). Note that the pattern is readily explained by source-side blockage at the Admiralty Islands. (b): Case of the 09:02 event. The absence at *T* waves at Monterey can be readily explained by blockage at the Ninigo Islands ("*N*" on frame (a)), but no structure can explain blockage along the path to TWGB (Taiwan).

7 Hz), and hence largely unknown. Figure 6 shows for example that the land paths can be regarded as a bandpass filter ($f \approx 4$ Hz) at ERM, as a low-pass filter at RAIO, but retain high frequencies of 7 to 10 Hz at both PET and VIB, these properties being uncorrelated with the length of the land path. Rather, it is probable that the retention of high frequencies in the *T* phase at PET, despite a land path reaching 280 km, expresses

favorable channeling through an efficient waveguide immediately below the Mohorovičić discontinuity.

The uncertainty surrounding the exact nature of the acoustic-to-station conversion and of the subsequent on-land propagation over distances reaching several hundred km, makes it impossible to invert *T*-phase arrival times in order to more precisely relocate the source of the 09:02 *T* phase. Rather, we simply verified that the observed arrival times were in all cases compatible with a legitimate conversion scenario. On the other hand, the duration of the various wavetrains analyzed on Figure 6 was found to be comparable to that measured on the WK31 record (Fig. 5), and thus to support the conclusion that the 09:02 source cannot be a simple dislocation of magnitude $m_b = 4.4$.

An additional intriguing observation is that we failed to detect 09:02 *T* phases at two Pacific sites which recorded powerful signals from the mainshock: Pin-lang, Taiwan (TWGB) and the Monterey Bay Ocean Bottom Seismometer (MCGR). The absence of signal at the latter is readily explained if we place the 09:02 source at the amphitheater, 42 km eastwards of the mainshock: this results in blockage by the Ninigo Islands (1.3°S; 144.4°E; "N" on Fig. 7a), which the *T* ray from the 08:49 epicenter just manages to avoid. This interpretation is confirmed by the weak amplitude of the 09:09 *T* waves at MCGR, which are expected to be generated mid-way between the 08:49 epicenter and the amphitheater, and whose ray should thus be grazing the islands.

On the other hand, selective blockage at Taiwan of *T* waves from the 09:02 event cannot be explained by an island or underwater ridge obstructing the path of the *T* wave (in particular, the main aftershock doublet at 09:09–09:10 generated *T* phases observed at TWGB). Rather, it must be attributed to the mechanism by which the acoustic energy is transferred to the water column in the epicentral area. A slump taking place inside the amphitheater would be expected to emit an acoustic signal directly inside the cavity into the water body of the ocean. Note in particular that the depth of the slump below sea level (1200 to 1800 m) is perfectly matched to that of the SOFAR channel, ensuring efficient excitation of acoustic energy into the wave guide. But this signal could leave the source only in the direction where the amphitheater is open to the high seas — to the north and northeast. To the northwest, in the azimuth of Taiwan, the western walls of the cavity, as mapped by SWEET and SILVER (2003), will essentially block propagation. This blockage would not be effective if the source was underground, since it could then illuminate seismically a larger section of the ocean floor, including presumably the seaside slope of the western wall of the cavity, and explaining the generation of a *T* phase in the azimuth of Taiwan by the main aftershock doublet.

4. Discussion

The principal characteristics of the *T*-wave signal recorded at the hydrophone station WK31 — duration and frequency content — cannot be reconciled with the reported body-wave magnitude of the event ($m_b = 4.4$). Rather, we have shown that the model of a slump taking place on the southern wall of the amphitheater mapped by TAPPIN *et al.* (1999) successfully explains these characteristics. Furthermore, these properties are generally supported by observations at other Pacific sites, instrumented with seismometers, in the geometry of a so-called *T*-phase station (OKAL, 2001a), and they are also generally comparable to the observations of the far-field hydroacoustic signatures of small underwater Hawaiian landslides reported by CAPLAN-AUERBACH *et al.* (2001).

We were also able to confirm several properties of the 09:02 event by studying its *P*-wave arrival at the seismic station WAKE, located on Wake atoll. From the epicentral area to the island receiver, the *P* wave travels as the oceanic P_n mode, identified as particularly efficient for regional-to-long distance propagation of high-frequency energy by WALKER (1977) (who called this phase "P_o") and TALANDIER and BOUCHON (1979). Figure 8 compares the spectrograms of the 09:02 and 09:40 *P* waves at the seismic station WAKE. The former features a duration of 40 s, which is in excellent agreement with the 47 s of the more dispersed acoustic *T* wave at the nearby hydrophone WK31; in contrast, the regular $m_b = 4.5$ aftershock has a P_o phase lasting no more than 13 s. We also confirm the occurrence of the highest frequency pulse (4.5 Hz for this record) about mid-way into the phase, in excellent agreement with the acoustic results.

We thus propose that the fresh slump identified inside the amphitheater by TAPPIN *et al.* (1999) and mapped by SWEET and SILVER (2003) actually took place at 09:02 GMT on 17 July 1998, thirteen minutes after the mainshock, and that this event was the source of the local catastrophic tsunami. As discussed by HEINRICH *et al.* (2000) and SYNOLAKIS *et al.* (2002), hydrodynamic models using the geometry and timing of the slump (with its estimated volume of 4 km^3) successfully predict the three characteristics of the observed run-up: its amplitude of 10 to 15 m along the coast, its rapid decay with distance along the coast, and the general timing of the arrival of the wave at the shore, neither of which can be explained by a dislocation occurring at the time of the mainshock.

How unique is the 09:02 event?

This interpretation of the 09:02 event as a slump responsible for the generation of the catastrophic tsunami requires a discussion of the uniqueness of its *T*-wave characteristics: in other words, we must establish that the event is indeed unique. For this purpose, we examined 15 hours of data at WK31, *i.e.*, the entire remainder of the GMT Julian day 1998:198, and in particular the *T* waves generated by the 40 aftershocks listed by the NEIC posterior to the main aftershock doublet at 09:09–09:10, the largest events having a reported magnitude $m_b = 4.7$ (at 18:17 and 18:50). Of those, six events (which may have had an on-land epicenter) did not generate *T* waves recordable at WK31, and one (at 11:28:51) corresponded to a gap in recording.

WAKE -- Wake Island Seismic Station

(a) P_o from 09:02 Event

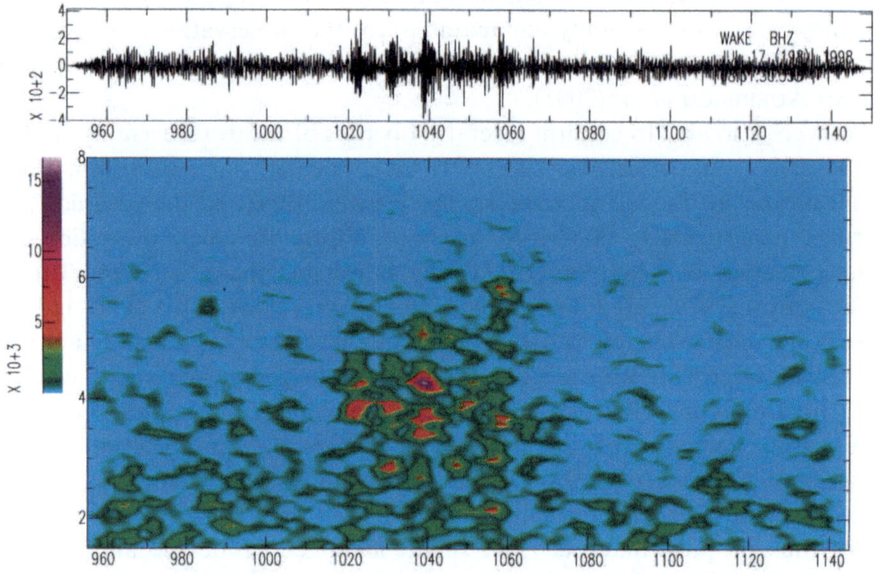

(b) P_o from 09:40 Aftershock

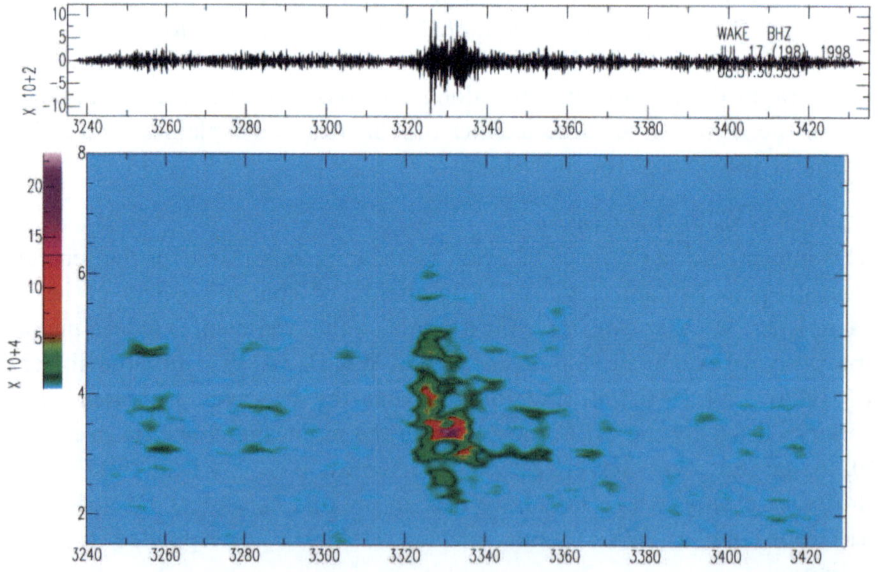

Of the remaining 33 sources, we could find only four earthquakes with *T* waves at WK31 lasting between 37 and 47 s, while all remaining 29 shocks (including the 09:06 aftershock, whose timing could have made it a potential candidate for the source of the tsunami) have *T* waves of "regular duration," *i.e.*, comparable to those of the 09:40 aftershock featured on Figure 5c (13 s). The four events with longer *T* waves all have spectral amplitudes at least one order of magnitude lower than those of the 09:02 event. Their locations are shown as triangles on Figure 2. They are:

* *Event at 09:29:14; 2.82°S, 141.82°E*

This earthquake relocates at the western end of the mainshock rupture. It is part of a series of puffs of energy lasting more than 300 s, but is the only one whose *T* waves last more than 20 s.

* *Event at 10:51:50; 2.85°S, 142.06°E (m_b = 4.3)*

This shock relocates essentially at the 09:02 epicenter; it could involve a rockslide or small slump on the Western wall of the amphitheater.

* *Event at 12:15:10 (m_b = 3.2)*

This small event could not be meaningfully relocated, with its error ellipse extending more than 150 km in the NNW-SSE direction. *T* waves at WK31 last 47 s, but at a level 100 times lower than for the 09:02 slump.

* *Event at 13:52:55; 3.00°S, 142.26°E (m_b = 4.5)*

This earthquake relocates North of Malol, with its error ellipse intersecting the canyon surveyed by R.V. Kairei (TAPPIN *et al.*, 1999). Its *T* wave at WK31 lasts 47 s, but with a spectral amplitude 30 times less than its 09:02 counterpart. This event may be interpreted as having triggered a small landslide or slump along the walls of the canyon.

In summary, we find that the 09:02 slump is clearly unique (among aftershocks, and discounting the large 09:09–09:10 doublet) in amplitude rather than in nature: we document that several events with comparable *T*-wave signatures but much lower amplitudes, did take place in the general area off the Sandaun coast on that day. We interpret them as minor underwater rockslides or slumps; their occurrence would certainly be expected in the aftermath of a sizable mainshock. Only the 09:02 slump was of a size sufficient to generate a locally destructive tsunami.

In conclusion, the present study identifies the 09:02 event as featuring singular source characteristics which can be reconciled with the occurrence of a major slump on the southern wall of the amphitheater identified by TAPPIN *et al.* (1999). In this respect, our work allows the timing of the occurrence of the slump mapped by SWEET and SILVER (2003), and which the shipboard surveys could only describe as "fresh."

◄

Figure 8

Spectrogram of the regional oceanic phase P_o (WALKER, 1977) recorded at the Wake Island seismic station from the 09:02 event (a) and the 09:40 aftershock (b). In both instances, we use a 200-s window analyzed in the frequency band $1.5 < f < 8$ Hz. Note the longer duration, higher frequency spectrum and lower amplitude of the 09:02 signal.

By suggesting that slumps may have specific and enhanced T-wave signatures while at the same time generating locally catastrophic tsunamis, we give a new impulse to the longstanding idea that T-wave and tsunami excitation may somehow be correlated. This was first suggested by EWING *et al.* (1950), based on their intuition that the excitation of both types of waves would require an extremely shallow source. However, WADATI and INOUYE (1953) countered this suggestion based on the observation of T waves from on-land sources, and we now know that the complex mechanics of T-wave generation can result in substantial T waves even from the deepest earthquakes (OKAL and TALANDIER, 1997, 1998). Also, TALANDIER (1966) has shown in the case of the Kuriles event of 20 October 1963 that a "tsunami earthquake," characterized by enhanced tsunami generation (relative to its surface waves) was actually deficient in T-wave excitation, a conclusion confirmed in the case of recent tsunami earthquakes such as Nicaragua (1992) or Peru (1996) (OKAL *et al.*, 2001). In summary, in the case of elastic dislocations, little if any correlation can be expected between the excitation of T waves (by the highest-frequency components of the source spectrum), and tsunamis (by their static or lowest-frequency components). On the other hand, a slump such as the 09:02 event is of course a completely different source, which being in contact with the water column, possesses the potential for direct excitation of acoustic energy into the SOFAR channel.

The idea that major tsunamis could be generated by underwater slumps is far from new: it goes back to such early visionaries as MILNE (1898) and MONTESSUS DE BALLORE (1907), with GUTENBERG (1939) arguing that slumps were indeed the main generator of tsunamis. By contrast with subaerial events such as the Mount St. Helens landslide (KANAMORI *et al.*, 1984), underwater slumps are generally impossible to witness directly; thus only a handful of such events have been documented and thoroughly studied, mostly on the basis of the observation of cable breaks or turbidity currents. For example, HOUTZ (1962) modeled the 1953 tsunami at Suva, Fiji as generated by an underwater slump estimated at 0.3 km^3, and HASEGAWA and KANAMORI (1987) modeled the 1929 Grand Banks earthquake and tsunami as involving a 500 km^3 submarine landslide. Similarly, AMBRASEYS (1960) proposed that the tsunami of 09 July 1956 at Amorgos in the Southern Aegean Sea emanated from a landslide, a conclusion based on reports of turbidity in the seawater, and upheld by recent marine surveys (PERISSORATIS and PAPADOPOULOS, 1999). Finally, SOLOV'YEV *et al.* (1992) identified tsunami waves on tidal gauge records in Southern Spain, following the land-based 1954 and 1980 Algerian earthquakes, the former having triggered in the Mediterranean Sea a turbidity current documented by numerous cable breaks (HEEZEN and EWING, 1955).

Apart from those few well-documented cases, several authors have speculated that landslides or slumps may have played a role in the generation of other significant tsunamis. These would include the still unresolved case of the 1946 Aleutian event (KANAMORI, 1985; PLAFKER *et al.*, 2002), the 1975 Kalapana earthquake on the southern coast of Hawaii (EISSLER and KANAMORI, 1987), and a few other intriguing events such as the 22 June 1932 aftershock of the large Colima–Jalisco earthquake of 03 June 1932.

It is probable that underwater landslides and slumps are ubiquitous phenomena on the ocean floor, and that the seismic and acoustic record of the past few decades includes literally myriads of them. As noted for example by AMBRASEYS (1960), the deformation field of a slump is dipolar in nature, and therefore, the tsunami that it raises affects mostly the near field. As such, only those slump events close enough to a populated shoreline may generate a detectable tsunami and thus be recognized. Yet, any assessment of the tsunami risk posed by underwater slumps requires accurate models of their population statistics, and possible recurrence patterns. Such investigations clearly represent a major challenge in the field of coastal hazards, which may constitute the scientific legacy of the catastrophic tragedy in Papua New Guinea.

Acknowledgments

I thank Costas Synolakis for arranging my participation in the Post-Tsunami Field Survey (under funding from the National Science Foundation), and for many subsequent discussions. I am grateful to an anonymous reviewer for constructive comments. *T*-wave records used in this study were obtained from the IRIS data management center, the POSEIDON data center, the Prototype International Data Center of the Comprehensive Nuclear Test Ban Treaty, or kindly provided by Cecily Wolfe (PELEnet), Jesse Williams (MBARI), and Bor-Shouh Huang (Taiwan network). Several figures were drawn using the GMT software (WESSEL and SMITH, 1991). Some aspects of this research were supported by the Defense Threat Reduction Agency, under contract DTRA01-00-C-0065.

REFERENCES

AMBRASEYS, N. N. (1968), *The Seismic Sea Wave of July 9, 1956 in the Greek Archipelago*, J. Geophys. Res. *65*, 1257–1265.

CAPLAN-AUERBACH, J., FOX, C. G., and DUENNEBIER, F. K. (2001), *Hydroacoustic Detection of Submarine Landslides on Kilauea Volcano*, Geophys. Res. Lett. *28*, 1811–1813.

DAVIES, H. L. (1998), *The Sissano Tsunami*, Univ. Papua New Guinea, Port Moresby.

DZIEWONSKI, A. M., EKSTRÖM, G., and MATERNOVSKAYA, N. (1999), *Centroid-moment Tensor Solutions for July–September*, 1998, Phys. Earth Planet. Inter. *114*, 99–107.

EISSLER, H. K. and KANAMORI, H. (1987), *A Single-force Model for the 1975 Kalapana, Hawaii Earthquake*, J. Geophys. Res. *92*, 4827–4836.

EWING, W. M., WOOLLARD, G. P., VINE, A. C., and WORZEL, J. L. (1946), *Recent Results in Submarine Geophysics*, Geol. Soc. Am. Bull. *57*, 909–934.

EWING, W. M., TOLSTOY, I., and PRESS, F. (1950), *Proposed Use of the T Phase in Tsunami Warning Systems*, Bull. Seismol. Soc. Am. *40*, 53–58.

GUTENBERG, B. (1939), *Tsunamis and Earthquakes*, Bull. Seismol. Soc. Am. *29*, 117–126.

HASEGAWA, H. S. and KANAMORI, H. (1929), *Source Mechanism of the Magnitude 7.2 Grand Banks Earthquake of November 18, 1929: Double-couple or Submarine Landslide?* Bull. Seismol. Soc. Am. *77*, 1984–2004.

HEEZEN, B. C. and EWING, M. (1955), *Orléansville Earthquake and Turbidity Currents*, Bull. Am. Assoc. Petrol. Geol. *39*, 2505–2514.

HEINRICH, P., PIATANESI, A., HÉBERT, H., and OKAL, E. A. (2000), *Near-field Modeling of the July 17, 1998 Event in Papua New Guinea*, Geophys. Res. Lett. *27*, 3037–3040.

HOFFMAN, I., SYNOLAKIS, C. E., and OKAL, E. A. (2002), *Systematics of the Distribution of Tsunami Run-up along Coastlines in the Near-field for Dislocation Sources with Variable Parameters*, Eos, Trans. Am. Geophys. Un. *83*, (22), WP54 (abstract).

HURUKAWA, N., TSUJI, Y., and WALUYO, B. (2003), *The 1998 Papua New Guinea Earthquake and its Fault Plane Estimated from Relocated Aftershocks*, Pure Appl. Geophys., *160*, 1829–1841.

HOUTZ, R. E. (1962), *The 1953 Suva Earthquake and Tsunami*, Bull. Seismol. Soc. Am. *52*, 1–12.

KANAMORI, H. (1985), *Non-double-couple Seismic Source*, Proc. XXIIIrd Gen. Assemb. Intl. Assoc. Seismol. Phys. Earth Inter. p. 425, Tokyo, (abstract).

KANAMORI, H., GIVEN, J. W., and LAY, T. (1984), *Analysis of Seismic Waves Excited by the Mount St. Helens Eruption of May 18, 1980*, J. Geophys. Res. *89*, 1856–1866.

KIKUCHI, M., YAMANAKA, Y., ABE, K., MORITA, Y., and WATADA, S. (1998), *Source rupture process of the Papua New Guinea earthquake of July 17, 1998 inferred from teleseismic body waves*, Eos. Trans. Amer. Geophys. Un. *79 (45)*, F573 (abstract).

MCCUE, K. F. (1998), *An AGSO Perspective on PNG's Tsunamigenic Earthquake of 17 July 1998*, Austral. Geol. Intl. *9*, 1–2.

MILNE, J., *Earthquakes and other Earth movements*, 376 pp. (Paul, Trench, Trübner and Co., London, 1898).

MONTESSUS DE BALLORE, F. La Science Séismologique (A. Colin, Paris, 1907), 579 pp.

OKAL, E. A. (2001a), *T-phase Stations for the International Monitoring System of the Comprehensive Nuclear-Test Ban Treaty: A Global Perspective*, Seismol. Res. Lett. *72*, 186–196.

OKAL, E. A. (2001b), *Converted T Phases Recorded on Hawaii from Polynesian Nuclear Tests: A Preliminary Report*, Pure Appl. Geophys. *158*, 457–474.

OKAL, E. A. and TALANDIER, J. (1989), *M_m: A Variable Period Mantle Magnitude*, J. Geophys. Res. *94*, 4169–4193.

OKAL, E. A. and TALANDIER, J. (1997), *T Waves from the Great 1994 Bolivian Deep Earthquake in Relation to Channeling of S-Wave Energy up the Slab*, J. Geophys. Res. *102*, 27,421–27,437.

OKAL, E. A. and TALANDIER, V. (1998), Correction to "*T Waves from the Great 1994 Bolivian Deep Earthquake in Relation to Channeling of S-wave Energy up the Slab*," J. Geophys. Res. *103*, 2793–2794.

OKAL, E. A., ALASSET, P.-J., HYVERNAUD, O., and SCHINDELÉ, F. (2003), *The Deficient T Waves of Tsunami Earthquakes*, Geophys. J. Intl. *152*, 416–432.

PERISSORATIS, C. and PAPADOPOULOS, G. (1999), *Sediment Instability and Slumping in the Southern Aegean Sea and the Case History of the 1956 Tsunami*, Mar. Geol. *161*, 287–305.

PLAFKER, G., OKAL, E. A., and SYNOLAKIS, C. E. (2002), *A New Survey of the 1946 Aleutian Tsunami in the Near-Field: Evidence for a Large Underwater Landslide at Davidson Bank*, Seismol. Res. Lett. *73*, 259 (abstract).

SOLOV'YEV, S. L., CAMPOS-ROMERO, M. L., and PLINK, N. L. (1992), *Orléansville Tsunami of 1954 and El Asnam Tsunami of 1980 in Alboran Sea (Southwestern Mediterranean Sea)*, Izv. Earth Phys. *28*, 739–760.

SWEET, S. and SILVER, E. A. (2003), *Tectonics and Slumping in the Source Region of the 1998 Papua New Guinea Tsunami from Seismic Reflection Images*. Pure Appl. Geophys. this issue.

SYNOLAKIS, C. E., BARDET, J.-P., BORRERO, J. C., DAVIES, H. L., OKAL, E. A., SILVER, E. A., SWEET, S., and TAPPIN, D. (2002), *The Slump Origin of the 1998 Papua New Guinea Tsunami*, Proc. Roy. Soc. London, Ser. A, *458*, 763–789.

TALANDIER, J. (1966), *Contribution à la prévision des Tsunamis*, C. R. Acad. Sci. Paris, *263B*, 940–942.

TALANDIER, J. and BOUCHON, M. (1979), *Propagation of High-frequency P_n Waves at Great Distances in the Central and South Pacific, and its Implication for the Structure of the Lower Lithosphere*, J. Geophys. Res. *84*, 5613–5620.

TALANDIER, J. and OKAL, E. A. (1998), *On the Mechanism of Conversion of Seismic Waves to and from T Waves in the Vicinity of Island Shores*, Bull. Seismol. Soc. Am. *88*, 621–632.
TALANDIER, J. and OKAL, E. A. (2001), *Identification Criteria for Sources of T Waves Recorded in French Polynesia*, Pure Appl. Geophys. *158*, 567–603.
TAPPIN, D. R. and 18 co-authors (1999), *Sediment Slump Likely Caused 1998 Papua New Guinea Tsunami*, Eos, Trans. Am. Geophys Un. *80*, pp. 329, 334 and 340.
WADATI, K. and INOUYE, W. (1953), *On the T Phase of Seismic Waves Observed in Japan*, Proc. Japan Acad. *29*, 47–54.
WALKER, D. A. (1977), *High-frequency P_n and S_n Phases Recorded in the Western Pacific*, J. Geophys. Res. *82*, 3350–3360.
WESSEL, P. and SMITH, W. H. F. (1991), *Free Software Helps Map and Display Data*, Eos, Trans. Am. Geophys. Un. *72*, pp. 441 and 445–446.
WYSESSION, M. E., OKAL, E. A., and MILLER, K. L. (1991), *Intraplate Seismicity of the Pacific Basin*, 1913–1988, Pure Appl. Geophys. *135*, 261–359.

Pure appl. geophys. 160 (2003) 1865–1877
0033–4553/03/111865–13
DOI 10.1007/s00024-003-2410-4

Pure and Applied Geophysics

Triggering Mechanisms of Slope Instability and their Relationship to Earthquakes and Tsunamis

S. G. WRIGHT[1] and E. M. RATHJE[1]

Abstract—Submarine and shoreline slope failures that accompany large earthquakes and large tsunamis are triggered by several mechanisms. Triggering mechanisms range from direct effects, such as inertial forces from earthquake shaking, to indirect effects, such as rapid drawdown that occurs when an earthquake-generated tsunami first approaches a shoreline. Soil shear strength also plays an important role in earthquake-related slope failures. Earthquakes change the shear strength of the soil by inducing excess pore water pressures. These excess pore water pressures change with time after the earthquake, resulting in changes in shear strength and slope stability with time. This paper reviews earthquake-related triggering mechanisms for submarine and shoreline slope failures. The variation in shear strength with time following an earthquake is examined and it is shown that delayed slope failures after an earthquake can occur as a result of changes in earthquake-induced excess pore water pressures and shear strength with time.

Key words: Slope stability, tsunamis, earthquakes, submarine slides, geotechnical engineering.

1. Introduction

Submarine and shoreline slope failures often accompany large earthquakes and are also associated with large, damaging tsunamis. The 1998 Aitape earthquake in Papua New Guinea represents one of the largest and most recent occurrences of a large earthquake accompanied by a large submarine slope failure and a large tsunami (KAWATA *et al.*, 1999; TAPPIN *et al.*, 1999). The more recent 1999 Kocaeli earthquake in Turkey produced reports of slope movements (lateral spreading) along the shoreline and possibly tsunamis (EERI, 2000). The slope failures induced by the Aitape and Kocaeli earthquakes are still the subject of ongoing research, but illustrate the close association that can exist between earthquakes, tsunamis and slope failures.

Several earthquake-related and tsunami-related mechanisms trigger slope instability. The role that each mechanism plays in causing instability varies from earthquake to earthquake. Thus, while slope failures are observed in most earthquakes, the underlying mechanisms for these slope failures may be quite

[1] University of Texas at Austin, Austin, Texas, U.S.A.

different. Furthermore, if a slope failure is associated with a tsunami, it is possible that the slope failure is either the cause or the result of the tsunami.

This paper examines the potential mechanisms for triggering a slope failure as a result of an earthquake. In any slope failure the shear strength of the soil is one of the most important considerations. Most slope failures that occur during earthquakes are caused by reductions in the soil shear strength, which lead to a reduced ability of the soil to resist the driving forces in the slope. Driving forces are principally due to gravity, but earthquakes and large water waves can also produce driving forces. Because the strength of soil is often reduced as a result of ground shaking, it is often assumed that the critical stability condition occurs at the end of the earthquake and that the stability of the slope increases with time after the earthquake. However, as shown in this paper, it is possible for the overall strength in the slope to decrease for a short time after the earthquake. Accordingly, the slope may reach its most critical stability condition either during, or some time after, the earthquake. Understanding the variation in soil strength with time during and after an earthquake is important to understanding the mechanics of slope failures.

2. Direct Triggering Mechanisms

2.1 Acceleration-induced Sliding

During strong ground motion from an earthquake the soils in a slope are subjected to horizontal and vertical accelerations (Fig. 1a). The inertial forces associated with these accelerations may momentarily reduce the factor of safety below 1.0, initiating downslope movement. If the accelerations are large enough or continue for an extended period of time, the slope may deform excessively or a complete failure of the slope may occur.

The impact of earthquake accelerations on slope stability is often evaluated using conventional limit equilibrium slope stability analysis procedures, where an

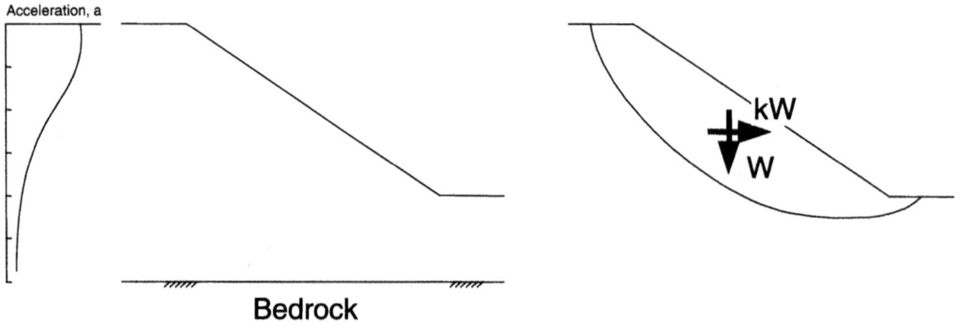

Figure 1
Simplified representation of earthquake accelerations used in pseudo-static slope stability analyses.

additional horizontal static force is applied to simulate earthquake shaking (Fig. 1b). Analyses that model the earthquake as an equivalent static force are commonly referred to as pseudo-static analyses. For pseudo-static analyses, the horizontal static force is calculated by multiplying the soil weight by a seismic coefficient, k, that represents the earthquake shaking. Seismic coefficients used in pseudo-static analyses are empirically derived to represent an equivalent seismic load. More rigorous analyses used to evaluate slope instability during earthquake shaking include decoupled Newmark-type sliding block analyses (NEWMARK, 1965; MAKDISI and SEED, 1978), coupled Newmark-type sliding analyses (*e.g.*, RATHJE and BRAY, 2000), and nonlinear dynamic finite element analyses (*e.g.*, PREVOST *et al.*, 1985). Newmark-type sliding block analyses calculate the permanent displacement in a slope by defining a yield acceleration (i.e., seismic coefficient that results in a factor of safety of 1.0) and using acceleration-time histories to evaluate when the yield acceleration is exceeded. In decoupled analysis, a seismic loading-time history for the slope (SEED and MARTIN, 1966) is calculated with a dynamic response analysis that ignores sliding. This seismic loading-time history is used as input into a Newmark sliding block analysis to calculate displacement. Because the dynamic response calculation is decoupled from the sliding displacement calculation, this analysis is termed decoupled. Alternatively, in a coupled analysis the dynamic response calculation and sliding displacement calculation are performed simultaneously. Finally, nonlinear dynamic finite element analysis can be used to evaluate earthquake-induced deformations throughout the slope. Internal forces from vertical accelerations may be included in any of these analyses; however, horizontal accelerations typically are the most important for slope deformations (YAN *et al.*, 1996).

Although earthquakes may produce substantial accelerations, it is unusual for the seismic forces alone to be high enough to cause large slope failures. Instead, the accelerations may serve as an initiation mechanism, which combines with other mechanisms to cause a slope failure.

2.2 Liquefaction-induced Sliding

The cyclic shearing produced by shear waves generated during earthquakes can weaken many soils. Earthquake-induced strength loss is particularly significant in saturated deposits of loose, cohesionless soils (silts and sands). These soils tend to compress during groundshaking, yet are unable to do so because there is not enough time for water to escape from the soil voids during the short duration of strong groundshaking (often less than 60 seconds). Earthquake-induced liquefaction is very likely in loose, cohesionless soils offshore because these soils are almost always saturated.

Liquefaction-induced sliding is initiated when strong ground motion reduces the shear strength of the soil below that which is required to resist the static and cyclic stresses imposed by the earthquake. A combination of reduced shear strength and earthquake-induced inertial forces probably contribute to most slope failures that occur during earthquakes.

2.3 Fault Rupture-induced Sliding

Noticeable changes in the ground surface profile can occur as a result of earthquakes when fault rupture extends to the ground surface (Fig. 2). Surface fault rupture was evident both onshore and offshore during the recent Kocaeli earthquake in Turkey (EERI, 2000). Even small changes in the surface profile of a slope can trigger sliding if the slope is marginally stable, and many natural slopes are formed by processes that leave the soil in a marginally stable condition. For example, the slopes found in actively advancing delta fronts, such as the Mississippi River Delta in the Gulf of Mexico, are only marginally stable. For these cases, any noticeable change in the surface profile due to fault rupture is likely to trigger sliding.

The extent to which fault rupture-induced sliding occurs is not well documented. Presumably, such sliding could be confined to the immediate area of the fault rupture or could extend over a much larger area, with the fault rupture only serving as a local trigger. Fault rupture-induced sliding can be evaluated with conventional slope stability analyses. Given the soil profile and slope geometry, along with the soil properties (shear strength, unit weight) and groundwater levels, stability can be evaluated using the post-rupture topography.

3. Indirect Triggering Mechanisms

Fault rupture, along with earthquake-induced accelerations and liquefaction-induced soil strength loss, represents direct effects of earthquakes on slope stability. Failure initiated by these mechanisms may be expected either during or immediately

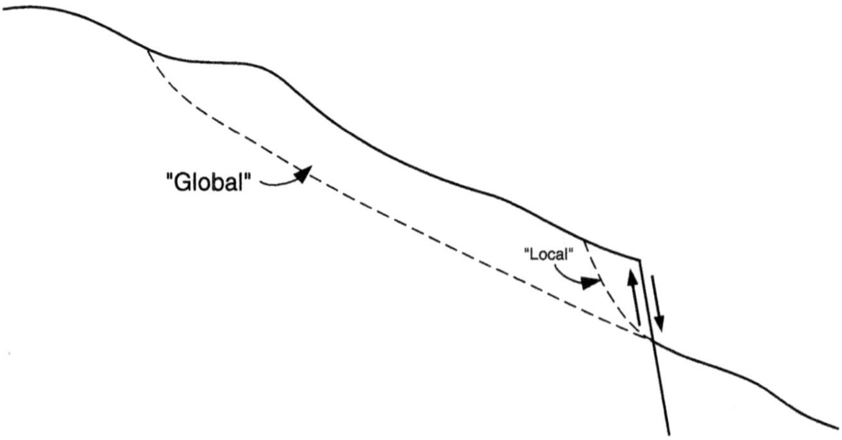

Figure 2
Fault rupture-induced slope instability.

following an earthquake. However, there are other mechanisms that may be an indirect result of the earthquake and may not occur until some time after the earthquake. These mechanisms are discussed below.

3.1 Water Wave-induced Sliding

Large ocean waves can generate significant changes in stress on the seafloor, particularly for water depths less than 100 m. Wave-induced stresses produce a disturbing moment (couple) that tends to shear, and possibly displace, the soil downslope (Fig. 3). This mechanism was the apparent cause of a large slide that resulted in the loss of at least one offshore oil drilling platform during hurricane Camille in the Gulf of Mexico (STERLING and STROHBECK, 1975). HENKEL (1970) proposed a simple, limit equilibrium model based on classical methods to predict instability caused by ocean waves. Henkel's model and more sophisticated finite element models were subsequently used to confirm that the observed slope failures during hurricane Camille in the Gulf of Mexico were caused by large, hurricane-generated waves (BEA *et al.*, 1983).

The advancing wavefront from a tsunami is capable of generating large hydrodynamic pressures on the seafloor, similar to large ocean waves, that may produce soil movements and slope instabilities. Suitable analytical models have been developed for computing the effect of such hydrodynamic pressures on the seafloor, once the wave hydrodynamics and seafloor soil properties are established (SCHAPERY and DUNLAP, 1977, 1978; WRIGHT and DUNHAM, 1972; WRIGHT, 1976).

3.2 Rapid Drawdown

When tsunamis approach the shoreline, water levels suddenly drop. The sudden lowering of water adjacent to a slope at the shoreline ("rapid drawdown") removes the resisting force exerted by the water and induces an undrained loading condition

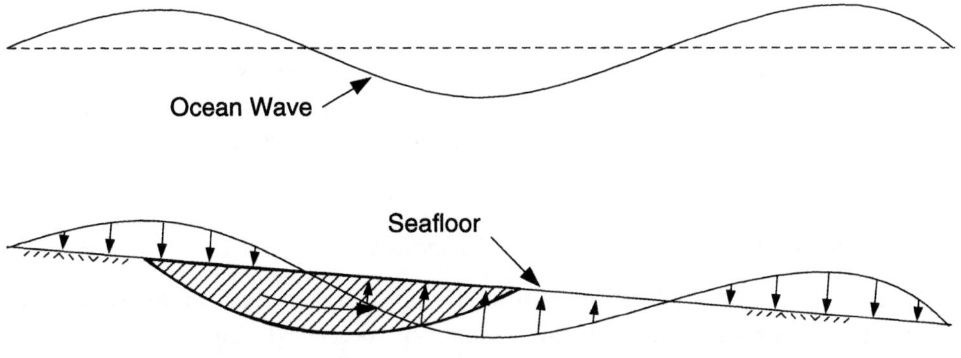

Figure 3
Rotational seafloor instability produced by ocean wave-induced hydrodynamic pressures.

because the soil has no time to drain. This type of loading is widely recognized as an important triggering mechanism for slope instability in earth dams. All earth dams and many riverbank slopes are susceptible to failure by rapid drawdown and are routinely designed to prevent such occurrences. A sudden drop of water level at the shoreline due to a tsunami also can be expected to produce slope instability, unless the slope has a relatively large margin of safety prior to the event. Any slope composed of fine-grained soil with a static factor of safety less than about 1.5 may be considered susceptible to failure if significant drawdown occurs as a result of a tsunami.

A large shoreline landslide was induced by drawdown from a tsunami at the Port of Nice in 1979 (SEED *et al.*, 1988). It is believed that a large, submarine landslide that occurred 15 km off-shore generated the tsunami that induced this landslide. The shoreline slope at the Port of Nice was stable prior to the tsunami, and the 3-m lowering of the sea level due to the tsunami increased the shear stresses in the slope only 1.5 to 2%. However, a continuous seam of loose, liquefiable sand was located at a depth of about 25 m. Parts of this sand seam had relatively large initial shear stresses compared to the normal stresses, making the soil susceptible to flow liquefaction under low levels of undrained loading. The small increase in shear stress induced by drawdown from the tsunami was enough to cause a strain-softening response in the sand and trigger a progressive failure of the slope. Other examples of shoreline landslides induced by drawdown are discussed by ANDERSEN and BJERRUM (1967) and EDGERS and KARLSRUD (1982).

3.3 Combined Triggering Mechanisms

All of the mechanisms and processes described above are known to produce slope failures both at the shoreline and on the seafloor. In the case of earthquakes and submarine slopes, it is certainly possible that several of these mechanisms may combine to cause slope failures. For example, earthquake-induced accelerations and liquefaction may trigger a large submarine slide. This slide may produce a tsunami, that in turn causes either rapid drawdown at the shoreline or significant wave-induced hydrodynamic pressures on the seafloor. Such occurrences may be overlooked because the interactions among several mechanisms are not typically considered in evaluating earthquake-induced slope failures, especially over relatively large areas where the different events and mechanisms may be separated by distances of several-to-hundreds of kilometers.

3.4 Shear Strength and Delayed Failure Mechanisms

Slope failures occur when the various driving forces exceed the resisting forces due to the soil strength. Most of the mechanisms described above are mechanisms that increase the net driving forces. Of the mechanisms discussed, only liquefaction represents a mechanism where the resisting forces of the soil (shear strength) are reduced. A second important failure mechanism that is triggered by a reduction in soil strength is related to dissipation of shear-induced excess pore water pressures after an earthquake.

3.4.1 Conceptual framework

Positive shear-induced excess pore water pressures are generated in loose, compressible soils, reduce the shear strength, and ultimately may cause liquefaction. After an earthquake, the strength of these soils increases as the shear-induced pore water pressures dissipate and the soil consolidates. In contrast to soils that compress, denser soils tend to dilate (expand) when they are sheared. When these soils are sheared suddenly without time to drain, the pore water pressures decrease. After an earthquake, these soils gradually expand and the pore water pressures increase back to their original, pre-shear values. This mechanism causes the soil shear strength to decrease with time after an earthquake.

Generally, soils that are loose or are subjected to relatively high confining stresses tend to generate positive excess pore water pressures when sheared without drainage. In contrast, pore water pressures decrease in soils that are either dense or subjected to low confining stresses when these soils are sheared without drainage. It is possible for both types of behavior to occur in a single slope because confining stresses vary significantly throughout the slope. This situation, where both compressive and dilative behavior are found, helps explain why slope failures often occur some time after an earthquake, even with no additional loading from aftershocks. This failure mechanism is illustrated by the hypothetical case described in the next paragraph.

Consider the slope and potential slip surface shown in Figure 4. Suppose that an earthquake occurs and the strong ground motion and cyclic shear stresses result in an increase in pore water pressures along the deeper portions of the slip surface (Zone B

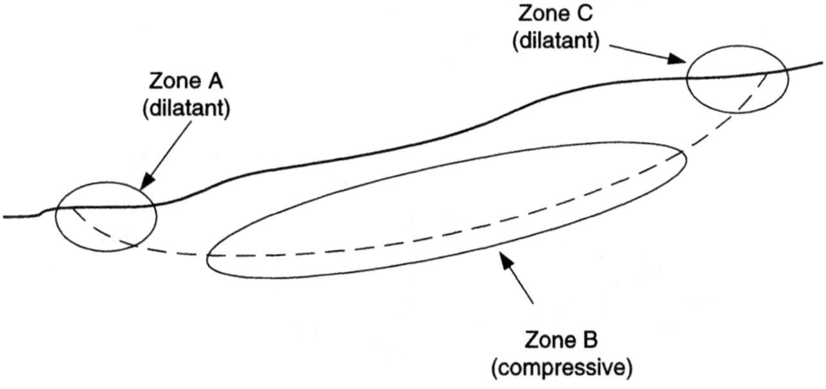

Figure 4
Compressive and dilatant shear zones along deep-seated failure surface.

in Fig. 4). The reduction in strength at depth causes the shear stresses in the soil along the shallower portions of the slip surface, identified as Zones A and C in Figure 4, to increase; particularly if the strength at depth becomes less than the initial driving shear stress. If the soil in the shallower zones is dilatant, the pore water pressures will decrease and the soil strength will increase near the surface. As time passes, the pore water pressures generated by shear will slowly dissipate as the soil drains. If most of the drainage and dissipation occurs due to water flowing into and out of the soil at the seafloor surface, the pore pressures will change, first primarily at the shallow depths, near the seafloor surface. Thus, the soil near the surface will lose strength, while the strength of the soil at depth remains essentially the same. The overall effect is a further reduction in strength along the slip surface. If the strength is reduced enough, failure may occur some time after the initial seismic event which induced the excess pore water pressures.

3.4.2 Illustrative example

To illustrate the variations of soil strength with time as a result of dissipation and redistribution of excess pore water pressure, consider the hypothetical submarine slope and deep potential sliding surface illustrated in Figure 5. For this illustration the potential sliding surface is shown extending to a depth of 300 meters, with dilatant behavior near the surface and compressible behavior at depth.

The distribution of excess pore water pressures generated by shaking is developed using estimates of excess pore pressures caused by earthquake shaking in the compressive zone, and active and passive shear in the dilatant zones (Fig. 5). We will assume that in the deepest one-third (100 m) of this deposit positive excess pore water pressures, equal to 80 percent of the original effective stress, are developed as a result of earthquake shaking. In the upper 100 meters of the soil deposit we have assumed the soil dilates during shear. Estimates of the change in pore water pressure due to undrained shear in this region were made using Skempton's pore water pressure coefficients, A and B (SKEMPTON, 1954). Assuming a value of -0.1 for A ($B = 1$ for saturated soil), the reduction in pore water pressure ($-\Delta u$) was estimated to be 1.3 times the original, pre-earthquake effective stress in the active zone along the upslope

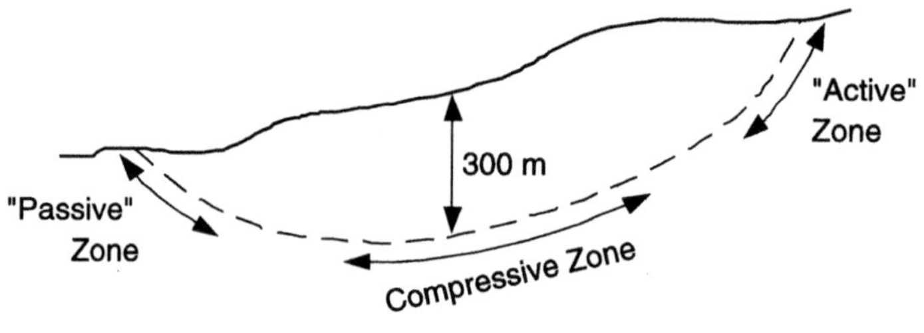

Figure 5
Hypothetical shear zone for deep-seated slope instability used in illustrative example.

portion of the shear surface. Assuming the same values for the pore pressure coefficients A and B in the passive zone of the shear surface near the toe of the slope, the reduction in pore water pressure was estimated as 0.4 times the original effective stress. On average, the reduction in pore water pressure near the two ends of the shear surface was estimated to be 0.8 times the effective vertical stress. These changes in pore water pressure represent the changes that would occur if the shear stresses increased at the two ends of the shear surface and the soils tended to dilate without time to drain. Such increases in shear stress and reductions in pore water pressure occur in the shallower soils as the deeper soils become weaker from earthquake-induced increases in pore water pressure. The changes in pore water pressure that occur in the upper portion of the deposit are not directly produced by cyclic loading, but rather are indirectly produced by static shear without drainage after the earthquake.

In the middle 100 meters of the hypothetical soil deposit the pore water pressures were assumed to vary linearly between the compressive and dilative zones. A submerged unit weight of 10 kN/m^3 was assumed to calculate the initial effective stress and the initial excess pore water pressures were calculated as described previously. The resulting distribution of initial excess pore water pressures is shown in Figure 6. The distribution of excess pore pressures shown in Figure 6 represents

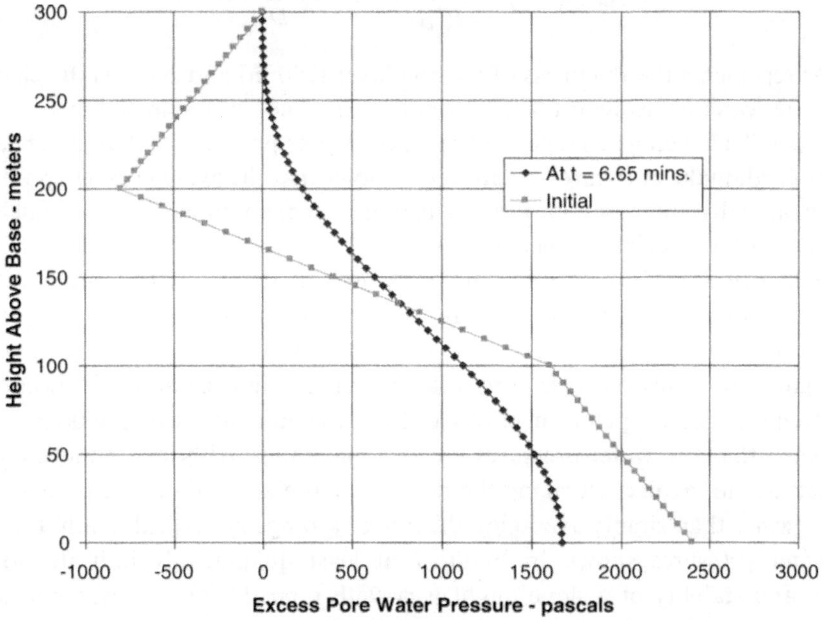

Figure 6
Variation in excess pore water pressures with depth in hypothetical deposit 300 meters thick – initial values and values 6.65 minutes after end of earthquake.

the excess pore water pressure that exists immediately after the earthquake and before any dissipation (drainage) occurs.

After an initial distribution of excess pore water pressures was established, the dissipation of pore water pressures was calculated using conventional one-dimensional consolidation (diffusion) theory and assuming drainage only occurs at the seafloor surface. A coefficient of consolidation, c_v, of 600 m^2/min was assumed based on an estimated compressibility of sand of 1×10^8 N/m^2 and a hydraulic conductivity of 0.1 cm/sec. The initial excess pore water pressure distribution shown in Figure 6 was assumed. A finite difference numerical scheme was used to solve for the excess pore water pressures as a function of time following the earthquake.

If the initial excess pore water pressures are all positive, the average pore water pressure decreases with time and the soil shear strength and stability increase. However, if some of the excess pore water pressures are negative, the average pore water pressure in the deposit may increase with time, resulting in a reduction in shear strength. To evaluate the variation of shear strength with time for the hypothetical case, the average change in excess pore water pressure was calculated as a function of time, t, after the earthquake. The average change in excess pore water pressure, $\Delta\bar{u}_{\text{avg}}(t)$, was calculated as follows,

$$\Delta\bar{u}_{\text{avg}}(t) = \frac{\int_0^D \Delta\bar{u}(t)\, dz}{\int_0^D dz} = \frac{\int_0^D \Delta\bar{u}(t)\, dz}{D}, \tag{1}$$

where D represents the thickness of the soil layer (300 m) and $\Delta\bar{u}(t)$ is the change in excess pore water pressure at time, t, from the initial condition immediately after the earthquake. The average change in excess pore water pressure is plotted versus time after the earthquake in Figure 7. This figure shows that the average excess pore water pressure actually increases at first, reaching a maximum at approximately 6–7 minutes, and then begins to decrease.

The distribution of excess pore water pressure with depth at 6.65 minutes after the earthquake is plotted in Figure 6 along with the initial excess pore water pressure. Figure 6 shows that all of the negative excess pore water pressures have dissipated, while significant positive excess pore water pressures still exist at depth. Because the shear strength of soil depends on effective stress, the minimum strength is expected to occur when the pore water pressures are at a maximum. Although a more rigorous approach would involve averaging the pore water pressures along the potential slide surface, rather than simply averaging them over a range in vertical depth, the excess pore water pressures shown in Figure 7 at least qualitatively indicate how the strength and stability of a slope might vary with time. These pore water pressures, which reach a peak positive value between 6 and 7 minutes, indicate that it is entirely possible that a slope may fail some time after the occurrence of a large earthquake. This phenomenon was also suggested by SEED (1979) in his Rankine lecture when he

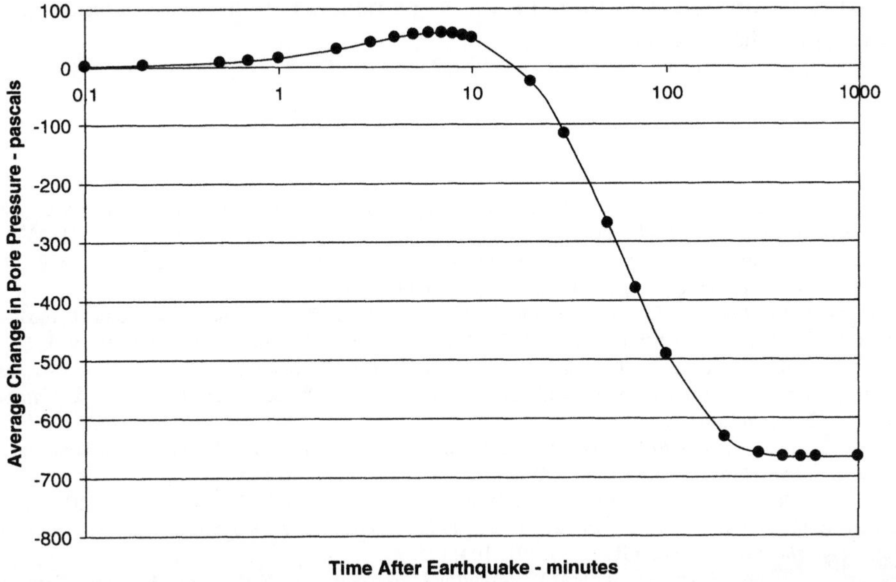

Figure 7
Variation in "average" excess pore water pressures (Δu_{avg}) in hypothetical deposit with time after end of earthquake.

observed that many slope failures associated with earthquake-induced liquefaction occurred some time after the earthquake. These observations are consistent with the hypothetical example and the changes in excess pore water pressure with time presented in Figure 7.

4. *Summary*

A number of mechanisms may contribute to earthquake-induced submarine and shoreline slope failures. Triggering mechanisms that are a direct result of an earthquake include loading from earthquake accelerations, fault rupture-induced sliding, and liquefaction and strength losses caused by undrained cyclic loading. Triggering mechanisms that are indirectly related to earthquakes include tsunami-induced rapid drawdown, hydrodynamic pressures on the seafloor from tsunamis, and dissipation and redistribution of excess pore water pressures induced by an earthquake. Depending on the mechanisms that cause sliding, failure may occur either during the earthquake or some time after the earthquake. Also, slope failures that are associated with a tsunami may either be the cause of the tsunami or represent an effect of tsunami waves. Identifying the appropriate triggering mechanisms applicable to any single earthquake and slope remains an important challenge when investigating such events.

References

ANDERSEN, A. and BJERRUM, L., *Slides in Subaqueous Slopes in Loose Sand and Silt.*, Marine Geotechnique (ed. A.F. Richards) (University of Illinois Press, Urbana, IL, 1967), pp. 221–239.

BEA, R. G., WRIGHT, S. G., SIRCAR, and NIEDORODA, A. W. (1983), *Wave-induced Slides in South Pass Block 70, Mississippi Delta*, J. Geotechn. Engin. *109*(4), 619–644.

EERI (2000), *The Kocaeli (Turkey) Earthquake of August 17, 1999*, Reconnaissance Report. Earthquake Engineering Research Institute, Oakland, California (Youd, T. L., Bardet, J.-P. and Bray, J. D., eds).

EDGERS, L. and KARLSRUD, K., *Soil Flows Generated by Submarine Slides – Case Studies and Consequences.* Proc. International Conference on Behavior of Offshore Structures, Vol. 2 (Cambridge, MA, 1982), pp. 425–437.

HENKEL, D. J. (1970), *The Role of Waves in Causing Submarine Landslides*, Geotechnique, Institutions of Civil Engineers, Great Britain, *20*(1), 75–80.

KAWATA, Y., BENSON, B. C., BORRERO, J. C., BORRERO, J. L., DAVIES, H. L., DE LANGE, W. P., IMAMURA, F., LETZ, H., NOTT, J., and SYNOLAKIS, C. E. (1999), *Tsunami in Papua New Guinea was as Intense as First Thought*, Eos Trans, AGU *80*(9), 101, 104–105.

MAKDISI, F. I., and SEED, H. B. (1978), *Simplified Procedure for Estimating Dam and Embankment Earthquake-induced Deformations*, J. Geotechn. Engin., ASCE *104*(7), 849–867.

NEWMARK, N. (1965), *Effects of Earthquakes on Dams and Embankments*, Geotechnique *15*(2), 139–160.

PREVOST, J. H., ABDEL-GHAFFAR, S. M., and LACY, S. J. (1985), *Nonlinear Dynamic Analysis of Earth Dams: A Comparative Study*, J. Geotechn. Engin., ASCE *111*(2), 882–897.

RATHJE, E. M. and BRAY, J. D. (2000), *Nonlinear Coupled Seismic Sliding Analysis of Earth Structures*, J. Geotechn. and Geoenvironm. Engin., ASCE *126*(11), 1002–1014.

SCHAPERY, R. A. and DUNLAP, W. A. (1977), *Wave-sea-bottom Interaction Study*, Texas Engin. Exp. Station Techn. Bull., July 4–13.

SCHAPERY, R. A. and DUNLAP, W. A. (1978), *Prediction of Storm-induced Sea Bottom Movement and Platform Forces*, Proc. Tenth Annual Offshore Technology Conference, Houston, Texas. Vol. 3, Paper No. OTC 3259, May 1789–1796.

SEED, H. B. (1979), *Considerations in the Earthquake-resistant Design of Earth and Rockfill Dams*, Geotechnique, Institutions of Civil Engineers, Great Britain, *29*(3), 215–263.

SEED, H. B. and MARTIN, G. R. (1996), *The Seismic Coefficient in Earth Dam Design*, J. Soil Mechanics and Foundations Div. ASCE *92*(3), 25–58.

SEED, H. B., SEED, R. B., SCHLOSSER, F., BLONDEAU, F., and JURAN, I. (1988), *The Landslide at the Port of Nice on October 16, 1979*, Report No. EERC 88-10, Earthquake Engineering Research Center, University of California, Berkeley, CA.

SKEMPTON, A. W. (1954), *The Pore Pressure Coefficients A and B*, Geotechnique, Institutions of Civil Engineers, Great Britain, *4*, 143–147.

STERLING, G. H. and STROHBECK, G. E. (1975), *The Failure of South Pass 70B Platform in Hurricane Camille*, J. Petroleum Technology, March 263–268.

TAPPIN, D. R., MATSUMOTO, T., WATTS, P., SATAKE, K., MCMURTRY, G. M., MATSUYAMA, M., LAFOY, Y., TSUJI, Y., KANAMATSU, T., LUS, W., IWABUCHI, Y., YEH, H., MATSUMOTU, Y., NAKAMURA, M., MAHOI, M., HILL, P., CROOK, K., ANTON, L., and WALSH, J. P. (1999), *Sediment Slump Likely Caused 1998 Papua New Guinea Tsunami*, Eos Trans, AGU *80*(30), 329, 334, 340.

WRIGHT, S. G. and DUNHAM, R. S. (1972), *Bottom Stability under Wave-induced Loading*, Preprints, Fourth Annual Offshore Technology Conference, Houston, Texas, Vol. 1, Paper No. OTC 1603, May 853–862.

WRIGHT, S. G. (1976), *Analyses for Wave-induced Seafloor Movements*, Proc. Eighth Annual Offshore Technology Conference, Houston, Texas, Vol. 1, Paper No. OTC 2427, May 41–53.

YAN, L., MATASOVIC, N., and KAVAZANJIAN, Jr., E. (1996), *Seismic Response of a Rigid Block on Inclined Plane to Vertical and Horizontal Ground Motions Acting Simultaneously*, Proc. Eleventh ASCE Engineering Mechanics Conference, Vol. 2, Fort Lauderdale, FL, 1110–1113.

Pure appl. geophys. 160 (2003) 1879–1894
0033 – 4553/03/111879 – 16
DOI 10.1007/s00024-003-2411-31

Pure and Applied Geophysics

Landslide-generated Tsunamis: Geotechnical Considerations

W. D. LIAM FINN[1]

Abstract—Landslide-induced tsunamis are particularly dangerous because the warning time is very short in the coastal area. Therefore it is critical to identify the potential for coastal landslides involving large masses of material. Almost invariably such large landslides will occur in material susceptible to liquefaction, particularly during earthquakes. This paper discusses characteristics of such materials, how they may be identified, and offers suggestions for estimating the potential for large deformations.

Key words: Flowslide, landslide, liquefaction, tsunami.

1. Introduction

This paper deals with a restricted class of tsunamis; those generated by submarine landslides. These tsunamis can produce very large wave heights and are particularly dangerous because of the very limited time between the landslide occurrence and the advent of the tsunami. The generation of tsunamis by submarine landslides during the Alaska earthquake has been discussed and illustrated by KACHADOORIAN (1965).

Very large landslides involving large volumes of material moving quickly over vast distances are required to generate highly damaging tsunamis. A catastrophic tsunami occurred in Papua, New Guinea on July 17, 1998. KAWATA *et al.* (1999), using methods developed by PELINOVSKY and POPLAVSKY (1997) and WATTS (1998), estimated that this tsunami could have been generated by a slide volume of approximately 10 km^3. Clearly, landslides with large volumes of material, moving considerable distances, are necessary for the generation of catastrophic tsunamis.

Many submarine landslides are most likely to be generated in slope materials that have a rather unstable structure which collapses when disturbed. This requirement is met by soils that have the potential to liquefy. The primary trigger for landslides under consideration here is groundshaking from an earthquake. In the context of the landslides being discussed here, liquefaction means the loss of a major proportion of the strength of the material as a result of the straining of saturated sands or cohesionless silts. The collapse of the soil structure due to the earthquake-induced

[1] Anabuki Chair of Foundation Geodynamics, Kagawa University, Takamatsu, Japan.

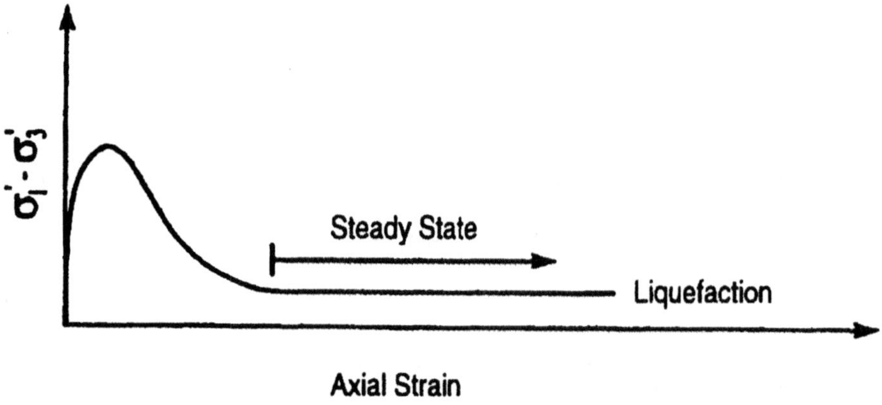

Figure 1
Stress-strain behavior of liquefiable soil.

shear strains causes very high residual porewater pressures to develop in the material. Transient normal stresses cause only transient porewater pressures which are balanced completely by the transient total normal stresses.

The stress-strain behavior of these materials is illustrated in Figure 1. After relatively small strains, peak undrained strength is reached and then the structure collapses. The shearing resistance drops to a low value that is maintained over a large range of strain. This low strength is called the residual strength or the steady-state strength. This strength controls the postliquefaction stability of the slopes. If the gravitational shear stresses substantially exceed the steady-state strength, very large post-liquefaction displacements will occur. Many submarine slopes failed by this mechanism during the Alaska earthquake. A notable example was the massive slide that occurred at Seward, which generated a very large local tsunami that caused extensive damage (SHANNON and HILTS, 1973). A description of this slide is presented below to illustrate some of the features dealt with in this paper.

2. *Landslide at Seward*

A huge submarine landslide occurred along the Seward waterfront during the 1964 Alaska earthquake. The coastal end of the slide, involving a strip of waterfront ranging in width from about 15 m to 125 m wide, and 1,250 m long, disappeared into the Bay. The south end of the breakwater, before the earthquake, was in 3 m of water; after the earthquake, the water at this location was about 40 m deep. A local tsunami was generated by the slide. The resulting waves deposited houses, boats and a tremendous amount of debris in the tidal lagoon and well above high tide level into the trees at the head of Resurrection Bay. Only about 50 m of the western part of the

Alaskan Railroad Wharf at the south end of town remained after the tsunami. All other facilities were destroyed, leaving little evidence that they had ever existed.

2.1 Mechanics of Landslide

SHANNON and HILTS (1973) made detailed studies of the Seward slide and reported on the likely mechanism as follows. "The first failure is believed to have been induced offshore and to have caused the observed recession of the water surface and the rapid drop of the Alaska Standard at her berth at the Standard Oil Dock. The slide then extended rapidly onshore. Whether the slide started as a shear slide or as a liquefaction slide is not known, but without a doubt it terminated as a liquefaction or flow slide, with the mass of material carried for great distances and spread out over the floor of Resurrection Bay as a thin layer that is now undetectable in the great depth of water. The mud boils observed at the surface of Resurrection Bay and the subsequent large waves were evidence of the extent of water turbulence caused by the slide."

2.2 Factors Contributing to the Submarine Slide

The earthquake triggered the submarine landslide. The sliding appeared to begin shortly after the start of strong groundshaking and to have ceased before the end of the earthquake.

The conditions in the submarine slopes at Seward were particularly conducive to a liquefaction flow failure. Alluvial deposits of loose sand and gravel overlie dense glacial deposits. The thickness of the overlying loose sands increases substantially towards the distal edge of the fan. According to SHANNON and HILTS (1973), these materials were probably deposited at the angle of repose and were of marginal static stability before the earthquake, particularly where the slopes were steepest. The stability was decreased further because the foreset bedding in some areas was parallel to the slopes. The submarine soils, consisting of alternating layers of sand and silt, increased in thickness to the north. The deposits contain lenses or layers of sand blanketed by relatively impervious silts. These loose sand layers are particularly susceptible to seismic liquefaction.

Another factor contributing to low stability before the earthquake is that nearly all the submarine slides resulting from the Alaskan earthquake were subject to seepage pressures from groundwater flow, from the neighboring mountain slopes, and valley drainage systems. In addition to this, the submarine slopes were subjected to seepage pressures from the ebb tide which was about 4 m below the previous high tide.

The prediction of landslide generated tsunamis, such as that at Seward, involves three important questions. Will liquefaction be triggered? Will large postliquefaction deformations occur? And if so, what is likely to be the magnitude of the resulting tsunami? The first two questions are in the realm of earthquake geotechnical engineering, and provide the basis for estimating the resulting tsunami threat by hydrodynamicists and oceanographers.

3. Triggering of Liquefaction

The primary factors controlling the liquefaction of saturated cohesionless sand and silt are the intensity and duration of earthquake shaking and the density of the soil. In engineering practice, the potential for liquefaction is assessed usually by using liquefaction assessment charts in which the *insitu* compactness of the soil is characterized by the Standard Penetration Resistance (SPT(N)), Cone Penetration Resistance, (CPT, q_c), or the *insitu* shear wave velocity (V_s). The standard charts are based on a magnitude 7.5 earthquake. Scaling factors are available for adjusting the charts to other magnitudes (NCEER, 1997).

The intensity of earthquake shaking is usually represented by the effective average cyclic stress ratio (CSR). SEED and IDRISS (1971) formulated the following equation for calculation of CSR,

$$\mathrm{CSR} = (\tau_{\mathrm{av}}/\sigma'_{\mathrm{vo}}) = 0.65(a_{\max}/g)(\sigma_{\mathrm{vo}}/\sigma'_{\mathrm{vo}})r_d, \tag{1}$$

where $a_{\max}$ is the peak horizontal acceleration at ground surface generated by the earthquake, g is the acceleration of gravity, σ_{vo} and σ'_{vo} are total and effective vertical overburden stresses, respectively, and r_d is a stress-reduction coefficient (see NCEER, 1997 for the latest values for r_d). The latter coefficient provides an approximate correction for flexibility of the soil profile. An approximate expression for r_d used in Japan is

$$r_d = 1.0 - 0.015H, \tag{2}$$

where H is the depth below the surface in meters of the location where the stress is being evaluated.

Updated liquefaction assessment charts recommended for engineering practice are given by NCEER (1997). The charts based on either the Standards Penetration Test or the Cone Penetration Test resistance are more appropriate for site specific studies, but are very expensive and time consuming for offshore regional studies. The most feasible approach for regional studies would involve the use of shear-wave velocity measurements to characterize the liquefaction potential of the submarine slopes. The liquefaction assessment chart based on shear-wave velocity is given in Figure 2, where V_{S1} is the shear-wave velocity normalized to an effective normal overburden pressure of 100 kPa (NCEER, 1997).

4. Determination of Residual Strength

4.1 Residual Strength from Field Data

A major challenge facing engineers is the selection of the appropriate residual strength of liquefied materials for use in analyses to assess the postliquefaction

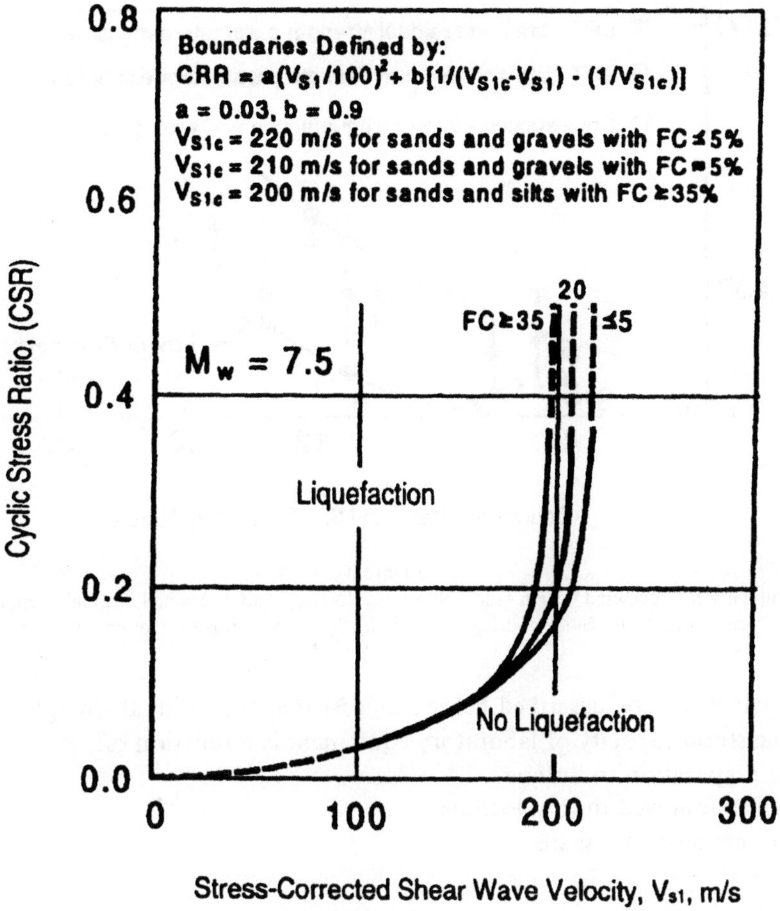

Figure 2
Liquefaction potential assessment chart based on shear wave velocity (NCEER, 1997).

stability of slopes and embankments. Seed back-analyzed a number of embankments in which significant displacements had occurred during earthquakes as a result of liquefaction. He published a chart linking the residual strength, S_r, to $(N_1)_{60\text{-}cs}$, the normalized penetration resistance for clean sands (SEED, 1987). A revised version of this chart by SEED and HARDER (1990) is shown in Figure 3. Use of this chart for estimating residual strength is widespread in engineering practice.

4.2 Residual Strength from Laboratory Tests

Extensive research has been conducted in the laboratory on the factors controlling the residual strength. Originally, it was thought to be a function only of the void ratio for a given sand (CASTRO, 1969; POULOS *et al.*, 1985). Research studies, since 1988,

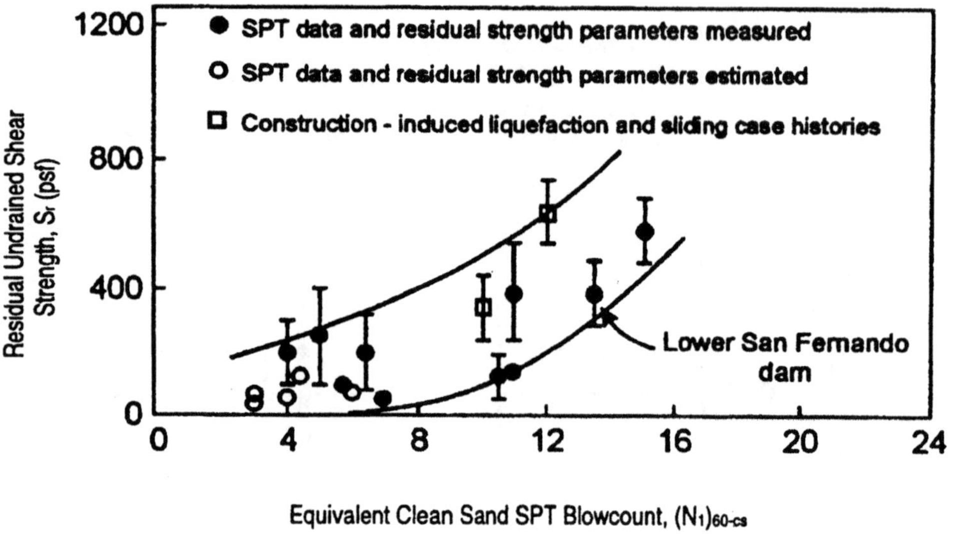

Figure 3
Relationship between corrected "clean sand" blowcount $(N_1)_{60\text{-}cs}$ and undrained residual strength, S_r, from case studies (after SEED and HARDER, 1990). Note that 20 psf is nominally 1 kPa.

some of which will be described below, suggest that the residual strength measured within the strain capacity of laboratory equipment is a function of:

- sample preparation technique,
- stress path followed during loading,
- effective confining pressure.

4.3 Effect of Sample Preparation

Many laboratory studies in liquefaction use samples prepared by moist tamping because it is the easiest way to form relatively loose samples. However, it frequently results in void ratios which are too high to be accessible to the same sand under deposition conditions in the field. Other methods in use are air pluviation and pluviation under water. VAID *et al.* (1999) has demonstrated that the residual strengths measured on samples prepared in different ways are quite different (Fig. 4).

VAID *et al.* (1999) also tested frozen field samples of two different Recent Sands to determine the residual strength. He then reconstituted the same samples to the same void ratio using pluviation in water. The reconstituted samples gave residual strengths very similar to the frozen field samples. These results are a strong argument for using pluviation under water to form representative samples of sands and silts which were originally deposited under water or were placed by hydraulic fill construction. The moist tamping method would seem to be more appropriate for forming representative samples of unsaturated compacted soils.

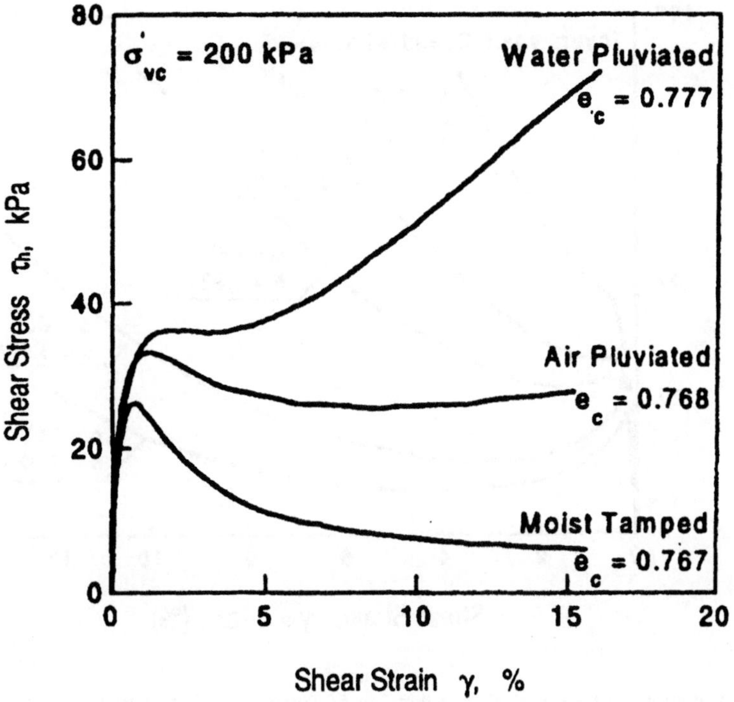

Figure 4
The effect of sample preparation on undrained simple response of Syncrude sand (VAID *et al.*, 1999).

4.4 Stress Path

UTHAYAKUMAR and VAID (1998) and YOSHIMINE *et al.* (1998) have explored the effects of stress path on residual strength over a wide range of stress paths defined by α, the inclination of the principal stress to the vertical axis of the sample, and the parameter $b = (\sigma_2' - \sigma_3')/(\sigma_2' - \sigma_3')$ which is a measure of the intermediate principal stress. The samples were tested using the hollow cylinder torsional shear device. Typical examples of this kind of data (YOSHIMINE *et al.*, 1998) are shown in Figure 5, which suggest that different residual strengths should be assigned to different parts of a liquefied zone in an embankment, depending on the predominant stress conditions in the sliding region. This selective use of shear strength for stability evaluation is not new. Bearing capacity under offshore structures in the North Sea is evaluated using compression, simple shear and extension strength data to suit stress conditions at different locations along potential sliding surfaces.

4.5 Residual Strength as a Function of Effective Confining Pressure

The practice of expressing the residual strength, S_r, as a fraction of the effective overburden pressure,p', has been followed in studies on several water-retaining and

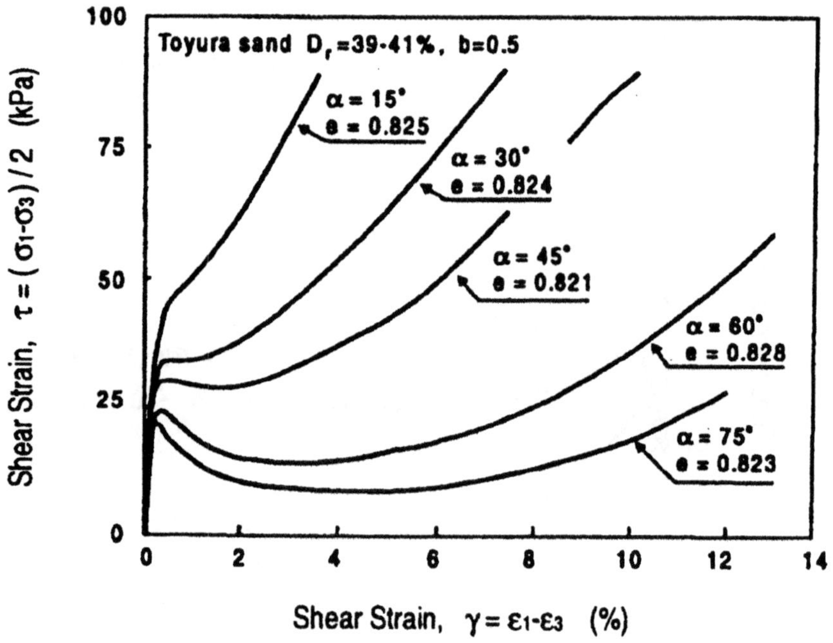

Figure 5
Effect of stress path on undrained behaviour of Toyoura sand (YOSHIMINE *et al.*, 1998).

tailings dams, since it was used on Sardis Dam in 1989 (FINN, 1990; FINN *et al.*, 1998). For the most part the ratio selected has been between 0.06 and 0.1. A value of $S_r/p' = 0.23$ was used for Duncan Dam, based on extensive testing of frozen samples (BYRNE *et al.*, 1994).

VAID and THOMAS (1994), using extension tests with frictionless end platens, determined the residual strength of Fraser River sand over a range of void ratios and the effective overburden pressures. The results clearly demonstrated the dependence of residual strength on effective overburden pressure. Their results are replotted in Figure 6; normalized with respect to the effective confining stress. Despite some scatter, the variation of Sr/p' is well represented by the curve shown (Fig. 6). Similar results were obtained for simple shear tests by VAID and SIVATHAYALAN (1996).

5. Postliquefaction Displacement Analyses

If liquefaction is triggered and the residual strength is less than the gravity induced stresses, the next challenge is to estimate the displacements on the slope and the volume of material that undergoes substantial movement. The calculation of the tsunami resulting from seismic dislocations of the seafloor are based on imposing the

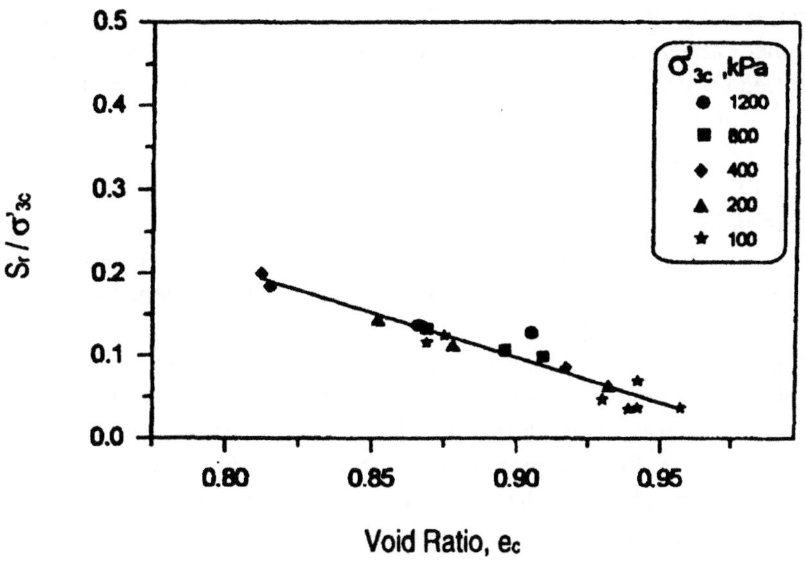

Figure 6
Relationship between the residual strength, normalized by the effective confining stress and void ratio in extension tests on Fraser River (after VAID and THOMAS, 1994).

tectonic displacements on the fluid mass and calculating the resulting wave forms. It is doubtful if this type of modeling is feasible for the more slowly moving landslides. The detailed geological and geotechnical description of the underwater slopes necessary to make even rough approximations to the deforming geometry of the slope are unlikely to be available in most cases. At present the prediction of changes in slope geometry in the case of massive flow slides moving quickly is unlikely to be accurate enough to make coupled analysis between slope and fluids reliable, but future developments could change this conclusion. Until then it seems more likely that some rough correlation with the total potential volume of the flow slide and initial depth will have to be relied upon to estimate the damage potential of the resulting tsunami. Therefore at present, the contribution of the geotechnical engineer to this aspect of the problem is likely to be limited to estimates of the total volume of potentially flowing material and whether the displacements are likely to be relatively contained or very large.

5.1 Large Displacement Analysis

The use of large displacement analysis in evaluating postliquefaction displacements has now become part of engineering practice. A computer program for large displacements analysis of flow deformations, TARA-3FL, was developed by FINN and YOGENDRAKUMAR (1989). It was first used to study the postliquefaction ground displacements in Niigata measured after the 1964 earthquake. The displacements had been mapped by HAMADA *et al.* (1988) on the basis of air photographs. Large displacement analysis was conducted over a distance of about half a kilometer on a line between Ogata school and the Tsusen River. The residual strengths were selected on the basis of the SEED (1987) lower bound correlation. The liquefiable layer and the measured and computed displacements are shown in Figure 7. The maximum measured displacements are in the range 7 m to 8 m. The predicted displacements agree reasonably well with the measured displacements. The largest discrepancies are in the area of maximum displacement. This discrepancy may be partly due to the fact that in the analyses an average slope was assigned to the liquefied layer rather than the actual undulating form.

The program has subsequently been used to study the postliquefaction performance of water retaining embankment dams and tailings dams. In the case of tailings dams, it has been used to track postliquefaction deformations with displacements as large as 50 m. The displacements in submarine slopes likely to cause tsunamis will probably move much more than that. It is my opinion that current methods for evaluating postliquefaction displacements will be useful in submarine analyses in the sense that they can demonstrate whether extremely large deformations are likely to

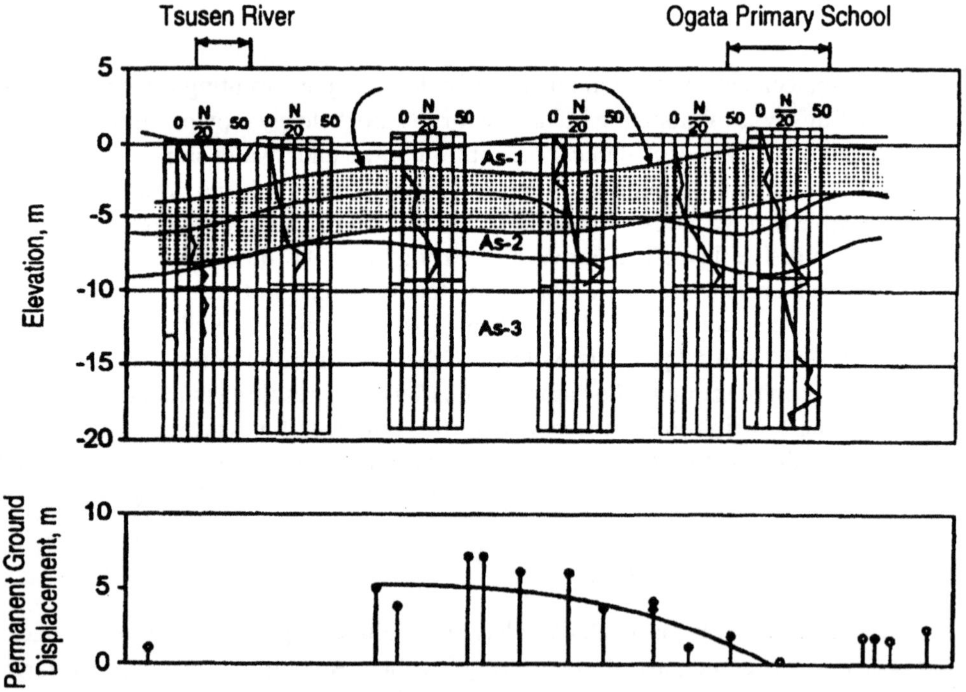

Figure 7
Comparison of computed and measured deformations in Niigata on a line from Ogata School to the Tsusen River (after FINN, 1990).

occur, however it is doubtful if these programs can track the displacements over the large distances involved. However, the initial displacement mass is probably the biggest factor in how catastrophic failure may be.

There has been little direct validation of postliquefaction displacement analysis slopes based on field data. The few case histories that have been available were poorly defined, both in terms of input motion and in the definition of the postliquefaction deformed shapes. Numerous embankment failures that occurred in Hokkaido during the Kushiro-oki earthquake of 1993 and the Nansei-oki earthquake of 1994, provided a good opportunity for validating postliquefaction deformation analysis by means of a series of blind tests in which predictions were checked against field data by an independent group. These case histories will be described below in some detail because of their importance for the validation of displacement analysis. However, it should be noted that displacements in these failures were considerably smaller than in typical tsunami generating landslides.

5.2 Postliquefaction Displacement Analysis of Hokkaido Dikes

Between 1995 and 1998, displacement analyses were conducted on many flood protection dikes in Hokkaido, Japan, which had been damaged during the Kushiro-oki and Nansei-oki earthquakes in 1993 and 1994, respectively. The objective was to develop a criterion based on potential postliquefaction crest settlement to prioritize the remediation of dikes against future earthquakes. A two-step strategy was adopted for the studies. First, failures of dikes in eastern Hokkaido would be simulated. If these simulations were satisfactory, then the analyses would be used to predict crest settlements in a number of dikes in western Hokkaido which had significant postliquefaction displacements. The displacement analyses were conducted in Vancouver, Canada (FINN, 1998), and the predictions were verified by engineers from the Advanced Construction Technology Center (ACTEC), Tokyo, and the Hokkaido Development Bureau on the basis of field data known only to them.

5.3 Simulation of Dike Failure at Section 9K850 in Eastern Hokkaido

The failure mode at a location on the left bank of the Kushiro river is shown in Figure 8. The height of the dike before the earthquake was about 7 m and the crest width was about 8 m. As a result of earthquake shaking, the crest of the dike settled about 2 m and movement of the slope of the order of 3 m took place towards the river. The ground was frozen to a depth of about 0.7 m at the time of the earthquake. The brittle nature of the frozen layer is probably responsible for the sharp step feature in the crest near the upstream slope. This frozen layer was taken into account during the simulation.

Appropriate input motions for seismic analysis were specified by Jishin Kogaku Kenkyusho, Tokyo. Relevant soil properties were provided by ACTEC. Dynamic analysis was conducted in the effective stress nonlinear mode using the program

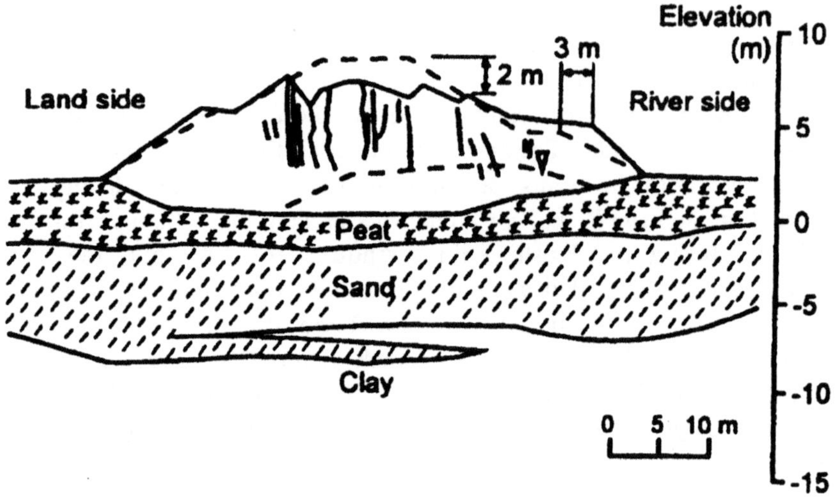

Figure 8
Mode of failure on left bank of Kushiro River at station 9K850.

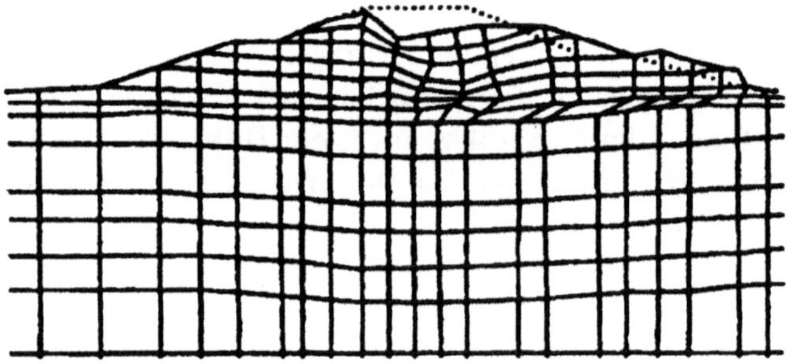

Figure 9
Computed post-liquefaction shape of the dike.

TARA-3 (FINN *et al.*, 1986). The large strain postliquefaction deformations were calculated using the program TARA-3FL (FINN and YOGENDRAKUMAR, 1989). This program allows the liquefied region to deform at constant volume and uses a Lagrangian updating scheme to handle large strains.

The computed deformed shape of the dike is shown in Figure 9. The sharp break in the surface shows the effect of the frozen ground. The deformed shape and the magnitudes of displacements agree fairly well with the displacements measured after the earthquake. The computed maximum settlement and horizontal displacement are 2.3 m and 2.7 m, respectively, compared to measured displacements of 2 m and 3 m.

The simulation studies were considered satisfactory and a major parametric study was approved to investigate the effects of some of the important cross-sectional parameters that control the consequences of liquefaction, such as the thickness of a non-liquefiable layer overlying the liquefied layer, the thickness of the liquefied layer itself, and the height and side slopes of the dikes. The effects of these parameters were characterized by the settlements of the crests of the dikes after liquefaction.

5.4 Estimation of Crest Settlements

The postliquefaction crest settlements S, were computed first for dikes with side slopes 1:2.5 and heights, H_D, as shown in Figure 10. The residual strength ratio was $S_r/\sigma'_{v0} = 0.10$. The thicknesses of the liquefied and nonliquefied layers, H_L and H_{NL} respectively, were varied and the resulting displacements after liquefaction are plotted in nondimensional form, S/H_D. in Figure 11 as a function of the nondimensional variable, β, defined by Eq. (3). The black data points represent the computed settlements.

$$\beta = \left[\frac{H_D}{H_{NL}} \frac{H_L}{H_{NL}}\right]. \tag{3}$$

The nondimensional computed crest settlements are matched by the curve shown in Figure.11. The equation of this curve is

$$\frac{S}{H_D} = 0.01e^{0.922\beta}. \tag{4}$$

This curve was adopted for predicting crest settlements.

Engineers from ACTEC and the Hokkaido Development Bureau compared the measured crest settlements from a wide variety of dikes in western Hokkaido which underwent noticeable displacements during the Nansei-oki earthquake in 1994 with

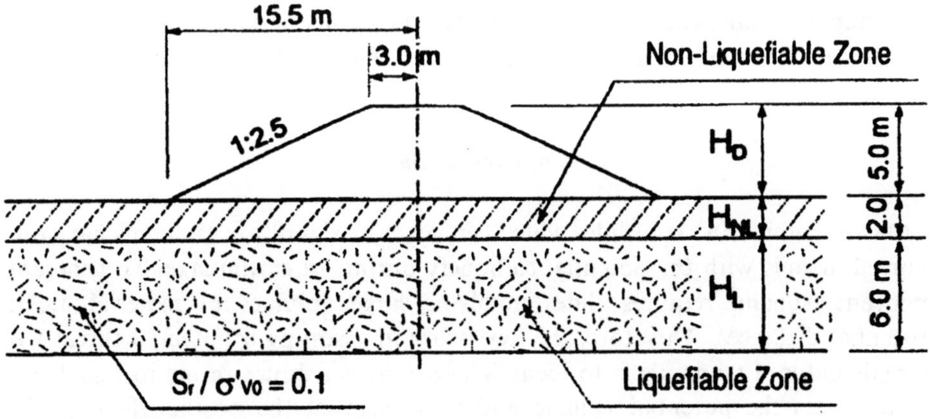

Figure 10
Typical cross-section of dike for analysis.

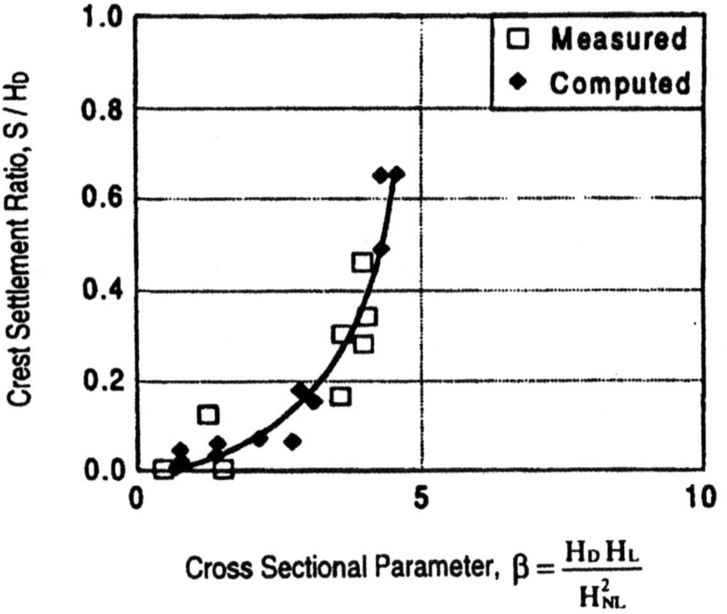

Figure 11
Comparison of crest settlements with prediction curve for dikes with slopes 1:2.5.

those predicted by Eq. (4). The black points represent the real cases corresponding to some of the idealized analyses done to develop the curve; the open points represent other dikes. The agreement was very good for dikes with slopes of 1:2.5. It can be seen that the criterion based on crest settlement can be useful for deciding which dikes should be remediated first. The larger the predicted settlement, the more urgent the need for remediation. The Hokkaido dikes study is an important case history because it is the only instance in which postliquefaction displacement analysis has been validated independently in blind tests involving a large number of earth structures undergoing different levels of postliquefaction displacements.

6. Conclusions

Massive submarine landslides which can generate significant tsunamis have often occurred in soils with the potential to liquefy during an earthquake. To generate a significant tsunami requires slides involving large volumes of material moving substantial distances. Therefore, the key issues in assessing a potential threat from landslide-induced tsunamis is to identify locations which are prone to liquefaction and to assess the potential volume and movement of the postliquefaction slide. Earthquake geotechnical engineers have very reliable tools for predicting the triggering of liquefaction. However, using these tools on a regional basis is prohibitively expensive with the possible exception of characterizing the *in situ* state of the soil deposits by shear-wave velocity measurements. Liquefaction potential can be determined following the procedures prescribed in NCEER (1997), on the basis of shearwave velocities.

The key variable controlling postliquefaction stability in any given slope and the potential for deformations is the residual strength. Correlations with field data are related to the normalized energy corrected standard penetration resistance, $(N_1)_{60\text{-}cs}$. The $(N_1)_{60\text{-}cs}$ values can be converted to cone penetration resistances although, in either case, it would be expensive to develop penetration data on a regional basis. Therefore, the definition of residual strength will be very approximate as it is likely to be based on a limited number of penetration resistances. Lower limits on the residual strength could be estimated on the basis of tests on water pluviated samples of material (VAID *et al.*, 1999) recovered from relatively crude samples, although the depth covered by such samples would be quite limited.

Existing large displacement analysis programs for postliquefaction analysis such as TARA3FL (FINN and YOGENDRAKUMAR, 1989), or FLAC (1996), can be used to estimate the potential for large deformations. These programs can track large deformations but not to the extremely large displacements developed in submarine flow slides. Therefore the main contribution that these programs can make to assessing the threat of a tsunami is in defining the depth and length of potential sliding (giving an indication of the volume of flowing material) and whether the displacements are likely to be very large.

The Workshop on the Prediction of Underwater Landslide and Slump Occurrence and Tsunami Hazards off of Southern California, held at the University of Southern California in March 2000, provided one of the few opportunities for geotechnical engineers to contribute to a team effort on assessing the threat from landslide generated tsunamis. Further cooperation between geotechnical engineers, geologists, seismologists, hydrodynamicists and oceanographers would lead to a greater shared understanding of the factors contributing to underwater landslides and the mechanism by which the resulting tsunamis are generated. This understanding should lead to a better blending of the skills available for dealing with landslide generated tsunamis and the development of better techniques for assessing the threat.

REFERENCES

BYRNE, P. M., IMRIE, A. S., and MORGENSTERN, N. R. (1994), *Results and Implications of Seismic Performance Studies for Duncan Dam.*, Canadian Geotechn. J. *31*, 979–988.

CASTRO, G. (1969), *Liquefaction of sands*, Harvard Soil Mechanics Series No.81, Pierce Hall, Cambridge, Massachusetts.

FINN, W. D. LIAM, YOGENDRAKUMAR, M., YOSHIDA, N., and YOSHIDA, H. (1986), *TARA-3: A Program for Nonlinear Static and Dynamic Effective Stress Analysis*, Soil Dynamics Group Report, University of B.C., Vancouver, B.C., Canada.

FINN, W. D. LIAM (1988), *Dynamic Analysis in Geotechnical Engineering*, Proc. *Earthquake Engineering and Soil Dynamics II —Recent Advances in Ground Motion Evaluation*, ASCE Geotechnical Engineering Division, Park City, Utah, 523–591.

FINN, W. D. LIAM, and YOGENDRAKUMAR, M., (1989), *TARA-3FL: A Program for Analysis of Flow Deformations in Soil Structures with Liquefied Zones*, Soil Dynam. Group Ret.,Department of Civil Engineering, University of British Columbia, Vancouver, B.C.

FINN, W. D. L., LEDBETTER, R. H., FLEMING, R. L., lr., FORREST, T. W., and STACY, S. T. (1998), *Stabilization of an Earth Dam Using Driven Prestressed Concrete Piles*, Int. J. Earthq. Eng. London *2*, 173–195.

FINN, W. D. LIAM (1990), *Analysis of Postliquefaction Deformations in Soil Structures*, Invited Paper, Proc., H. Bolton Seed Memorial Symposium, University of California, Berkeley (ed. J. M. Duncan), (Bi-Tech Publishers, Vancouver, Canada,) May 9–11, *2*, 291–311.

FINN, W. D. LIAM. (1998), *Estimating Postliquefaction Displacements in Embankment Dams and Prioritizing Remediation Measures*. Proc., 41 1nternational Conference on Case Histories in Geotechnical Engineering (ed., Shamsher Prakash), CD ROM ISBN 1-887009-03-5, Continuing Education, University of Missouri-Rolla.

ITASCA (1996), FLAC (Version 3.3). Itasca Consulting Group Inc., 708 South 3rd Street, Suite 310, Minneapolis, Minnesota 55415.

KAWATA, Y., BENSON, B. C., BORRERO, I. C., BORRERO, I. L., DAVIES, H. L., DE LANGE, W. P., IMAMURA, F., LETZ, H., NOTT, I., and SYNOLAKIS, C. E. (1999), *Tsunami in Papua New Guinea was as Intense as First Thought*, EOS, Transact., Am. Geophys. Union, March 2nd, 80: 101, 104–105.

NCEER (1997), Proc. NCEER Workshop on Evaluation of Liquefaction Resistance of Soils (ed. T. Leslie Youd and Izaat Idriss), Technical Report, NCEER-7-0022, December 31.

PELINOVSKY, E. N. and POPLAVSKY, A. A. (1997), *Simplified Model of Tsunami Generation by Submarine Landslides*, Phys. Chem. Earth. *21*, 13–17.

POULOS, S. I., CASTRO, G., and FRANCE, W. (1985), *Liquefaction Evaluation Procedure*, J. Geotech. Eng. Div., ASCE *111*, 772–792.

SEED, H. B. and IDRISS, I. M. (1971), *Simplified Procedure for Evaluating Soil Liquefaction Potential*, J. Soil Mechanics and Foundations Div., ASCE *97*, 1249–1273.

SEED, H. B. and IDRISS, I. M. (1982), *Ground Motions and Soil Liquefaction During Earthquakes*, Earthq. Eng. Res. Institute Monograph, Oakland, California, EERC.

SEED, H. B. (1987), *Design Problems in Soil Liquefaction*, J. Geotech. Eng., ASCE *113*(8), 827–845.

SEED, R. B. and HARDER, L. F., Jr., *SPT-Based Analysis of Cyclic Pore Pressure Generation and Undrained Residual Strength*. In Proc., H. Bolton Seed Memorial Symposium, May (Bi-Tech Publishers Ltd., 1990), pp. 351–376.

SHANNON, W. L. and HILTS, D. E., *Earthquake-Caused Sub-Marine Landslide at Seward, Alaska. In The Great Alaska Earthquake Of 1964: Engineering*. NAS Pub. 1606. (Washington, National Academy of Sciences, 1973).

UTHAYAKUMAR, M. and VAID Y. P. (1998), *Static Liquefaction of Sand Under Multiaxial Loading*, Canadian Geotech. J. *35*, 273–283.

VAID, Y. P., SIVATHAYALAN S., and STEDMAN, D. (1999), *Influence of Specimen Reconstituting Method on the Undrained Response of Sand*, Geotech. Testing J., ASTM *22*(3), 187–195.

VAID, Y. P. and SIVATHAYALAN, S. (1996) *Static and Cyclic Liquefaction Potential of Fraser Delta Sand in Simple Shear and Triaxial Tests*, Canadian Geotech. J., *33*, 281–289.

VAID, Y. P. and THOMAS, J. *Post-liquefaction Behavior of Sand*. Proc., I3th International Conference on *Soil Mechanics and Foundation Engineering* (New Delhi 1994), *3*, 1305–1310.

WATTS, P. (1998), *Wavemaker Curves for Tsunamis Generated by Submarine Landslides*, J. Waterways, Port and Coastal Ocean Eng. *124*, 127–137.

YOSHIMINE, M., ISHIHARA, K., and VARGAS, W. (1998), *Effect of Principal Stress Direction and Intermediate Principal Stress on Undrained Shear Behavior of Sand.*, Soils and Foundations *38*(3), 179–188.

Pure appl. geophys. 160 (2003) 1895–1922
0033–4553/03/111895–28
DOI 10.1007/s00024-003-2413-1

Pure and Applied Geophysics

The Aitape 1998 Tsunami: Reconstructing the Event from Interviews and Field Mapping

H. L. Davies,[1] J. M. Davies,[1] R. C. B. Perembo,[1] and W. Y. Lus[1]

Abstract—On the evening of 17 July 1998, on th e Aitape Coast of Papua New Guinea, a strongly felt earthquake was followed some 10–25 minutes later by a destructive tsunami. The tsunami comprised three waves, each estimated to be about 4 m high. The second of the three waves rose to a height of 10–15 m above sea level after it had crossed the shoreline and caused most damage. Maximum wave heights and the greatest damage were recorded along a 14-km sector of coast centered on Sissano Lagoon. In this sector the wavefronts moved from east to west along the coast; all structures were destroyed, and in the two main villages 20–40 percent of the population was killed. Partial destruction extended 23 km to the southeast and 8 km to the northwest, and effects of the tsunami were felt as far as 250 km to the west–northwest, beyond the international border. More than 1600 people are known to have died, with some estimates as high as 2200; 1000 were seriously injured, and 10,000 survivors were displaced. This paper presents information from interviews with eye-witnesses and from mapping of damage and inundation, and includes new information on the height, shape and timing of the waves; on the possible escape of petroleum and other gases from beneath the seafloor before and during the tsunami; on unusual sound effects that preceded the waves, and lighting effects that followed; on possible deep circulation (to 250 m) of sea water in the waves; on subsidence of the order of 50–70 cm at the coastal sand barrier; and on the resilience and potential protective capacity of certain species of trees. Eye-witness accounts indicate that the tsunami reached the shore at between 09:00 and 09:08 UT, which is earlier than is proposed in published models of the timing and location of the source of the tsunami.

Key words: Tsunami, Papua New Guinea, eyewitness, wave-height, wave-shape, timing.

1. Introduction

In the early evening of Friday 17 July 1998, following soon after a strongly felt earthquake, three great waves struck the coast west of Aitape, Sandaun Province, Papua New Guinea (PNG; Fig. 1). The waves came almost without warning and there was little people could do to save themselves, other than struggle to survive in the turbulent water. Two entire villages were destroyed almost without trace, and survivors were carried inland into lagoons, mangrove and sago swamp, and forest (Fig. 2). People died by drowning or by impact with hard objects. In all, more than 1600 people lost their lives, with some estimates of fatalities as high as 2200, another 1000 were seriously injured, and 10,000 were displaced to new villages.

---

[1] University of Papua New Guinea, University Post Office NCD, Papua New Guinea.

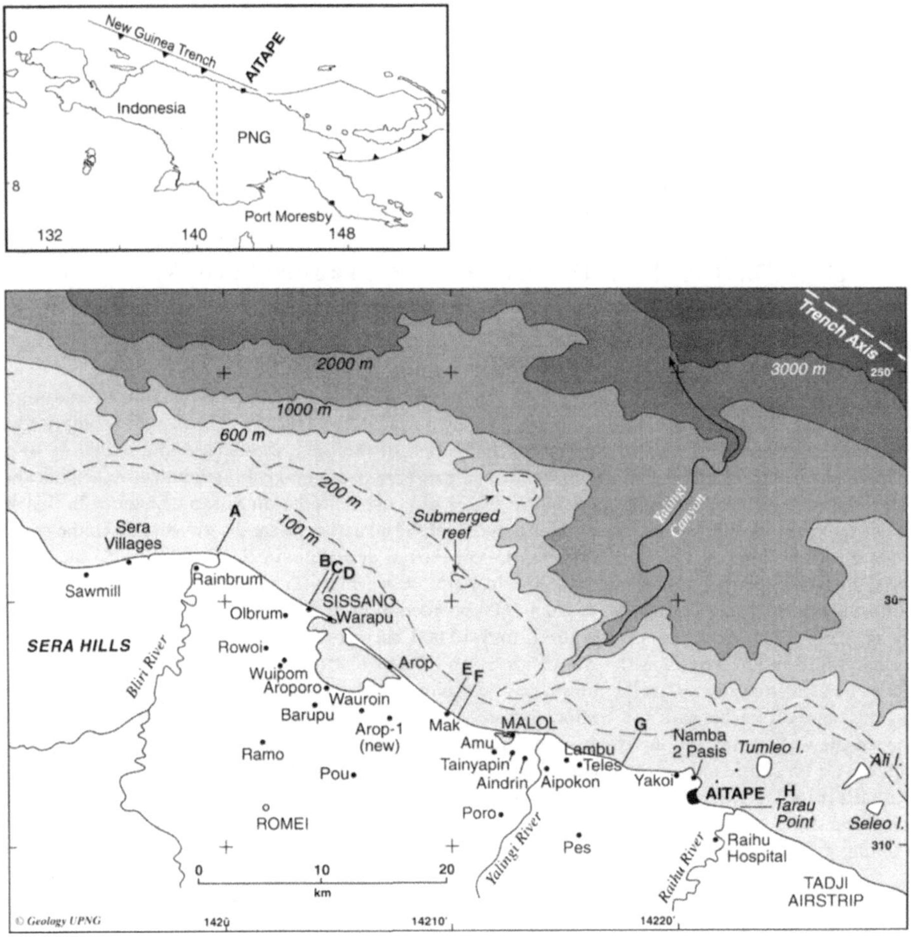

Figure 1

Locality map and map of the Aitape coast. Bathymetry is from multibeam survey (MATSUMOTO and TAPPIN, this volume: solid lines) and from spot depths on a navigation chart supplementary to chart Aus 389: dashed lines. The sunken reefs are at depths of 460 m and 82 m, respectively. Sissano Lagoon is the body of water inland from Warapu and Arop. The Sissano Government Station was at D and Nimas village (the easternmost Sissano village, mentioned in the text) at C. Sissano Mission was at B. Resettlement villages are Rainbrum, Olbrum, Rowoi, Wuipom, Barupu, Wauroin, Arop-1 (new), Amu, Tainyapin, Aindrin, Aipokon, Teles and Lambu. The original Teles and Lambu villages were on the coast directly seaward of the new villages of the same name. Aitape jetty is just south of the letter A of 'AITAPE'. The main focus of the wave was on the 14-km sector of coast from Mak to just beyond Warapu, DE. Here the wave heights were 10 m or more, all buildings were destroyed, and destruction extended as far as 500 m inland. From the mouth of the Bliri River to Sissano (AB), and from Mak to Tarau Point (FH), wave heights were not more than 4 m above sea level. In sectors AB and GH, only structures close to the water's edge were affected by the waves.

The first scientific investigation of the tsunami was carried out by an international team that entered the area on 1 August 1998. The team remained in the field for a week and reported their findings to the authorities in Aitape, and then in Port Moresby, the national capital (ITST, 1998). Other investigations followed, including a second visit by an international team in September 1998, and four major marine surveys. Results of the investigations have been presented by KAWATA *et al.* (1999a), KAWATA *et al.* (1999b), TAPPIN *et al.* (1999, 2001), SYNOLAKIS *et al.* (2002) and MATSUMOTO *et al.* (this volume), amongst others.

From experience at the time of the Rabaul seismic crisis in 1983 and the eruption that followed in 1994, we were aware of the need to get reliable information to the survivors and disaster managers, not only because of concerns for their safety, but also for peace of mind. At their briefing in Port Moresby, the international team confirmed that there was such a need at Aitape. Accordingly we mounted a public information program, on a small scale, starting 13 August 1998, and continuing today. While carrying out the program there was opportunity to interview many of the survivors and, in later months, to spend time in geological investigations (DAVIES, 1998a,b; J.M. DAVIES *et al.*, 1999; H.L. DAVIES *et al.*, 2000, 2001). Information from the interviews and field investigations is brought together in this paper.

2. The Setting

The island of New Guinea (Fig. 1) lies athwart the northern margin of the Australian continent; is 2500-km long and up to 800-km wide; and has a central mountainous spine with peaks to just under 5000 m. The mountains are flanked on the south side by broad plains, and on the north by riverine plains and coastal ranges. The western half of the Island is the Indonesian province of Papua (was Irian Jaya); and the eastern half of the Island, together with adjacent islands, forms the Independent State of Papua New Guinea (PNG).

The Island was formed by successive collisions between the northward-advancing Australian craton and exotic and para-autochthonous terranes. Currently, in the border region, the Pacific and Australian plates are converging at a rate of 11.1 cm/yr on an azimuth of 070° (PEGLER *et al.*, 1995). Crustal shortening is accommodated by oblique subduction at the New Guinea Trench (Fig. 1), by faulting in the hanging wall plate, and by faulting and folding in a foreland thrust belt on the south side of the central ranges.

Aitape is on the north coast, 160 km east of the international border. This is an area of low-lying coastal plain interrupted by fault-uplifted blocks of Oligocene volcanic arc rocks and associated limestone. Inland are foothills and the steep-fronted Bewani and Torricelli ranges. The coastal plain is bounded seaward by a coastal sand barrier that stands 1–2 m above sea level and is typically a few hundred meters across (Fig. 3). The sand barrier is highest at the beachfront and slopes gently

Figure 2
An artist's impression of the arrival of the tsunami. Wall painting by Lucas Rawah of Aitape, reproduced by permission.

Figure 3
All buildings on the sand spit at Arop were swept away, and the people carried into the lagoon and the mangrove swamps that fringe the lagoon. Photographed in September 1998 by H. Davies.

downwards away from the sea – a common morphology for coastlines where sea level is rising relative to the land, and where there is a steady supply of sand distributed along the coast by longshore drift (e.g., see KOMAR, 1998; Figs. 2–22 and 2–25). Much of the sand barrier is planted with coconut palms and there are occasional large trees (kalopilam *Callophylum inophyllum*; breadfruit or kapiak *Atocarpus altissima*; talis *Terminalia catappa*; and mango *Magnifera indica*) and thickets of yar (*Casuarina equisetifolia*).

Surf beaches of grey sand form the front of the sand barrier. Seaward of the beaches, sandbanks extend for 100–130 m, and beyond the sand banks a gently-sloping shelf extends out to sea for 10–15 km. Beyond the shelf edge an irregular broken slope descends to the floor of the New Guinea Trench, exceeding 4000 m in depth (Fig. 1). A submarine canyon cuts through the shelf in the east.

The beaches in the sector of maximum destruction are not protected by reefs or islands and for much of the year waves 1–2 m high break on the near-shore sandbanks. The surf is stronger in the Northwest Monsoon season, December to March, when ocean swells roll in from the open ocean, and is calmer in the season of Southeast Tradewinds, May to September. There is a longshore drift from the northwest in the Monsoon season, and from the southeast in the other season.

Aitape is a small commercial center, district administrative headquarters and port, served by regular air services and accessible by road from Wewak in the east, and by 4-wheel-drive track and road from Vanimo in the west, when the weather is dry. Minor roads connect to some of the nearby villages and the coastal villages are accessible by powered dinghy.

Before the tsunami, about 12,000 people lived in the coastal villages west of Aitape, from Malol to Sissano (Fig. 1). Most houses were of traditional materials, and most were within a few hundred meters of the waterfront on land that was not more than a few meters above sea level. Each village extended for a kilometer or more along the coast.

3. The Earthquakes of 17 July 1998

The events of 17 July 1998 began with a strongly felt earthquake of M_W 7.0 at 6.49 p.m. local time (0849 UT), followed by a series of aftershocks. There was a mainshock at 08:49:13 Universal Time and a widely felt aftershock, itself composed of two events at respectively 09:09:32 (m_b = 5.6) and 09.10:02 (m_b = 5.9). In between, two small aftershocks are documented by the National Earthquake Information Center at respectively 09:02:06 (m_b = 4.4) and 09:06:03 (no magnitude reported) (SYNOLAKIS *et al.*, 2002). The epicenter of the mainshock was about 4 km north of the Sera villages (Fig. 1) and the epicenters of the aftershocks lay along a line east from the mainshock, the first due north of Sissano-Warapu and the second due north of Malol (HURUKAWA *et al.*, this volume). The focal mechanism of the main shock can be interpreted as either a steep reverse fault dipping to the north (*e.g.*, GEIST, 2000), or a low-angle thrust dipping to the south–southwest (HURUKAWA *et al.*, this volume).

The mainshock was sufficiently vigorous and prolonged that at Malol, Arop and Warapu people left their houses and moved into open space. At Arop and Warapu cracks opened in the ground, and water squirted upwards, house foundation posts shook and water rose around the posts, and there was a smell of hydrogen sulphide. At Sissano Mission the earthquake caused minor damage to the 62-year-old church, and in the nearby villages some houses collapsed. At Malol the shaking was strong enough to cause concern that the water tanks at the Mission might collapse. Even at Vanimo, 140 km from the epicenter, the earthquake was described by one long-term resident as stronger and more prolonged than any he had experienced.

A weak foreshock that preceded the mainshock by about ten minutes was felt by observers at Malol, Arop, Warapu and Sissano, but was not recorded by observatories at Jayapura and Port Moresby. The foreshock had a gentle sideways motion, and people referred to this as a 'normal' earthquake, unlike the sharp shaking and fast up-and-down movement of the mainshock that followed.

Surveys of the felt intensity of the earthquakes by Y. TSUJI and others (unpublished data) and RIPPER *et al.* (1999) confirmed that the felt intensities were such as would be expected from a M_W 7.0 earthquake. In the Aitape area, and inland from Aitape at Nuku, the aftershocks were felt more strongly than the mainshock.

4. The Tsunami

The mainshock was followed, some minutes later, by a loud boom, as though of thunder; this was heard from Sissano to Malol. A few minutes or up to five minutes later there was a roaring sound, variously described as the noise of a low-flying heavy jet plane, the approach of a large ship, or as the woop-woop-woop of a heavy helicopter. The sound progressed eastward along the coast then back again to the west, and was heard all along the coast from Sissano in the west to Aitape High School (near Tadji airport, Fig. 1) in the east.

Although the sun had set at 6.37 p.m., there was still sufficient daylight that the day's activities were continuing. Men were painting a canoe, young people were playing touch football and their elders were moving around in the villages. People went to the beach to investigate the unusual noise and observed that the sea was 'boiling' or bubbling, and had receded by 50 m or so, exposing the nearshore sea bed. They then saw a wave develop in the distance, as a dark line on a sea surface that otherwise reflected the light of the sky. The wave approached and, when 200–300 m from the beach, started to break, rolling from the top. 'Smoke' or haze rose from the top of the wave, and many saw a red glow in the top of the wave.

One observer (John Sanawe, a former Colonel in the PNG Defence Force) reported that he first saw the sea on the skyline rise and explode, sending spray high in the air where it caught and diffracted the late afternoon sunlight into rainbow colors. He then heard a sound like distant thunder. He wondered at hearing thunder on a day when the sky was clear, then linked this sound to the explosion. Then there was a sound like a heavy helicopter, or such as can be heard when air bubbles escape from a bottle that is held under water, and the sea started to retreat from the shore. The rhythm of the helicopter noise slowed as the retreat of the sea slowed. Then there was silence for 4–5 minutes, followed by the noise like that of a low-flying jet aircraft. Sanawe looked to sea and saw that a wave had formed at or near the site of the explosion. The wave then approached at great speed.

People ran from the approaching waves but most were caught. A few escaped by climbing trees, or pushing their boats into the lagoon. People in the waves were vigorously tumbled and turned in water that was laden with sand and debris. They were stripped of their clothing, lost skin by sand abrasion, were battered by hard objects and some cut or impaled by timber and metal objects. Those who were fortunate were carried into the lagoon and were able to cling to floating debris. An infant was deposited miraculously on the floating roof of a house. Those less fortunate were carried into swampland or into the mangroves that fringe the lagoon where some were impaled or were buried under piles of logs and debris. Some who had survived the initial impact were swept out to sea as the waters receded. Most had ingested water from the waves.

By 7.20 or 7.25 p.m. the water had retreated, though much standing water remained. At this moment, according to survivors, the sky was filled with a yellow or yellow-red glow that provided sufficient light for people to start searching for family members. They said that without the glow this would not have been possible (see also 5.6 *Unusual Lights*).

At Sissano Lagoon a low haze had advanced with the wave and this now blotted out the stars, so that it became pitch dark, so dark that people moving inland, away from the lagoon, held on to each other to maintain contact.

Rescue began that night, the survivors helping each other. The first outside help arrived 16 hours after the event on the Saturday morning, and a major rescue effort began a day later, 40 hours after the event. By Monday evening (72 hours) all surviving injured were in hospitals at Vanimo, Raihu (Aitape) and Wewak (TAYLOR *et al.*, 1998; HOLIAN and KEITH, 1998; DAVIES, in prep.).

4.1 The Shape of the Waves

The shape of the tsunami waves has been reconstructed from interviews with survivors, and from mapping the damage and the trails of debris left by the wave (our unpublished data). The mapping made use of low level aerial photographs that were flown three weeks after the tsunami (NATIONAL MAPPING BUREAU, 1998).

The tsunami was seen by observers as an initial lowering of sea level followed by three large waves. It caused damage along 45 km of coastline (AH in Fig. 1), but maximum energy was focussed on a 14-km sector (DE in Fig. 1). Here wave heights, on shore, were 10–15 m above sea level and there was extensive damage for distances of up to 500 m from the coast. Damage was less on either side of the 14-km sector. Towards the periphery, in sectors AB and GH, wave heights were lower and only structures close to the water's edge were damaged.

Three waves, about 2-m high, caused minor damage at towns and villages along the coast to the west–northwest for a distance of 250 km (NELSON JOKU, unpublished data) and smaller waves were recorded 170 km to the east–southeast at Wewak, and 500 km to the east–northeast at Bipi Island, Manus Province. A wave with a peak-to-peak amplitude of 7 cm was detected on the tide gauge in Rabaul harbor, 1000 km to the east, and smaller waves were detected on tide gauges in Japan (TANIOKA, 1999).

Within the 14-km sector the frontal wave was observed to approach the coast obliquely from the east, striking the coast first at or near Mak (E in Fig. 1; Figs. 4, 5)

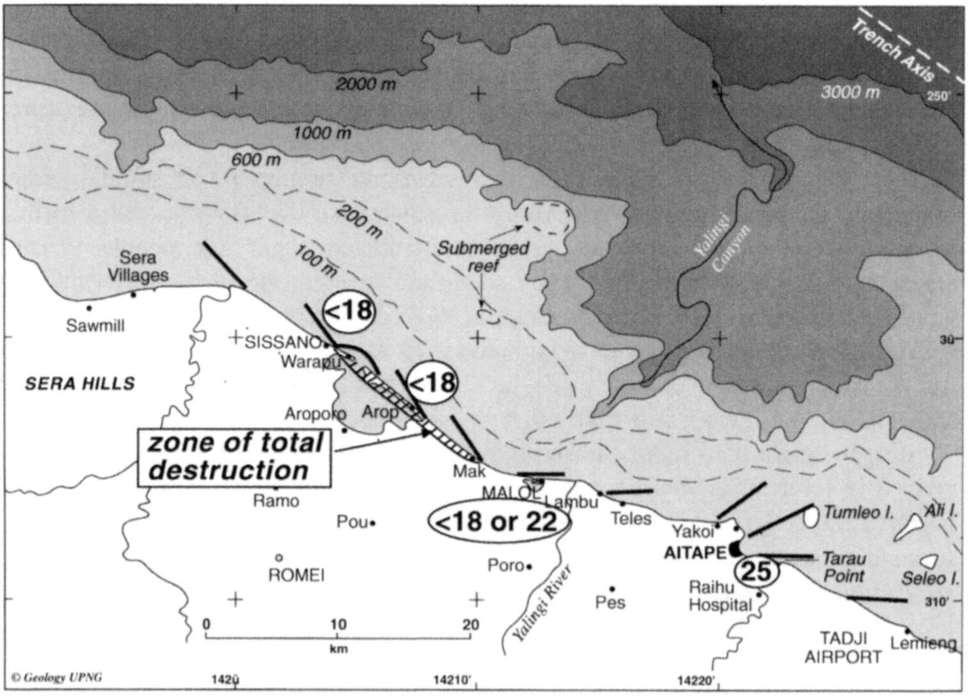

Figure 4

Progress of the wave along the Aitape coast as determined from interviews and damage mapping. The heavy lines indicate the approximate alignment of the wave-front as it intersected each section of the coast. The numbers indicate the time the wave arrived at the coast, shown as minutes after the earthquake of 08:49 hours Universal Time.

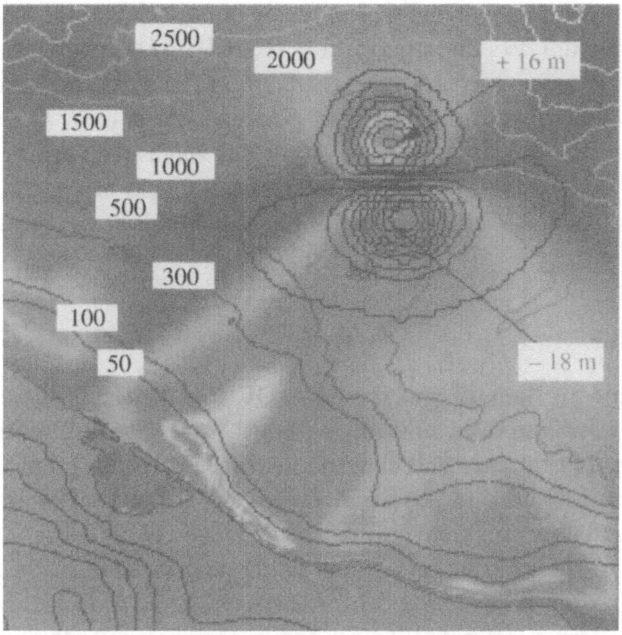

Figure 5
Computer model of the wave front from SYNOLAKIS *et al.* (2002), reproduced by permission of The Royal Society.

and progressing to the west, so that the wave struck Arop before it struck Warapu (Fig. 1). At Warapu the wavefront was curved, or C-shaped in plan view, with the result that the western and eastern ends of the village were hit before the center (Fig. 4).

People in a powered dinghy returning from Aitape saw the wave approach from the east, reaching the beach behind them. They accelerated to maximum speed of perhaps 35–40 km/hr, trying to outrun the wave but, at Arop, the wave loomed and broke 200–300 m out to sea, caught the dinghy and carried it ashore, killing all but one occupant.

Eye-witnesses at Arop and Warapu described how the first wave came ashore as a broken wave or bore, about 1-m high. People were knocked over by the wave and by the cushion of air moving in front of the wave, but houses were only slightly damaged. Jerome Kinisa, who was on low land just east of the mouth of the lagoon, described how the wave broke offshore then rushed at speed across the peninsula, as a 1-m-high mass of foaming water, sand and logs. He was carried by the wave into the middle of the lagoon. At Arop and Warapu, the second wave reached the shoreline before the first had withdrawn, loomed to tree-top height and then crashed down upon the villages. This was the wave that caused most destruction. At Warapu the third wave followed closely behind the second, and a weak fourth wave followed after that.

West of the 14-km sector of most extensive damage the wave progressed westward. East of the 14-km sector the wave progressed eastward. This pattern: the wave moving outward away from 14-km sector in both directions, suggested that the source of the tsunami was directly offshore from DE (DAVIES, 1998a, p. 28).

At Aitape the three waves approached around the Aitape headland from the northwest. The first wave came ashore at Namba 2 Pasis (Fig. 1) and was reflected back to meet the second wave at the headland. The waves then progressed to east and southeast. The wave crests were estimated to be 500–600 m apart (observations by Dickson Dalle). As observed here, and at Sissano Mission, the three waves were of about equal height.

4.2 The Damage

The predominant damage was in sector DE (Fig. 1). Here the villages of Arop and Warapu were removed almost without trace, leaving only the concrete foundation slabs of churches and classrooms (Fig. 6). House foundation posts were lifted out of the ground and removed, except for a few at Warapu, and massive concrete blocks at Warapu were carried for distances of 50–60 m.

In contrast to the destruction of man-made objects, mature coconut palms and well established, deep rooted trees with extensive lateral rooting, such as the many casuarinas (*Casuarina equisetifoli*) and the occasional callophylum (*Calophyllum inophyllum;* Fig. 7), kapiak or breadfruit (*Atocarpus altissima*), talis (*Terminalia catappa*), and mango (*Magnifera indica*), withstood the waves. A robust callophylum at Warapu provided some protection for Camillus Kanuvo and family from the worst of the first wave. People who climbed trees and palms at Warapu and at Arop-2 escaped the wave. The less salt-tolerant of these trees, notably the mango trees, died some months later.

The beachfront was eroded by the wave so as to create a low escarpment 1–2 m high at the front edge of the vegetated sand ridge. This effect was seen from Malol in the east to Sissano in the west. The low cliffs were modified by the return flow of the water which carved channels through the beachfront berm.

At the back of the sand ridge, on the lagoon side, the waves caused scouring to depths of several meters, and scours were then partly filled with sand (Fig. 8). The scouring occurred along the line where the second wave dropped on the houses, and reflects the great turbulence at this point. Scouring was most severe where the ground already had been disturbed, *e.g.*, where house posts had stood, or in the graveyard near the Warapu church; or where there was any steepening in the slope of the sand ridge back towards the lagoon. The scouring and its causes have been discussed by MATSUTOMI *et al.* (2001).

Further west at Nimas, the easternmost of the Sissano villages (CD in Fig. 1), the damage was marginally less severe. Although most houses were destroyed, there was scouring in only a small area at the seaward end of the village and house posts remained standing, all canted inland by the force of the water. Four houses, 400 m from the coast, were preserved intact. In the other Sissano villages, at BC, many

Figure 6
Immediately west of Warapu, the Sissano District Office was swept away by the wave, leaving only the former foundation stumps and the base of a concrete-brick water tank. Photograph H. DAVIES.

Figure 7
The callophylum is a deep-rooted and salt-tolerant tree that can provide protection from tsunamis. This tree is on the west coast of Tumleo Island, where blocks of stone broken from the fringing coral reef were carried ashore by the tsunami. Photograph H. DAVIES.

Figure 8
Where the waves crashed down on to the villages at Arop and Warapu they scoured out the sand, leaving low escarpments that faced toward the lagoon. Photograph H. DAVIES.

houses remained standing, but at the western end of BC the waves destroyed buildings at Sissano Mission (B in Fig. 1), including the great church that had stood since 1926 (Fig. 9). At the Mission, buildings more than 400 m from the coast were not damaged.

Dramatic evidence of the power of the waves was preserved in the low country inland from Arop and Mak, at the eastern end of DE, where clumps of wild sago palm were lifted and overturned by the waves, and tall trees were uprooted and transported inland for hundreds of meters, like so much match wood (Fig. 10). This swathe of destruction tailed off east of Mak, in sector EF.

At Malol, seaward of Malol Lagoon, houses within 50 m of the water's edge were destroyed but most other houses were left standing. The mission station and the adjacent villages, 200–300 m from the beach, were not damaged; this included Tainyapin village, which is on a low-lying Island in the lagoon. East of the Yalingi River, the villages of Lambu and Teles were mostly destroyed except the westernmost houses, which were protected by the headland of the Yalingi river mouth, and houses on the inland side of the road, which were more than 150 m from the water's edge. Lambu and Teles were on a low, poorly vegetated sand ridge, less than a meter above high water mark, that slopes downward away from the beachfront berm.

From the Waipo River to Tarau Point (GH in Fig. 1), and from Sissano Mission to the mouth of the Bliri River (AB in Fig. 1), the energy of the wave was less and only houses close to the water's edge were damaged.

Figure 9
The church at Sissano Mission, built by German missionaries in 1926 (top), was swept away by the wave, leaving only some concrete slabs at ground level (bottom). Photographs by Fr. Eugene McKinnon in 1962 (top) and H. DAVIES.

4.3 Height of the Waves

Wave height information was collected by the international team (ITST, 1998) and reported by KAWATA *et al.* (1999a, 1999b). They found that wave heights reached 10–15 m in the area of maximum devastation (DE in Fig. 1). These extreme wave heights were confirmed by survivors at Arop and Warapu, who spoke of waves

Figure 10
East of Arop village the tsunami cut a swathe for 500 m towards Arop Community School. Mature trees were uprooted and stacked like drift wood. Photograph H. DAVIES.

as high as the tops of the coconut palms. At Arop and Warapu it was the second wave that reached these heights.

However, contrary to what has been generally accepted from the ITST results, we think it unlikely that the waves were 10–15 m high when they were still approaching the beach. Rather the maximum wave height while still at sea was of the order of 4 m. Heights of 10–15 m were only achieved when the broken waves were deflected upwards, after they had reached the beach. This point cannot be proven, but accords with descriptions by eye-witnesses who watched the waves approach the shore. They described the waves as about 4-m high, as did the sole survivor of the powered dinghy that was caught by the wave offshore from Arop.

This is an important point because it bears upon the classification of the tsunami in terms of tsunami magnitude (see under Discussion), and because it changes a key element in the modeling of the source and path of the tsunami. The published models have sought to explain a wave that reached a height of 15 m as it approached the shore, not 4 m (*e.g.*, see Fig. 6 in TAPPIN *et al.*, 2001; and Fig. 9 in SYNOLAKIS *et al.*, 2002, reproduced as Fig. 5 in this paper).

According to eye-witnesses the first wave started to break, rolling from the top, at a point 200–300 m from shore and reached the shore as a broken wave. The first wave then swept through Arop and Warapu villages and the lagoon mouth as a bore about 1 m high (which equates to about 2 m above sea level). It was the second wave that reared to great height and crashed down on the two villages.

Although we do not know for sure what happened, one can imagine that the second wave, travelling fast on the back of the flooding caused by the first wave, was deflected upwards when it reached the shore, because of the change in slope at the beach and the beachfront berm. The first wave had progressed across the sand barrier and into the lagoon, and so had not retreated before the second wave arrived. The effect of the flooding by the first wave would have been to reduce bottom friction and provide an elevated platform on which the second wave approached the beach. One eye-witness (John Sanawe of Arop) described the second wave as being deflected upwards when it reached the beachfront.

It is unlikely that the second wave, while at sea, was inherently bigger than the other waves. All eye-witnesses who saw the three waves as they approached the shore described them as being approximately equal in height.

East and west of DE, the sector of greatest damage, wave heights recorded by the international team were lower, at 2–4 m above sea level (KAWATA *et al.*, 1999b). At only one location outside sector DE did these authors record a wave height of 10 m. This was at Uyang, the westernmost part of the Malol villages. Our investigations suggest that this was either an anomalous local effect, or an incorrect reading of the field evidence; a conclusion supported implicitly by MATSUTOMI *et al.* (2001).

4.4 The Timing of the Waves

Information regarding when the first wave hit different parts of the coast (Fig. 4) can help to define the location of the source of the tsunami. However, precise information is not easily obtained, because people in the villages generally do not concern themselves with knowing the time to the minute but, rather, take the time from the sun. In the late afternoon, village activities continue until after sundown – people moving around, young people playing – and when the light becomes insufficient to move around freely people sit down to the evening meal. There is no street lighting in the villages and, in fact, no electric light.

We sought information about the timing of the wave using two reference points: (1) did the wave arrive before or after the strongly-felt aftershocks (which occured at 09:09 and 09:10 hours UT), and (2) how much daylight remained when the wave arrived. Of these, the first is the less subjective.

Our inquiries revealed that the wave reached villages in sector BE (Fig. 1) before the aftershocks. Any number of eye-witnesses concur on this. A survivor from Warapu Elizabeth Bade felt the aftershocks while being carried in the wave from eastern Warapu to the islands just inshore from the lagoon mouth. The priest at Sissano Mission, Fr Otton, felt the aftershocks after having run 300 m to escape the wave.

At Malol, too, the waves may have arrived before the aftershocks (several reliable witnesses) although there are conflicting accounts. A group of young men, interviewed at Malol in October 1998, recalled that the wave reached Malol just as the shaking from the strongly-felt aftershocks ceased. At the time of that interview there was no opportunity to cross-check with other Malol survivors because all had dispersed to inland camps. The information from the youths was published (DAVIES, 1998a, p. 26) and was adopted as a key argument in at least one published paper (TAPPIN *et al.*, 2001), but could be in error. More recent inquiries at Malol have not resolved the question. A number of survivors recalled that the wave reached Malol some minutes before the strong aftershocks, however others maintain otherwise.

From the observers in sector BE (Fig. 1), we know that the waves reached all of the villages in this sector before the strongly felt aftershock at 09:09 hours UT. From other sources we know that the waves progressed westward along the coast, beginning at Mak, or possibly at Malol (Fig. 4). We do not know how much time was needed for the wave to progress from Mak to Sissano Mission, but do know that some significant time elapsed, because people saw the wave as it approached from the east, and had time to try to outrun the wave in a powered dinghy (in one instance), and, at Arop-2 and Warapu, to climb trees and to jump into canoes that were pulled up on the shore of the lagoon. If the progress from E to B took 5 minutes or more, as seems likely, then the wave first came ashore at Mak (or Malol) at or before 09:04 hours UT.

Possibly the wave that reached Malol was part of a separate and later wavefront. We know that the waves at Mak struck the coast while travelling obliquely from the east. In contrast, the youths at Malol described the waves as approaching square-on to the coast, *i.e.*, heading due south.

Further confirmation that the waves reached Arop and Warapu, and probably Malol, some minutes before the aftershocks is provided by people's recollections of how much light remained in the sky, at the time they saw the waves approaching. At all three localities people observed the waves as dark features on an ocean surface that otherwise was lit by reflection of light from the sky. At this time of the year (July), and in similar weather to that of the night of the tsunami, with scattered cumulus cloud's, there is little or no reflection of light from the sky onto the surface of the sea after 09:05 UT (7:05 p.m. local time), and none at all by 09:09 UT.

Fifteen km east of Malol, at Aitape jetty (Fig. 1), the arrival of the waves was recorded by several observers independently at 09.14 or 09:15 UT (7.14 or 7.15 p.m.), 2–3 minutes after the aftershocks. At Tumleo Island and at Pro (near Lemieng, Fig. 4) observers estimated that the waves arrived "about 5 minutes" after the aftershocks.

4.5 What was in the Wave?

People who saw the broken wave approach the mouth of the lagoon described it as a dirty mass of broken water filled with dark sand, logs and other debris. The sand was deposited across much of the devastated area as a layer about 10-cm thick, thinning to a few cm close to the waterfront and towards the inland limit of deposition (JAFFE *et al.*, 1998; and our unpublished data). The sand deposit is coarse at the base and fine at the top, and includes rip-up mud clasts. Locally the deposits are much thicker, especially in areas where the ground surface had a steeper than normal negative gradient (sloped more steeply away from the waterfront), and where the waves had caused scouring (see Fig. 9; DAVIES, 1998a, pp. 22–23).

Probably much of the sand was lifted from the beachfront and the immediate offshore sand banks as the wave came ashore (see also MATSUTOMI *et al.*, 2001). At Teles-Lampu the sands included pebbles up to 1-cm across that presumably were carried from the mouth of the Yalingi River.

The wave also carried considerable marine life and deposited this on the shore or in the lagoon, including sharks, quantities of fish, and some sea turtles. The fish species included Red Emperor (*Lutjanus sebae*), a bottom-feeding snapper species that normally inhabits depths of 30–50 m, and living specimens of a species of foraminifera that normally lives at depths of 250–300 m: a keeled form of *Globorotalia*. Foraminifera were found only in sands collected at Teles (Fig.1).

4.6 Hot and Smelly

All survivors in the Arop-Warapu area reported that the water in the wave was hot, and many reported that it had a bad smell and stung the skin. Some described the smell as a 'chemical' smell, and others as 'rotten eggs' (hydrogen sulphide). Others, particularly at Arop, noted the smell of kerosene or petrol or oil. People speculated that the impurities that caused the stinging effect also caused wounds to become infected very quickly, and caused the skin on bodies floating in the lagoon to turn black. However, medical authorities advise that the blackening of the skin is only to be expected. Survivors further to the east (*e.g.* at GH, Fig. 1) did not mention any unusual temperature or smell of the water in the wave.

The reports of oil in the water pointed to the possibility that an explosive escape of gas may have been a factor in the tsunami (DAVIES, 1998a, p. 16, and see John Sanawe's observations under 4, and discussion under 5.2 and 5.3, below).

4.7 Other Phenomena

Possible ground deformation. John Kimene of Nimas was one of a group that was fishing at a drowned reef 8–10 km from the coast at about 10 a.m. on Thursday, 16 July 1998. This probably is the reef marked on the map (Fig. 1) just inside the 200-m isobath, which stands at 82-m depth. As the party trolled 1–2 km west of the submerged reef they were surprised to run into a succession of 2–3 m waves that loomed and steepened as though about to break. They took this to be evidence of a new shoaling of the water in an area that previously had been quite deep. There was a smell of dead fish.

Escape of gas. On Friday, 17 July 1998 at about noon Tom Kaisiera, a teenager from Nimas, paddled to the same general area and was surprised to find the sea bubbling with odorless gas. The area of bubbling was large, perhaps 100–200 m across. The canoe was drawn toward the center of bubbling area and it was only by paddling strongly that he could escape (DAVIES, 1998a, p. 16).

Unusual lights. Three unusual lighting effects were reported. Many observers saw a red light on the horizon before the tsunami developed: "After the first earthquake a long streak of red light like fire appeared just above the ocean on the northern horizon. It flashed and then disappeared, then within seconds there was a loud bang." Also, many observers described a red glow or "fire" in the top of the wave.

After the wave passed, observers at widely separated locations (Warapu, Malol and Raihu) saw a yellow or yellow-red glow in the sky over the sea. "The sky lit up after the wave had destroyed the villages" (observer at Malol) and "after I climbed down from the tree I saw a big light over Arop and in the direction of Aitape" (observer on an island near the lagoon mouth). The Sisters at Malol recall that after the waves had passed they looked seaward and saw a calm golden sea. Warapu survivors recall that the yellow glow in the sky helped light their search for survivors.

4.8 Subsidence

There has been 50–70 cm of subsidence in an area from Nimas (C in Fig. 1) to Arop, and possibly beyond Arop towards Mak. This is evidenced in the flooding and erosion of the tip of the Warapu Peninsula at the mouth of the lagoon (Fig. 11), and in the high water table at the former village sites of Warapu and Arop. Subsidence is thought to be the cause of the dying back of some vegetation a few hundred meters

Figure 11
The sand spit at the mouth of the Sissano Lagoon, previously occupied by Warapu village, has subsided and in November 2000 was being eroded. Photograph by H. Davies.

inland from Nimas. The subsidence may have been partly coseismic and perhaps mostly postseismic.

While in the field in the first week of August 1998 the international team searched for evidence of subsidence but found none. We too saw no proof of subsidence on several visits in August and September 1998. Another team that entered the area in August saw and photographed, on 17 August 1998, the partly submerged bole of a tree on the lagoon shore at Arop, and grass that had been flattened by the high tide, and cited this as proof of subsidence (Goldsmith *et al.*, 1999; McSaveney *et al.*, 2000). Because no other observers saw this, we suggest that the flooding seen on 17 August 1998 may have been an ephemeral effect due to an influx of river water into the lagoon and restricted outflow through the lagoon mouth. (Flow through the lagoon mouth was restricted by newly-formed sand banks for five to six months following the tsunami.)

The present Sissano Lagoon was formed by an earlier, coseismic subsidence event, in 1907 (Neuhauss, 1911; Welsch, 1998).

5. *Discussion*

The Aitape tsunami was the worst natural disaster in the history of the Independent State of Papua New Guinea, and compares with the catastrophic eruption of Mount Lamington in January 1951, in which 3000 people were killed. Since the beginnings of written history in this region, the Aitape tsunami is exceeded in impact by only the 1888 tsunami, triggered by the collapse of Ritter Island volcano (EVERINGHAM, 1977).

The tsunami developed at a time when the nation was ill prepared. There had been no major tsunami in Papua New Guinea since 1930 (EVERINGHAM, 1977), sufficiently long ago that recollections had been lost or dulled. Recent catastrophic tsunamis in Indonesia that should have triggered a state of increased awareness, including Flores in 1992 and Biak in West Papua in 1996, had passed without drawing much attention in the media or in disaster management circles. In addition, the promotion of tsunami awareness by the Government had lapsed in recent decades, since the efforts of EVERINGHAM (1977), and the government agency responsible for disaster management was undermanned and temporarily without telephones.

Despite these disadvantages, a response to the disaster was cobbled together, and was followed by a strong recovery program. Vital to the response and the recovery were the contributions of individuals and groups within the country and overseas, the generous donations of funds, goods and services, and the concerted efforts of key people on the ground.

The scientific investigation of the tsunami, similarly, was an amalgam of international and locally-based contributions. Key amongst these was the arrival of the international team two weeks after the disaster, their rapid investigation of the situation, and their reporting back to the authorities at the end of the third week. The members of the team made all information available to the PNG-based scientific community, as photocopies of field data initially and as a bound report later in the year (ITST, 1998; KAWATA *et al.*, 1999a). After their departure they continued to provide support in the way of advice and encouragement.

The other major contribution to the investigation of the tsunami was the funding of four research cruises in 1999–2001 by the Japanese Government, through the Japan Marine Science and Technology Corporation (JAMSTEC; TAPPIN *et al.*, 2001; MATSUMOTO and TAPPIN, this volume). The cruises were carried out at the request of the South Pacific Applied Geoscience Commission (SOPAC), acting on behalf of the PNG Government.

For PNG-based scientists the first objective was to provide reliable information to the public, especially those in the disaster area. There was no umbrella organization nor budget for this, and some of those involved in management of the disaster did not accept the need for the program. Therefore it was necessary to begin the program without funds, to establish the credibility of, and need for, the program, and then to seek funds so it could continue. Because of the low profile of the program initially there were difficulties arranging transport in the field and time was wasted. Over several months the program gained support, particularly amongst the managers and victims in the field at Aitape, sponsorship was found, and operations became more efficient.

In the course of the public information program it was inevitable that information would be gleaned from the survivors and that, over time, a considerable store of information about the tsunami would be accumulated. However, these inquiries were never the main thrust of the program and were never pursued exhaustively. More could have been learned with a concerted program of inquiry. Also, stories could have been cross-checked and independently verified. The uncertainty about the time of the arrival of the wave at Malol is a case in point.

The PNG-based scientists had limited financial resources, but had the advantage that, unlike their overseas counterparts, they could make repeated visits to the field area (more than 18 visits in 3 years), and gather information in a more leisurely and systematic way, long after the emergency had eased. This resulted in, for example, a better understanding of the timing and path of the wave and better documentation of the damage. The repeated visits also were vital to the success of the public information program because they provided an opportunity to build linkages with community leaders and to identify and respond to community concerns. A prime concern was to persuade the communities that the tsunami was a natural event, and not man-made. In this way, the program has been a factor in helping the communities return to normalcy.

There are still a number of issues about the tsunami that are unclear or unresolved as are discussed below.

5.1 The Cause of the Tsunami

Because the wave heights of 10–15 m were greater than should be expected from an earthquake of magnitude 7, members of the international team suggested that there may have been an augmentation of the wave by some secondary process, such as a submarine landslide (as reported in DAVIES, 1998a, p. 7). Subsequently the discovery of submarine landslide deposits (SWEET and SILVER, this volume) and of extensional fissures in the seafloor sediments (TAPPIN *et al.*, 2001; MATSUMOTO and TAPPIN, this volume) has provided supporting evidence. The hypothesis that the Aitape tsunami was generated by a slump or landslide of seafloor sediments has become widely accepted (TAPPIN *et al.*, 1999, 2001; SYNOLAKIS *et al.*, 2002).

SYNOLAKIS *et al.* (2002) concluded that the slump occurred at 09:02 hours UT (7:02 p.m. local time), at a location where fissures in the seafloor had been mapped, about 30 km north–northeast of Arop. The slumping generated a *T*-wave signal that was picked up by hydroacoustic and seismic stations across the Pacific to the north and northeast of Aitape. Computer modeling shows that a wave generated at this source would deliver maximum energy to the devastated area, DE in Figure 1, and would approach the coast obliquely from the east. Depending on the dimensions of the slump block and the distance that the slump moved, the wave would have a maximum elevation of 10–15 m at the coast (Fig. 5).

The model matches the field data with the exception that it requires the wave to reach Arop at 09:12 hours UT (7.12 p.m. local time), which is after the aftershocks and is 4–5 minutes later than our best estimate of the arrival time, based on survivor accounts. This raises two possibilities: (a) that the 09:02 UT seismic event does not mark the initiation of the tsunami; or (b) that the source of the 09:02 UT seismic event was much closer to the shore, perhaps as little as 10–12 km from Arop. The seafloor at this possible source area has not been mapped.

5.2 The Possible Involvement of Gas

An escape of gas may have accompanied or even initiated the tsunami. The evidence includes the vigorous bubbling of gas encountered by a lone fisherman on the day of the tsunami; John Sanawe's description of what appeared to be an explosion in the wave that threw spray high in the air; the recollections by a significant number of the survivors that there was a smell of kerosene, petrol or oil in the wave; and Sanawe's observation that there was a film of light oil on vegetation at Arop on the Saturday morning.

The sediments that underlie the Aitape coast are known to contain petroleum. Seeps have been mapped in the Bewani and Torricelli Mountains (NORVICK and HUTCHISON, 1980), and weak gas seeps are present at one location in Sissano Lagoon and at another offshore from the Waipo River (offshore from G in Fig. 1; information recorded during our investigations). In addition, there is a newly-recorded oil seep offshore from Lemieng (Fig. 4; information from staff of the Diocese of Aitape, January 2002). Gas seeps also were found in water depths of 1000–2000 m during the post-tsunami marine surveys; the seeps are marked by chemosynthetic faunal communities (MATSUMOTO and TAPPIN, this volume).

Gas escape on a more modest scale caused the 'boiling' of the ocean that was reported by many eye-witnesses. This was the escape of gas that is normally held in equilibrium in the seafloor sediments. The gas escapes when pressure at the seafloor is reduced at the time of the extreme lowering of sea level that precedes the arrival of the positive tsunami wave (M. HOVLAND, pers. comm. 1999; HOVLAND and JUDD, 1988).

SOTER (1999) has suggested that earthquakes may be triggered by gas ascending from depth. Escape of some of this gas to the surface would explain why gas venting might precede an earthquake.

5.3 Hot and Smelly

One of the most puzzling features of the Aitape tsunami was that most or all witnesses in the Sissano Lagoon area described the water in the waves as unusually hot. This remains unexplained. Possibly this too was linked to the escape of gas.

The heating effect of escaping gas was demonstrated in the Gulf of Corinth in the eastern Mediterranean where HASIOTIS *et al.* (1996, cited by SOTER, 1999) recorded that water temperatures in the Bay of Patras increased by as much as 6°C in several pulses before the 1993 Patras earthquake. They attributed the heating events to the escape of gas from mapped vents on the sea bed. SOTER (1999) cited other references to the heating of ocean water at the time of the 1817 Aigion earthquake and tsunami, in the same area.

Water in the deeper, more stagnant parts of Sissano Lagoon – parts that are not flushed by tides or by fresh water from the rivers – also can become uncomfortably hot, as we found while free diving from the shoreline at old Arop. The heat presumably is due to the decay of vegetable matter. However, this warm water probably is not present in sufficient quantity to explain the warming of the large volume of water that was in the waves. Nor would it explain warm water in waves that arrived directly from the sea.

Hydrogen sulphide in the tsunami wave is more easily explained. Hydrogen sulphide that could be smelled during the first strong earthquake was released from subsurface sediments as they started to liquefy. H_2S is widely present in the lagoonal mudflats and in the sediments just below the floor of the lagoon (our observations), and probably also is present in the sediments below the seafloor which are, at least in part, mudstones that accumulated in a lagoonal environment, before subsidence and shoreline retreat (e.g., see KOMAR, 1998, Figs. 2–22, 2–25).

Another puzzle is that the water stung the skin. This remains unexplained. Did the water become weakly acidic, and if so what might have caused this? Perhaps carbon dioxide gas escaping from the sediments beneath the seafloor or beneath the lagoon caused acidification.

5.4 The Magnitude of the Tsunami

Although there was clear evidence for maximum wave heights of 10–15 m onshore in sector DE (ITST, 1998; KAWATA, *et al.*, 1999a,b), it is almost certain that the height of the waves as they approached and reached the beach was about 4 m (see under 4.3, above). The wave heights of 10–15 m were likely caused by upward deflection after the wave reached the coast. This has implications for computer modelling of the source, and for the classification of the tsunami. A maximum wave height of 4 m indicates a tsunami magnitude of 2, rather than magnitude 3 for a wave height of 10–15 m (Imamura-Iida scale of tsunami magnitude; IIDA, 1958, cited by SHUTO, 1999).

5.5 The Thunderous Boom and the Roaring Noise

Most observers in the Sissano Lagoon area recall a thunderous boom or "pairap" (Tok Pisin) as though of thunder, that was heard after the mainshock, and that this was followed some minutes later by a roaring noise as though of a low-flying heavy jet plane or a large ship. Some heard a woop-woop sound, as of a low-flying heavy helicopter, before the roaring noise. The jet-like roaring sound travelled eastward initially (it was heard as far east as Aitape High School near Tadji (Fig. 4) and possibly beyond), then returned to the west. At the high schools east and southeast of Aitape it was heard as a whistling sound that progressed to the east, then to the west. Sanawe observed that the woop-woop sound coincided with the retreat of the sea from the shoreline, and slowed as the retreat of the water slowed.

Loud explosions have been heard at the time of other earthquakes and may be the effect of the *P* wave reaching the surface and releasing sound waves at frequencies greater than 15 cycles per second (BOLT, 1999). The delay between feeling the earthquake and hearing the sound can be explained because shock waves travel through rock faster than sound waves travel through air. Alternatively, as was suggested by John Sanawe (see under 4), the thunderous boom may have been caused by the gas-driven (?) explosion that he had witnessed.

SHUTO (1997) reviewed cases where roaring sounds had been heard at the time of tsunamis in Japan and concluded that the sound can be caused by the breaking wave, or by the rolling of pebbles as the wave retreats. Neither solution seems apt for the roaring sound that preceded the tsunami at Aitape. The source of the sound remains a mystery, as does the fact that the sound travelled along the coast to the east and then back to the west. Perhaps the sound was generated by displacement of air at the sea surface by the fast-moving low-amplitude tsunami wave, while it was propagating in deep water.

Survivors of the Aitape tsunami interpreted the thunderous boom and the roaring sound to be evidence that the tsunami was caused by a bomb that had been dropped by a low-flying jet plane. Persuading people against "the bomb theory" was one of the prime objectives of the public information campaign, because the hypothesis worked diametrically against people accepting the tsunami as a natural phenomenon, and so prolonged the period of uncertainty and rumour that followed the tsunami.

5.6 Unusual Lights

Three kinds of light effects were reported by different observers at different points. Firstly, many observers saw a flash of red light along the seaward horizon at or near the time of the mainshock. This may have been a sunset effect caught on the lower edge of a cumulus cloud, as was suggested by some observers. Later, as the wave approached, many saw a red glow in the top of the wave. This was seen not only at the area of most destruction (DE in Fig. 1) but also as far away as Aitape. This may have been bioluminescence, as suggested by KAWATA, *et al.* (1999b), though eye-witnesses dispute this. Finally, after the waves had passed and all was quiet, many saw a luminous yellow glow in the sky. What might have caused this?

Unusual lights at the time of an earthquake have been widely reported (*e.g.*, SOTER, 1999; BOLT, 1999, p. 202) but remain unexplained. SOTER (1999) suggested the possibility that friction amongst particles entrained in a discharge of gas might cause a glowing electrostatic discharge.

5.7 Water Circulation in the Wave

The tsunami sands deposited at Teles included living foraminifera of a species that characteristically lives at 250-m water depth. This appears to indicate a surprisingly great depth of water circulation in the tsunami wave.

5.8 How Dangerous is this Coast?

Since written records began, at the time of the first European settlement in 1896, there has been no major tsunami on this coast other than the 1998 event. According to contemporary records (the account by NEUHAUSS (1911) and the diaries of A.B. Lewis (WELSCH, 1998)) there was no tsunami at the time of the 1907 earthquake, and a tsunami described by MCCARTHY (1963) at the time of a major earthquake in 1935 was not severe and no deaths were recorded. On the evidence of that relatively short record, the recurrence interval of major tsunamis may be at least 100 years and possibly more. Currently we are investigating the prehistoric record.

The 1998 tsunami demonstrated that for villages in sector DE any tsunami is potentially dangerous, both because there are inadequate escape routes and because any tsunami generated near here will become focussed on this sector of coast, by virtue of the focussing effect of the shape of the seafloor (TAPPIN *et al.*, 1999, 2001; SYNOLAKIS *et al.*, 2002; MATSUMOTO and TAPPIN, this volume). Plantings of deep-rooted, salt tolerant trees (Fig. 7) will provide some protection from future tsunamis on this coast and elsewhere in PNG.

6. Conclusions

Information from our investigations has built upon and modified information collected by the international team soon after the tsunami. In particular, we present new information about the timing and westward progress of the waves, which arrived at the coast 4–7 minutes earlier than current models allow; and the maximum height of the waves while still offshore. The extremely high waves that left traces of their passage at 10–15 m above sea level onshore, as recorded by the international teams, are thought to have developed when a 4-m broken wave crossed the beach and was deflected upwards. Given that the coastal topographic profiles are known (our unpublished data), it should be possible to test this idea in the laboratory. The revised wave height must be taken into account in models of the source of the tsunami. The escape of gas may have played a part in the tsunami and can be investigated by searching for gas blow-out structures in the as yet unmapped areas of the seafloor.

Acknowledgements

Field operations were facilitated by the Catholic Diocese of Aitape and specifically Bishop Austen Crapp, Father Tim Elliott, Balthasar Maketu, the Presentation Sisters, Philip Turner, and driver Oderich; by the Controller of the Emergency, the late Ludwick Kembu and the Deputy Controller, Vincent Tutu with Major John Sehi; Robert and Margaret Parer, Brother Stephen Garry and Stephen Matawe; District Office staff Dickson Dalle, Tuatureme Bodde and Martin Selmatin; and members of the Local Government Council. Rondi Davies drafted the map (Figs. 1 and 4) and Simon Saulei advised on species and names of trees. Members of the international tsunami science community provided essential advice at different times during the investigation, notably Fumihiko Imamura, Frank Gonzalez, Costas Synolakis and Yoshinobi Tsuji. Professor Imamura provided seed funding for field operations, and Professor Synolakis, funding to attend the 1998 Fall Meeting of the American Geophysical Union. The paper was improved by critical reviews from Eli Silver and Professor Imamura. Orogen Minerals Limited provided the initial financial support for field operations, and the National Disaster Committee provided funds subsequently.

References

Bolt, B. A. Earthquakes (New York, Freeman 4th edition 1999).

Carey, S. W. (1935), *Preliminary Notes on a Recent Earthquake in New Guinea*, Australian Geographer *2*(8), 8–15.

Davies, H. L. (1998a), *Tsunami PNG 1998 – Extracts from Earth Talk*, Waigani, University of Papua New Guinea.

Davies, H. L. (1998b), *Sissano Tsunami: A Reconstruction of the Events of July 17, 1998, Based upon Interviews with Survivors*. 1998 Fall Meeting American Geophysical Union, Supplement to EOS, Transactions AGU *79*(45), p. F572.

Davies, H. L., Davies, J. M., Lus, W. Y., Perembo, R. C. B., Joku, N., Gedikile, H., and Nongkas, M. (2000), *Tsunami at Aitape – "Is it Safe to Go Back?"* (abstr.). Proc. 15th Australian Geological Convention, Geological Society of Australia Abstracts 45.

Davies, H. L. (in prep.), *Report on the Aitape 1998 Tsunami*, Waigani, PNG National Disaster Management Office.

Davies, H. L., Davies, J. M., Lus, W. Y., Perembo, R. C. B., Joku, N., Gedikile, H., and Nongkas, M. (2001), *Learning from the Aitape Tsunami*, Proc. Internati. Tsunami Symp. ITS2001, Seattle.

Davies, J. M., Davies, H. L., Joku, N., and Gedikile, H. (1999), *The Aitape Tsunami – reconstructing the events and planning for the future at PARTIC*, 1999 Fall Meeting Am. Geophys. Union, Supplement to EOS, Transactions AGU 80 p. F750.

Everingham, I. B. (1977), *Preliminary Catalogue of Tsunamis for the New Guinea/Solomons Region, 1768–1972*, Bureau of Mineral Resources, Australia, Report 180.

Geist, E. L. (2000), *Origin of the 17 July 1998 Papua New Guinea Tsunami: Earthquake or Landslide?*, Seismol. Res. Lett. *71*, 344–351.

Goldsmith, P., Barnett, A., Goff, J., McSaveney, M., Elliott, S., and Nongkas, M. (1999), *Report of the New Zealand Reconnaissance Team to the Area of the 17 July 1998 Tsunami at Sissano Lagoon, Papua New Guinea*, Bull. NZ Soc. Earthq. Engin. *32*, 102–118.

HASIOTIS, T., PAPATHEODOROU, G., KASTANOS, N., and FERENTINOS, G. (1996), *A Pockmark Bed in the Patras Gulf (Greece) and its Activation during the 4/7/93 Seismic Event*. Marine Geology *130*, 333–344.

HOLIAN, A. C., and KEITH, P. P. (1998), *Orthopaedic Surgery after the Aitape Tsunami*, Medical J. Australia *169*, 606–609.

HOVLAND, M., and JUDD, A. G., *Seabed Pockmarks and Seepages* (London, Graham and Trotman 1988).

HURUKAWA, N., TSUJI, Y., and WALUYO, B. (2003), *The 1998 Papua New Guinea Earthquake and its Fault Plane Estimated from Relocated Aftershocks*, Pure appl. geophys. *160*, 1829–1841.

IIDA, K. (1958), *Magnitude and Energy of Earthquakes Accompanied by Tsunami, and Tsunami Energy*, J. Earth Sciences, Nagoya University *6*, 101–112.

ITST (1998), *Preliminary Report on the Tsunami Caused by the Earthquake at Sissano, West Sepik, PNG, on 17 July 1998*, International Tsunami Survey Team unpublished report, September 1998.

JAFFE, B., GELFENBAUM, G., BENSON, B., DAVIES, H., and NONGKAS, M. (1998), *Sedimentation, Erosion and Flow in the July 17, 1998 Papua New Guinea Tsunami*, 1998 Fall Meeting Am. Geophys. Union, Supplement to EOS, Transactions AGU *79*(45), p. F572.

KAWATA, Y., TSUJI, Y., MATSUTOMI, H., IMAMURA, F., MATSUYAMA, M., and TAKAHASHI, T. (1999a), *Field Survey of the 1998 Tsunami in the Northwestern Area of Papua New Guinea*, Report to Ministry of Education, Science, Sports and Culture, Japan.

KAWATA, Y., BENSON, B. C., BORRERO, J. C., BORRERO, J. L., DAVIES, H. L., DELANGE, W. P., IMAMURA, F., LETZ, H., NOTT, J., and SYNOLAKIS, C. E. (1999), *Tsunami in Papua New Guinea was as Intense as First Thought*, EOS Transactions of the Am. Geophys. Union *80*, pp. 101, 104–105.

KOMAR, P. D., *Beach Processes and Sedimentation*. (New Jersey, Prentice Hall, 1998), 2nd edition, 543 pp.

MCCARTHY, J. K. (1963), *Patrol Into Yesterday*, Canberra, Cheshire, p. 156.

MCSAVENEY, M. J., GOFF, J. R., DARBY, D. J., GOLDSMITH, P., BARNETT, A., ELLIOTT, S., and NONGKAS, M. (2000), *The 17 July Tsunami, Papua New Guinea: Evidence and Initial Interpretation*, Marine Geology *170*, 81–92.

MATSUTOMI, H., KAWATA, Y., SHUTO, N., TSUJI, Y., FUJIMA, K., IMAMURA, F., MATSUYAMA, M., TAKAHASHI, T., MAKI, N., and HAN, S. S., 2001, *Flow Strength on Land and Damage of the 1998 Papua New Guinea Tsunami*. In *Tsunami Research at the End of a Critical Decade* (G. T. Hebenstreit, ed.), (Dordrecht, Kluwer 2001)

NATIONAL MAPPING BUREAU (1998), *Aitape Disaster Area West Block and East Block, Aerial Photographs at 1:10,000 scale*. Port Moresby, National Mapping Bureau.

NEUHAUSS, R., *Deutsch-Neu-Guinea*, 3 vols. (Berlin: Dietrich Reimer 1911).

NORVICK, M. and HUTCHISON, D. S. (1980), *Aitape-Vanimo 1:250,000 Geological Series Map and Explanatory Notes, Geological Survey of Papua New Guinea SA/54-15 and SA/54-11*.

PEGLER, G., DAS, S., and WOODHOUSE, J. H. (1995), *A Seismological Study of the Eastern New Guinea and the Western Solomon Sea Regions and its Tectonic Implications*, Geophys. J. Internat. *122*, 961–981.

RIPPER, I. D., LETZ, H., and MOIHOI, M. (1999), *Felt Effects of the 17 July 1998 Sissano Lagoon Tsunami Earthquakes*, Geological Survey of Papua New Guinea Report 99/13.

SHUTO, N. (1997), *A natural warning of tsunami arrival*. In *Perspectives on Tsunami Hazard Reduction* (G. Herbenstreit ed.), (Dordrecht, Kluwer 1997), pp. 157–173.

SHUTO, N. (1999), *A Short History of Tsunami Defense Works in Japan*. In *Study on Regional and Historical Change of Tsunami Disasters – In Commemoration of the 1993 Hokkaido Nansei-oki Earthquake Tsunami*. Research Project, Grant-in-aid for Scientific Research (B) no. 09044126 (1998–1999), Faculty of Policy Studies, Iwate Prefectural University, pp. 35–67.

SOTER, S. (1999), *Macroscopic Seismic Anomalies and Submarine Pockmarks in the Corinth-Patras Rift, Greece*, Tectonophysics *308*, 275–290.

SWEET, S., and SILVER, E. A. (2001), *Seismic reflection images of the source region of the 1998 Papua New Guinea tsunami*. In: *Prediction of Underwater Slump and Slide Hazards*. (Watts, P., Synolakis, C. E., and Bardet, J. P., eds), (Rotterdam, Balkema 2001).

SYNOLAKIS, C., BARDET, J.-P., BORRERO, J., DAVIES, H., OKAL, E., SILVER, E., SWEET, S., and TAPPIN, D. (2002), *The Slump Origin of the 1998 Papua New Guinea Tsunami*, Proc. Roy. Soc. London A *458*, 763–789.

TANIOKA, Y. (1999), *Analysis of the Far-field Tsunamis Generated by the 1998 Papua New Guinea Earthquake*, Geophys. Res. Lett. *26*, 3393–3396.

TAPPIN, D. R., MATSUMOTO, T., WATTS, P., SATAKE, K., McMURTRY, G. M., MATSUYAMA, M., LAFOY, Y., TSUJI, Y., KANAMATSU, T., LUS, W., IWABACHI, Y., YEH, H., MATSUMOTO, Y., NAKAMURA, M., MOIHOI, M., HILL, P., CROOK, K., ANTON, L., and WALSH, J. P. (1999), *Sediment Slump Likely Caused the 1998 Papua New Guinea Tsunami*, EOS Transact. Am. Geophys. Union *80*, pp. 329, 334, 340.

TAPPIN, D. R., WATTS, P., McMURTRY, G. M., LAFOY, Y., and MATSUMOTO, T. (2001), *The Sissano, Papua New Guinea Tsunami of July 1998 – Offshore Evidence of the Source Mechanism.* Marine Geology *175*, 1–23.

TAYLOR, R. P., EMONSON, D. L., and SCHLIMMER, J. E. (1998), *Operation Shaddock – The Australian Defence Force Response to the Tsunami Disaster in Papua New Guinea*, Medical J. Australia *169*, 602–606.

WELSCH, R. L. (1998), *An American Anthropologist in Melanesia. A.B. Lewis and the Joseph N. Field South Pacific Expedition 1909–1913.* 2 vols. (Honolulu: University of Hawaii Press 1998).

Pure appl. geophys. 160 (2003) 1923–1943
0033–4553/03/111923–21
DOI 10.1007/s00024-003-2414-0

Pure and Applied Geophysics

Possible Coseismic Large-scale Landslide off the Northern Coast of Papua New Guinea in July 1998: Geophysical and Geological Results from SOS Cruises

TAKESHI MATSUMOTO[1], DAVID R. TAPPIN[2] and SOS Onboard Scientific Party

Abstract—The epicentral area of the 1998 Papua New Guinea Earthquake and Tsunami off Sissano Lagoon, northern coast of Papua New Guinea ($M = 7.1$; 17 July, 1998) was surveyed by the Research Vessel *Kairei* in January 1999. Precise bathymetric survey by use of SEABEAM-2112 and other geophysical surveys (bottom reflectivity, sub-bottom profiling, surface ship gravity and geomagnetic surveys) were carried out and four piston core samples were collected during the nine days' survey. The area was also surveyed by use of the Japanese deep sea research *ROV Dolphin-3K* and research submersible *Shinkai-2000* after the regional geophysical survey, in order to locate the possible seismic faults and/or underwater landslides as the source of tsunami and to study the process which took place off Sissano Lagoon and the driving force of the event.

The study area is characterised by substantial fan sediment supply from Sissano Lagoon. Straight small-scale submarine canyons and valleys are eroding the shelf slope constructed by the fan sediment. Topographic features of arcuate slumps caused by landslides are recognised at numerous sites of the study area. Most of them are old and the most recent is located 25 km northeast off the Sissano Lagoon. Two major topographic depressions on the shelf were located off Sissano Lagoon. The western is a depression of about 10-km width and the eastern is a meandering deep-sea canyon.

Amphitheatre topographic features with slumps caused by landslides that were discovered by *Kairei*'s cruise were extensively surveyed in the study area. Six dives by *Dolphin-3K* and seven dives by *Shinkai-2000* revealed that a fresh crack of about 15 km in total length is located along the slope of the amphitheatre. The crack is apparently the upper slope of a large-scale slumping on the amphitheatre, characterised by tensional stress on it.

The easternmost part of the fresh crack is accompanied by living chemosynthetic organisms such as mussels and tube worms. The chemosynthetic community is apparently associated with cold seepage along the crack suggesting that the crack was constructed very recently. The areas other than the crack were rather old with bioturbation.

Key words: The 1998 Papua New Guinea Tsunami, Sissano Lagoon, underwater landslide, amphitheatre.

[1] Marine Science Department, Nippon Marine Enterprises, Ltd. 14-1 Ogawacho, Yokosuka 238-0004, Japan.

[2] British Geological Survey, Kingsley Dunham Centre, Keyworth, Nottingham, England NG12 5GG, United Kingdom.

1. Introduction

The northern coast of Sissano Lagoon and Town of Aitape, Sandaun Province of Papua New Guinea enormously suffered from an $M = 7.1$ earthquake which occurred at 08:49 (UTC), July 17, 1998 (RIPPER *et al.*, 1998; KAWATA *et al.*, 1999). A large-scale tsunami also occurred just after the earthquake and more than 2,200 people were killed due to the runup of the tsunami in the Aitape and Sissano areas.

The area is characterised by a plate convergent margin along the Wewak Trench (COOPER and TAYLOR, 1987), with clear distribution of the Benioff zone subducting from the North Bismarck Sea Plate towards mainland Papua New Guinea (TREGONING *et al.*, 1998). The estimated epicentre of the mainshock and the possible origin of the tsunami are also included in this area. Preliminary hypocentres (latitude, longitude, depth) of the mainshock were obtained by three different organisations as follows: (i) 2.42°S, 142.38°E, 18.5 km by the Earthquake Research Institute, University of Tokyo; (ii) 2.78°S, 142.57°E, 15.0 km (centroid of rupture) by Harvard University; (iii) 3.084°S, 141.762°E, 6 km (apparently on land) by United States Geological Survey. The study area was specified to cover these results.

The main objectives of the present project are: (1) to detect the nature of the "1998 Papua New Guinea Tsunami," (2) to detect the seismic fault, possible underwater landslide and the mechanisms of tsunami occurrence, and (3) to detect the process which took place in the epicentral area and how it generated these phenomena, in the context of regional tectonics.

The South Pacific Applied Geoscience Commission (hereafter SOPAC), with headquarters in Suva, Fiji, took this natural disaster very seriously and requested JAMSTEC to establish a research project on the geology and geophysics of the epicentral area. JAMSTEC officially decided to implement a geophysical and geological survey in order to clarify the nature of the earthquake and of the subsequent tsunami.

The first scientific research cruise in this project was SOS-1, a topographic and geophysical survey by *R/V Kairei* (Cruise ID: KR98-13). Its main objective was to locate the earthquake faults and/or underwater landslides in the survey area through precise topographic and bottom reflectivity surveys by use of the SEABEAM 2112 multibeam bathymetric survey system. The result of the topographic survey was also used as a basic bathymetric data set for numerical simulation of the propagation and inundation of the tsunami in order to replay the process from the initiation of the tsunami off the northern coast of Papua New Guinea to its propagation towards and inundation on the land mass, especially at Sissano Lagoon and its surroundings. The gravity and geomagnetic data together with the topographic data were also of great importance to define the regional tectonics which should form the background of the series of hazardous geological phenomena. Piston core sampling was also carried out at four stations, in order to investigate the present and past occurrence of turbidity currents induced by large-scale landslides, their character, age and their physical properties.

After analysing the results from this first cruise, a second cruise (SOS-2) was launched using the Remotely Operated Vessel (ROV) *Dolphin-3K* and her mothership *R/V Natsushima* (Cruise ID: NT99-01). It involved a rather broader survey, especially regarding visual observation and geological sampling in the study area. Later, a third cruise (SOS-3) involving a more focused, precise survey was carried out using the deep sea submersible *Shinkai 2000* and her mothership *R/V Natsushima* (Cruise ID: NT99-11). Through these cruises, the following investigations took place: (i) identification of the possible collapse, underwater landslide, and seismic faulting mechanism which took place on the occasion of the 1998 tsunami; and (ii) estimation of the processes involved in the regional area around the epicentre and their potential in terms of such phenomena, in the context of regional tectonics.

2. *Topographic Features*

The swath bathymetric and geophysical survey area is located off Sissano Lagoon and its surroundings (Fig. 1). The final survey box was between 3°20′S and 2°00′S, and between 141°00′E and 143°30′E. The area of 19000 km^2 was almost fully covered by the swath survey of the SEABEAM 2112 system. Results of the present study define several topographic features which characterise the regional tectonics and potential tsunami events (Figs. 2a,b).

Wewak Trench

The Wewak Trench is located about 50 km north of the coastline. The general trend of the trench is N45°W. However, the landward side of the trench is characterised by topographic lineations of EW escarpment and the other parallel to sub-parallel to the general trend of the trench axis. The eastern end of the trench is located about 40 km NE of Aitape, with a water depth of about 3000 m. Towards the NW, the width and water depth of the trench increase gradually. Maximum water depth in the study area is 4,200 m about 50 km NNE of Sissano Lagoon. Several seamounts on the ocean floor are colliding against the trench on the northern side from the North Bismarck Sea Plate. Some are not a typical volcanic cone but form an elongated feature perpendicular to the general lineament of the trench. The trench floor is divided into several basins. The water depth of those basins is 4100–4200 m in most cases.

Amphitheatres

Four major "amphitheatres" are located from 141°40′E to 142°20′E off Sissano and Aitape, about 30 km off the coast as for the western three and about 15 km off as

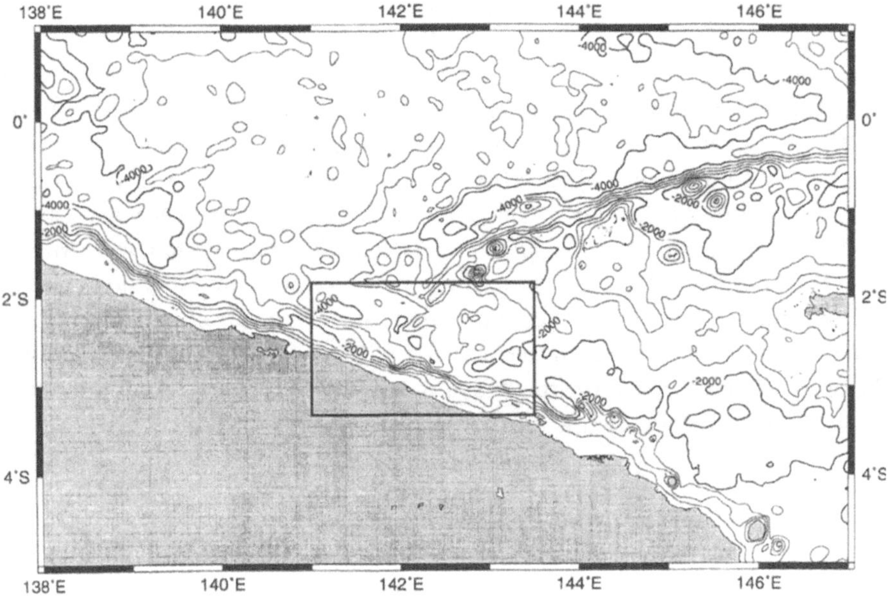

Figure 1
Index map of the study area off the northern coast of Papua New Guinea.

for the easternmost. These are characterised by a topographic depression edged by an arcuate steep slope (headwall) and a flat basin off the slope. The average gradient of the upper slope is 14 to 17 degrees. A few occurrences of inverted topography (convex-concave topography) suggesting recent landslides are located on the steep slopes. As for the western two, the topographic features of erosion with some small-scale channels and a wide basin just off the steep slope, apparently due to sediment deposition, are predominant. These topographic features show that no large-scale deformation which may be the origin of tsunami took place on these two amphitheatres recently.

The amphitheatre NNE of Sissano Lagoon is associated with a ENE-WSW-trending knoll just off the basin (Pop-up Block, "PUB"). The western half of the amphitheatre is characterised by erosion by small channels perpendicular to the topographic contour line and the eastern half by a sediment deposition lobe parallel to the strike of the steep slope. There is a topographic high with a subsided coral reef

Figure 2a
Topographic contour map obtained by the SEABEAM-2112 multibeam bathymetric survey system on board *Kairei* during the SOS-1 cruise.

Figure 2b
Close-up topographic map of the study area.

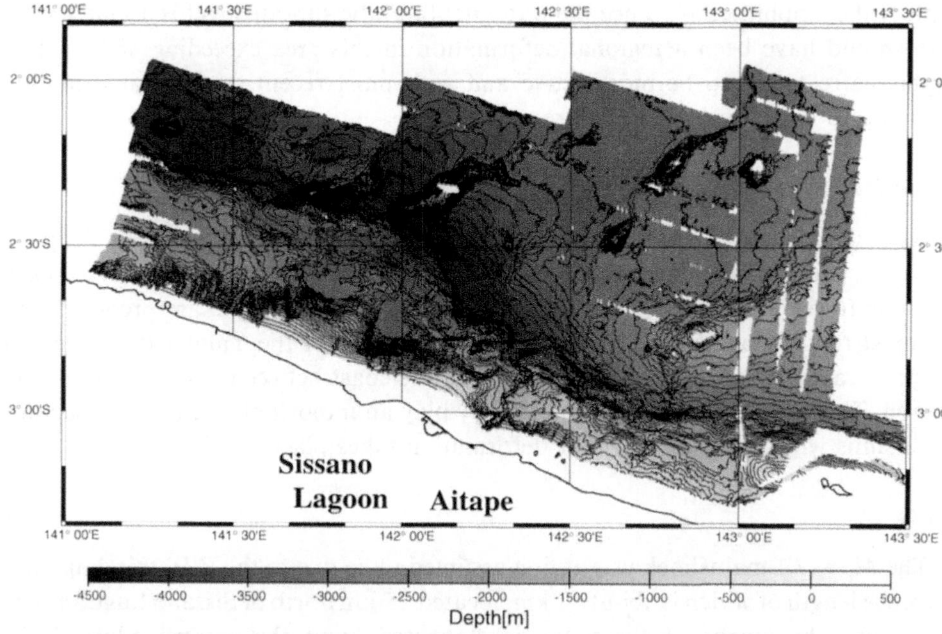
141° 00'E
141° 30'E
142° 00'E
142° 30'E
143° 00'E
143° 30'E
2° 00'S
2° 30'S
3° 00'S
Sissano
Lagoon
Aitape
-4500
-4000
-3500
-3000
-2500
-2000
-1500
-1000
-500
0
500
Depth[m]

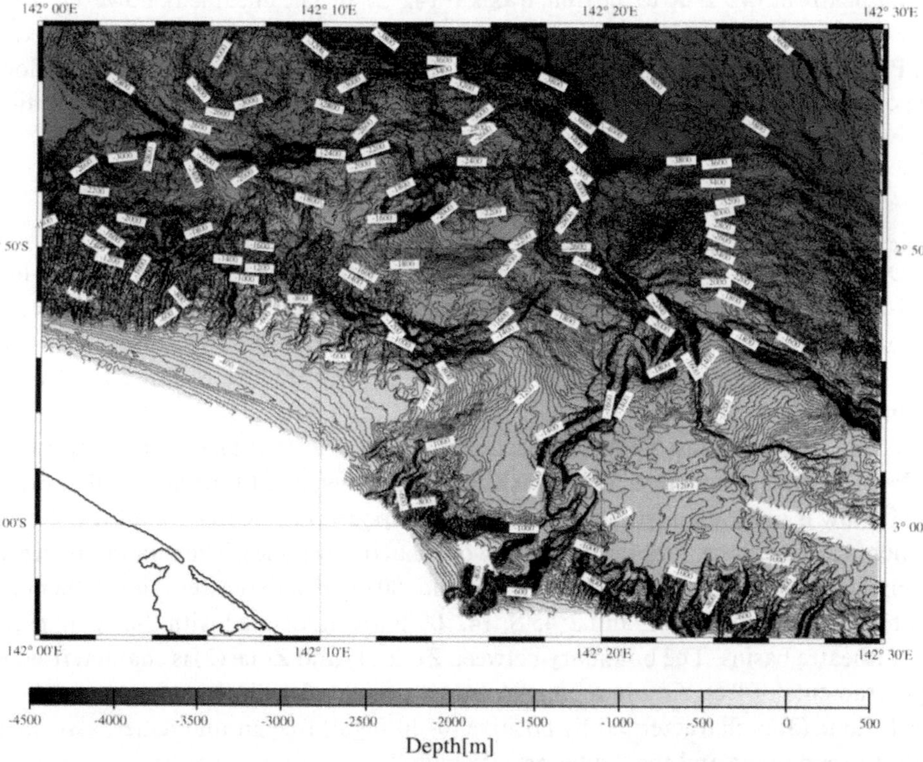
142° 00'E
142° 10'E
142° 20'E
142° 30'E
2° 50'S
3° 00'S
-4500
-4000
-3500
-3000
-2500
-2000
-1500
-1000
-500
0
500
Depth[m]

south of the amphitheatre. Since the water depth of the subsided reef is about 460 m, there should have been a regional deformation in this area exceeding 400 m. This amphitheatre seems to be most active and with most recent seafloor deformation according to the topographic feature.

Submarine Canyons and Valleys

As described above, the area is also characterised by the development of deep sea canyons, channels and topographic depressions. Small-scale straight valleys apparently due to the erosion on the shelf slope shallower than 1000 m are predominant. The most remarkable is the meandering channel north of the Yalingi River west of Aitape, located in the amphitheatre 25 km off the coast between Aitape and Sissano Lagoon. These topographic depressions may play an important role as a propagator of tsunamis when a certain seafloor deformation takes place nearby.

Faults

The $M = 7.1$ mainshock was at first assumed to be due to the E-W trending fault scarp, the length of which is about 40 km, located 25 km north of Sissano Lagoon. The topographic lineament of the scarp is elongated from the eastern edge of the amphitheatre at 142°E up to the trench axis at 142°27′E. The lineament, however, seems to be split into two by PUB and no sign of the overriding of the fault scarp is observed on the PUB. Therefore, the mainshock is not corresponding to the fault on the seafloor. The origin of the tsunami should be another fault or some other seafloor deformation.

3. *Gravity*

On-board surface-ship gravity measurements were carried out during the whole SOS-1 cruise by use of the Bodenseewerke KSS-31 gravity metre. The theoretical normal gravity based on the Geodetic Reference System 1967 was subtracted from each gravity measurement to get the free air anomaly. Figure 3 shows the free air anomaly map of the study area.

The study area is classified into three zones sub-parallel to the shoreline: (1) positive, aside the shore; (2) negative, along the topographic low; and (3) positive, off-shore. A gravity low is associated with the lineament of topographic depression of the Wewak Trench. However, the zone is split into two isolated low zones. The minimum free air anomalies are – 100 mgal in the western side and -80 mgal in the eastern side. Other lows located at 2°35′S, 141°45′E and 2°45′S, 142°00′E are associated with the "collapsed" amphitheatre basins. The boundary between Zone (1) and Zone (2) is characterised by high gradients of free air anomaly, which are associated with the steep slope off the coast. Zone (3) is characterised by positive (0–40 mgal) free air anomalies, associated with the seamounts and the "outer gravity high."

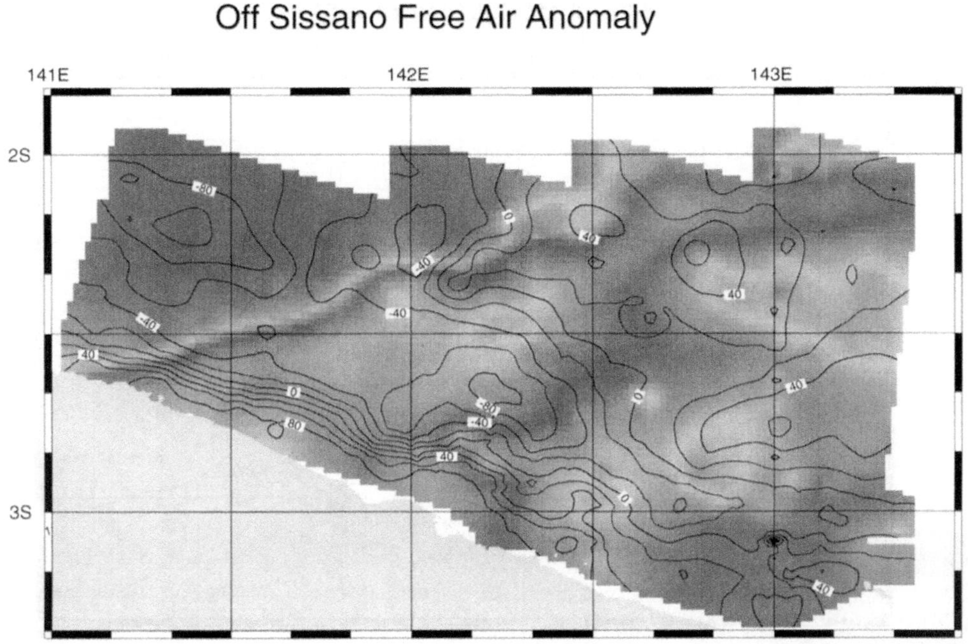

Figure 3
Free air anomaly map of the study area. Data are from the gravity measurements during the SOS-1 cruise.

Bouguer anomalies were calculated by adding the Bouguer correction to free air anomalies. The density of crust is taken as a constant, 2,670 kg/m^3. A three-dimensional calculation of terrain correction by use of FFT was applied. Figure 4 shows the result thus calculated. The area north of 2°30′S is characterised by a maximum Bouguer anomaly, associated with typical oceanic crust.

4. *Magnetic Anomaly*

The geomagnetic anomalies, shown on Figure 5, were mapped by proton-precession and 3-D vector magnetometres. The total force anomaly shows that a negative anomaly is predominant in the coastal zone and a positive one off the coast. The large intensity positive and negative pair in the northeastern part of the survey area are associated with the seamount chains located in this area. These seamounts are considered to be of volcanic origin and are colliding against the plate convergence margin off Sissano.

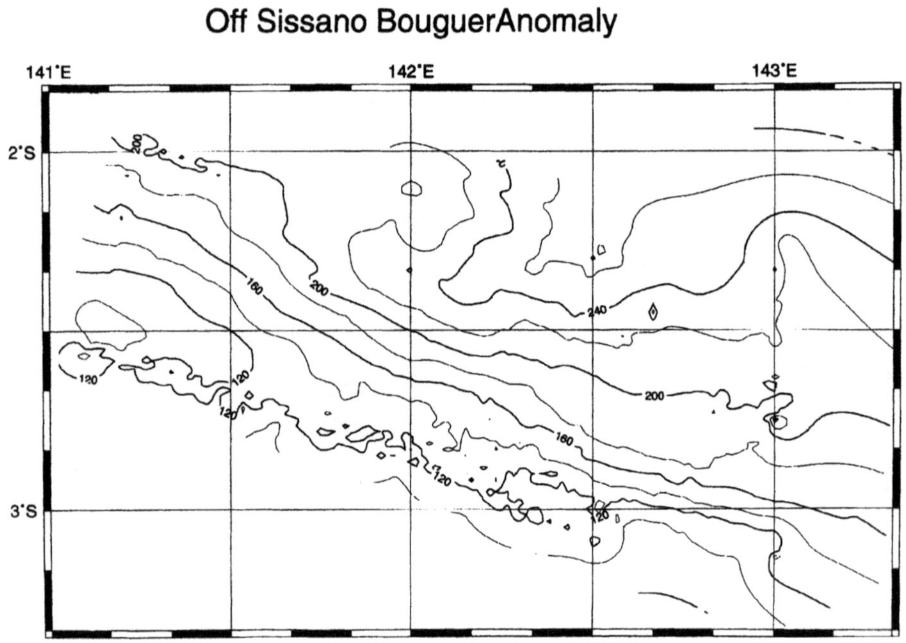

Figure 4
Bouguer anomaly map of the study area derived from the SOS-1 gravity and topography data.

5. *Observation by ROV Dolphin-3K and the Deep Sea Submersible Shinkai 2000*

The SOS-2 and SOS-3 cruises were focussed on the precise seafloor observation by ROV Dolphin-3K (SOS-2) and Manned Submersible SHINKAI 2000 (SOS-3). Total 13 dives over the potential tsunamigenic deformational area were completed in order to identify the origin of the 1998 event (Figs. 6, 7).

Cracks and Fissures along the Headwall of the Amphitheatre (cf. Fig. 8a,c)

Cracks and fissures on the sediment with sharp edges on both sides along the headwall of the amphitheatre were first discovered during Dive 3K#398 at the eastern to central portion of the amphitheatre, 1500 to 1530 m in water depth. The width of the crack is a couple of metres and the depth seems to be more than a couple of metres. Large cracks were found to exist from the lower portion up to a depth of 1400 m. These cracks apparently indicate the tensional stress applied recently on the headwall. The general orientation of the cracks is perpendicular to the steepest gradient.

The cracks were traced westward during Dive 3K#402. The orientation of the crack was WNW-ESE along the track, almost perpendicular to the strike of the steepest gradient of the headwall. The crack was found to be not single, but rather consisting of *en échelon* minor cracks some 100 m in length, indicating a shear stress component on the headwall. Along the track of the same dive, some relics of old cracks

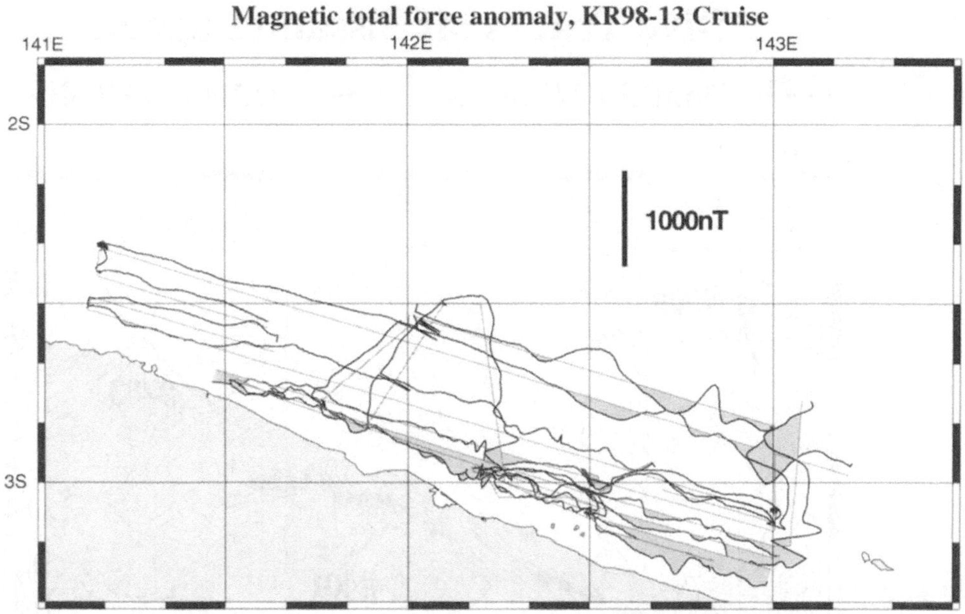

Figure 5
Total force magnetic anomaly obtained by the proton precession magnetometre during the SOS-1 cruise.

covered with sediment were observed on the terrace at 2°53′S, 142°13.5′E, in water depth shallower than 1400 m. Fresh angular rock debris and breccia apparently from the upper slope were also observed on the foot of the steep slope edging the terrace.

Dive 2K#1159 took place on the eastern part of the amphitheatre on the sediment deposit lobe off the foot of the headwall towards the arcuate slope. From the landing point at 1664 m to the base of the headwall there was an undulating seabed. The foot of the headwall comprised a steep scarp slope with a gradient of ~25 to 40 degrees. The mid-part of the headwall comprised an undulating slope of shallow gradient. The upper headwall again was of a steep gradient. The dive identified recent slumping at the bottom and top of the amphitheatre headwall slope. At the base the slump included both unlithified, but cohesive sediment, and limestone rock. At 1662 m there was an extensive and prominent sinuous fissure trending ~2 m wide, metres deep and tens of metres long with a steep southward dip. Above 1654 m there were further fissures and sediment slumps. The slump was associated with chemosynthetic communities of mussels and tube worms as well as bacterial mats over 40 metres of seabed, in water 1600 m deep, apparently derived from cold water seepage. The upper headwall slumping involved only sediment. For most of the dive transect the seabed was formed of a thin layer of soft yellowish brown sandy mud underlain by slightly more cohesive olive green clay. Over most of the dive the seabed

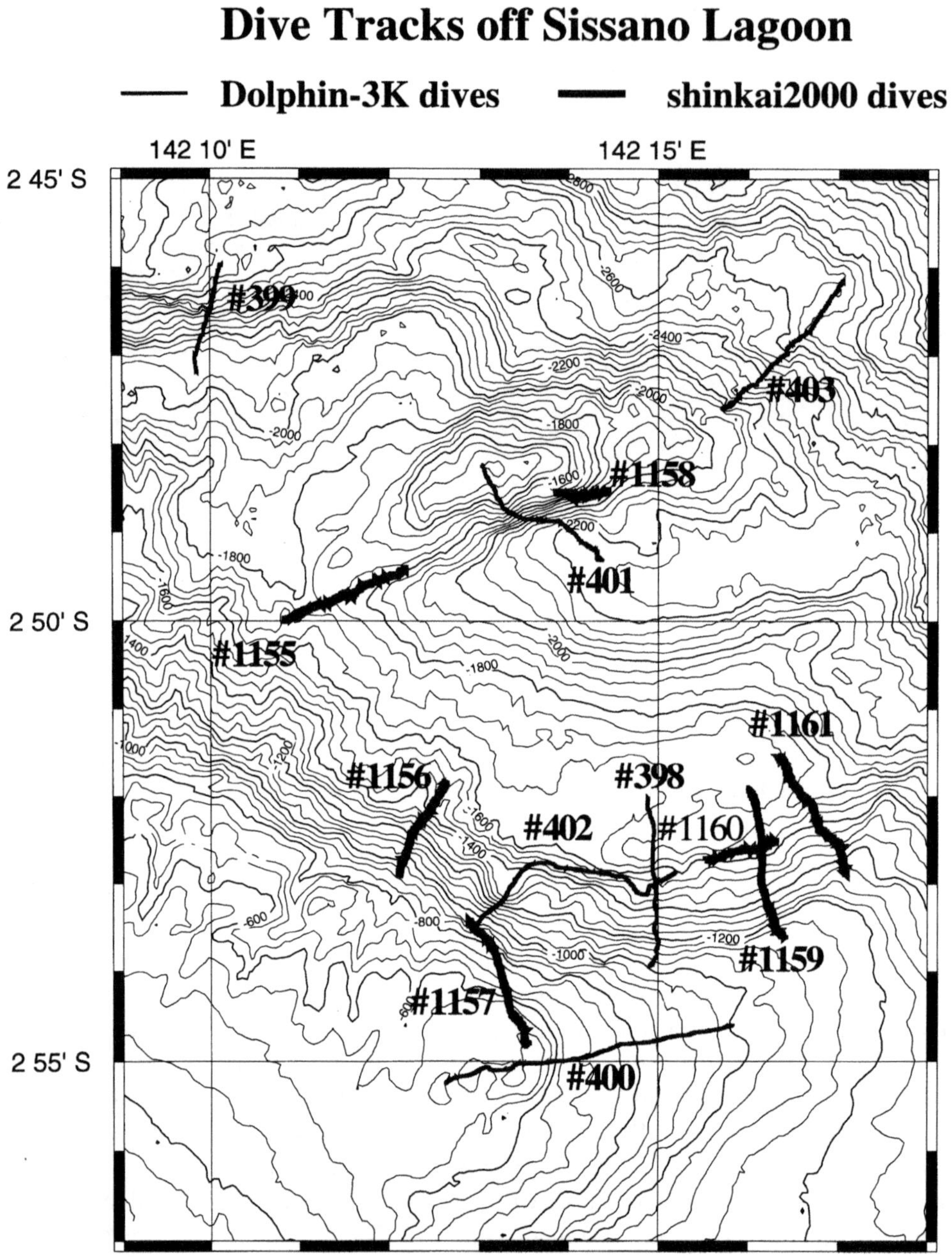

Figure 6
Dive tracks by the ROV *Dolphin-3K* (#398 to #403, SOS-2 cruise, February 1999) and Submersible *Shinkai 2000* (#1155 to #1161, SOS-3 cruise, November 1999) in the amphitheatre and Pop-up Block area.

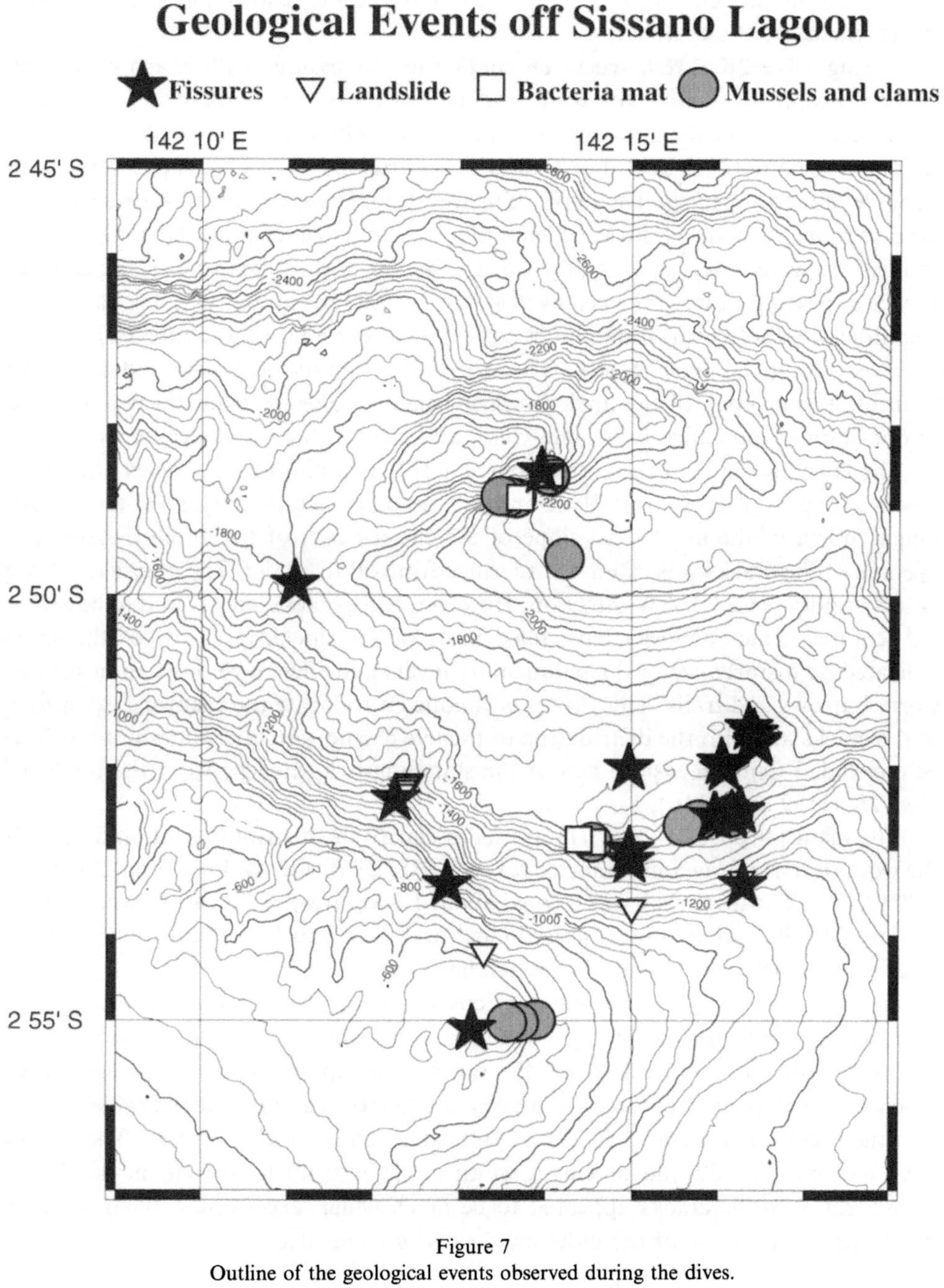

Figure 7
Outline of the geological events observed during the dives.

displayed a faunal activity. Trails and tracks covered the seabed and were the result of movement by these fauna.

During Dive 2K#1160, recent *en échelon* cracks/fissures with sharp edges and minor slumping associated with the major cracks were observed along the mid-wall of the eastern section of the amphitheatre, along depth contours 1630–1650 m, from the first landing point (2°52.02′S, 142°16.282′E) towards the west. The cracks were about a couple of metres wide and a couple of metres deep. However, the western section of the dive route showed very little evidence of sea bottom disturbances. No major cracks and fissures were observed along the western border of the dive route. These features may be appearing either at the headwall above the 1600 m contour or at the base of the amphitheatre. Biological communities consisting of live mussels and tube worms were observed along the fresh cracks. The area also consists of gullies and ridge-like features. The geology of the area comprises mainly cohesive sediment deposit. The composition of the sediment is mainly clay and coarse grain sand structures.

The Dive 2K#1161 at the eastern portion of the amphitheatre showed a series of cracks at depths of 1666 m, 1671 m, and 1683 m on the southern slope of the deposit tongue north of the northward slope of the eastern part of the amphitheatre. The opening of the cracks was estimated to range from 60 to 80 cm. The cracks are more or less trending in a NE-SW direction. However, these cracks were found to be rather old features. Finer deposits have filled most of the openings. Most of the areas indicated no disturbance of the sedimentary layers and effect of bioturbation such as worm burrows and trails. This dive was conducted to locate the eastward extensions of the cracks and also the distribution of the live chemosynthetic community. It is to be concluded that the eastern end of the slumping is located between the tracks of Dive 2K#1159 and #1161.

The main purpose of Dive #1156 was to clarify the distribution of landslides in the western part of the amphitheatre between 2°52.17′S, 142°12.126′E and 2°52.88′S, 142°12.11′E. The dive track was located about 1.8 km west of the most western point of Dive 3K#402, where cracks in sediment and rock had been found, but without locating a landslide or slump. At the landing site the seabed was relatively flat and covered by a thick, muddy sedimentary cover. Sediments were undisturbed and no evidence of rock debris or outcrop was found during the dive. Old slumps of cohesive lumps of sediment were found on the northward dipping slope. Although steep, much of the slope was covered by muddy sediments. The most remarkable feature was the cracks at a depth of 1191 metres, at a strike of 120 degrees. The cracks measured between 30 and 50 cm in width and about 40 to 100 cm in depth. At greater distance the cracks appeared to be much wider, exceeding 150 cm. A fresh fissure was seen in one of the older cracks and documented.

Southern Escarpment of the Pop-up Block (cf. Fig. 8b)

The southern slope of the Pop-up Block (PUB) was surveyed during Dives 3K#401 and 2K#1158. The faulted surface of the strike which is east to northeast may represent the footwall fault surface of the E-W fault at the southern base of the PUB. Evidence of fresh cracks associated with a chemosynthetic community on the basement rocks suggests accommodation of recent movement on the fault surface. It is unclear if the fresh cracks are derived from the July 17, 1998 earthquake activities and the tsunami. Landslide debris along the traverse route consists of not pure cohesive clay but a mixture of rock, shell fragments, and proportions of soft clay sediment. This suggests that marine landslide material on the PUB is different from cohesive clay dominated material at the southern wall of the amphitheatre.

The main purpose of Dive #1155 was to clarify whether or not the cliff of the southern slope of PUB extends to the western end of the west flank of the block. The cliff is possibly formed by the fault dislocation which caused the 1988 Aitape Earthquake. However, no evidence of the extension of the cliff or the fault activity in the area of the western flank of the PUB was observed.

Subsided Coral Reef (cf. Fig. 8c)

The topographic survey during the SOS-1 cruise revealed the existence of a submerged coral reef drowned to 465 m and centred at 2°55′S, 142°13′E. Dive

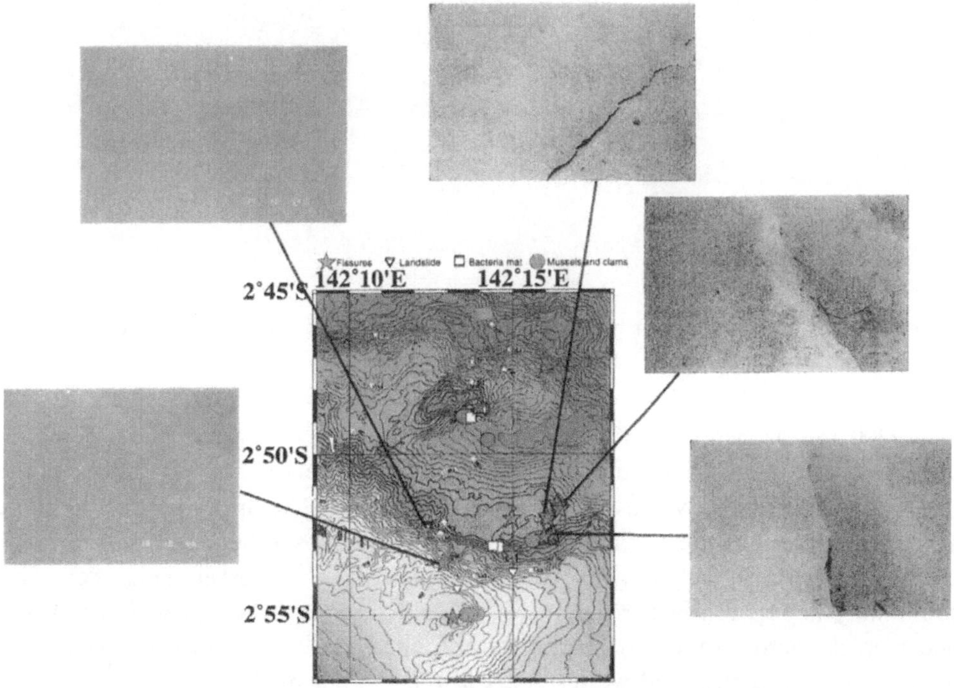

Figure 8a
Photographs of geological features observed during dives—cracks and fissures along the headwall of the amphitheatre.

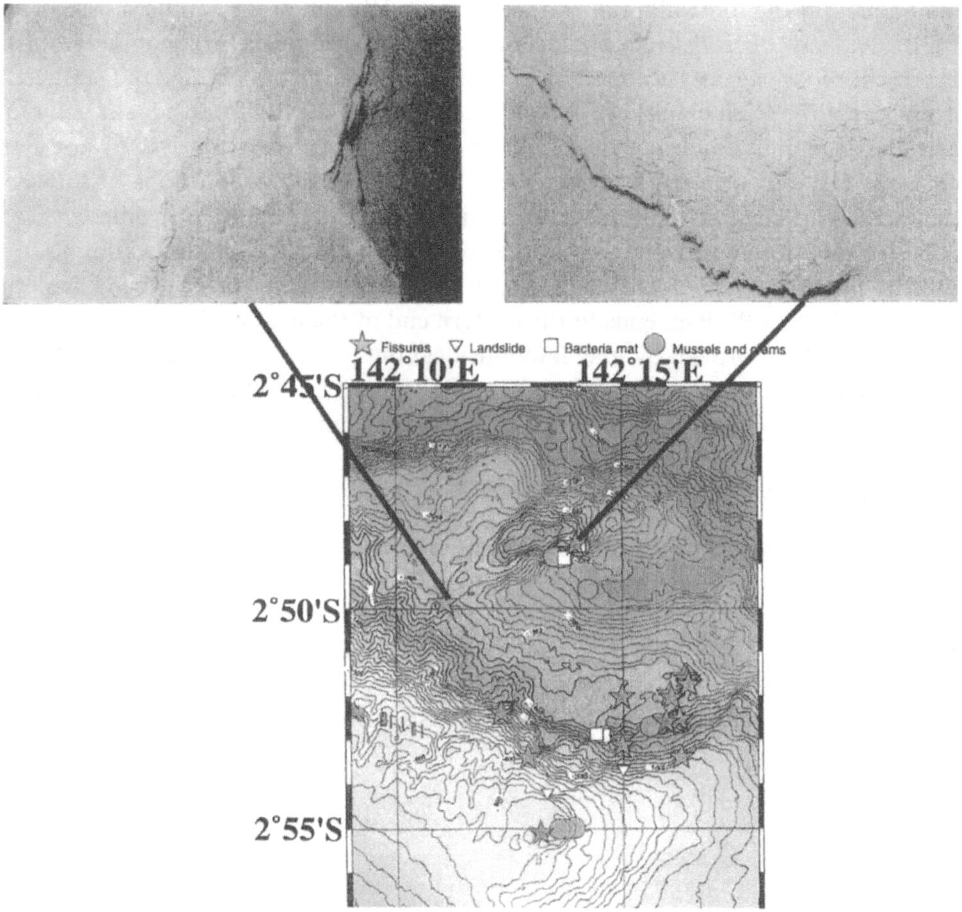

Figure 8b
Photographs of geological features observed during dives—southern escarpment of the Pop-up Block.

D3K#400 approached the reef from about 4000 m ENE (1000 m in water depth) in order to detect the mechanism and process of the subsidence. The lower slope of the reef was characterised by heavily bioturbated sediment. The upper part of the slope at the eastern foot of the reef is rather steep and a chemosynthetic community made up of tube worms, calyptogena sp. and mussels was observed (Fig. 8c). This community is apparently derived from the cold seepage from the seafloor. Many linear cracks across outcrops, gaping several cm wide, spaced metres apart, were observed on the central and western part of the summit of the reef. The outcrop is bioclastic limestone with ferro-manganese crust. These outcrops apparently are coral limestone.

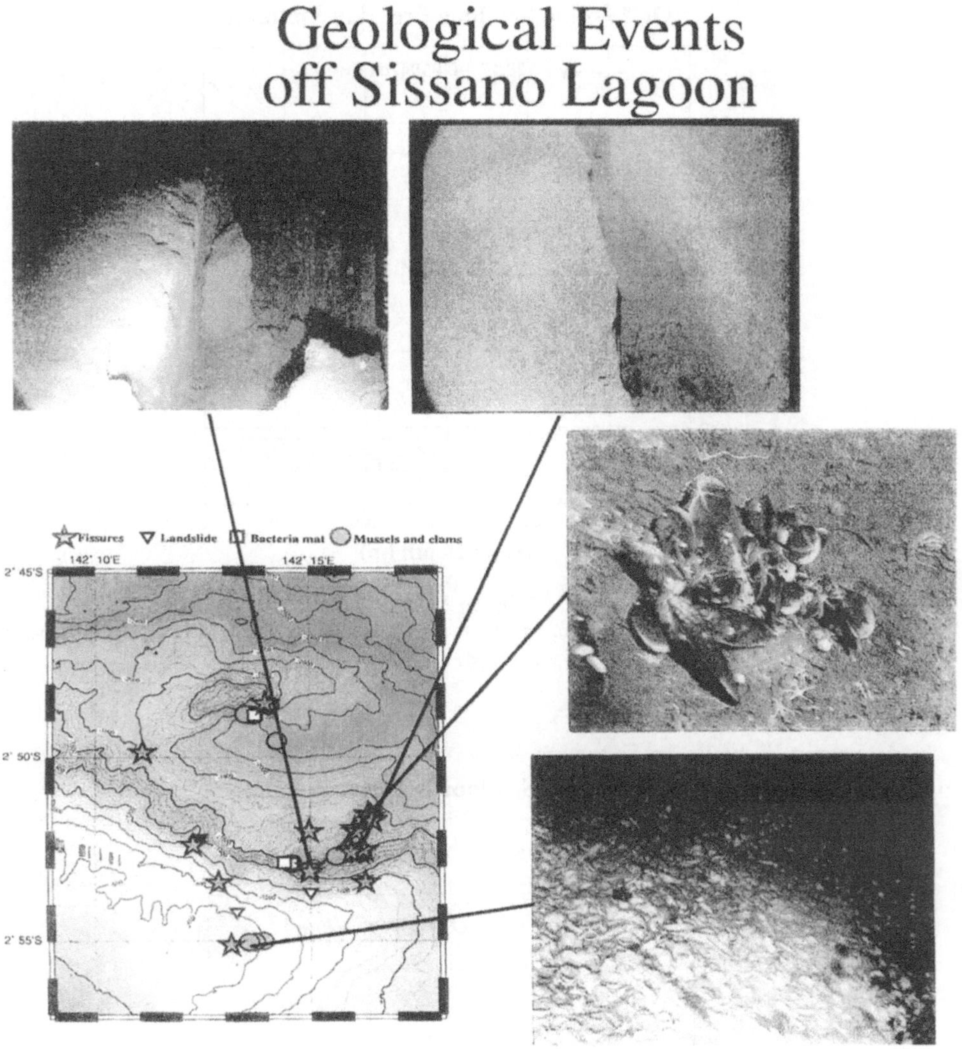

Figure 8c
Photographs of geological features observed during dives—chemosynthetic communities.

Dive 2K#1157 started near the end point of Dive D3K#402 and ended at a site about 300 m north of the chemosynthetic community site to conduct geological mapping along the N-S direction. It may be concluded that the area of the chemosynthetic site is very limited within a couple hundred metres in horizontal scale.

Seafloor gravity measurements were carried out at two stations: the deepest and the shallowest points during the *SHINKAI 2000* Dive Number 1157 on the northern slope of the subsided atoll offshore the Sissano Lagoon by use of the Scintrex CG-3M

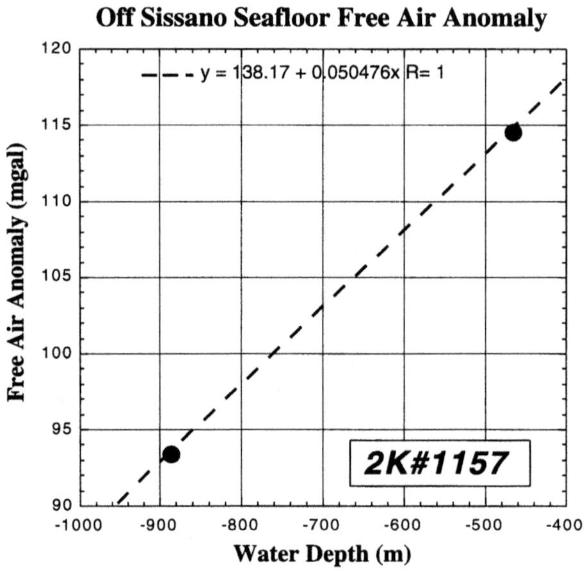

Figure 9a
Free air anomaly obtained from seafloor gravimetry at the upper slope of the headwall near the subsided reef.

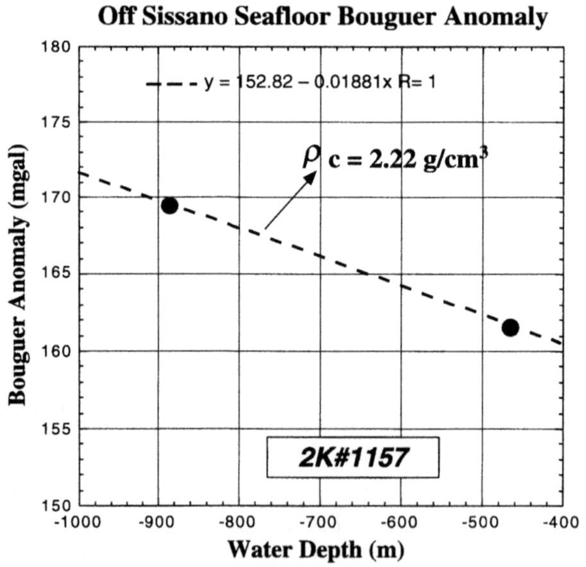

Figure 9b
Bouguer anomaly obtained from measured seafloor gravimetry at the upper slope of the headwall near the subsided reef.

gravity metre. The obtained gravity data were reduced to the values on the mean sea surface by the following equation:

$$\Delta g' = g_{obs} - \gamma - (\beta - 4\pi G\rho_w)d \tag{1}$$

where $\Delta g'$ is free air anomaly reduced to the mean sea surface, g_{obs} is observed gravity, γ is normal gravity, β is gravity gradient (assumed to be 0.3086 mgal/m near the surface of the earth), G is the gravitational constant, ρ_w is the density of sea water (1.03 g/cm^3) and d is water depth at the gravity station. The Bouguer anomaly was calculated by three-dimensional terrain correction with the assumed crustal density of 2.67 g/cm^3. Figure 9a shows the free air anomaly and Figure 9b the Bouguer anomaly along the submersible track. The result shows that the ratio of Bouguer anomaly to depth curve is −0.0188 mgal/m. The estimated average density of the bedrock between the two stations is 2.22 g/cm^3 and corresponds to the collected limestone sample.

6. *Piston Coring*

During the SOS-1 cruise, piston coring was carried out at four stations off Sissano Lagoon to ascertain any possibilities of an occurrence of turbidity current induced by possible underwater landslides and to understand the timing and emplacement processes of the sedimentary layers (Fig. 10). Piston core samples also indicate the previous events of the turbidity currents and may suggest if underwater landslides take place periodically at the stations. Location of piston core cast was decided after detailed sub-bottom profiling survey by the SEABEAM-2112 system on board *Kairei*. Core samples were cut into 1-m long short cores for the convenience of analyses. The cores were then vertically split into two halves; a working half and an archive half.

Piston core PC-1 was cast at 07:30–10:30 Local Standard Time (LST, GMT + 10), 8 January 1999, at 2°38.9890′S, 142°13.5186′E, at 4161-m water depth. The total length of recovered sediment core was 802.5 cm. The major lithology is moderately greyish olive calcareous clay composed of biogenic fragments such as foraminifera and shell. The calcareous clay displays massive and sometimes faint mottled structures in visual inspection. Several thin black sand layers are intercalated in all intervals of this core. The intervals between 0 and 50 cm of Section 7 show the convex shape of sand layer. The interval between 0 and 84 cm of Section 8 shows the upward grading structure. In Section 9, considerable alteration of clay and thin sand layer (several cm thick) is observed. As a whole, the station is characterised by the occurrence of frequent turbidity current in the past. However, no evidence of the present turbidite was observed.

Piston core PC-2 was cast at 14:05–15:05 LST, 8 January, at 2°54.7098′S, 142°15.3413′E, at 913.0 m water depth, 5 km east of the subsided atoll, and 20 km

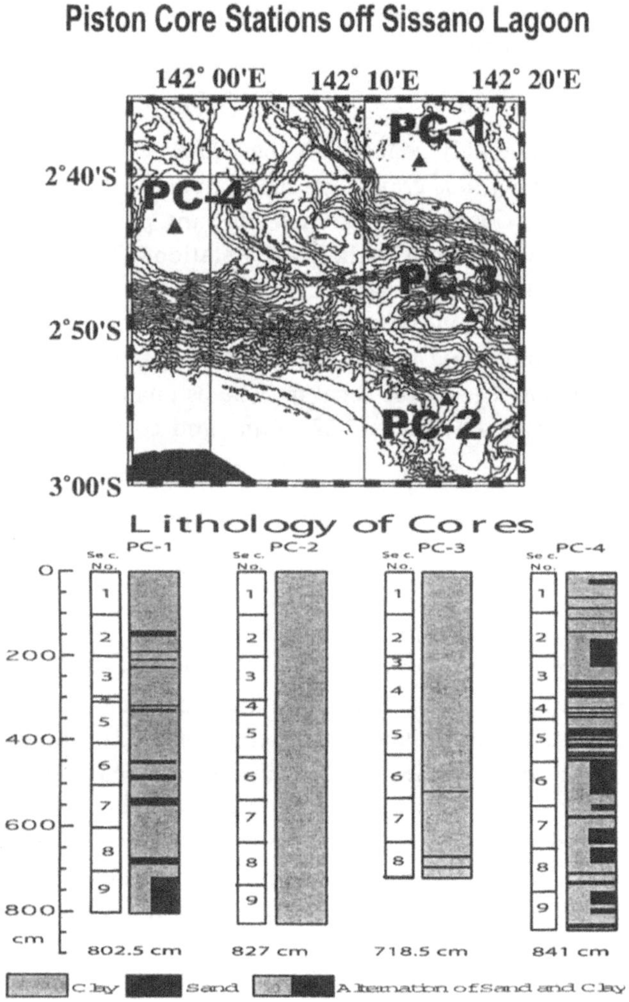

Figure 10
Location of the piston coring sites during the SOS-1 cruise, and preliminary results. cruise.

NE of Sissano Lagoon. The total length of recovered sediment core is 827 cm. The principal lithology is clay composed of biogenic fragments such as foraminifera and shell. The calcareous clay reveals massive and cohesive structures in visual inspection.

Piston core PC-3 was cast at 08:00–10:30 LST, 9 January, at 2°49.1080′S, 142°16.9336′E, at 2282 m water depth, at the center of the amphitheatre 30 km NE of Sissano Lagoon, between the southern slope and the pop-up seamount. Total length of the recovered sediment core was 718.5 cm. The major lithology is moderately greyish olive calcareous clay composed of biogenic fragments such as foraminifera and shell. The calcareous clay shows massive and sometimes faint mottled structures in visual inspection. Sandy layers appeared in the lower part of the core. Frequency of sandy layer appearance is less than that of PC-1 and PC-4. No signs of present turbidity current were observed from the core sample.

Piston core PC-4 was cast at 12:00–15:08 LST, 9 January, at 2°43.3438′S, 141°57.7814′E, 3253 m water depth at the centre of the western amphitheatre basin, 30 km off the coast. Total length of the recovered sediment core was 841 cm. The major lithology is a turbidite-rich carbonate ooze. It is moderately greyish olive calcareous clay composed of biogenic fragments such as foraminifera and shell. The calcareous clay shows massive and sometimes faint mottled structures in visual inspection. Seventy-eight thin black sand layers are intercalated in the entire interval of this core. The intervals between 58 and 70 cm of Section 3 show upward grading.

The lithology of the stations in and off the amphitheatre shows no signs of present turbidity current associated with the 1998 event. Since it is clear that a large-scale landslide took place on the amphitheatre from the observation by the submersible, the underwater landslide might be the type without surface collapse of the soft sediment layer. The interpretation is that the sediment is cohesive, suggested by its origin in the northern coast of Papua New Guinea Island and the condition of sedimentation off Sissano. This is why the deformation is rather solid with a long fresh crack of more than 20 km associated with the major aftershock on 17th July 1998; no disturbance of the surface of the sediment is observed and even bioturbation was proved to be predominant on locations other than the cracks. Stratigraphic study of these cores will be of great help in determining their origin and calculating the frequency of turbidite deposition that continues to the present offshore islands.

7. *Discussion*

Remarkable tension cracks were observed along the mid-slope of the amphitheatre through the dive surveys during the SOS-2 and SOS-3 cruises. The crack extends to 15 km. However, this is not a single one but is composed of *en échelon* cracks the length of which is several tens of metres. This is apparently associated with the upper part of an underwater landslide along the amphitheatre and derived from the tensional stress of the mass wasting.

Observation by *Dolphin-3K* and *Shinkai 2000* showed chemosynthetic communities on the submerged atoll, along the tension crack on the amphitheatre and along the southern fault escarpment of the Pop-Up Block. As for the tension crack, the distribution of the live community is limited in the eastern part of the amphitheatre. This suggests that cold seepage is taking place in the eastern part of the tension crack along the amphitheatre, apparently due to recent fracturing. On the other side, no chemosynthetic community was observed in the western part of the tension crack.

Observation by *Dolphin-3K* and *Shinkai 2000* also established that the tension crack on the amphitheatre has a sharp edge in its eastern to central part, suggesting that it was produced recently. The western part of the crack has a round edge and is covered with sediment, suggesting that this part is rather old.

The contrast between the eastern and western part of the crack is also described by topographic features. The western part of the amphitheatre is characterised by a couple of small-scale channels derived from the erosion of the seafloor. Therefore, this part did not move recently. The flank of the eastern part of the amphitheatre is rather flat, suggesting that mass wasting took place very recently. However, it is noted that the area other than the tension crack is characterised by bioturbation such as benthic fauna and their trails in both the eastern and western part of the amphitheatre. This is suggesting that "solid" deformation of the cohesive sediment should be taking place in the eastern part. This is supported by the result of piston core sampling. No recent (corresponding to the uppermost layer) turbidite layer was observed in all the core samples. PC-1 (trench axis) and PC-4 (western basin) are characterised by frequent turbidity current in the past. However, PC-2 (east of the submerged atoll) and PC-3 (eastern part of the amphitheatre) are characterised by no or scarce occurrence of recent turbidity current. The contrast between the two cases derives from the nature of the sediment. The sediment in the PC-4 area is from the deep sea fan with the erosion on it located at 141°55′E off the coast. Considering that the free air anomaly in the basin of the PC-4 area is remarkably low and comparable to that of the trench, continuous sediment supply and the consequent subsidence of the basement may be taking place. The sediment in the PC-3 area is cohesive due to a certain process of compaction or some specificity of the origin of the sediment derived from the nearest on-land area.

Acknowledgements

The authors would like to thank the crew members of *R/V Yokosuka* for their skillful operation and excellent technical support during the whole cruise. The authors also thank the SOPAC secretariat, the Government of Papua New Guinea and the JAMSTEC Directorate for giving us a good opportunity to carry out such an important project, not only for Japan and SOPAC member countries, but also for all the countries which experience numerous earthquake and tsunami hazards. We thank the Japanese tsunami research group for participation in the planning stages of this project. The operating expense for the cruises by the JAMSTEC research fleet was supported by the Japanese Science and Technology Agency through the research project fund of JAMSTEC.

References

COOPER, P., and TAYLOR, B. (1987), *Seismotectonics of New Guinea: A Model for the Arc Reversal Following Arc-continent Collision*, Tectonics *6*, 53–67.

TREGONING, P., LAMBECK, K., STOLZ, A., MORGAN, P., MCCLUSKY, S. C., VAN DER BEEK, P., MCQUEEN, H., JACKSON, R. J., LITTLE, R. P., LAING, A., and MURPHY, B. (1998), *Estimation of Current Plate Motions in Papua New Guinea from Global Positioning System Observations*, J. Geophys. Res. *103*, 12,181–12,203.

KAWATA, Y., BENSON, B. C., BORRERO, J. C., BORRERO, J. L., DAVIES, H. L., DE LANGE, W. P., IMAMURA, F., LETZ, H., NOTT, J., and SYNOLAKIS, C. (1999), *Tsunami in Papua New Guinea Was Intense as First Thought*, EOS, Trans. Am. Geophys. Un. *80*, 101 and 104–105.

RIPPER, I. D., LETZ, H., and MOIHOI, M. (1998), *Pre-tsunami Large Earthquakes of the Aitape Region North Coast Mainland of Papua New Guinea*, Papua New Guinea Geolog. Survey Rep. *98/7*, Port Moresby.

Pure appl. geophys. 160 (2003) 1945–1968
0033 – 4553/03/111945 – 24
DOI 10.1007/s00024-003-2415-z

Pure and Applied Geophysics

Tectonics and Slumping in the Source Region of the 1998 Papua New Guinea Tsunami from Seismic Reflection Images

S. SWEET[1] and E. A. SILVER[1]

Abstract—In September 1999, we collected seven high resolution seismic reflection profiles along the northern continental margin of Papua New Guinea, which targeted the source region of the 1998 tsunami that inundated Sissano Lagoon. We utilized swath bathymetry collected by the JAMSTEC/SOPAC groups in January 1999. The seismic profiles image several faults, bottom simulating reflectors, and a large rotational slump. The slump has a head scarp of 100 m vertical extent, coinciding with the headwall and tension cracks observed previously by submersible at the southern edge of the amphitheater. The central, back-rotated part of the slump is coherent with parallel reflections. The interpreted basal failure plane has a maximum depth of 760 m below the seafloor, and it crops out at a steep escarpment, about 100 meters high, located 4.5 km north of the head scarp. This escarpment separates the slide toe from a series of seafloor-parallel, coherent reflections that are top-lapped by basin deposits at the foot of the amphitheater to the north. The cross-sectional area of the displaced mass is 2.3 km^2. From the bathymetry, the width is approximately 2.5–3 km, yielding a total volume (assuming parabolic shape) of 3.8–4.6 km^3. Based on these interpretations, the slump was restored to its undeformed position. This technique yields a center of mass vertical drop of 380 m, horizontal movement of 840 m and slip of 980 m along the slide plane.

Key words: Tsunami, submarine slides, tectonics, seismic reflection, gas hydrates.

1. Introduction

On July 17, 1998, the northern continental margin of Papua New Guinea, in the region of Sissano Lagoon, experienced a magnitude 7 earthquake, followed shortly thereafter by a tsunami that inundated the coast, killing over 2000 residents (Fig. 1). An International Tsunami Survey Team visited the area two weeks after the event and documented wave runup heights along a 40 km stretch encompassing the area of greatest devastation. Runup heights averaged 10 m (peaking at 15 m) along Sissano Lagoon spit and adjacent westward beaches, and decayed quickly to 2 m away from this area (KAWATA *et al.*, 1999). This geographically concentrated runup pattern suggests a non-dislocative source for the tsunami such as a submarine slump (DAVIES, 1998; TAPPIN *et al.*, 1999; and SYNOLAKIS *et al.*, 2002).

[1] University of California Santa Cruz, Santa Cruz, CA, U. S. A.

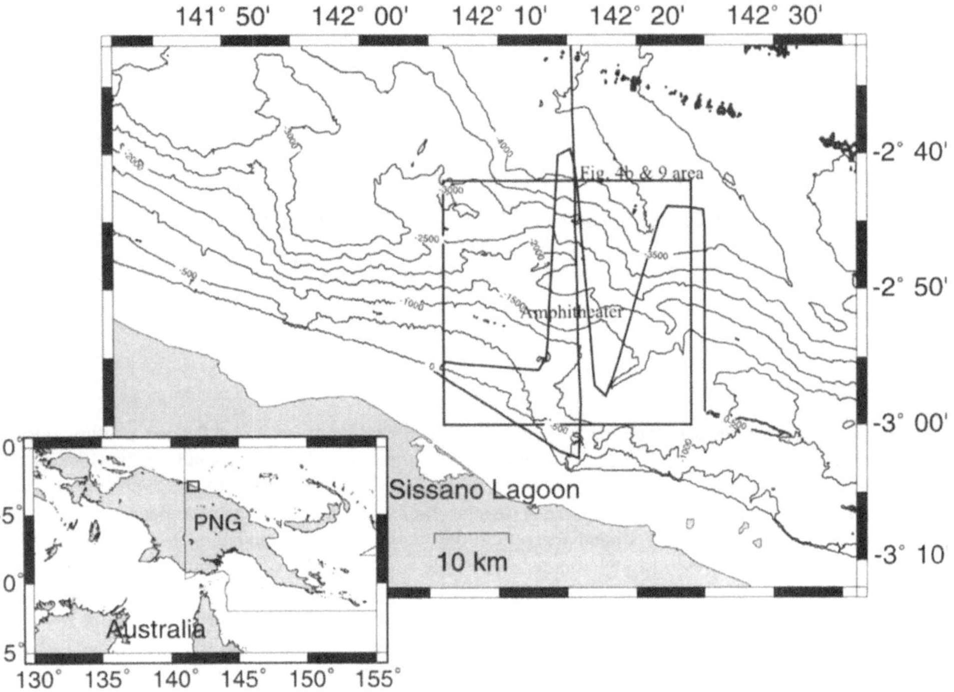

Figure 1
Location map of survey area. Sissano Lagoon is the area of greatest tsunami devastation. Seafloor bathymetry collected by JAMSTEC/SOPAC groups is shown. Seismic profile track lines are shown. Area of Figures 4b and 9 are designated by box. Inset map shows regional location.

In early 1999, scientists representing the Japan Marine Science and Technology Center (JAMSTEC), the South Pacific Applied Geoscience Commission (SOPAC), the University of Papua New Guinea, and the University of Southern California visited the region to collect seafloor bathymetry and perform ROV and submersible dives. Using swath bathymetry they identified faulting and an amphitheater-shaped feature, and they characterized the margin as undergoing subduction-erosion and subsidence. Subsequent submersible dives revealed fresh tensional cracks at the southern scarp of the amphitheater and talus deposits within the amphitheater suggesting recent movement (TAPPIN *et al.*, 1999).

In September 1999, aboard the *R/V Maurice Ewing*, we collected seven high resolution seismic reflection profiles in the identified source region of the tsunami. We utilized the bathymetry collected by the JAMSTEC / SOPAC groups to select our track lines. As shown in Figure 1, we crossed the identified amphitheater with three north-south trending seismic profiles. Two additional profiles cross the continental margin east of the amphitheater and the final two profiles west of the amphitheater are near-shore and coast-parallel. The purpose of collecting high-resolution seismic reflection data was to ascertain whether the internal structure of the tsunami source region was consistent with a slump, and if so to determine the vertical and lateral dimensions for use in refining tsunami generation models. We collected data more regionally than the likely limits of the slump zone (Fig. 1), to see whether young faults might be present in the source area to provide an alternate explanation for the tsunami.

Here we discuss the seismic data, the tectonic setting and our interpretation of slump structure and dimensions, including a balanced reconstruction of the inferred slump. We also show evidence for several faults that displace the seafloor, though they are located beyond the most likely area of generation of the tsunami.

2. Data Acquisition and Processing

In total, we collected 218 kilometers of seismic reflection data (Fig. 1). In order to image the shallow subsurface with the greatest detail, we utilized the *Ewing's* high resolution seismic reflection system in our data collection (see Table 1 for parameters). Initial on-board processing yielded 24-fold common depth points, spaced at 12.5 meters, using SIOSEIS processing software (HENKART, 1981). Subsequent processing and interpretation was done at UCSC using Landmark Graphics software. We followed a typical processing sequence to gain a final product of migrated, depth sections. The processing parameters are shown in Table 1.

Table 1

Data acquisition and processing parameters

Data Acquisition	Processing*
Source: Single GI airgun, 45 in^3	Common depth point sort-24 fold, 12.5-m spacing
1200-m digital streamer	Velocity analysis
48 channels with 25-m spacing	Progressive mute analysis
25-m shot spacing	Brute stack
8-second record length	Spherical divergence (1/distance)
Boat speed: 9 km/hr	Decibels/second amplitude adjustment
	Bandpass filter
	Mute
	Predictive deconvolution
	Normal moveout correction
	Stack
	Automatic gain control adjustment
	F-K migration (Memory Stolt)
	Depth conversion

* Using Landmark Graphics packages

3. Tectonic Environment

3.1 Regional Tectonic Environment

The Papua New Guinea (PNG) region displays a complex interaction of active tectonic plates (Fig. 2). PNG has been characterized as three distinct tectonic zones which include a young fold-and-thrust belt at its center, a zone of major left-lateral strike-slip faults in the north and west, and a zone of subduction at the New Guinea Trench (KREEMER *et al.*, 2000; HAMILTON, 1979). The island of New Guinea is split by a political boundary into Irian Jaya (Indonesia) to the west and PNG to the east, and lies on the northern margin of the Australian Plate. In northwestern PNG, the Pacific and Australian Plates are converging to form the New Guinea Trench at a highly oblique angle of approximately 60° (KREEMER *et al.*, 2000) (Figs. 1 and 2).

Major east-trending left-slip faulting in the western region of Irian Jaya includes the Sorong and Tarera fault systems (HAMILTON, 1979; ABERS and MCCAFFREY, 1988). Offshore, the motion between the Bismarck Plate and the Pacific Plate to the north has 130 mm/yr of relative left-lateral strike-slip motion (TREGONING *et al.*, 1998), based on Global Positioning System (GPS) studies. The boundary is well defined by earthquake-epicenter locations and seafloor morphology. COOPER and TAYLOR (1987) called this boundary the Bismarck Sea Seismic Lineation and

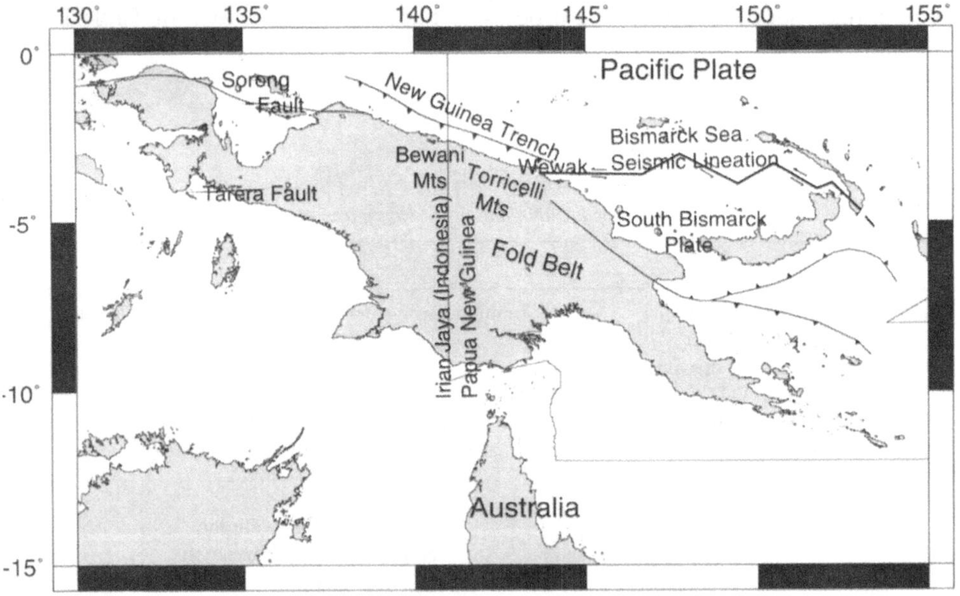

Figure 2
Tectonic environment after HAMILTON (1979), ABERS and MCCAFFREY (1988), and COOPER and TAYLOR (1987), showing fold and thrust belt, New Guinea Trench, Bismarck-Pacific Plate boundary and relative motion, the Torricelli and Bewani Mountains, and the Sorong and Tarera Faults.

investigated its offshore and onshore expression. They found that fault planes of the western-most portion of the boundary are oriented E–W. This trend continues on land as a series of faults that cut westward through the Torricelli and Bewani mountains (COOPER and TAYLOR, 1987; RIPPER *et al.*, 1980). These mountain ranges trend E–W, as far eastward as the coast of Wewak. These faults likely partition some of the left-lateral component of oblique convergence across the New Guinea Trench.

A recent GPS and seismicity study of strain partitioning in western PNG (KREEMER *et al.*, 2000) determined that the trench-normal component of convergence is small and most deformation is taken up in an E–W trending zone of almost pure strike-slip deformation. Velocity field modeling relative to Australia indicates a direction which is consistent with NUVEL-1A Pacific Plate-Australian Plate convergence of 250°, but with a velocity slightly less than the NUVEL-1A estimate (DEMETS *et al.*, 1994).

3.2 Overview of Continental Margin

Seismic reflection data (Fig. 3) shows that the continental margin off NW PNG is essentially sediment starved, with 500–1000 m of unconsolidated sediment concentrated near shore, thinning dramatically seaward. The thickest sediment imaged near shore is contained in a basin surrounding the head of Yalingi Canyon (herein referred to as the near shore basin). The continental slope dips 10–12° with little unconsolidated terrigenous sediment on the slope. Beneath the existing sediment are

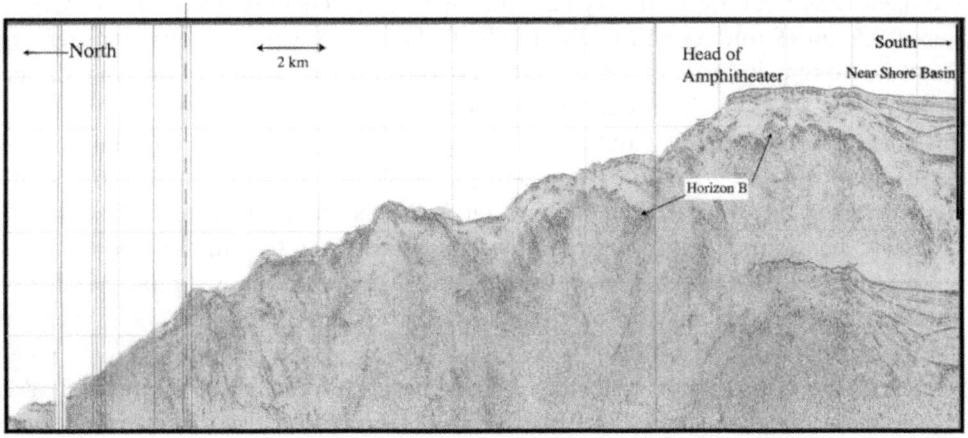

Figure 3
Full Line 5 seismic depth profile with 4 × vertical exaggeration. Presented as an overview of the continental margin. Showing 'near shore basin', Horizon B, and the amphitheater.

faulted, high-amplitude, discontinuous reflections. Very little structure is resolvable below this set of reflections, labeled 'Horizon B', which is displaced at least 700 m as it enters the amphitheater from the south (Fig. 3). This will be discussed in conjunction with faulting in the next section. The seaward edge of the continental margin ends in a small deformed toe in most profiles. The subducting Pacific Plate enters the trench with up to a kilometer of unconsolidated sediment above an acoustic basement. The combination of this entering sediment pile and the lack of resolved accretionary structure implies a nonaccreting margin. A drowned coral reef complex on the upper slope (TAPPIN *et al.*, 1999) implies subsidence by subduction erosion.

Complex faulting is prevalent throughout the margin due to the oblique subduction environment. Both extensional and compressional features are evident from slumping, normal faulting and uplifted blocks. Some sediment has been displaced from faulting or slumping in all profiles crossing the margin.

3.3 Left-slip Faulting on the Northern Margin of PNG

The bathymetry and seismic profiles show extensive faulting which is consistent with this strike-slip environment. Figure 4a is a shaded relief image showing faulting interpretations. We identified two major linear faults with significant left-slip motion cutting through the continental margin (Faults A and B), which terminate at Basins A and B. Fault A corresponds to the 40 km fault noted by TAPPIN *et al.* (1999). We infer a large left-slip component on these faults as shown by westward offset of the inner trench slope on the north side of the faults. Fault A may step over to Fault B at 142°05'E, 2°45'S, where a ridge trends northwest and may be a restraining bend linking the two faults. This ridge creates a basin that appears to trap terrigenous sediment (Basin A). Basin B may be a similar feature. These faults offset the margin about 9.5 km (Fault A) and 6 km (Fault B). Sediment deposits cover the trace of Fault B and may indicate that this fault is no longer active. Along Fault A we do not see such deposits, which may indicate ongoing activity.

There is a zone of E–W faulting in the region of Fault A that is imaged in Lines 4 and 5 (Figs. 4a and b). The most prominent are two normal faults that dip south approximately 30° which are shown on Line 5. We see similar features in Line 4, that appear to be normal faults, but are not well resolved at depth. This may be a zone of splay faults that join into a large fault at depth. These faults show significant vertical displacement of the surface in this region, implying there may be a vertical component of slip to Fault A.

Line 6 may also show disturbance from Fault A. The trend of the fault line intersects the trench in Line 6. The Pacific Plate enters the trench at a steeper angle than along other profiles and there is no deformed toe in the trench, possibly due to material removed by the fault. We image a decollement cropping out at the toe of the margin and dipping under it at an angle of $\sim 10°$. Offset is not resolvable along this fault.

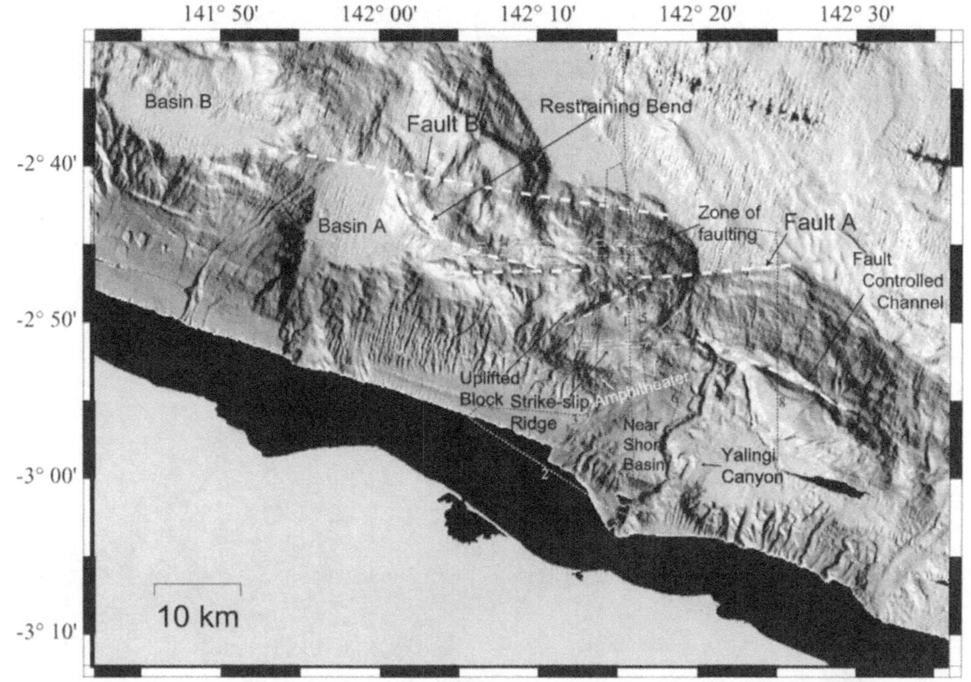

Figure 4a
Shaded relief image of the bathymetry showing regional faulting interpretations. Fault A and Fault B are two major faults showing left-slip that cuts through the continental margin and terminate at Basins A and B. Restraining bend and basins are identified. The zone of faulting associated with Fault A (imaged in Lines 4 and 5), the amphitheater, uplifted ridge, and flower structure areas are identified. The fault controlled channel that diverts the Yalingi Canyon is identified.

To the east of Fault A we see a linear channel, 24 km long, trending parallel to the trench, which intersects and diverts Yalingi Canyon to the northwest (Fig. 4a). A deep, v-shaped valley that cuts 300 m into sediment and into Horizon B in seismic profile 8. No visible offset of Horizon B is shown across this valley, therefore, it is interpreted as a strike-slip fault controlled channel (Fig. 5).

3.3.1 Amphitheater morphology

One of the most prominent formations creating this feature is the uplifted ridge, identified by TAPPIN *et al.* (1999), which bounds the northern extent of the amphitheater (Figs. 4a and b). This ridge is bounded to the north and south by normal faults. The southern fault dips approximately 20° in Lines 5 and 1 and nearly 40° in Line 4. The fault displaces Horizon B between 1000–1500 m along the ridge, creating a basin which has accumulated 400–700 m of sediment. The ridge is bounded to the north by a fault dipping 40° north with a surface offset of approximately 300 m. The fault is imaged only in Line 5 and is not well resolved at depth. This fault appears to terminate to the northeast at Fault A.

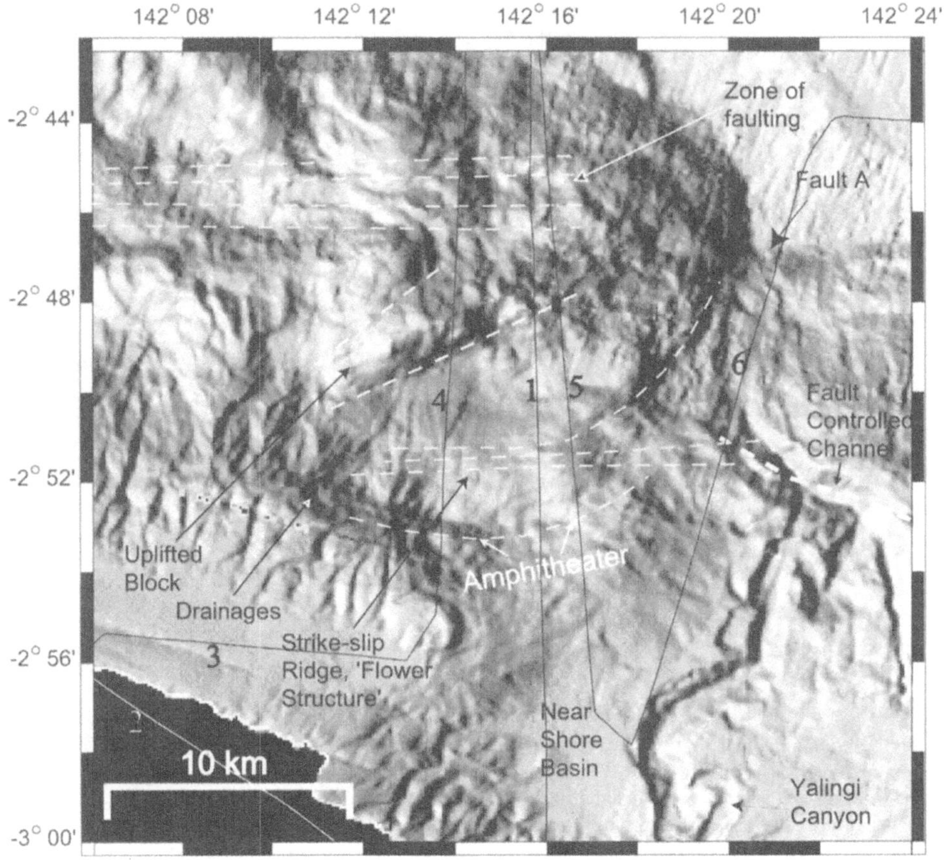

Figure 4b
Blowup of Figure 4a, extent shown on Figure 1. Amphitheater, uplifted ridge, zone of faulting along Fault A, and zone of strike-slip faulting (flower structure) are shown. Drainages in the western portion of the amphitheater are identified.

The center of the amphitheater is a complexly faulted ridge, with indications of flower structure that is characteristic of strike-slip faults. Both normal and reverse offset along faults is noted in the profiles. From examination of the bathymetry and profiles, this appears to run parallel to Fault A with a possible splay fault that extends to Fault A. In Line 4 at the ridge, Horizon B approximates a parabolic form and is predominately reverse faulted. Line 1 also shows a parabolic-shaped ridge that is up to 1 km wider and is faulted with both reverse and normal offsets on high angle faults. Line 5 shows a similar ridge that is reverse faulted, adjacent to an area that is more heavily deformed and relatively uplifted. Horizon B in this area has a hummocky appearance which is reflected on the ocean floor. Faulting shows both normal and reverse offsets. This set of faults with varying displacement may be the

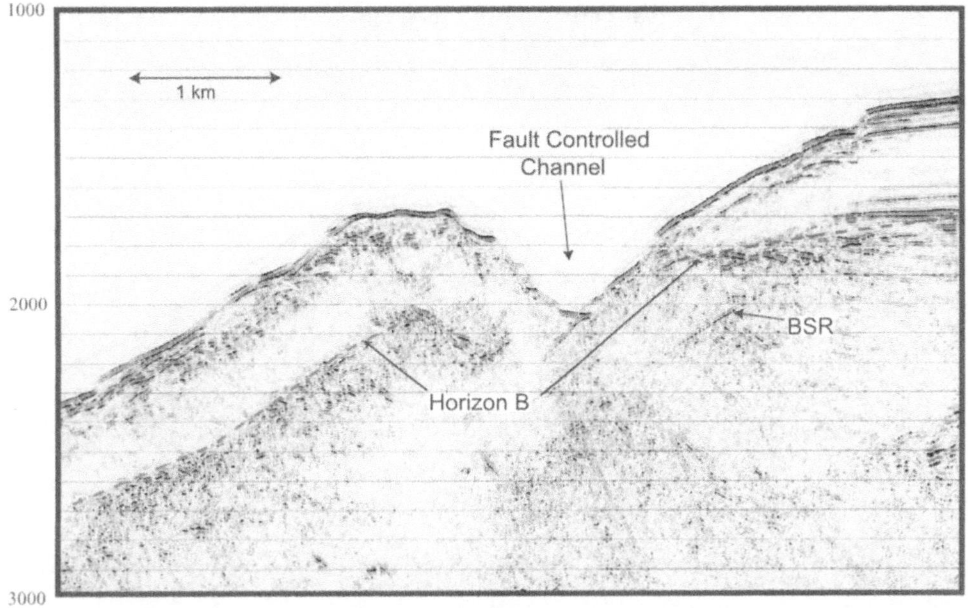

Figure 5
Line 8 seismic depth profile on the continental margin showing the strike-slip fault controlled channel and BSR. Horizon B is labeled.

upward branches of a strike-slip fault, or a flower structure (CHRISTIE-BLICK and BIDDLE, 1985; ALLEN and ALLEN, 1990, pp. 120–122).

The arc-shaped scarp south of this ridge defines the head of the amphitheater. Referring back to Figure 3, the full cross-section of the margin, Horizon B is offset up to 1 km along this scarp. Line 1 (Fig. 6) shows a linear normal fault dipping 30° to the north, which is discussed further in Section 4.2.

These features support a strike-slip environment with some associated compression and extension. The arc-shaped features may extend to Fault A and be part of a larger system. Uplifted areas noted create an environment for basin filling and mass failures.

4. Submarine Slumping and Sediment Distribution

The entire mass at the base of the amphitheater-shaped feature has previously been interpreted as a slump (TAPPIN *et al.*, 1999). Although slide material is interpreted in all three lines along the southern wall of the amphitheater, only Line 5 images a large coherent slump mass. Line 1 is approximately 1.6 km west of this line as it crosses the slump, but Line 1 images a fault-bounded basin covered by a small

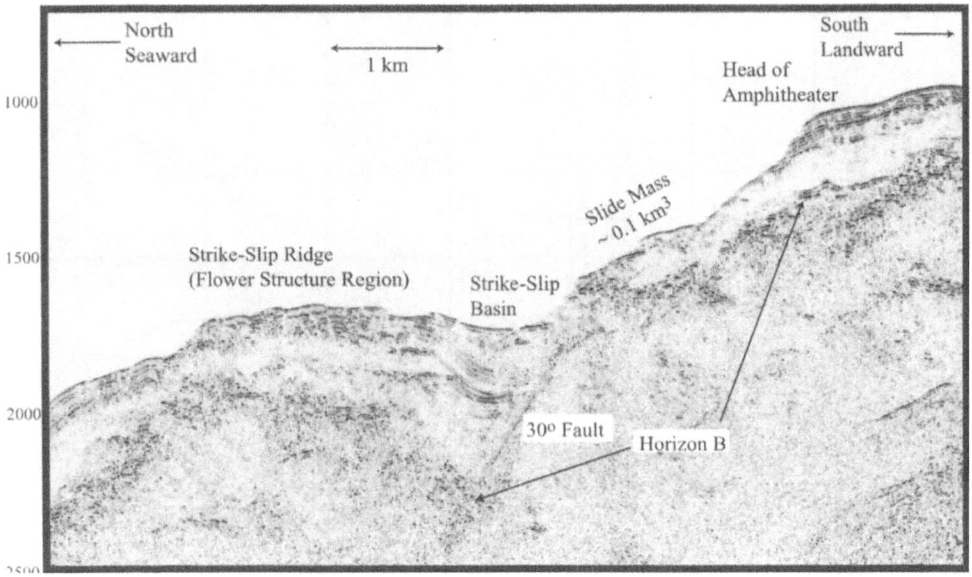

Figure 6
Line 1 seismic depth profile of amphitheater showing faulting, basin fill, and slide mass with 2 × vertical exaggeration.

amount of slide material along the south wall. Line 4 also crosses a small slide mass on the southern amphitheater wall, but as noted from the bathymetry, the eastern half of the amphitheater is covered with undisturbed drainage patterns. Therefore this portion of material at the base of the amphitheater did not accumulate as one coherent slump, but more likely it accumulated over time.

4.1 Slump Interpretations

Figure 7 is a blow up of Figure 3 showing Line 5 in the amphitheater (2 × vertical exaggeration). We interpret the southern portion of this area as a slump based on several lines of evidence. The southern extent of the amphitheater in this profile is a scarp of 100 m vertical extent. To the north, Horizon B is offset approximately 700 m vertically to a location within the amphitheater where we see a block of parallel strata (Block A), that is tilted south approximately 14°. This is the main mass of a large rotational slump. We map the basal slide plane at the lower extent of back-tilted reflections, yielding a maximum depth of 760 m below the seafloor. On the slope between the Block A and the head scarp, the seafloor is normally faulted. Reflections are back-tilted toward the scarp, indicating that this material has rotated down slope as smaller individual blocks. The slide plane is interpreted to be the base of back-tilting that crops out at the headscarp. South of the Block A is the area of hummocky reflections interpreted as a flower structure of a strike-slip faulted ridge, labeled the

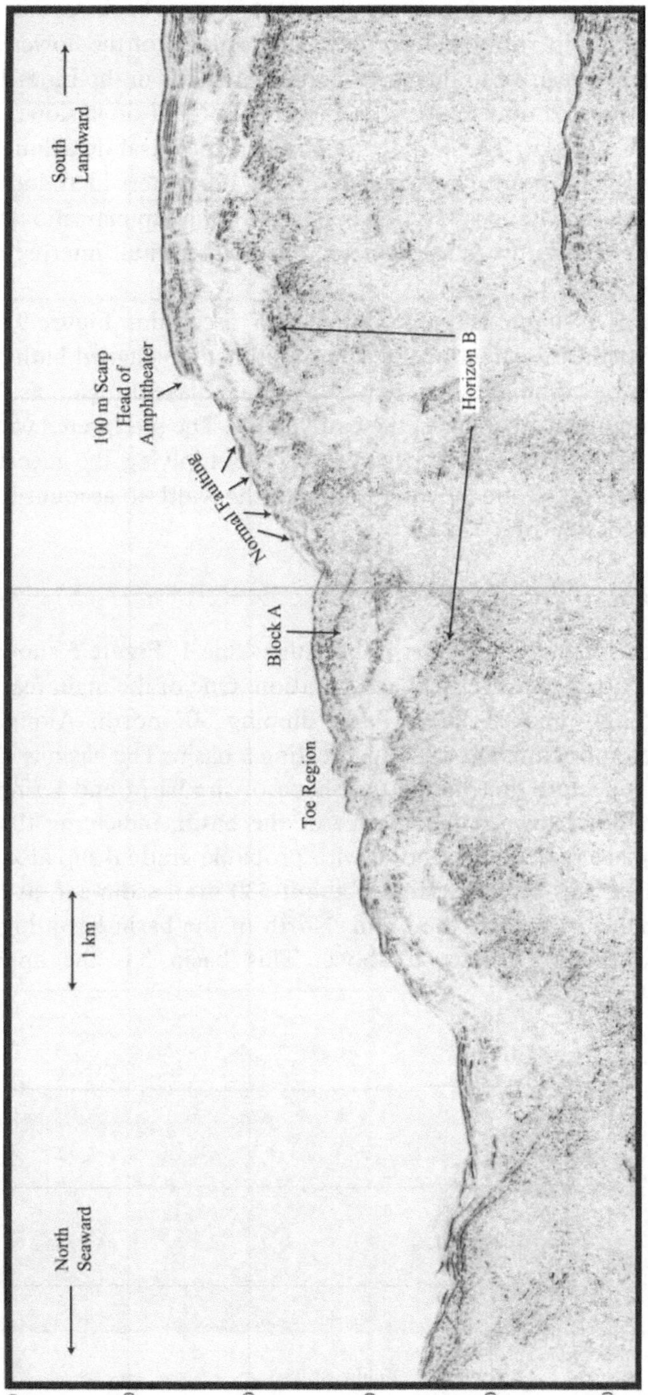

Figure 7

Blowup of Figure 3 showing Line 5 slump and amphitheater features with 2 × vertical exaggeration. Block **A**, 'toe region', head scarp, normal faulting and horizon **B** are shown.

'toe region' on Figure 7. This area may have acted as a backstop for the upper coherent block, halting its rotation. The toe region portion of the flower structure is uplifted and deformed relative to the corresponding regions in the Lines 1 and 4 and was probably compressed and uplifted as Block A rotated downslope. This is the extent of displaced material. The seafloor outcrop of the basal detachment near the toe region coincides with a steep escarpment, about 100 meters high, located 4.5 km north of the head scarp. The cross-sectional area of the slump mass along this profile is 2.3 km^2. Figure 8 shows the basal plane and associated fault interpretations with no vertical exaggeration.

The width of the slump is more difficult to ascertain. Figure 9 shows our interpretations of the areal extent of the slump outlined on shaded bathymetry. The width is delineated based on sediment deposits in the adjacent region as discussed in the next section and lateral changes in the bathymetry. The interpreted width is 2.5 to 3 km, which yields a volume of 3.8 to 4.6 km^3, multiplying the measured cross-sectional area of Line 5 (2.3 km^2) by two-thirds of the width to account for probable parabolic shape at depth.

4.2 Sediment Deposits

The areal extent of the slump does not include Line 1. Figure 6 shows Line 1 in the amphitheater with 2 times vertical exaggeration. One of the main features in this profile is the strongly imaged linear fault, dipping 30° north. Along this fault, Horizon B is offset approximately 1000 m, creating a basin. The basin is identified by concave-bending reflectors that mimic the shape of the basin and terminate at the fault. Bending shallows toward the surface of the basin, indicating that sedimentation has taken place over a long period with probable gradual slip along the fault. Above Horizon B in this basin, we image about 530 m of sediment, at least 100 m more sediment than just outside the basin. North of the basin lies a highly-faulted ridge or flower structure, discussed above. This basin has the appearance of

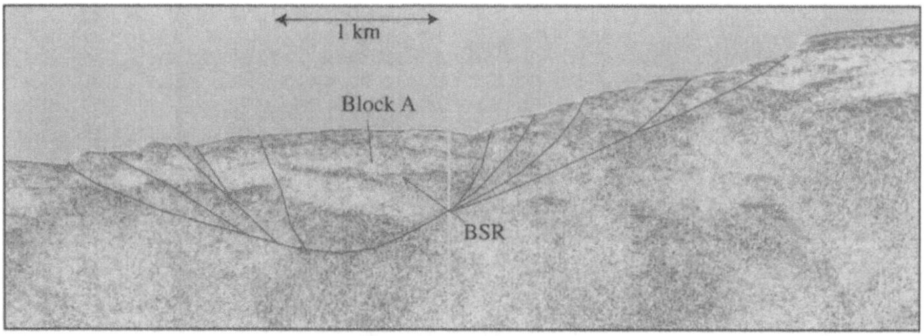

Figure 8
Seismic profile of slump area in Line 5 showing fault and slide plane interpretations. No vertical exaggeration. BSR is shown.

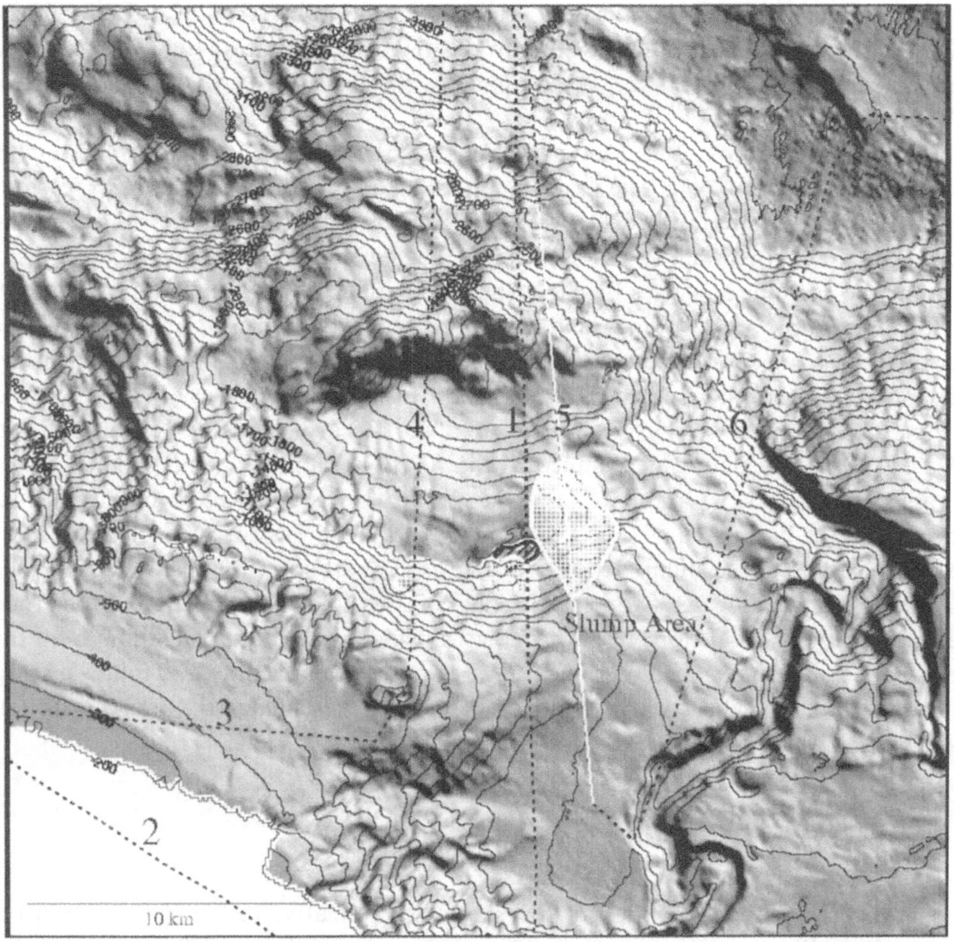

Figure 9

Close up of survey area, showing amphitheater. Area extent is shown in Figure 1. Bathymetry shown with shaded relief overlay and contour interval of 100 m. Sun angle is 30° from an azimuth of 40° Track lines of seismic profiles are shown. Line 5 is highlighted in white. The slump area is outlined in white. The small slide blocks of Lines 1 and 4 area are also outlined in white and the strike-slip basin of Line 1 is outlined in black.

documented strike-slip basins (MIALL, 1984; ALLEN and ALLEN, 1990, pp. 252–253), such as a deep, but narrow geometry and rapid deposition.

The fault plane creates a likely plane of weakness for landsliding as seen in Line 5. In this profile there is some slide material on the upper slope that is resting on the basin and obscuring the fault outcrop. From examination of the bathymetry this constitutes an individual slide volume of approximately 0.1 km^3.

Drainages cover the western portion of the amphitheater (Fig. 4b). Thus it is likely that no recent coherent mass movement has occurred here. In Line 4 we see surface-parallel reflections indicating gradual deposition by series of turbidity flows. The southern wall of the amphitheater in this region also appears to have small slide blocks similar to Line 1 and approximately the same size, which are noted on Figure 9.

As discussed above, lines 1 and 4 do not cross the slump of line 5, thus limiting its size and westward extent. The eastern end of the amphitheater is the likely eastern extent of this slump and northern and southern extents can be determined by the seismic Line 5. In order to more closely delineate the lateral extent of the Line 5 slump, we examine traverses of the bathymetry. Figure 10 is a blow up of Figure 9 with the slump extent outlined and nine intersecting bathymetric traverses. We see evidence of a lateral scarp or gradual rise in each traverse along the proposed slump boundary. The southern extent is determined by the 100 m scarp as imaged in Line 5 (Fig. 7). We continue it northeast along an obvious scarp, shown in Traverses F and G. The northern extent is determined by the 100 m offset in the 'toe region' shown in cross section (Fig. 7). On the western edge, in Traverses A and B, we note a gradual rise out of the basin area of Line 1 to Block A of Line 5. Traverses C, D, and E cross a scarp leading to the slump deposit. Traverses H and I show the eastern edge of the slump. We have indicated two possible edges on Figure 10. The location of the inside extent is on a gradual slope in both traverses, which is an interpolated line from the northern extent to the head scarp near G. The outer extent is drawn to follow a scarp leading from G northeastward. Volume calculations presented in the previous section are based on the inside extent.

5. *Gas Hydrates*

A high amplitude reflection of reverse polarity (relative to the water bottom) is seen within Block A, an average of 260 m below the seafloor (1770 m water depth) (Fig. 8). The reflection continues for 800 m and cuts across sediment reflections within the rotated block. A reflection of polarity opposite of the ocean floor indicates a velocity inversion, which may represent the base of a gas hydrate. Gas hydrates parallel or simulate the ocean floor and so are referred to as bottom simulating reflectors (BSRs). They gradually deepen with increasing water depth, consistent with their equilibrium pressure and temperature stability fields (HENRY *et al.*, 1999). We identify several such BSRs in the profiles.

We image an extensive BSR in the near-shore basin west of Yalingi canyon, which continues north toward the amphitheater scarp. Within the amphitheater sediments, we image BSRs discontinuously. We also see a BSR on the western side of the canyon, leading into the strike-slip channel discussed in conjunction with Line 8 faulting (Fig. 5).

Figure 11a is a graph of seafloor water depth verses BSR depth below the seafloor at several locations along each BSR identified in the seismic depth sections. Best fit

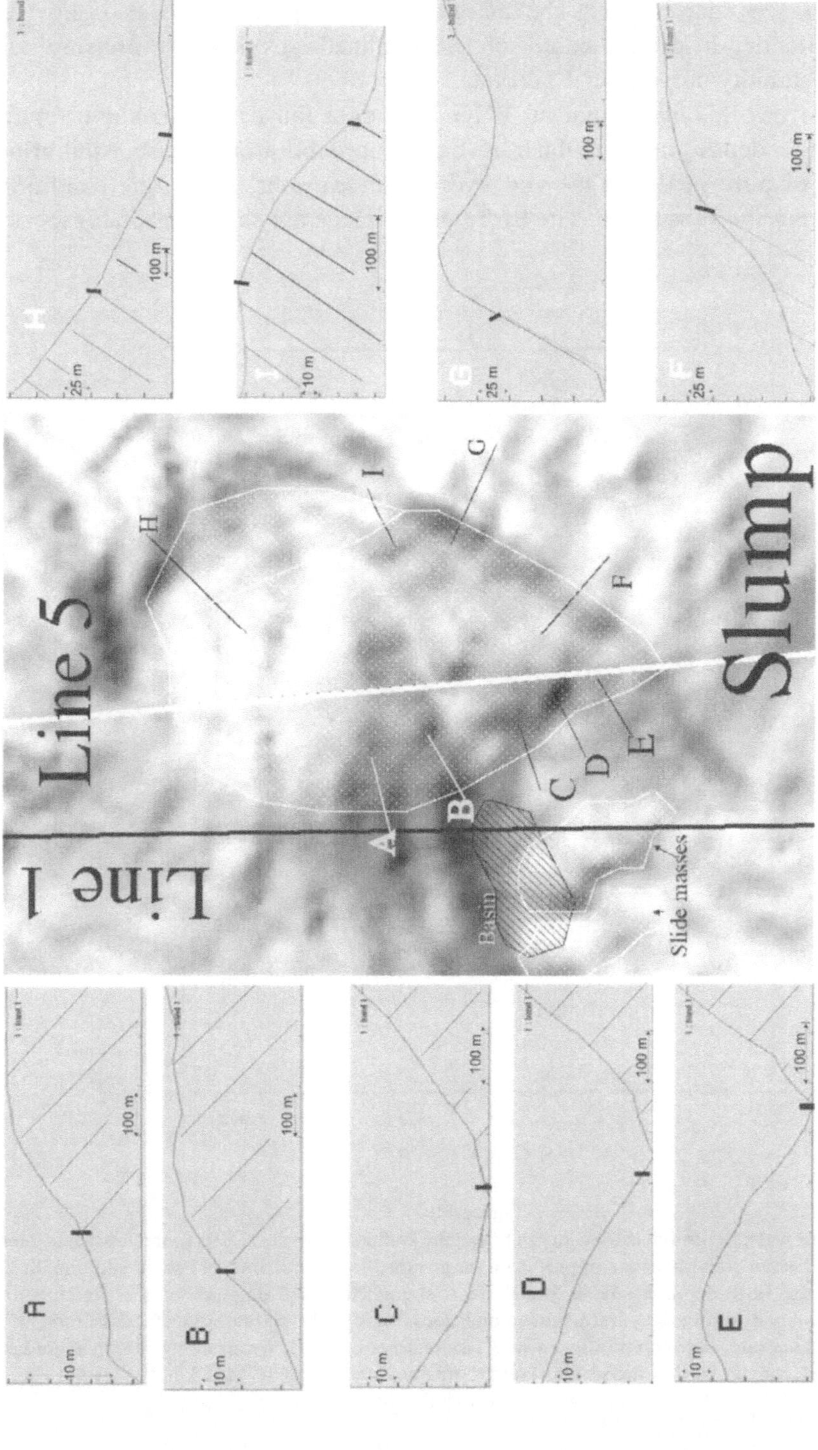

Figure 10

Closeup of slump extent on bathymetry showing 9 traverses across the edge of the slump to see lateral changes. Hash marks and diagonal lines on the traverses show the location of slump boundary and slump material, respectively, as outlined in white. Also shown is the extent of the basin imaged in Line 1 and the slide deposits that top-lap this basin, interpreted from bathymetry and seismic profiles.

trendlines are applied to data from each line and allow one to observe the local variability in BSR depth along the margin. For comparison, we also show an idealized BSR depth profile based on a simplified gas hydrate pressure and temperature stability curve after CLENNELL *et al.* (1999).

We note from this figure that all PNG BSR data fall below the curve for the 'idealized BSR depth profile' which is based on laboratory phase equilibrium experiments of pure methane gas and hydrate in seawater, simulating conditions found in the marine subsurface. The hydrates we image are shallower than expected

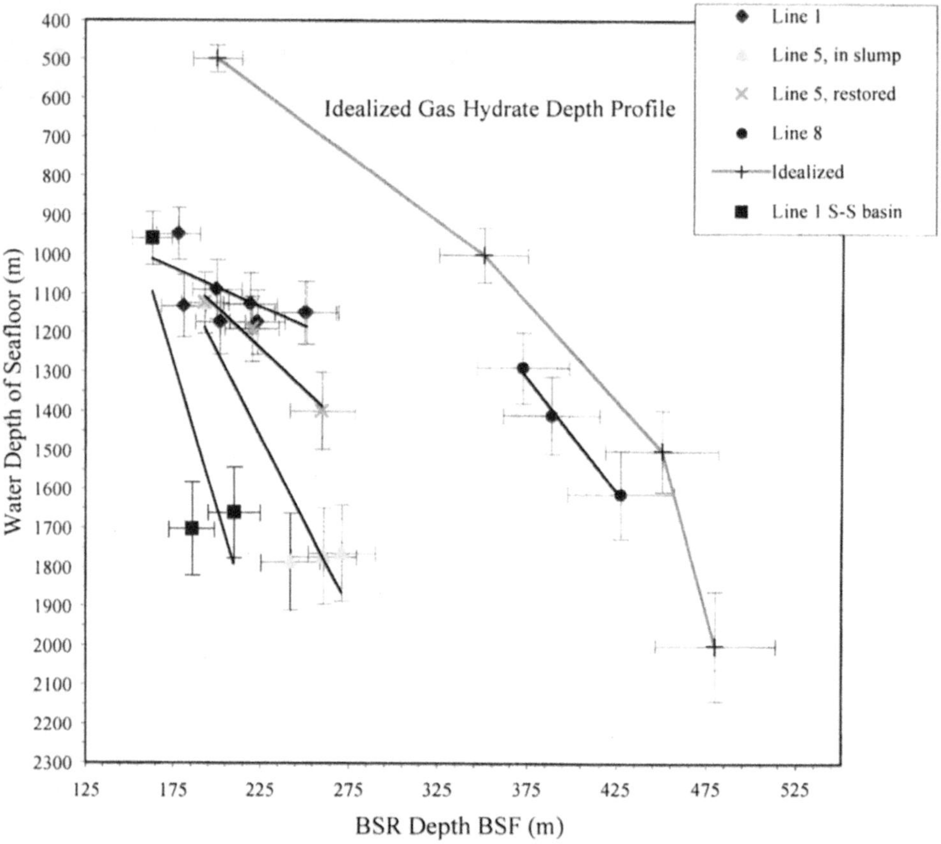

Figure 11a
Graph of seafloor water depth versus BSR depth below the seafloor of several BSRs found in seismic depth profiles. Linear best fit trend lines are applied to data grouped by seismic line and some additionally by location. Idealized BSR depth profile is based on a simplified phase diagram of the pressure and temperature stability field for gas hydrates under conditions found in the marine subsurface assuming pure methane gas and seawater fluid composition, after CLENNELL *et al.* (1999). We apply a geothermal gradient of 40 °C km^{-1} (starting at 4 °C at the seafloor) and a pressure gradient of 10 Mpa km^{-1} to stability curve and convert to depths.

for their water depths. One explanation may be that this margin has a high-temperature gradient. The idealized BSR depth profile is based on a 40 °C km^{-1} temperature gradient which is a typical temperature gradient for deep sea sediments. CLENNELL *et al.* (1999) cite a range of 20–80 °C km^{-1} as typical temperature gradients for deep-sea sediments. KVENVOLDEN and BARNARD (1983) use a temperature gradient of 27.3 °C km^{-1} to present an idealized BSR depth profile based on the phase equilibrium of pure methane gas and hydrate in pure water. Both use a seafloor temperature of approximately 4 °C on continental slopes within the approximate depth range of 1000–1700 m and a pressure gradient equivalent to 10 Mpa km^{-1}, assuming hydrostatic and lithostatic pressure.

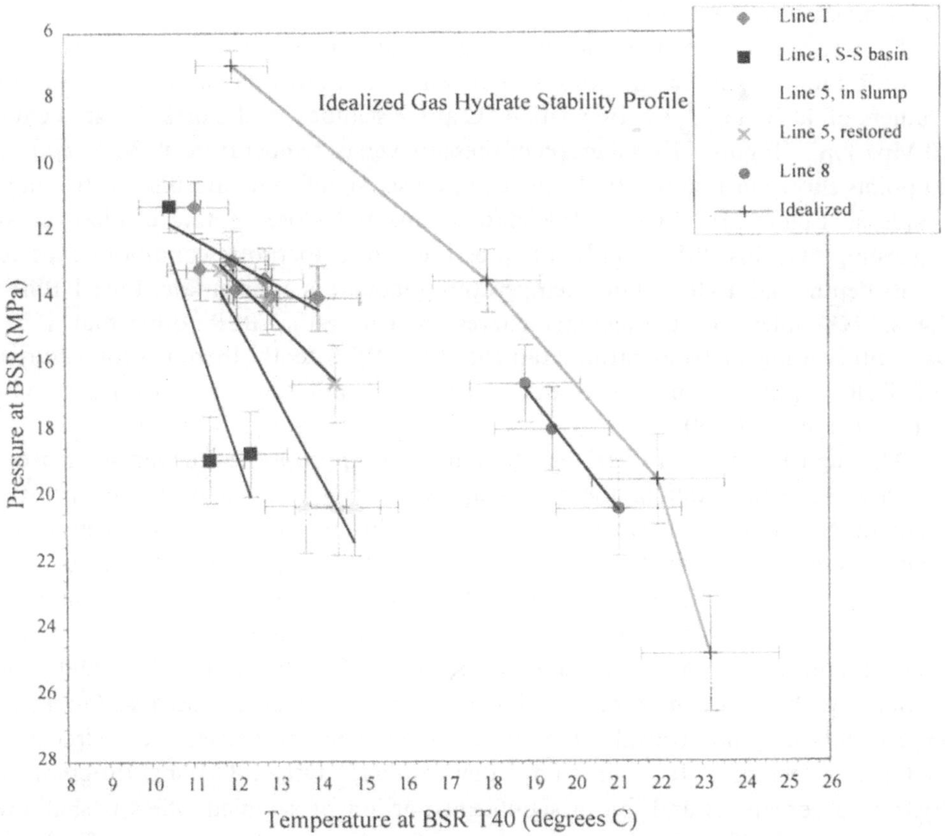

Figure 11b

Graph of BSR data from 11a in *T-P* environment. Converted water depth and BSR depth to pressure and temperature using a 10 Mpa km^{-1} pressure gradient and a 40 °C km^{-1} temperature gradient, starting at 4 °C at the seafloor. (*s-s* basin is strike-slip basin. Line 1 data includes near shore basin BSR. Line 5 restored BSR depths based on Figure 12).

Line 8 BSR data points fall within 7% error bars (used to account for possible velocity processing errors) of the idealized stability curve, but Lines 1 and 5 have significantly shallower BSRs for their water depths. The BSR data series labeled Line 1 is approximately 145 m shallower than the methane/seawater stability curve shown and approximately 85 m shallower than the methane/pure water stability curve. We have delineated BSR measurements taken within the amphitheater of line 1 (in the strike-slip basin) and line 5 (in the slump) from other measurements along these profiles. In the strike-slip basin, the BSR is as shallow as other Line 1 BSRs found, but seafloor depths are 500–600 m greater. This is near an active fault which may influence the fluid circulation at depth and localized temperature gradient. In the slump, the BSR is up to 90 m deeper than in the strike-slip basin, but is still 160–210 m shallower than Line 8 and the methane stability curves. This hydrate may be a residual from pre-slump pressure and temperature conditions.

To further assess possible temperature gradient or pressure variability, we convert the BSR data to likely temperature and pressure conditions. We apply a temperature gradient of 40 °C km^{-1}, starting with 4 °C at the seafloor, and a pressure gradient of 10 Mpa km^{-1}. Figure 11b is a graph of pressure versus temperature at BSR depth for all points shown in Figure 11a. Here again, we see significant variability with similar trends as Figure 11a. Line 8 BSR data points fall close to the idealized curve, suggesting that this BSR is under the pressure and temperature conditions expected for its depth, and a 40 °C km^{-1} temperature gradient is appropriate. Line 1 BSR is 2.4–4.5 °C cooler than the methane curves, indicating that BSRs other than Line 8 have either a higher temperature gradient at the BSR depth, there is subpressure at the BSR depth, or there is some mechanism inhibiting gas hydrate stability (Clennell *et al.* 1999).

Although the data are variable, they may be appropriate for various reasons. Activity along the fault in the strike-slip basin may cause a higher temperature gradient which would drive the hydrate equilibrium to a shallower depth. The BSR in the slump may be shallow because it has not yet re-equilibrated after mass failure as the seismic data were collected only one year after the tsunami. In this case the BSR within this slump block should restore to a depth similar to those found in the adjacent undeformed area. Clennell *et al.* (1999) work emphasizes the need for *in situ* studies of gas hydrate stability because laboratory experiments may not consider the effects of sediment properties, fluid circulation and active tectonics. In their study, they evaluate Ocean Drilling Program gas hydrate observations and find a significant portion of sampled BSRs is shallower than predicted. They further suggest that fine grain sediment may affect pore space in such a way as to inhibit hydrate stability in these cases. In spite of this localized variability in BSR depth and/or temperature gradient, all BSRs display subbottom depths acceptable for gas hydrates in continental margins (Kvenvolden and Barnard, 1983).

6. *Restoration of Slump*

We estimated the amount of slump displacement by restoring it to an undeformed state. Based on faulting and slide plane interpretations, we employed a commonly used extensional restoration methodology of vertical/oblique slip using GeoSec software. This method restores fault bounded blocks along a single slip direction based on the interpreted faulting. It assumes no material has moved into or out of the plane of the cross section, including displacement by strike-slip faulting. Restoration was performed on the main slump mass (Block A) and upper rotated blocks. The horizons and fault interpretations were modified as necessary to improve the restoration and preserve area balance of the cross section.

We restored each block shown on Figure 8 individually to get a cumulative movement of Block A so that horizons match those found in the undeformed sediment south of the scarp. We present the maximum displacement allowed without over-rotation of the parallel reflections of Block A. The material in the toe region is smoothed along one horizon essentially to fill in the area missing by moving Block A. In doing so, we smooth the uplifted and hummocky Horizon B in the toe region to show a gradual slope. The reconstruction shows that some material is missing, which may be due to out-of-plane motion during the slump. Figure 12 is an overlay of the restoration results on the slump interpretations.

The center of mass is determined in Block A by averaging the amount of movement of each reflection in this block from a restored to a deformed state. The restoration shows the following center of mass displacements: 380 m of vertical drop, 840 m of horizontal movement and slip of 980 m along the slide plane.

An important consideration in the restoration is the location of the restored BSR. If the BSR in Block A did not re-equilibrate from pre-slump temperature and pressure conditions to post-slump conditions, it should restore to a depth comparable to adjacent undeformed areas of Line 1. We plot this restored BSR on Figures 11a and b, and convert to temperature and pressure by the same techniques described above. From an examination of the figures, we note the restored BSR falls close to Line 1 BSR (in the near-shore basin). It is approximately 170 m shallower and 5.6 °C cooler than the methane/seawater curve shown and approximately 100 m shallower and 3.5 °C cooler than the methane/pure water curve. This variation is similar to Line 1 and suggests the BSR restored to an appropriate depth for pre-slump temperature and pressure conditions.

7. *Discussion*

Tsunamis may be generated by fault motion during seismic events or by landslips that may or may not be associated with seismicity. Other causes include volcanic eruptions and collapse of volcanic edifices, and bolide impact. Seismic profiling in the

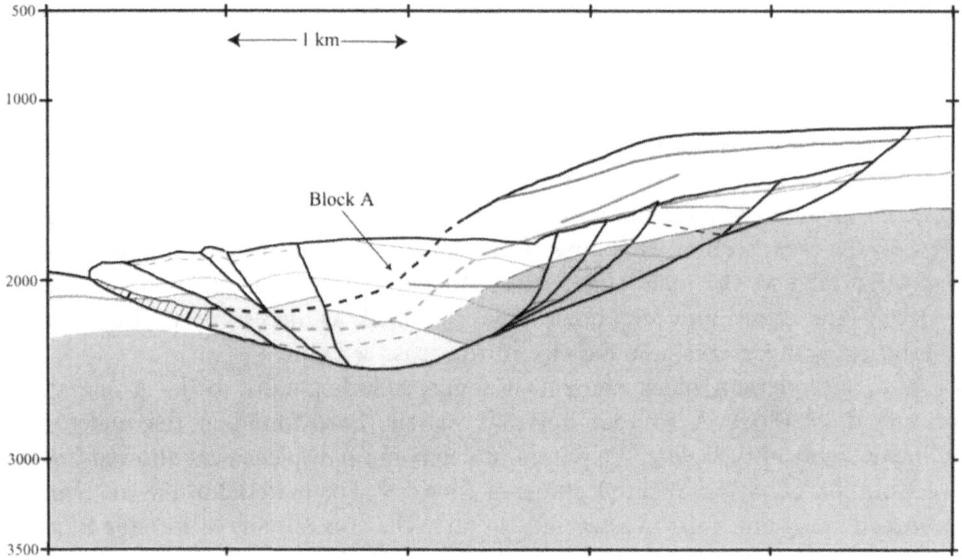

Figure 12
Line drawing of seismic section showing slump and restoration. Deformed state is shown with faulting interpretations shown in Figure 8. Restoration surface is shown in heavy line. Deformed and restored location of BSR is shown. See text and Figures 11a and b for depths of BSR. Horizons are dashed where uncertain. Missing area of restoration shown with diagonal lines.

source area of the 1998 PNG tsunami has shown evidence of both faulting and submarine slumping, and the region is volcanically active. Thus a number of potential tsunamigenic mechanisms occur in this region. On July 17, 1998, no volcanic eruption or bolide impacts occurred, but a $M = 7$ earthquake was followed by a large aftershock just offshore of the zone of greatest tsunami damage. In addition, the potential source region shows evidence for submarine slumping.

The mainshock of the earthquake occurred at 6:49 PM local time. A large aftershock 20 minutes later was felt by a number of people shortly before the tsunami struck, and a number of others involved in the tsunami did not feel the aftershock (DAVIES, 1998). Tsunami propagation models show that the wave would have struck the coast about 10 minutes after initiation, implying that the wave was not generated at the time of the mainshock, nor could it have been generated by the aftershock. In addition, the earthquake magnitude was very small compared to the size of the wave that struck the coast (IMAMURA, 1937; KAWATA *et al.*, 1999; TAPPIN *et al.*, 1999). Some earthquakes are "slow," generating significantly larger tsunamis than expected from their magnitudes (KANAMORI, 1972). However, the July 17, 1998 event was not a slow earthquake (NEWMAN and OKAL, 1998).

Local tsunami maxima could be generated from slip on a high angle fault breaking the surface in relatively shallow water (FUKAO, 1979). We have mapped several faults that show apparent vertical displacement of the seafloor along their trace, but are at depths of 1500 m or greater. These faults could be capable of generating a significant tsunami, depending on the magnitude of vertical displacement during a slip event. However, several lines of evidence suggest that these faults were not likely to be the cause of the July 17, 1998 tsunami. First, the distribution of aftershocks (TAPPIN *et al.*, 1999) covers a broad area, implying that the main event was a low angle thrust, rather than a high angle reverse fault (the only two alternatives given by the first motion results). Second, our seismic data suggest that the faults along this margin are dominantly strike-slip, rather than reverse and are not in shallow water. Third, the timing of the earthquake does not match the timing of the wave, as discussed above.

Other lines of evidence favor a submarine slump as the likely cause of the tsunami. It is more likely that a slump will fail 10 minutes after the causative earthquake than the seafloor along the fault plane breaking 10 minutes after the fault starts to slip. Secondly, Synolakis *et al.* (2002) documented *T*-phase arrivals at Wake Island for a small, relatively long duration event 13 minutes after the main shock. This event could be the seismic signal for the slump. Thirdly, we have imaged a major submarine slump in the likely source area of the tsunami, whose size, location, vertical drop and water depth are consistent with the parameters necessary to generate the observed tsunami.

We utilized a high-resolution source for the seismic data that gave returns rich in the higher frequencies and a sharp signature that needed little deconvolution. Depth of penetration was sufficient to image the slump mass. The key elements of the slump include a central mass over 700-m thick which is rotated southward about 14 degrees, a set of smaller rotated slump blocks with normal surface offset on the upper part of the feature, and a deformed toe with 100 m of surface relief.

We have restored the slump to a reasonable pre-slump condition using GeoSec software. This reconstruction provides quantitative constraints on the total volume of the slump and the vertical and horizontal components of displacement. Modeling of the slide geometry and kinematics shows that this slump is capable of generating the observed tsunami (SYNOLAKIS *et al.*, 2002). Assisting our reconstruction is the geometry of the BSR in seismic Line 5. Comparing the reconstructed depths with those observed in nearby profiles that are not associated with slumping yields good agreement. Line 1 BSR is an average of 18 m deeper than the restored BSR of Line 5 when compared to the methane stability curves and Line 8 BSR. This is within a 10% error of the measure BSR depths.

We discovered a significant difference in the implications for slumping between Lines 1 and 5, which are spaced about 1.6 km apart where they cross the amphitheater. Line 1 is interpreted to cross a strike-slip basin showing growth folding of the shallowest layers, suggesting slow rather than sudden movement of material. Line 5 does not show signs of progressive growth, suggesting rapid movement. Examination of bathymetric profiles from the digital bathymetry shows a gradual rise traversing from the strike-slip basin of Line 1 to Block A of Line 5 and a scarp between the lines on the amphitheater south wall. This is interpreted as the western extent of the slump.

TAPPIN *et al.* (1999) suggested that this margin was probably undergoing subduction erosion, based on the nature of the bathymetry and their discovery of a drowned coral reef complex at 500 m depth. Our seismic data allow us to concur with this interpretation. The lower slope region has very little sediment cover and many of the lines crossing the trench show little or no sediment accretion at the front of the margin. Seismic indications of strike-slip faults can be mapped and continued laterally using the swath bathymetry, which we enhanced using ERMapper imaging software. The front part of the margin is undergoing subduction of the Pacific Plate, and the low-angle thrust plane is visible on the seismic records. In addition, significant left-lateral strike slip motion has disrupted the margin, displacing the slope into a series of arcs and straight segments with sharp corners formed where they meet. Large unfilled basins exist on the margin; evidence that tectonics is occurring much more rapidly than sedimentation. This environment is ideal for generating major slumps, and one such slump is likely to have generated the destructive tsunami of July 17, 1998.

8. Conclusions

We performed a seismic reflection survey in the source region of the 1998 PNG tsunami to investigate sediment deposition characteristics. A previously identified, amphitheater-shaped feature is a likely location for a slump that may have generated this geographically concentrated tsunami. We crossed this region with three profiles and imaged a slump of 3.8 to 4.6 km^3 within the identified amphitheater. Although the other two profiles cross some slide material, only a small portion of the sediment load in the amphitheater is slump material. We restored the slump in Line 5 to an undeformed state and balanced the cross section. This technique yielded a center of mass vertical drop of 380 m, horizontal movement of 840 m and slip of 980 m. A BSR imaged in the slump mass restores to an appropriate position based on other PNG BSRs presented and published data. Although we have imaged a slump in the likely location of the tsunami source area, we did not constrain the age of this mass movement.

Acknowledgments

We would like to thank the *Ewing* crew and other ship-board scientists for on-board support of data acquisition and processing, Paul Henkart for extensive support of SIOSEIS seismic processing software, Landmark Graphics for use and support of software, Sara Foland and Paradigm Geophysical for use of GeoSec software. Location maps were created using GMT and ERMapper. We further acknowledge Gretchen Zwart and Ed Boring for UCSC computer support, and Hugh Davis, Casey Moore, Laura Wallace and Phil Watts for very helpful advice. Supported by Addendum to NSF Grant #EAR-9725792.

References

ABERS, G. and MCCAFFREY, R. (1988), *Active Deformation in the New Guinea Fold-and-thrust Belt: Seismological Evidence for Strike-slip Faulting and Basement-involved Thrusting*, J. Geophys. Res. *93*, 13,332–13,354.

ALLEN, P. A. and ALLEN, J. R., *Basin Analysis Principles and Applications* (Oxford, Blackwell Scientific Publications, 1990), 451 pp.

CHRISTIE-BLICK, N. and BIDDLE, K. T. (1985), *Deformation and basin formation along strike-slip faults*. In *Strike-slip Deformation, Basin Formation and Sedimentation* (eds., K. T. Biddle and N. Christie-Blick), Spec. Publ. Soc. Econ. Paleont. Mineral. *37*, 1–34.

CLENNELL, M. B., HOVLAND, M., BOOTH, J. S., HENRY, P., and WINTERS, W. J. (1999), *Formation of Natural Gas Hydrates in Marine Sediments 1. Conceptual Model of Gas Hydrate Growth Conditioned by Host Sediment Properties*, J. Geophys. Res. *104*, 22,985–23,003.

COOPER, P. and TAYLOR, B. (1987), *Seismotectonics of New Guinea, A Model for Arc Reversal Following Arc-continent Collision*, Tectonics *6*, 53–67.

DAVIES, H. L. (1998), *The Sissano Tsunami*, Univ. Papua New Guinea, Port Moresby.

DEMETS, C., GORDON, R. G., ARGUS, D. F., and STEIN, S. (1994), *Effect of Recent Revisions of the Geomagnetic Reversal Time Scale on Estimated of Current Plate Motions*, Geophys. Res. Lett. *21*, 2191–2194.

FUKAO, Y. (1979), *Tsunami Earthquakes and Subduction Processes near Deep-sea Trenches*, J. Geophys. Res. *84*, 2303–2314.

HAMILTON, W. (1979), *Tectonics of the Indonesian Region*, U.S. Geol. Surv. Prof. Pap. *1078*, 345 pp.

HENKART, P. (1981), SIOSEIS, Scripps Industrial Associates and National Science Foundation.

HENRY, P., THOMAS, M., and CLENNELL, M. B. (1999), *Formation of Natural Gas Hydrates in Marine Sediments 2. Thermodynamic Calculations of Stability Conditions in Porous Sediments*, J. Geophys. Res. *104*, 23,005–23,022.

IMAMURA, A. *Theoretical and Applied Seismology* (Maruzen, Tokyo, 1937), 358 pp.

KANAMORI, H. (1972), *Mechanism of Tsunami Earthquakes*, Phys. Earth Planet. Interiors *6*, 346–359.

KAWATA, Y., BENSON, B. C., BORRERO, J. L., DAVIES, H. L., de LANGE, W. P., IMAMURA, F., LETZ, H., NOTT, J., and SYNOLAKIS, C. (1999), *Tsunami in Papua New Guinea was as Intense as First Thought*, EOS. Trans. Am. Geophys. Un. *80*, 101, 104–105.

KREEMER, C., HOLT, W. E., GOES, S., and GOVERS, R. (2000), *Active Deformation in Eastern Indonesia and the Philippines from GPS and Seismicity Data.*, J. Geophys. Res. *105*, 663–680.

KVENVOLDEN, K. A. and BARNARD, L. A. (1983), Hydrates of natural gas in continental margins. In *Future Petroleum Provinces of the United Stated* (eds., J. S. Watkins and C. L. Drake), AAPG Memoir *34*, 631–640.

MIALL, A.D., *Principles of Sedimentary Basin Analysis* (New York, Springer-Verlag, 1984), 490 pp.

NEWMAN, A. V. and OKAL, E. A. (1998), *Teleseismic Estimates of Radiated Seismic Energy: The E/M_o Discriminant for Tsunami Earthquakes*, J. Geophys. Res. *103*, 26,885–26,898.

RIPPER, I., *Seismicity, earthquake focal mechanisms and tectonics of the Indo-Aus/SolomonSea plate boundary* (Dept. of Min. and Energy Rep. 1980), 8015, 55 pp. Geol. Surv, Papua New Guinea, Port Moresby.

SYNOLAKIS, C., BARDET, J., BORRERO, J., DAVIES, H., GRILLI, S., OKAL, E., SILVER, E., SWEET, S., and TAPPIN, D., (2001), *The Slump Origin of the 1998 Papua New Guinea Tsunami*, The Roy. Soc. Proc.: Math., Phys. and Engin. Sci. *458*, 763–789.

TAPPIN, D.R. *et al.* (1999), *Sediment Slmp Likely Caused 1998 Papua New Guinea Tsunami*, EOS. Trans. Am. Geophys. Un., *80*, 329, 334, 340.

TREGONING, P. *et al.* (1998), *Estimation of Current Plate Motions in Papua New Guinea from Global Positioning System Observations*, J. Geophys. Res. *103*, 12,181–12,203.

Pure appl. geophys. 160 (2003) 1969–1999
0033 – 4553/03/111969 – 31
DOI 10.1007/s00024-003-2416-y

Pure and Applied Geophysics

Erosion and Sedimentation from the 17 July, 1998 Papua New Guinea Tsunami

GUY GELFENBAUM[1] and BRUCE JAFFE[2]

Abstract—This paper describes erosion and sedimentation associated with the 17 July 1998 Papua New Guinea tsunami. Observed within two months of the tsunami, distinct deposits of a layer averaging 8-cm thick of gray sand rested on a brown muddy soil. In most cases the sand is normally graded, with more coarse sand near the base and fine sand at the top. In some cases the deposit contains rip-up clasts of muddy soil and in some locations it has a mud cap. Detailed measurements of coastal topography, tsunami flow height and direction indicators, and deposit thickness were made in the field, and samples of the deposit were collected for grain-size analysis in the laboratory. Four shore-normal transects were examined in detail to assess the shore-normal and along shore distribution of the tsunami deposit. Near the shoreline, the tsunami eroded approximately 10–25 cm of sand from the beach and berm. The sandy layer deposited by the tsunami began 50–150 m inland from the shoreline and extended across the coastal plain to within about 40 m of the limit of inundation; a total distance of up to 750 m from the beach. As much as 2/3 of the sand in the deposit originated from offshore. Across most of the coastal plain the deposit thickness and mean grain size varied little. In the along-shore direction the deposit thickness varied with the tsunami wave height; both largest near the entrance to Sissano Lagoon.

Key words: Tsunami deposit, Papua New Guinea, coastal sedimentation, erosion, flow direction indicators.

Introduction

Tsunami waves approach land with sufficiently high velocities and bed shear stresses to suspend and transport large quantities of sediment. Despite the destructive power of a tsunami, in which trees are uprooted and structures destroyed, a tsunami can leave behind a clearly identifiable deposit. If the sediment deposited by a tsunami is buried and preserved, then a geologic record of that tsunami will be created. Using a sedimentological analysis, geologists may be able to read the geologic record to infer the occurrence of past tsunamis. The recognition of deposits from past tsunamis allows geologists to extend the relatively short or non-existent historical record of

[1] U.S. Geological Survey, 345 Middlefield Road, MS999, Menlo Park, CA 94025, U. S. A. E-mail: ggelfenbaum@usgs.gov

[2] U.S. Geological Survey, 1156 High Street, University of California, Santa Cruz, CA 95064, U. S. A. E-mail: bjaffe@usgs.gov

tsunamis in an area, thus improving assessment of tsunami hazard (ATWATER, 1987; BOURGEOIS and MINOURA, 1997; GOFF *et al.*, 2001; JAFFE and GELFENBAUM, 2002).

The interpretation of palaeotsunamis is improved by studies of sedimentary deposits associated with modern tsunamis. An important goal of these studies is to determine which characteristics of a tsunami, if any, can be interpreted from the sedimentological record that is left behind. For instance, do trends in the grain-size distribution or thickness of a tsunami deposit relate to the height or speed of a tsunami wave? Questions such as these can only be answered from a more complete, detailed analysis of modern tsunami deposits and the tsunami waves that produced them.

Since 1992 there have been 10 major tsunamis worldwide, including Nicaragua and Flores in 1992, Okushiri in 1993, East Java, Shikotan, and Mindoro in 1994, Irian Jaya and Peru in 1996, Papua New Guinea in 1998, and most recently Peru in 2001. Recent studies of these events has led to an improved understanding of tsunami generation, propagation, and inundation. In particular, studies of the sedimentary deposits from the Okushiri, Japan tsunami (SATO *et al.*, 1995; NISHIMURA and MIYAJI, 1995), and the Flores Island, Indonesia tsunami (DAWSON, 1994; SHI *et al.*, 1995) describe the pattern of sedimentation associated with a tsunami. For example, SHI *et al.* (1995) find that the grain-size fines landward in the deposit on Flores Island. The purpose of this manuscript is to describe the erosion and sedimentation associated with the 17 July 1998 Papua New Guinea tsunami, thus adding to the observations of modern tsunami sedimentation that will ultimately improve the identification and interpretation of palaeotsunamis in the geologic record.

Description of 1998 PNG Tsunami

On the evening of July 17, 1998 a magnitude 7.0 earthquake was followed by a series of tsunami waves that devastated villages on the north coast of Papua New Guinea (PNG). As later confirmed by field surveys, eyewitnesses reported waves at least 7–10 m high at the coast. The confirmed death toll was over 2200, and the coastal villages of Warapu and Arop, located on barrier spits around Sissano Lagoon, were totally destroyed (Figs. 1 and 2). Thousands of survivors were forced from their homes and villages and set up temporary camps further inland or on nearby hills.

The tsunami consisted of three large waves, the first making land fall approximately 20 minutes after coastal residents felt the earthquake (DAVIES, 1999). The second came within 5 minutes of the first wave and the third wave came within 5 minutes of the second wave. According to eyewitnesses, the second and third waves flowed over the low-lying coast before the first wave completely receded. Waves ripped large trees out of the ground by the roots and moved both traditional wooden and modern brick and mortar structures off of their pilings or foundations tens to hundreds of meters away into the lagoon. Preliminary analysis of seismographs and other field evidence put the location of the seafloor disturbance that caused the

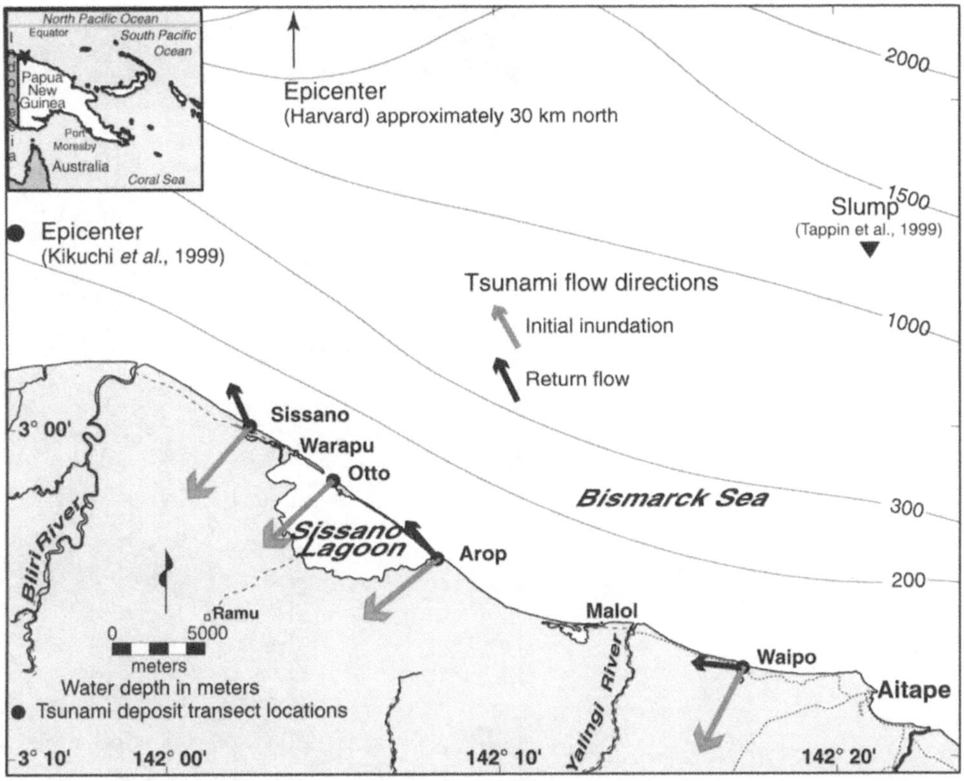

Figure 1
Location map of the north coast of Papua New Guinea struck by the 1998 tsumani. The gray shaded arrow indicates the flow direction of the main tsunami inundation; the black arrow indicates the direction of return flow after inundation.

tsunami approximately 20 km offshore the Sissano Lagoon. Initial tsunami generation and propagation model results suggested that an earthquake of this size alone could not be responsible for the large wave heights observed on the coast (KAWATA *et al.*, 1999), thus motivating a search for an additional seafloor disturbance. A marine geophysical survey later identified a submarine failure which may have contributed to the generation of the tsunami (TAPPIN *et al.*, 1999). A debate persists regarding whether the tsunami that devastated the north coast of PNG was generated by an earthquake, a landslide, or a combination of the two mechanisms (GEIST, 2000; OKAL and SYNOLAKIS, 2001; GEIST, 2001; TAPPIN *et al.*, 2001).

An international team of tsunami researchers, referred to as the 1st International Tsunami Survey Team (ITST), surveyed the area within a few weeks of the tsunami. The team of scientists and engineers from Japan, the United States, Australia, and New Zealand installed seismographs, measured water levels, and interviewed eyewitnesses to learn more about the tsunami and its effects (KAWATA *et al.*, 1999).

b
a

Soon after the 1st ITST completed their survey, a second group of international scientists returned to PNG and retrieved seismographs, collected more water-level and velocity indicator data, assessed damage to buildings and structures, documented societal effects, and examined the sediments left behind by the tsunami. The 2nd ITST arrived in Aitape, Papua New Guinea on September 29, 1998, and included representatives from Japan, the United States, Korea, and Papua New Guinea.

Historical Tsunamis in Northern Papua New Guinea

Northern Papua New Guinea lies near the boundary of the Australian and Pacific plates. The plate boundary is complex and includes subduction, transform faulting, and thrusting (TAPPIN *et al.*, 2001). Overall, convergence is at an oblique angle and at a rate of about 100 mm/yr. This rapid convergence gives rise to numerous large earthquakes. In the past 100 years there have been 16 earthquakes with magnitude over 5.7. The largest, in 1935 and 1971, had a magnitude of 7.9 (RIPPER *et al.*, 1998).

Most of these earthquakes, however, occurred on land and failed to generate tsunamis. Moreover, for northern Papua New Guinea there is only a short historical record of tsunamis impacting the coast. In the 20th century, a magnitude 7 earthquake generated a tsunami that hit the area west of Aitape on December 15, 1907 (RIPPER *et al.*, 1998). Prior to the earthquake, there was a smaller lagoon, called Warupu Lagoon with two small islands inhabited by the Warupu clan. Coseismic subsidence of about 100 km^2 of coast formed a larger lagoon in the location of the present Sissano Lagoon (MCSAVENEY, *et al.*, 2000). A second tsunami generated by a 7.9 magnitude earthquake occurred on September 20, 1935 near the northern shore of PNG in an area close to the 1998 event. However, no tsunami intensity or runup heights are available for that event. A third tsunami in northern Papua New Guinea occurred on October 31, 1970 near Madang to the east of the present event. This tsunami was triggered by a magnitude 7.0 earthquake and created 4-m high waves.

The area of greatest impact from the 1998 tsunami lies between the Bliri River to the west and the Yalingi River to the east (Fig. 1). These rivers, and several smaller ones, drain the Torricelli Mountains that lie 10–15 km from the coast. The rivers supply sand to the coast and both rivers have built large sandy deltas. This sediment supply feeds the beaches and has built sandy barrier spits that separate the low-lying coastal plain and lagoons from the ocean. The coastal plain in this region is heavily wooded, low-lying, and swampy. Although only a few meters in elevation, the high ground along the spits was the site of many of the coastal residences.

◄

Figure 2
Photographs of debris torn off nearby structures by the tsunami and wrapped around standing trees.

Tsunami Deposit Transects

Measurements of land elevation, flow direction, flow depth, and tsunami deposit thickness and character were made approximately two months after the tsunami. Topography was measured with a rod and level and values are reported relative to mean sea level. The small tide range and steep beach slopes permitted field estimates of mean sea level to within approximately 0.25 m. Flow direction was estimated in the field from bent or broken palms, exposed roots and grass aligned parallel on their sides, and most commonly, debris wrapped around or piled against immobile objects like palm trees left standing. Flow direction was interpreted from the orientation of the feature and measured with a compass. Flow depth was estimated by measuring the height of broken limbs, debris found in trees, debris found piled on the ground, or water marks on buildings (Fig. 2). In general, estimates of flow depth from these indicators represent minimum values.

Small pits were dug to examine the sedimentary deposits left by the 1998 tsunami. Tsunami deposits were common and identified as gray sand above a brown, rooted soil (Fig. 3). The deposits were measured, photographed, and described in the field

Figure 3
Photograph of sand deposited by the 1998 tsunami overlying a buried soil.

and samples were taken for laboratory analyses. In the field, bulk samples of the deposit were taken, and, in a few locations, samples were taken at 1-cm intervals down through the deposit.

Grain size was analyzed for all of the sediment samples collected. Samples were wet sieved at 4 phi to separate silt and clay from the sand-sized material. The sand was run through a setting tube and the results reported in ¼-phi intervals. The fine sediment was run through a laser-diffraction particle analyzer and also reported in ¼-phi intervals. Grain-size statistics were calculated using methods described in CARVER (1971).

Detailed measurements were made along four cross-shore transects in the vicinity of Sissano Lagoon (Fig. 1). The transects were located, from east to west, at Waipo Village, Arop School, Otto near the west end of East-Sissano spit, and at Sissano Village.

Waipo Transect

Setting and topography

The Waipo transect is adjacent to the inlet of a small lagoon near the village of Waipo, about half way between Aitape and Malol (Fig. 4). An interview with a local farmer who survived the tsunami provided a first-hand account of the event. The eyewitness reported feeling a large earthquake and seeing the trees swaying. He reported that it was difficult to stand, but that the village huts were not extensively damaged by the earthquake. Within approximately 15 minutes after the earthquake the sea first lowered, followed shortly thereafter by an approaching wall of water. The first wave was approximately 1.5-m high at a location about 200 m from shore (Fig. 4). Surveying later showed this site to be 1.5 m above sea level for a water elevation of 3 m above sea level. The first wave swept the eyewitness off his feet and carried him inland. Before the water from the first wave could withdraw, a second wave approached, riding over the water from the first wave. A short time later, a third wave approached, again riding atop the water from the first two waves. The maximum water depth at this site 200 m from the shore, as determined by information from the eyewitness, was 4.5 m above the ground, giving a total water elevation at this site of 6 m above sea level. The tsunami destroyed most of the village huts at Waipo. In addition, a steel and wooden bridge that crossed the river was ripped off its foundation and transported several tens of meters inland (Fig. 4).

The topography along the Waipo transect undulates across sandy ridges and low-lying troughs (Fig. 5a). Adjacent to the shoreline is a steep beach face and a berm at an elevation 2 m above sea level. Beyond the berm is a back beach trough that is sloped down to the west allowing water to flow parallel to shore and drain through the trough. Landward of the trough is a 1–4 m high erosional scarp with exposed roots. An eyewitness reported that the scarp was formed by the tsunami. It is not clear if a smaller scarp existed prior to the tsunami, or if the scarp formed entirely as

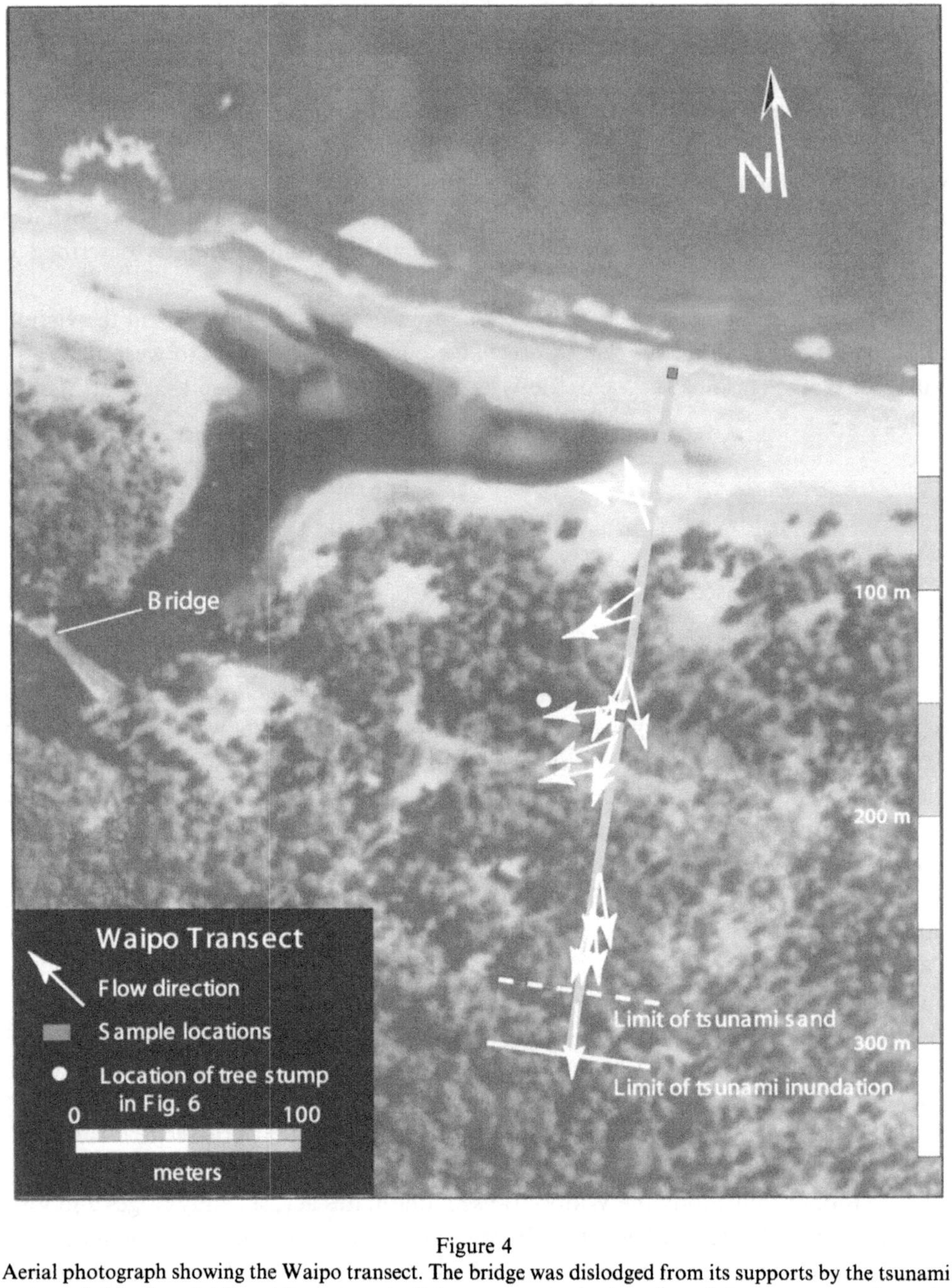

Figure 4
Aerial photograph showing the Waipo transect. The bridge was dislodged from its supports by the tsunami and carried landward. The white dot west of the transect is the location of the tree stump shown in Figure 6. Photograph taken August 8, 1998, three weeks after the tsunami, by the PNG National Mapping Bureau.

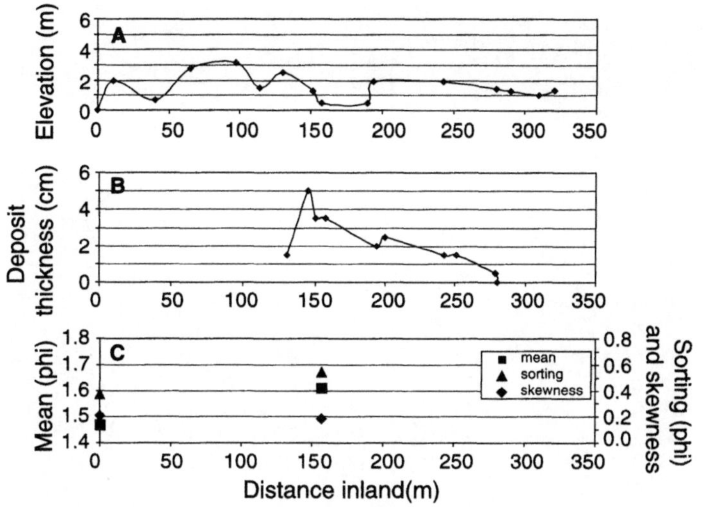

Figure 5
Topography, tsunami deposit thickness, and grain-size statistics of the beach sand and of the tsunami deposit along the Waipo transect.

a result of the return flow of tsunami water. The profile has two more sandy ridges, with elevations of 3.2 m and 2.5 m above sea level. Approximately 150 m from the shoreline is a 40-m wide trough with an elevation less than 0.5 m above sea level. This trough is intermittently wet and dry. Landward of this trough the profile flattens out at an elevation of between 1 and 2 m.

Flow direction and depth indicators

Flow at the Waipo transect traveled in two main directions, 280 degrees (west alongshore) and 210 degrees (southwest onshore) (Fig. 4). At a few locations along the transect flow indicators suggest flow in both of these directions. For example, at about 160 m from the shoreline, the flow at different times was in the cross-shore direction and perpendicular to that direction. The two directions suggest that the initial inundation of the tsunami was in the cross-shore direction, but that the water drained the coastal plain alongshore to the west to the nearby river channel.

The 1st ITST, noting broken branches and debris in trees, estimated wave heights of 4–7 m in the area around Waipo (KAWATA *et al.*, 1999). A few months later, few flow depth indicators remained along the Waipo transect, nevertheless one that did, approximately 160 m from the shoreline, was a tree stump capped nearly 1 m above the ground with a thin layer of sand (Fig. 6). This sand had likely settled out of suspension from within the tsunami and had been deposited on the stump. If so, the tsunami crested at least 1 m high at this location and carried sand-sized material in suspension at least 1 m above the bed.

Figure 6
Photograph of sand atop a tree stump 1m above the ground. The sand was deposited out of suspension from the tsunami.

Deposit

No tsunami deposit was observed over the area from the shoreline inland for 120 m. Exposed roots and debris on the ground indicated the tsunami had eroded this area. The depth of erosion was indeterminable. More than 130 m from the shoreline, a layer of light-colored sand rested on a rooted and compacted dark-colored sandy soil. The deposit thickened from 1.5 cm to a maximum of 5 cm almost 150 m from the shoreline (Fig. 5b), then thinned to the most inland deposit 280 m from the shoreline. Vegetation (trees and underbrush) was densest beyond the landward limit of the tsunami deposit at this transect. The limit of tsunami inundation was found about 40 m further landward than the limit of the tsunami deposit.

Sedimentology

The tsunami deposited a thin sedimentary layer that grades upward from coarse sand at the base of the deposit to fine sand above. This fining upward was observed only in deposits more than a few centimeters thick. The thinner deposits either were not vertically graded or the grading was not detected. The tsunami deposit overlay a compact sandy soil and the contact between the deposit and the soil was typically abrupt. The tsunami deposit was too thin along this transect to collect several vertically-spaced samples for laboratory analysis of the fining-upward sequence. Grain-size analyses were conducted on bulk samples from two sites along the Waipo transect. The mean grain size of the beach sand at the shoreline was 1.5 phi, slightly coarser than the mean size of the tsunami deposit 157 m from the shoreline which was 1.6 phi (Fig. 5c). Sorting and skewness are likewise similar at the two sites. The mode of both samples is 1.25 phi. The size distributions suggest that the sand at the shoreline could be the source of the sandy layer deposited by the tsunami further inland.

Arop Transect

Setting and topography

The Arop transect, located about 500 m east of Sissano Lagoon, traversed the Arop school complex (Fig. 7). The Arop village was one of the areas hit hardest by the tsunami, with over 800 people killed and all of the village huts destroyed. Much of the village had been located on the narrow sandy spit between the ocean and Sissano Lagoon (Fig. 1). Close to shore, the tsunami completely uprooted full-sized palm trees and carried them inland. The Arop school, located east of the Lagoon and approximately 600 m south of the shoreline, suffered damage to some of its buildings.

The Arop transect crosses a steep beach face backed by a broad low-lying plain (Fig. 8a). The beach berm is 2.2 m high and is the highest elevation along the entire 700-m long transect. Landward of the berm the profile drops to less than 0.5 m above mean sea level and gradually rises to 2 m above sea level 700 m inland.

Flow direction and depth indicators

Flow direction indicators along the Arop transect suggest two main directions of flow (Fig. 7). Many of the indicators record onshore flow roughly perpendicular to the shoreline (240 degrees). Other indicators imply return flow parallel to the shoreline (320 degrees) through a broad topographic low (approximately 80 to 350 m from the shoreline). Landward of 350 m from shore, where the elevation gradually rises from 0.5 to 2 m above sea level, all evidence indicates onshore flow only. In general, the return flow was toward the lagoon and not back toward the beach.

The 1st ITST estimated wave heights of 10–15 m along the coast near Arop (KAWATA *et al.*, 1999). These estimates came from broken branches near the tops of trees and debris found in the trees within 100 m of the beach. We estimated water levels further inland along the transect. At 200–450 m inland, water 1.7–4.5 m high piled debris against trees and broke branches in standing trees. Between 500–700 m inland, water levels decreased from 1.3 m to 0.3 m within 50 m of the limit of tsunami inundation. These estimates contain uncertainty because the level at which debris is piled up against an obstacle like a tree can be either higher or lower than the water level at the time of inundation.

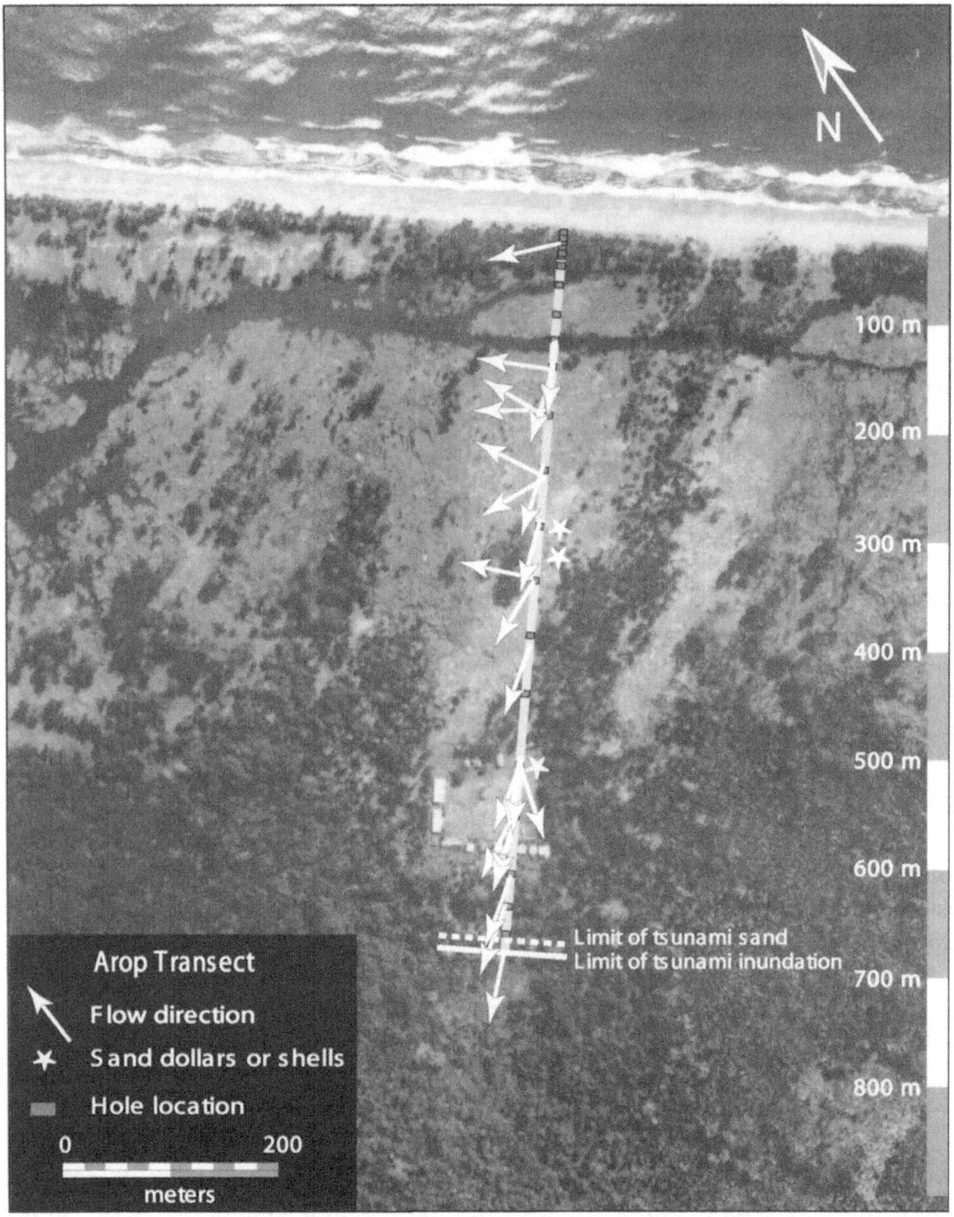

Figure 7
Aerial photograph showing the Arop transect. Photograph taken August 8, 1998 by the PNG National Mapping Bureau.

Deposit

The beachface and berm showed no evidence of deposition from the tsunami. On the berm, exposed roots and scour at the base of some palm trees indicated erosion of

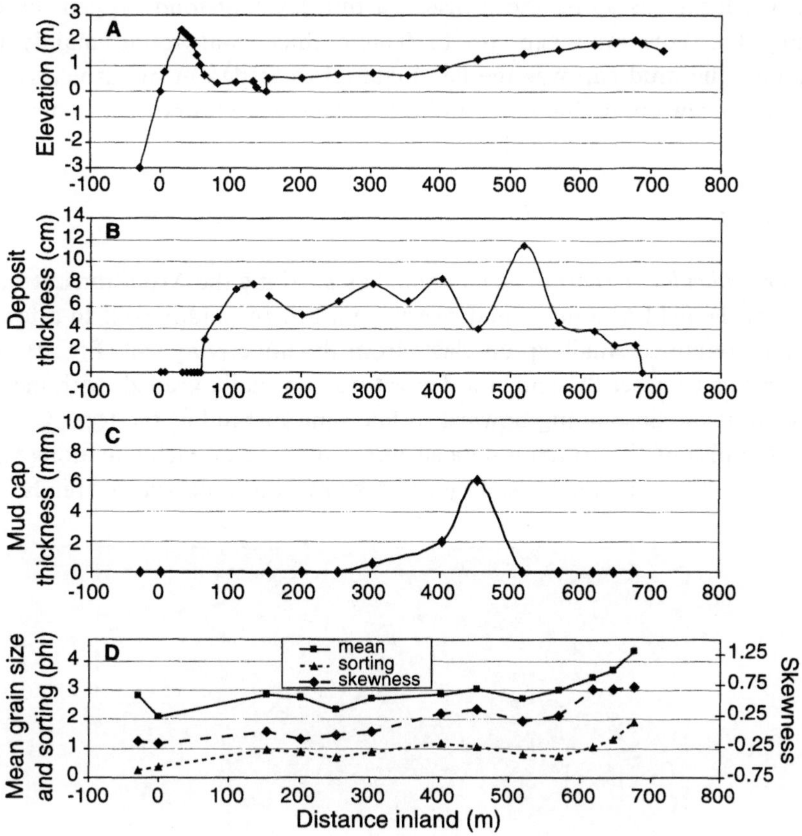

Figure 8
Topography, tsunami deposit thickness, and grain-size statistics of the beach sand and of the tsunami deposit along the Arop transect.

an estimated 20–30 cm of sand. The zone of erosion extended to the landward side of the berm, about 50 m from the shoreline.

At about 70 m from the shoreline and within the span of a few tens of meters, the deposit thickness increased to about 8 cm (Fig. 8b). The tsunami deposit was easily identified as a gray sand overlying a brownish muddy soil. The deposit thickness remained relatively constant from 100 to 500 m from shore, varying by only a few centimeters. The deposit thickness peaked at 11 cm approximately 520 m from the shoreline. From there, the deposit thickness decreased in the landward direction over the next 150 m. The deposit further tapered to near zero thickness at a distance of 680 m from the shoreline. The limit of tsunami inundation, as defined by a small debris line, was 720 m from the shoreline.

At some locations along the transect, a thin layer of mud capped the tsunami sand (Fig. 8c). This mud cap, up to 0.6-cm thick, was found mainly in local depressions. The mud cap was the last material to settle out of suspension. Whole sand dollars (a few cm in diameter) and small shells were found on the surface of the deposit as far as 500-m inland (Fig. 7).

Sedimentology

Trends perpendicular to shore The tsunami deposit along the Arop transect was easily identified in the field as a gray sand overlying a brown muddy soil. In a few places, the deposit contained small rip-up clasts from the underlying soil (Fig. 9). At most sites along this transect the tsunami deposit was normally graded, with medium to coarse sand at the base fining upward to fine sand or mud at the top.

Bulk samples of the tsunami deposit were collected at each site along the Arop transect for laboratory analysis. Figure 8d shows the cross-shore trends in mean

Figure 9
Photograph of the tsunami deposit layer overlying a buried soil. The deposit is normally graded with medium sand near the base and fine sand toward the top. Within the deposit is a rip-up clast of muddy soil.

grain size, sorting, and skewness across the Arop transect. Sand samples taken from the shoreline and from -3 m depth only 30 m offshore show local variability in grain size of the beach material. The beach sand is well sorted and the mean size varies from 2 to 3 phi. Combining both the offshore (-3m depth) and shoreline sites, sand is found in all size classes from 1 to nearly 4 phi.

Along the transect from approximately 150–520 m from the shoreline the mean grain size and the sorting of the sand in the tsunami deposit remain relatively constant. From about 520 m from the shoreline to the limit of the tsunami deposit at 680 m the deposit gets finer. The mean grain size decreases to about 3.25 phi. The decrease in mean size is due almost entirely to an increase in the amount of silt and clay-sized material. The grain-size trend at the Arop transect does not follow the simple model of landward fining. While it is true that the grain size at the beach is coarser than the grain size at the limit of deposit inundation, for the majority of the transect at Arop, the grain size of the tsunami deposit does not vary with distance inland.

Vertical trends At a site approximately 400 m from the shoreline, several sub-samples were taken to quantify the vertical variations in grain size within the deposit. At this site the tsunami deposit was 8-cm thick. Eight samples were taken at 1-cm intervals. The vertical distribution in mean grain size confirms the fining upward sequence that was observed in the field (Fig. 10). The mean grain size smoothly

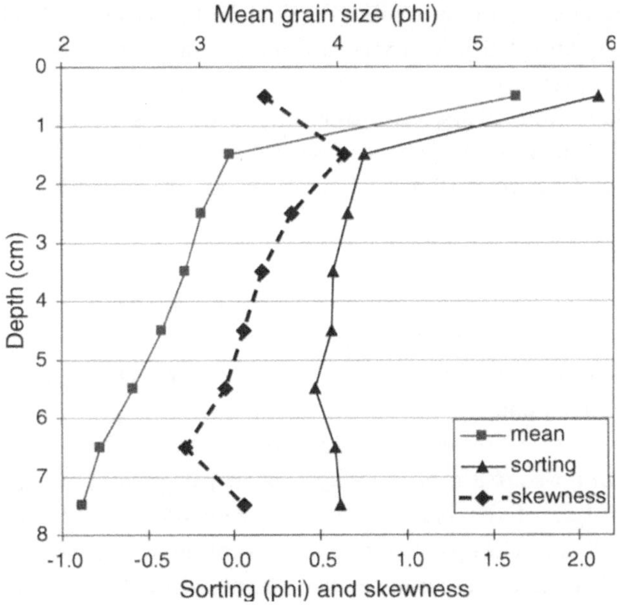

Figure 10
Mean grain size, sorting and skewness at 1-cm intervals of the tsunami deposit in Figure 9.

decreases from 2.25 phi at the base of the deposit to 3.25 phi just below the surface. The surface layer contained mud that decreased the mean size of the top 1 cm to 5.3 phi. The origin of the mud, either from erosion of the underlying soil or from an offshore source, could not be determined.

Otto Transect

Setting and topography

The Otto transect was located near the west end of the sandy barrier spit that extends to the west from Arop and fronts Sissano Lagoon. The width of the spit along this transect, as well as along much of the over 5 km length of the spit, is narrow, extending less than 200 m across from the ocean to the lagoon (Fig. 11). The spit was sparsely to densely populated with vegetation, primarily with a few different varieties of palm trees and low grasses. The highest elevation along the Otto transect is merely 0.73 m above sea level at the primary beach berm located 15 m from the shoreline (Fig. 12a). The elevation drops steadily from a maximum at the berm down to zero at the edge of the lagoon a distance of 160 m from the shoreline.

Flow direction and depth indicators

The entire barrier spit fronting Sissano Lagoon, including the area at the Otto transect, was inundated by the tsunami. All flow direction indicators point landward, roughly perpendicular to the shoreline (Fig. 11). Indicators included fallen palm trees, laid-over grasses, and debris wrapped around obstacles such as standing trees. No flow indicators representing return flow were found along the spit near this transect.

Few flow depth indicators were found along the west end of the eastern barrier spit fronting Sissano Lagoon. Those observed by the 1st ITST suggested maximum tsunami heights of between 10 and 15 m (KAWATA *et al.*, 1999).

Deposit

As at Waipo and Arop, the beach face and berm at Otto showed no evidence of a recent tsunami deposit. Along the first 40–50 m across the top of the berm, exposed roots of grasses and scour around trees indicate erosion. The amount of this erosion is estimated at 10–20 cm.

Starting at a distance of about 60–70 m from the shoreline, about 10 cm of well-sorted gray medium-coarse sand overlay a brown organic-rich soil of muddy sand. Just below the base of the tsunami deposit were root clumps. Fresh grasses still connected to their root clumps were laid over pointing inland within the lower part of the gray sand. The sandy tsunami deposit increased in thickness to a maximum of about 26 cm at a distance of 90 m from the shoreline, then remained nearly constant for at least several tens of meters (Fig. 12b).

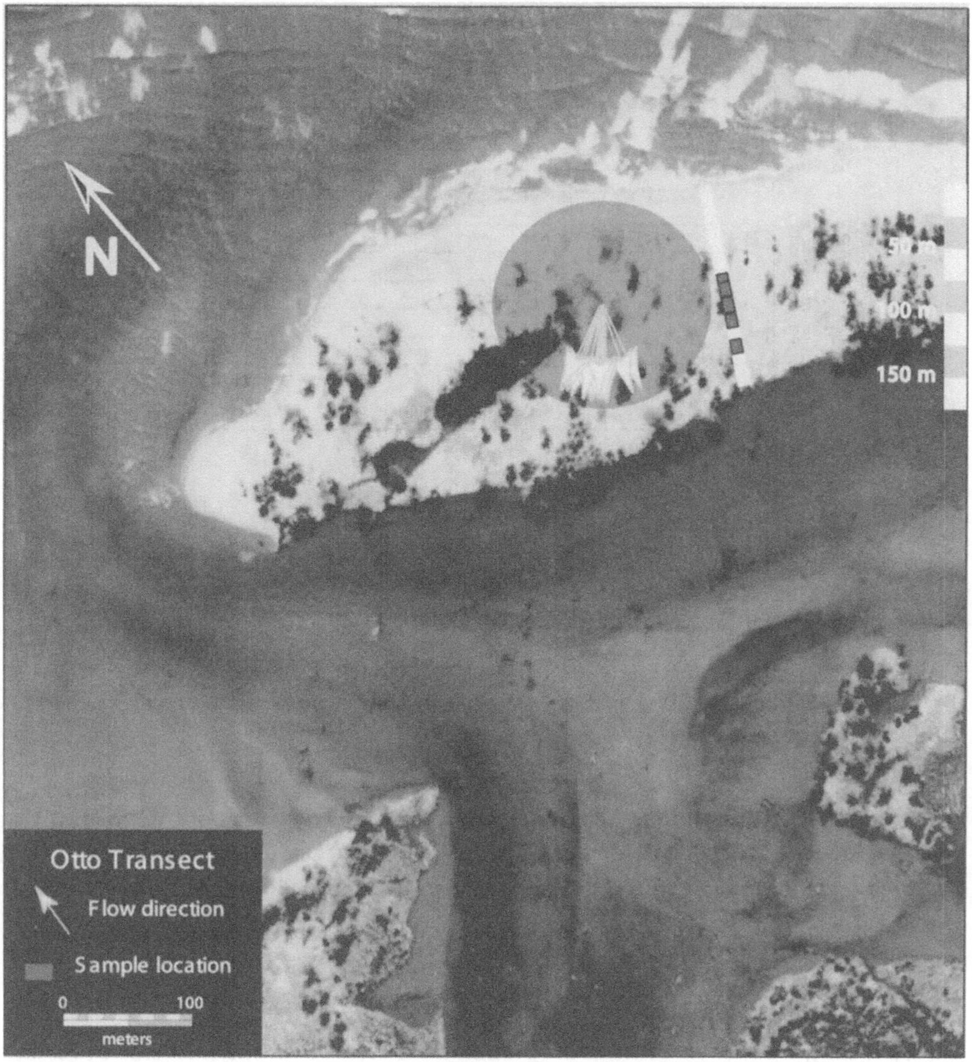

Figure 11
Aerial photograph showing the Otto transect. Photograph taken August 8, 1998 by the PNG National Mapping Bureau.

Sedimentology

Sediment samples of the tsunami deposit were taken from five sites along the Otto transect for laboratory analysis. These sites span only a distance of about 50 m along the center of the spit. In this distance the mean grain size, sorting, and skewness all remain nearly constant across the transect (Fig. 12c). This distance is probably too short to demonstrate horizontal trends in grain size.

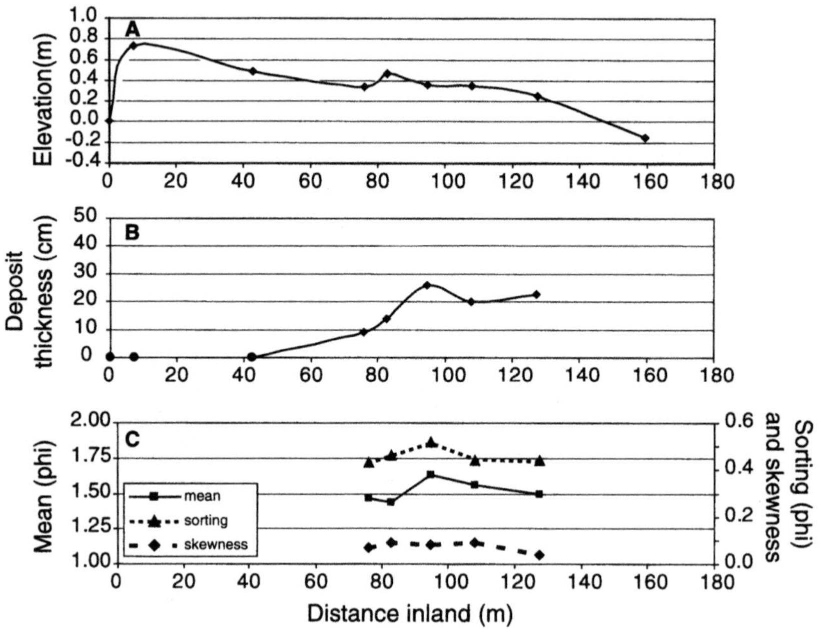

Figure 12
Topography, tsunami deposit thickness, and grain-size statistics of the tsunami deposit along the Otto transect.

Field observations along the Otto transect revealed a tsunami deposit with a massive lower layer from 10–15 cm thick below a thinly laminated upper layer 5–10 cm thick. The laminae showed no signs of ripple cross bedding.

Sissano Transect

Setting and topography

The villages of Sissano and Warapu had been located on the west side of Sissano Lagoon. Warapu, located on the narrow spit between the ocean and the lagoon, was totally destroyed by the tsunami. Sissano was further to the west and set back from the ocean (Fig. 1). Over 1000 people from these two villages lost their lives as a result of the tsunami, and many more were displaced.

The Sissano transect crossed the Sissano village about 3 km west of Sissano Lagoon (Fig. 13). The transect was approximately 600-m long and ended at a small river that flows parallel to the shoreline and is connected to the lagoon.

We interviewed two eyewitnesses who had been in their homes in the Sissano village about 450 m from the shoreline when the earthquake and tsunami occurred. They reported seeing a highly turbulent wave approach that inundated the area all the way to the river. At their house, the wave height was about 1.4 m. The first

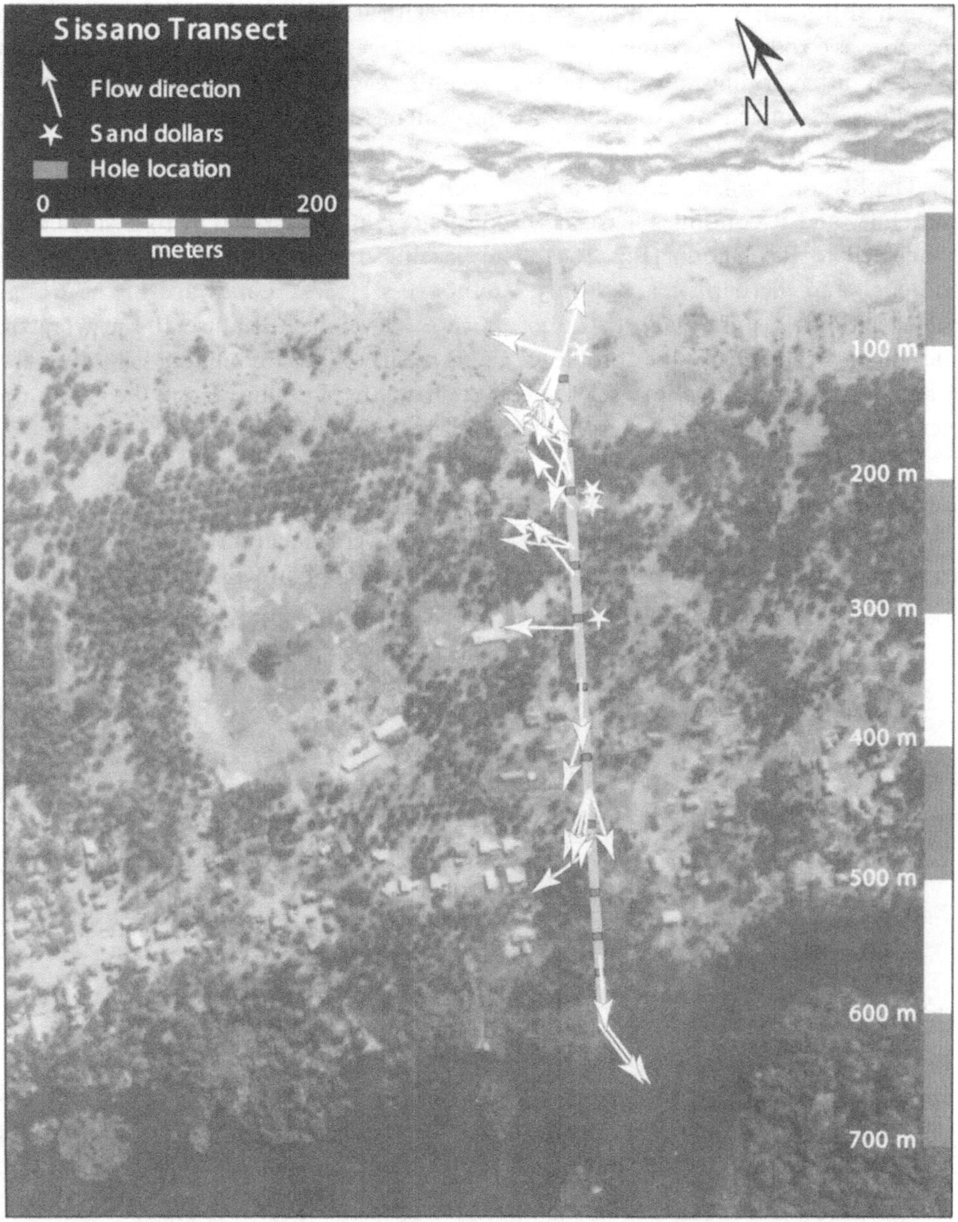

Figure 13
Aerial photograph showing the Sissano transect. Photograph taken August 8, 1998 by the PNG National Mapping Bureau.

wave had little or no return flow associated with it. Soon after the first wave, the second wave rode in over the already flooded village. Likewise, the third wave came in soon after, riding on top of the second wave. After the third wave, the eyewitnesses noticed a return flow as the waters in the flooded village started to recede (Fig. 14).

The Sissano transect crosses a low beach berm and narrow trough close to the shoreline, and it continues across a broad flat plain almost 600 m to a small river connected to the lagoon (Fig. 15a). The beach berm and trough are close to the shoreline and the berm reaches only 1 m above sea level. Landward of the trough the elevation rises gradually to 2.5 m above sea level at a distance 100 m from the shoreline. From there the profile remains nearly flat, undulating slightly, and gradually rising to 3 m above sea level 500 m from the shoreline. From the maximum elevation of 3 m the profile descends toward sea level at the river.

Flow direction and depth indicators

Flow direction indicators were measured at several locations along the Sissano transect and are plotted on the aerial photograph in Figure 13. Indicators suggest

Figure 14
Photograph near the Sissano transect showing two flow directions. The fallen palm trees indicate the direction of the main tsunami inundation (1) and the bent grasses indicate the direction of subsequent return flow (2).

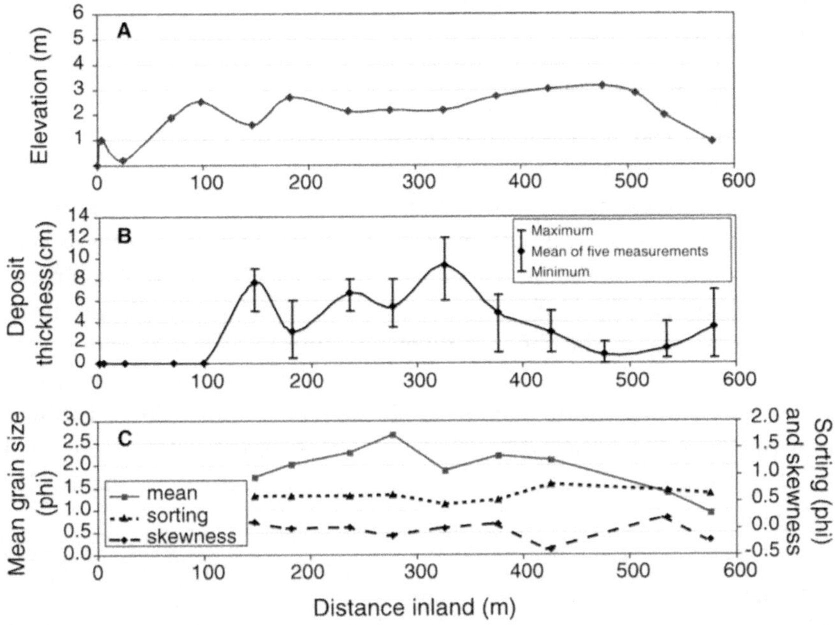

Figure 15
Topography, tsunami deposit thickness, and grain-size statistics of the tsunami deposit along the Sissano transect.

flow in several directions along the Sissano transect. Numerous fallen trees within 200 m of the shoreline represent the primary flow and are all aligned in approximately the southwest direction at 230°. Bent-over grass and piles of debris wrapped around standing trees suggest a secondary flow in the northwest direction at 340°. Other secondary flow indicators were aligned toward the northeast at 40° or toward the north at 350°. We ascribe the primary flow to initial high-energy inundation of the tsunami, and the secondary flow to backwash or return flow. Secondary flow indicators like grasses are easily aligned by even small flows, so they probably indicate the direction of the last flow to which they were subjected. A site 185 m from the shoreline contains both primary (220°) and secondary (350°) flow indicators suggesting two phases of flow at different times (Fig. 14). The return flow indicators all along the transect are aligned toward local topographic lows.

The 1st ITST estimated wave heights along the coast near Sissano of 5–10 m (KAWATA *et al.*, 1999). These estimates came from broken branches near the tops of trees and debris found in the trees. Inland along the transect we estimated the flow depth by measuring the height of debris found in trees and the height of debris piled up against trees or pilings. At a distance of 350 m from the shoreline we observed flow depth indicators at 1.8 m above the ground. Around 475 m from the shoreline we measured depth indicators at 0.8 m, 0.9 m, 1.2 m, and 1.5 m above the ground.

Deposit (small-scale lateral variability)

Within the first 100 m of the shoreline along the Sissano transect we found no evidence of tsunami deposition. By 150 m from the shoreline the tsunami deposit was almost 8 cm thick on a rooted and compact sandy soil. Variability in the tsunami deposit thickness was determined by measuring the thickness along the transect and by measuring the thickness 5 and 10 m away along a line perpendicular to the main transect (Fig. 15b). From 150 to 425 m inland along the Sissano transect the variability in the shore-parallel direction is similar to the variability in the shore-normal direction. In this zone, the average thickness is about 6 cm, the minimum is 0.5 cm, and the maximum is 12 cm. The tsunami deposit thins from a maximum of 12 cm thick at about 350 m to a minimum of 0–2 cm thick at a distance of 474 m from the shoreline. Approaching the river, the tsunami deposit thickens again slightly.

Sedimentology

Field observations revealed more variability and complexity in the sedimentology of the tsunami deposit along the Sissano transect than along the Otto, Arop, or Waipo transects. At Arop in particular, the tsunami deposit is a single fining upward layer. Along the Sissano transect the deposit in some places is a single fining upward layer, but in other places the deposit is composed of multiple layers. At 150–275 m from the shoreline we noted only one layer. Along that stretch, the mean grain size of the deposit decreases landward (Fig. 15c). At 320 m and beyond, the deposit contains two recognizable layers, both of which commonly fine upwards. Mean grain size increases landward along this part of the transect. Except at the location 320 m from the shoreline, only one depth-integrated sample was collected for grain-size analysis at each location. If there were multiple layers deposited from different waves, then a simple trend in grain size of the composite would not be expected.

At the site 320 m from the shoreline, where the deposit was 6-cm thick, samples were collected at 1-cm intervals from the top of the deposit to the base. The mean, sorting, and skewness of the sediment at each depth in the deposit is shown in Figure 16. While the sorting and skewness remain nearly constant with depth in the deposit, the mean grain size clearly shows two fining-upward sequences. Because the mean grain size is reported at the depth of the center of the sample, Figure 16 does not properly show the actual transition between the two layers. In fact, the transition between the two layers is abrupt. The abrupt transition in grain size may or may not represent an erosional surface. We do not know whether or not a part of the lower layer was eroded before deposition of the upper layer.

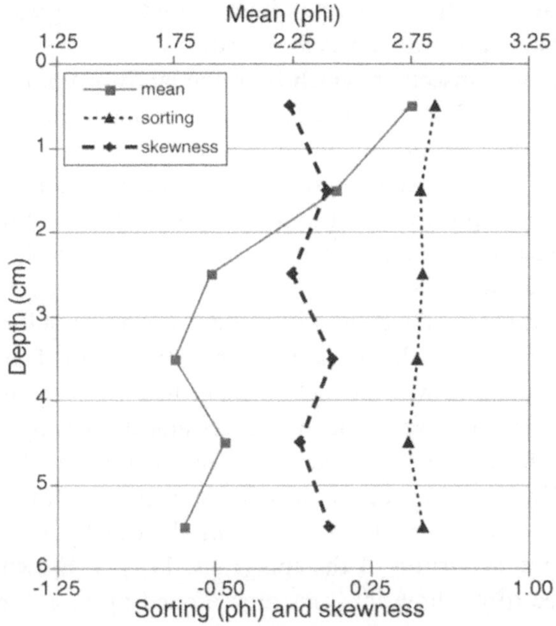

Figure 16
Mean grain size, sorting and skewness at 1-cm intervals of the tsunami deposit on the Sissano transect 320 m from the shoreline. The deposit here contains two fining-upward sequences suggesting deposition out of suspension from two waves.

Discussion/Summary

Shore-normal Trends in Deposit

The tsunami waves that struck PNG approached land with sufficiently high velocities to suspend and transport large quantities of sand. Despite their destructive power, in which trees were uprooted and structures destroyed, the waves left behind a sandy deposit. This sandy layer was deposited as a continuous sheet that varied in thickness depending, in part, on local topography, the presence of obstructions, and the density of vegetation.

A tsunami deposit was found at every site we visited along the northern PNG coast in the vicinity of Sissano Lagoon extending from approximately 100 m from the shoreline across the coastal plain to very near the limit of inundation, a distance of up to 750 m. At Waipo the deposit tapers inland from a maximum thickness near the shoreline. At Arop and Sissano, however, the deposit is more nearly tabular, with a rapid increase in thickness near the coast, a long nearly flat middle section, and then a gradual decrease on the landward side. The maximum thickness is not near the coast, but farther inland. Along the Otto transect the spit is so narrow that it was completely inundated by the tsunami, which flowed a long ways into the lagoon. Therefore, a deposit shape was not determined.

Along each of the transects in which both the landward limit of sand deposition and the landward limit of the tsunami could be identified, the landward limit of the deposit fell short of the limit of tsunami inundation. The sand stopped 40–50 m short of the landward limit of the tsunami at the Arop and Waipo transects. At Otto, the landward limit of the deposit was not found, and at Sissano, neither limit was found. DAWSON (1994) measured similar distances between the deposit and the maximum inundation for the Java tsunami.

Along all four transects, the areas with net erosion as a result of the tsunami were limited to portions of the beach and berm close to the ocean. The width of that zone varied from 50–150 m and was located adjacent to the shoreline. The thickness of sand eroded by the tsunami was difficult to determine accurately, but we estimated it at 10–25 cm using the depth of exposed roots. Only at Warapu did erosion extend far from the beach. There, on a narrow western spit between the ocean and Sissano Lagoon (Fig. 1) erosion occurred not only along the beach and the top of the berm, but also on the central portion of the spit (Figs. 17, 18). Eyewitness accounts and scour exposing tree roots show that this area eroded up to 1.3 m for many tens of meters in the shore-normal direction. Perhaps the wave broke over the top of the spit

Figure 17
Photograph of a tree with exposed roots near Warapu. The depth of the exposed roots and eyewitness accounts suggest 1–1.5 m of erosion of sand from the barrier spit.

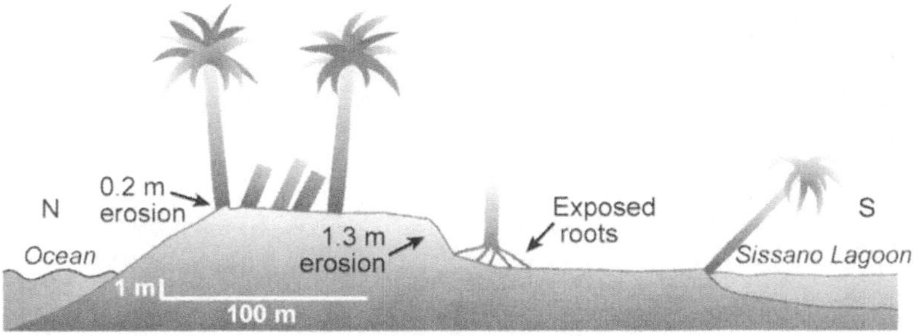

Figure 18
Sketch of profile near Warapu showing erosion on the back side of the barrier spit.

and plunged to the ground resulting in the added scour, or an alongshore flow originated from the lagoon that scoured into the spit, or maybe a hydraulic jump formed over the berm.

The source of the sand transported by a tsunami can be difficult to identify. Several clues help identify the source for the sediment deposited in PNG. First, along both the Sissano and the Arop transects the deposit contains many marine shells, including small sand dollars. The sand dollars must have been transported from at least several tens of meters offshore since sand dollars typically live outside the normal wave breaking zone. Second, too little sand was eroded from the sub-aerial beach to account for the total volume deposited. The volume of sand deposited per unit distance in the alongshore is calculated along the Arop transect to be 36 m^3/m. If the average scour depth was 25 cm, the amount of sand eroded landward of the shoreline was only 14 m^3/m. This is likely a high estimate because the maximum scour across the berm was probably 25 cm, the average would be less. This analysis suggests that sand must have been supplied from offshore. In fact, more than twice as much sand came from offshore as came from the subaerial beach and berm. This result for PNG differs from the observations by SATO *et al.* (1995), who found that deposits of the 1993 and 1983 Japan Sea tsunamis came mainly from the sub-aerial beach and berm.

Alongshore Trends in Deposit

The zone of inundation from the 1998 tsunami covered only about 40 km in the alongshore direction. Within this 40-km long stretch of coast the maximum wave levels varied from 2–3 m near Aitape in the east to nearly 15 m along the spit fronting Sissano Lagoon and back down to 2–3 m near the Bliri River to the west. A section of coast spanning about 20-km long sustained water levels over 10-m high. This focusing of wave energy may reflect several factors, including the proximity of the source disturbance and the focusing nature of the bathymetry on the shelf and slope (MATSUYAMA *et al.*, 1999).

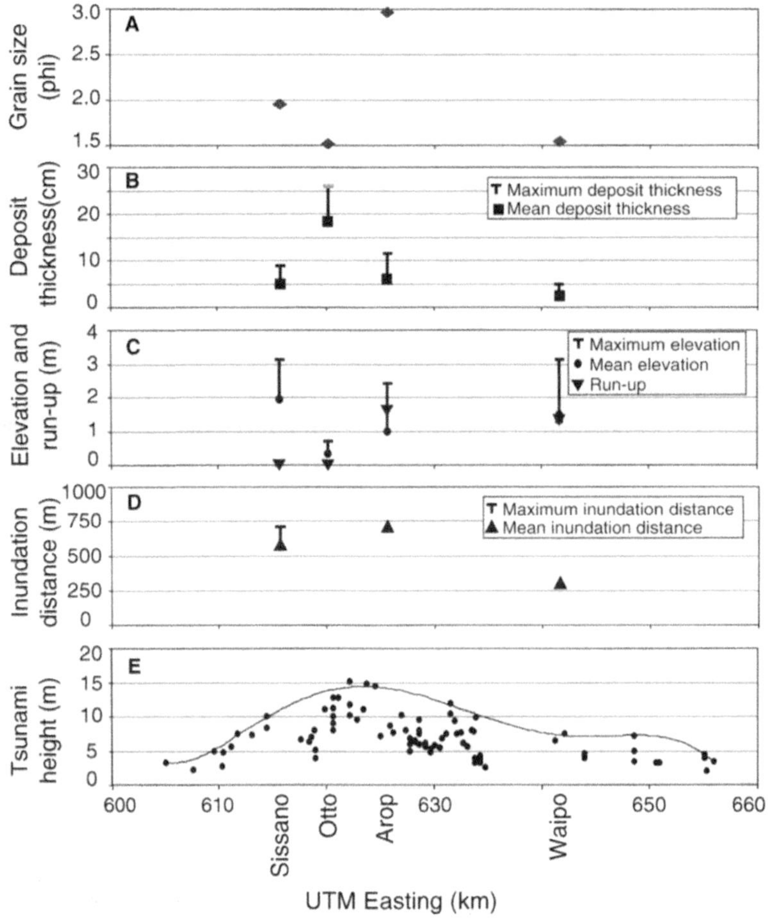

Figure 19

Alongshore variation in A) mean grain size of the tsunami deposit from each transect; B) mean and maximum tsunami deposit thickness; C) average and maximum elevation, and runup; D) inundation distance; and E) tsunami height as reported by KAWATA *et al.* (1999). The smooth curve is a spline through the maximum wave heights.

Tsunami inundation distance, or the distance measured from the shoreline to the landward limit of inundation, varied in a similar manner to the wave height, with the maximum inundation near the lagoon (Fig. 19). The maximum inundation distance that we observed over land, at the Arop transect, measured about 750 m. The tsunami penetrated even farther where the waves traveled over the lagoon. Eyewitnesses reported that debris carried by the tsunami washed up against the southern shores of the lagoon up to 4–5 km from the spit, or nearly five times farther over water than over land.

Tsunami runup, or the elevation relative to sea level at the landward limit of inundation, does not appear related to tsunami wave height (Fig. 19). By contrast, for short-period wind waves on beaches, runup is found to depend on beach slope, wave height, and wave period (BATTJES, 1974; HOLMAN and SALLENGER, 1985). Moreover, based on laboratory and numerical studies, SYNOLAKIS (1991) found that the runup from unbroken long-period tsunami waves on a plane beach depends on beach slope and wave height. In Papua New Guinea, runup of the tsunami was complicated by, among other things, irregular topography; none of the sites have a simple topographic profile of a monotonically increasing elevation from the shoreline to the limit of inundation. In other words, none of the sites consist entirely of a plane beach. Along much of the coast the maximum elevation is at the beach berm or along a beach ridge, then the elevation decreases again in the landward direction. In fact, even though the wave height was over 10 m at the Otto and Sissano transects, the runup is close to 0 m because the tsunami inundation extended to the far side of a lagoon and a river leading to the lagoon, respectively. The complicated topography, varying roughness across sand, mud, and vegetation, and varying degree of infiltration into the soil below result in no clear relationship between wave height and runup (Fig. 19).

On a regional scale, tsunami deposit thickness does appear related to wave height (Fig. 19). Deposit thickness (both the maximum across the transect and the mean) is smallest at Waipo where the wave height was smallest and greatest at Otto and Arop, where the wave height was greatest. The thickness of the sand layer deposited by the tsunami varies locally as well as regionally in the alongshore direction. Locally, the tsunami deposit thickness varies about twofold in response to variations in local topography. In general, deposit thickness increases slightly in local depressions and decreases over local highs in the topography.

Mean grain size of the tsunami deposit seems unrelated to deposit thickness or wave height (Fig. 19). The grain size of the deposit more likely depends on the grain size of the sediment available to be transported. Based on estimates of the water velocity in the tsunami from the 1st ITST (1998), all of the sediment grain sizes on the PNG coast within the study area are easily transported.

Vertical Trends in Deposit

Sediment transport and depositional processes can be inferred from the detailed vertical trends in the tsunami deposit. Most of the deposits examined are vertically graded, meaning the grain size varies with depth in the deposit. Most sites had one continuously graded bed with the coarsest material at the base and fining upward. A few sites had multiple normally graded beds one atop another. Each graded bed implies deposition from suspension as flow velocity decreases. Normal grading is a common sedimentary structure associated with deposition from a turbulent suspension. For example, normal grading is a common feature of turbidites, the deposits formed by deposition from a turbidity current (SANDERS, 1977). In a turbulent flow there is a sediment grain-size gradient, with coarse material proportionally concentrated near the bed and finer material more evenly distributed throughout (SMITH, 1977; GELFENBAUM and SMITH, 1986). As the flow wanes and the sediment settles out of suspension, the coarse material will settle first and the finer material will follow. At Arop, the normally graded deposit suggests deposition from a single event or wave. At Sissano, the two layers of normally graded sediment imply deposition from two tsunami waves.

At both locations, the number of graded beds is less than the number of waves reported by eyewitnesses. Throughout the region, eyewitnesses consistently reported three waves inundating the coast. Apparently, each wave that struck the coast did not deposit a sandy layer. Or else, each wave deposited a sandy layer that was subsequently resuspended with the passage of the next wave. Unfortunately, there does not appear to be a one-to-one relationship between the number of waves inundating the coast and the number of sandy layers in a deposit.

Conceptual Model of Tsunami Sedimentation

The detailed observations of erosion and deposition from the PNG tsunami can be used to develop a conceptual model of tsunami sedimentation. In general, the thickness and grain-size distributions found in tsunami deposits are the result of the sediment transport during tsunami inundation. Based on conservation of mass, the rate of change in bed elevation (erosion or deposition) is related to the rate of change of storage of sediment in the water column and to the horizontal gradient of sediment transport, and can be expressed by

$$\frac{\partial \eta}{\partial t} = -\frac{1}{C_b}\left(d\frac{\partial \overline{C}}{\partial t} + \frac{\partial Q}{\partial x}\right) , \tag{1}$$

where η is the bed elevation, C_b is the sediment concentration in the bed (≈ 0.6), d is the flow depth, $\overline{C}$ is the mean suspended-sediment concentration, and Q is the sediment flux in the shore-normal (x) direction (MIDDLETON and SOUTHARD, 1984). The sediment flux is the product of the velocity and the sediment concentration. The erosion of the beach and berm along the seaward side of each of the transects suggests that the sediment flux within the tsunami was increasing in the landward direction all the way to where deposition began, at a point some 50–150 m inland from the shoreline. This transition from erosion to deposition most likely coincides with the location of the maximum horizontal velocity along each transect.

Deposition of sediment from the tsunami results from a combination of both terms in Equation (1). Both processes, a decrease in sediment storage in the water column as sediment settles out of suspension and convergence of the horizontal sediment flux, take place to varying degrees at different times and locations. Isolating the dominant term for any given situation is not simple. If the horizontal convergence of sediment flux is the dominant process leading to deposition, then the grain size in the deposit should decrease with landward distance. If the decrease in sediment storage in suspension over time controls deposition, then the deposit should be vertically graded with a fining upward sequence. Because the tsunami deposit at both the Arop and the Sissano transects vary little in grain size in the shore-normal direction, and because fining upward sequences are common, settling from suspension was the dominant process leading to deposition at these sites in Papua New Guinea.

Conclusions

Shore-normal transects show signs of flow, erosion, and deposition from the Papua New Guinea tsunami of 17 July, 1998:

1. A distinct deposit was left by the tsunami. The deposit was distributed continuously across the coastal plain and varied in thickness. The average thickness along the transects was 8 cm and the maximum thickness was 26 cm. Rip-up clasts of mud and soil indicate erosion of the underlying soil.
2. Upward fining sediments suggest sediment fell out of suspension before the water retreated. Some deposits show multiple graded beds suggesting multiple waves.
3. The tsunami accelerated across the beach and berm. Erosion of the beach and berm extended 50–150 m inland from the shoreline.
4. As much as 2/3 of the tsunami deposit came from offshore, as shown by clam shells and sand dollars in the deposit and by comparison between volumes of sediment deposited and eroded on the sub-aerial beach and coastal plain.
5. The incoming tsunami flowed nearly perpendicular to shore, but a slower backwash, or return flow, was directed obliquely to the shore, toward local lows in topography.
6. Tsunami deposit thickness and inundation distance varied along the coast. They were greatest where the tsunami wave height was greatest and decreased with decreasing wave height.

Acknowledgements

Funding came from the U.S. Agency for International Development, Office of Foreign Disaster Assistance and from the USGS Coastal and Marine Geology Program. Laura Kerr of the USGS carried out laboratory analysis and prepared the figures. Members of the 2nd ITST included Dr. Yoshinobu Tsuji, University of Tokyo; Dr. Norio Maki, Earthquake Disaster Mitigation Research Center; Dr. Hideo Matsutomi, Akita University; Han She-Sub, Korea Meteorological Administration; and Dr. Russell Perembo, University of Papua New Guinea. Professor Hugh Davies, University of Papua New Guinea sponsored our trip and assisted with logistics and important historical information. Michael Nongkas, University of Papua New Guinea did untiring work in the field. David Freyne, with the National Mapping Bureau, obtained the aerial photography. Chris Amunty and Willie Martin safely ferried us back and forth from Aitape to Sissano Lagoon. The Aitape Tsunami Disaster Control Officials generously provided hospitality under difficult circumstances. Finally, Brian Atwater and Eric Geist provided thorough and constructive technical reviews of this manuscript.

References

Atwater, B. F. (1987), *Evidence for Great Holocene Earthquakes along the Outer Coast of Washington State,* Science *236*, 942–944.

Battjes, J. A. (1974), *Surf Similarity*, Proc. 14th Coastal Engineering Conference, Am. Soc. Civil Engrs., 466–479.

Bourgeois, J., and Minoura, K. (1997), *Palaeotsunami studies—Contribution to mitigation and risk assessment.* In *Tsunami Mitigation and Risk Assessment* (ed. V. K. Gusiakov) pp. 1–4.

Carver, R. E. *Procedures in Sedimentary Petrology* (New York, John Wiley and Sons 1971) 653 pp.

Davies, H. (1999), *Tsunami, PNG 1998*, University of Papua New Guinea Press, Port Moresby, 49 pp.

Dawson, A. G. (1994), *Geomorphological Effects of Tsunami Run-up and Backwash*, Geomorphology *10*, 83–94.

Geist, E. L. (2000), *Origin of the 17 July 1998 Papua New Guinea Tsunami: Earthquake or Landslide?* Seismol. Res. Lett. *71* (3), 344–351.

Geist, E. L. (2001), *Reply to Comments by E. A. Okal and C. E. Synolakis on "Origin of the 17 July 1998 Papua New Guinea Tsunami: Earthquake or Landslide?"* Seismol. Res. Lett. *72* (3), 367–372.

Gelfenbaum, G. and Smith, J. D. (1986), *Experimental Evaluation of a Generalized Suspended-Sediment Transport Theory.* In Shelf Sands and Sandstones (Knight, R. J. and McLean, J. R., eds.), Canadian Society of Petrolium Geologists, Memoir, pp. 133–144.

Goff, J., Chague-Goff, C., and Nichol, S. (2001), *Palaeotsunami Deposits: A New Zealand Perspective*, Sedimentary Geology *143*, 1–6.

Holman, R. A. and Sallenger, A. H. (1985), *Setup and Swash on a Natural Beach*, J Geophys. Res *90*, 945–953.

Jaffe, B. E. and Gelfenbaum, G. (2002), *Using Tsunami Deposits to Improve Assessment of Tsunami Risk*, Solutions to Coastal Disasters '02 Conference Proceedings, 836–847.

Kawata, Y., Benson, B., Borrero, J. Davies, H., de Lange, W., Imamura, F., Letz, H., Nott, J., and Synolakis, C. E. (1999), *Tsunami in Papua New Guinea was as Intense as First Thought*, EOS, Trans. Am. Geophys. U. *80*, 101–105.

Matsuyama, M., Walsh, J. P., and Yeh, H. (1999), *The Effect of Bathymetry on Tsunami Characteristics at Sissano Lagoon, Papua New Guinea*, Geophys. Res. Lett. *26* (23), 3513–3516.

McSaveney, M. J., Goff, J. R., Darby, D. J., Goldsmith, P., Barnett, A., Elliott, S., and Nongkas, M. (2000), *The 17 July 1998 Tsunami, Papua New Guinea: Evidence and Initial Interpretation*, Marine Geology *170*, 81–92.

Middleton, G. V. and Southard, J. B. (1984), *Mechanics of Sediment Movement*, SEPM Short Course Number 3, 2nd Edition, Providence, 401 pp.

Nishimura, Y. and Miyaji, N. (1995), *Tsunami Deposits from the 1993 Southwest Hokkaido Earthquake and the 1640 Hokkaido Komagatake Eruption, Northern Japan*, Pure Appl. Geophys. *144* (3/4), 719–733.

Okal, E. A., and Synolakis, C. E. (2001), *Comment on "Origin of the 17 July 1998 Papua New Guinea Tsunami: Earthquake or Landslide?" by E.L. Geist*, Seismol. Res. Lett. *72* (3), 362–366.

RIPPER, I. D., LETZ, H., and MOIHOI, M. (1998), *Pre-tsunami Large Earthquakes of the Aitape Region North Coast Mainland of Papua New Guinea*, Papua New Guinea Geological Survey Report 98/7.

SANDERS, J. E. (1977), *Primary Sedimentary Structures Formed by Turbidity Currents and Related Resedimentation Mechanisms*, SEPM Special Publication *12*, 192–219.

SATO, H., SHIMAMOTO, T., TSUTSUMI, A., and KAWAMOTO, E. (1995), *Onshore Tsunami Deposits Caused by the 1993 Southwest Hokkaido and 1983 Japan Sea Earthquakes*, Pure Appl. Geophys. *144* (3/4), 693–717.

SHI, S., DAWSON, A. G. and SMITH, D. E. (1995), *Coastal Sedimentation Associated with the December 12th, 1992 Tsunami in Flores, Indonesia*, Pure Appl. Geophys. *144* (3/4), 525–536.

SMITH, J. D. (1977) *Modeling of sediment transport on continental shelves*. In *The Sea*, Vol. 6, (E. D. Goldberg, I. N. McCave, J. J. O'Brien, and J. H. Steele, eds.) (Wiley-Interscience, New York) pp. 539–577.

SYNOLAKIS, C. E. (1991), *Tsunami Run-up on Steep Slopes: How Good Linear Theory Really Is*, Natural Hazards *4*, 221–234.

TAPPIN, D. R., MATSUMOTO, T., WATTS, P., SATAKE, K., MCMURTRY, G. M., MATSUYAMA, Y., LAFOY, Y., TSUJI, Y., KANAMATSU, T., LUS, W., IWABUCHI, Y., YEH, H., MATSUMOTU, Y., NAKAMURA, M., MAHOI, M., HILL, P., CROOK, K., ANTON, L., and WALSH, J. P. (1999), *Sediment Slump Likely Caused 1998 Papua New Guinea Tsunami*, EOS, Trans. Am. Geophys. U. *80*, 329–340.

TAPPIN, D. R., WATTS, P., MCMURTRY, G. M., LAFOY, Y., and MATSUMOTO, T. (2001), *The Sissano, Papua New Guinea Tsunami of July 1998—Offshore Evidence on the Source Mechanism*, Marine Geology *175*, 1–23.

Pure appl. geophys. 160 (2003) 2001–2031
0033–4553/03/112001–31
DOI 10.1007/s00024-003-2417-x

Pure and Applied Geophysics

Mitigation Lessons from the July 17, 1998 Papua New Guinea Tsunami

LORI DENGLER[1] and JANE PREUSS[2]

Abstract—The July 17, 1998 tsunami killed over 2,100 people, injured at least 800 severely enough to require hospitalization, permanently displaced over 10,000 and disrupted the social and economic framework of the coastal communities of eastern Saundaun Province, Papua New Guinea. Initial response to the disaster was delayed 16 hours because of the failure to communicate the magnitude of the impact outside the affected area. Once the scope of the disaster was known, international assistance was rapid and substantial. Medical teams, supplies, air transport and mobile field hospitals were provided to assist national medical personnel and facilities. Seven care centers were established for the displaced survivors. Nineteen countries and 17 NGOs (Non-Governmental Organizations) donated money, relief and rebuilding supplies valued at over 6 million US$ in addition to substantial contributions from national recovery funds and the Catholic Diocese. The three Malol villages, two Arop, four Sissano and the Warupu village were permanently abandoned and new villages established inland. Closure of the lagoon and restriction of coastal fishing impacted adjacent villages as well as tsunami survivors. By the second anniversary of the tsunami construction of schools, water and sanitation systems and some roads had been completed. Survivors were provided tools and building supplies to construct new homes, canoes and fishing equipment. Relief and recovery efforts were complicated by coordination difficulties among the many responding agencies and organizations, the disruption of daily routines for both survivors and villages hosting the care centers, and the intrusion of outside aid workers in a region that had been isolated before the disaster. Adaptation to the new village sites has been difficult due to their inland locations that are hotter, more insect-infested and have water and sanitation difficulties. The high number of casualties from the tsunami was the result of several factors:

Population concentrated in the area of maximum tsunami impact.
Date and time of occurrence.
Siting of villages on vulnerable sand spits.
Failure of residents to self-evacuate after feeling the earthquake.
Substantial delay in mobilizing response.

However, further losses were likely reduced because of the rapid deployment of air transport and medical teams, and massive international relief support. The tsunami has affected long-term tsunami mitigation in the Sissano region through the relocation of villages, heightened awareness and education programs. It has also had a significant effect on mitigation in other countries by illuminating the potential dangers of landslide-generated tsunamis and in media coverage of tsunami hazards.

Key words: Impacts, response, relief, recovery, mitigation.

[1] Department of Geology, Humboldt State University, Arcata, CA 95521, U.S.A.
[2] GeoEngineers, Redmond, WA 98052, U.S.A.

Introduction

On July 17, 1998, a magnitude M_w 7.0 earthquake occurred near the coast of eastern Saundaun Province, Papua New Guinea (Fig. 1). It was closely followed by a tsunami that reached heights of 10–15 m, uprooting trees and damaging structures over 500 m inland. The tsunami destroyed or damaged villages along a 45-km stretch of coastline in the area of Sissano Lagoon, killing at least 2,100 people and displacing nearly 10,000, making it Papua New Guinea's third most deadly natural disaster in the 20th century (EM-DAT, 2001). This tsunami is important because of the large loss of life, the relatively moderate magnitude of the triggering earthquake compared to the size of the waves (discussed elsewhere in this volume) and the impact it has had on tsunami hazard assessment and mitigation. This paper examines results of post-tsunami field studies in Papua New Guinea and reports from international aid agencies to describe strengths and weaknesses in emergency response, relief and recovery efforts and to identify factors that contributed to the high number of casualties. It also looks at how the event has effected mitigation both within Papua New Guinea and in other tsunami-prone areas.

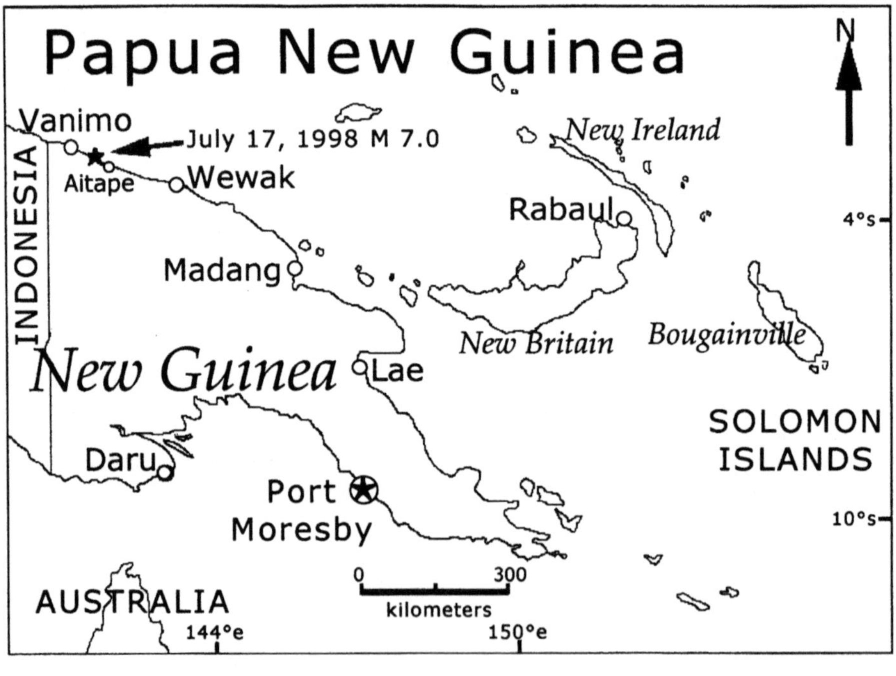

Figure 1
Location map. Star is the epicenter of the July 17, 1998 earthquake.

The Setting: Physical, Social and Political Context

The unique physical setting and the cultural and political history of Papua New Guinea influenced the magnitude of the disaster, the nature of response and continue to affect the recovery process. The country's present political situation is strongly influenced by its colonial past. Portuguese, Spanish, Dutch, British and German colonial interests all made claims on parts of the island of New Guinea. In the 20th century, Australia gained jurisdiction over the eastern half of the island, taking over administration of the British colonies and occupying the German colonies during World War I. Papua New Guinea became a country when Australia granted independence to its former colonies on September 16, 1975.

At present there are just over 5 million people in Papua New Guinea and 24,000 expatriates, 7,500 of whom are refugees from Irian Jaya living in camps in the western provinces. About 6,500 expatriates are Australian; many of whom work on short- and intermediate-term contracts for international companies, at the University and as government advisors. Significant portions of the expatriate population are missionaries. Mission workers have been in Papua New Guinea for over a century and represent a broad spectrum of Christian denominations. Churches and missionaries have a major role in providing health and education services, particularly in the more remote parts of the country.

Papua New Guinea is an independent democracy within the British Commonwealth. The parliament tends to change substantially at each election. Bill Skate, who became Prime Minister in 1997, resigned two years later following allegations of loan improprieties. Urban unemployment is high and the currency, devalued 12% in 1993, has been declining ever since. Economic and political considerations led to the loss of all permanent positions in several government agencies including the National Disaster and Emergency Services (NDES), the agency responsible for disaster management at the national level, in May 1998. It was operating with temporary staff throughout the immediate response and relief periods of the Sissano tsunami. NDES had a 24-hour radio alert system that had been curtailed before the tsunami due to budget issues (DENGLER and PREUSS, 1999).

Saundaun (formerly West Sepik) Province is one of the least developed areas of Papua New Guinea. Only the provincial capital of Vanimo (population 8000) and the town of Aitape (population 4000) have infrastructures including power, permanent roads, telecommunications and hospitals. The economy of the coastal villages is based on fishing, hunting and gardening, and harvesting crops such as sago and cassava, taro and yams grown in garden plots. Villages exchange food and raw materials among themselves and other villages in the region (WELSCH and TERRELL, 1998) but are isolated from the provincial or national economy. Most of the coast of Saundaun Province is sparsely populated but people concentrated in the vicinity of Sissano Lagoon because of its abundant and diverse community of fish and benthic invertebrates. The lagoon supplied most of the protein requirements of the estimated

12,000 people living in the region (JENKINS, 1997). Although the Sissano Lagoon communities might be described as underdeveloped or primitive from the perspective of outsiders, the village society was complex, functional, culturally rich and economically self-sufficient (WELSCH and TERRELL, 1998).

The coastal land near Sissano Lagoon is generally flat, low-lying and densely-vegetated with sago and coconut palms and casuarina trees near the coast, and mangroves on the lagoon shore. The water table is at or very near the surface, especially during the rainy season. Mosquitoes and malaria have been an endemic problem in the region. Homes are constructed of traditional materials on elevated posts. Four kinship groups, Sissano, Warupu, Arop and Malol, were settled in coastal villages before the 1998 tsunami (Fig. 2). The Arop villages were on the south spit, an 8-km long, 150-m wide strip of sand and the Warupu were on the much shorter north spit; other villages were near the beach within 400 m of the ocean on land only 1 to 3 m above sea level. Transportation in the region is by foot and by boat (often in hand-carved canoes) using both ocean and the Sissano and Malol lagoons. Grass landing strips provide access to small planes at Sawmill, Sissano and Romei villages. Radio communications were provided by the Catholic missionaries at the village sites and a Sub-District Headquarters in Sissano and were not monitored on a 24-hour basis. The Catholic Diocese provided schools and teachers, funded by the government, through missionary facilities at Sissano, Arop, Warupu

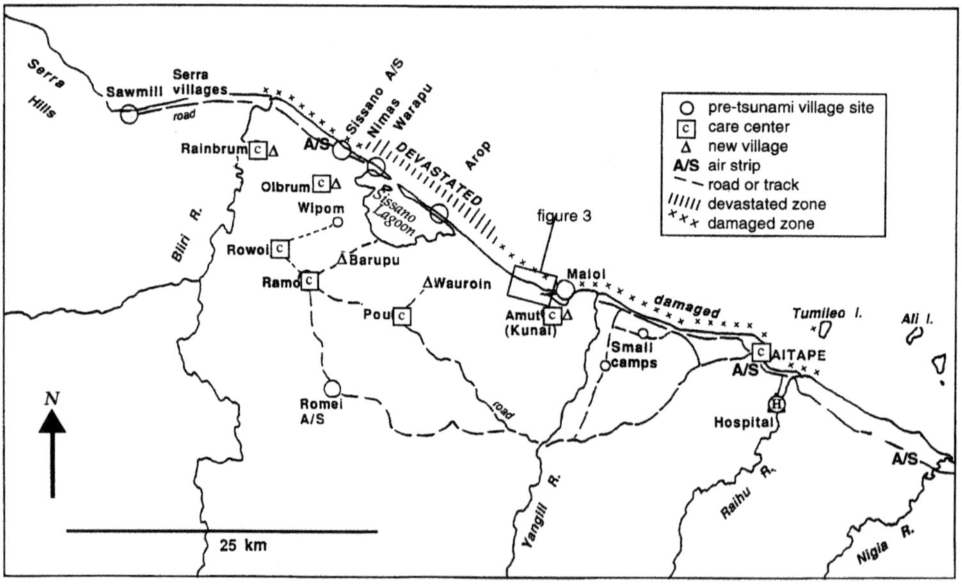

Figure 2
Location map of villages affected by the July 1998 tsunami. After Hugh DAVIES, 1998. Rectangle near Malol is where the Air Photo in Figure 3 is taken.

and Malol. The Diocese also ran the closest Basic Level Services (BLS) hospital, Raihu, in Aitape. The hospital had been closed the afternoon of the tsunami due to delays in receiving funding and all the patients had been moved out before July 17.

Natural and human disasters are common in Papua New Guinea. At the time of the 1998 earthquake and tsunami, International Relief Agencies were responding to two disaster situations: an El Niño drought followed by La Niña flooding that killed 70 and affected 1.2 million people in 1997–1998, and refugees from the Bougainville rebellion that began in 1989 (EM-DAT, 2001). Landslides in 1991 killed 200, an earthquake in 1993 killed 53, floods in 1992 affected 90,000 people and an eruption in 1994 displaced over 150,000 people and destroyed the town of Rabaul (EM-DAT, 2001). A permanent International Federation of the Red Cross (IFRC) supply depot was operating in Lae at the time of the Sissano tsunami.

The Earthquake

The earthquake occurred at 08:49 UTC (6:49 PM local time) July 17, 1998. It was felt widely in coastal communities in Saundaun and East Sepik provinces from at least Vanimo to Wewak. In the Aitape region, it was described as strong enough to make walking difficult and to knock objects over (RIPPER *et al.*, 1999). Some observers in Arop reported cracks forming along the beach and sand blows with water squirting out of the sand (DEPT. OF THE ARMY, 1998; WELSCH, personal communication). Statues in front of the Sacred Heart Church at Sissano were broken and the shaking caused some damage to a roof brace and cracks in walls (DAVIES, 1998a,b). The earthquake was perceived as strong by the population but all adults had felt similar earthquakes in an area where 21 earthquakes of MMI intensity V or greater have occurred since 1900 (RIPPER *et al.*, 1998). The mainshock was followed by two strong aftershocks at 09:09:35 ($m_b = 5.6$) and 09:10:02 ($m_b = 5.9$), likely perceived as a single event by persons in the area. Few in the area connected groundshaking with the likelihood of a tsunami. The exception were some elderly people who reportedly paddled canoes to the back of the lagoon or climbed trees because of their own past experiences or stories they had heard relating earthquakes and tsunamis (DENGLER and PREUSS, 1999).

The Tsunami

Effects of the tsunami were observed along a 60-km long reach of the coast from as far west as Wuntung on the border with Irian Jaya to Tumleo Island offshore of Aitape (KAWATA *et al.*, 1999a,b). Impacts were concentrated along the 40-km zone from Aitape to the Sissano villages (Fig. 2). Because the closest tide gage was 640 km away, the near-source characteristics of the tsunami can only be determined from field studies and by interviews with survivors. A number of international tsunami field teams visited the region in the months after the tsunami. The first International Tsunami Survey Team (ITST) visited the region two weeks after the tsunami to measure runup heights from available high-water marks (KAWATA *et al.*, 1999a,b). A reconnaissance team from the New Zealand Society for Earthquake Engineering visited the area a month after the tsunami to examine the effects of the tsunami on structures and landforms and to analyze sediment deposits (MCSAVENY *et al.*, 2000; GOLDSMITH *et al.*, 1999). A UNICEF provincial team visited the region August 18–25 to assess the situation of children and families affected by the disaster (UNICEF, 1998). A second ITST arrived in late September to study sedimentation, erosion and flow characteristics of the tsunami (GELFENBAUM *et al.*, 2001). Robert Welsch, an anthropologist, had studied the peoples of the Sepik coast over the past decade and revisited the area three times since the tsunami (WELSCH and TERRELL, 1999; WELSCH, personal communication). An Earthquake Engineering Research Institute (EERI) team spent 10 days in November 1998 interviewing survivors, government personnel and responders to examine societal response and issues that exacerbated impacts (DENGLER and PREUSS, 1999). A Japanese team visited the area in March 1999 to examine response and recovery issues (MAKI, personal communication). Hugh Davies of the University of Papua New Guinea made repeated trips to the Aitape region, the affected villages and the care centers to speak with survivors (DAVIES *et al.*, 2001). Professor Davies has published many of the survivor accounts in "Earth Talk," a weekly column in the Port Moresby newspaper The National (DAVIES, 1998a). The summary of the tsunami characteristics given below has been drawn from the survey teams' results and Davies' reports.

About 20 minutes elapsed between the earthquake and the arrival of the first wave at most coastal locations. Only in the Arop area did survivors not report feeling the strong aftershock before the tsunami hit. Many survivors described a significant initial drawdown of the water surface, exposing 50 to 100 m of sea floor. As in many past tsunami disasters, some people went out onto the exposed platform to gather marine life. The first positive wave was preceded by extremely loud sounds, variously described as a jet plane, helicopter or a bomb. Many people went down towards the beach to determine what the sounds were. At Arop village, some survivors described a blast of air strong enough to knock people over preceding the wave. Survivors from the Warupu and Arop villages described the subsequent positive wave as a breaking wave or bore that broke down on top of the villages, sweeping over the spit and carrying debris into the lagoon. Most survivors described three significant waves. Many declared the first as the strongest but some said the second was larger and produced more damage. Most agreed that the third wave was the smallest. People described the second wave as riding on the back of the first and the entire duration of damaging wave activity was probably less than 30 minutes.

The waves uprooted trees, snapped tree trunks and piled debris as far as 500 m from the coast on land (Fig. 3). Damage extended further from the coast in the

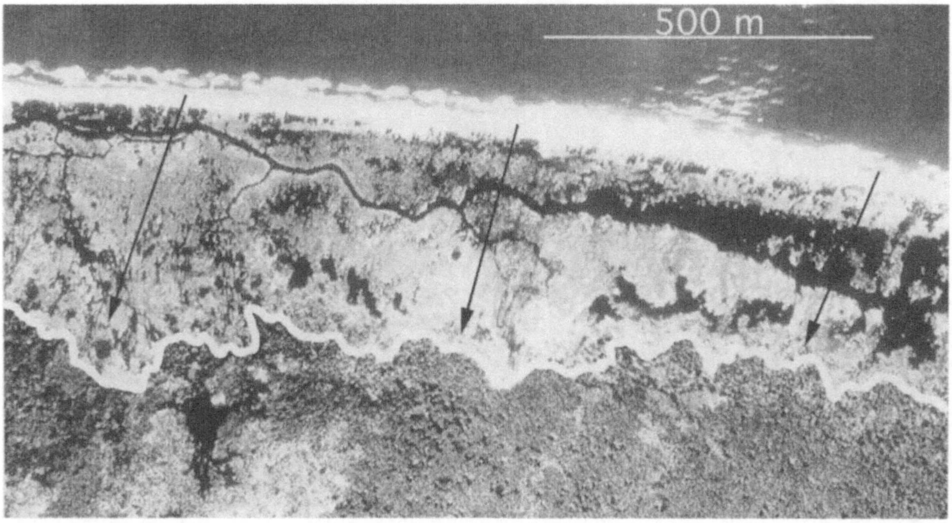

Figure 3
Air photo taken between Arop villages and Malol a month after the tsunami. Before the tsunami, transportation from the Arop villages to Malol and Aitape was mainly by boat through the lagoon at the right. The tsunami penetrated about 0.5 km inland in this area. The white line shows the limit of uprooted vegetation; black arrows are the direction of incoming waves. Photo: National Mapping Bureau of Papua New Guinea.

lagoon where mature mangroves were snapped off 1–2 m above the water level at distances as far as 1,300 m from the coast (Fig. 4). Water heights estimated from watermarks and debris left in trees suggest an average flow depth of about 10 m along the sand spit separating Sissano Lagoon from the ocean. At several locations, peak heights approached 15 m (KAWATA *et al.*, 1999a,b; BORRERO, 2001). The zone of very high elevations was narrowly focused along the spit and heights rapidly diminish away from this zone. The lagoon side of the spit was as dangerous as the ocean side; 1 to 2 meters of material were eroded, probably by a combination of turbulent tsunami wave action and high current velocities within the lagoon. Tree roots were exposed and undermined causing many to topple. People described being swept into the lagoon along with trees, houses and animals. Many of the survivors were carried to the far side of the lagoon where they became entangled within and sometimes impaled on broken mangroves. Although there are no photographs or videos of the tsunami, post-event drawings done by survivors vividly depict a chaotic debris-filled mix of trees, houses and animals (Fig. 5).

Major tsunamis have been rare in Papua New Guinea in historic times. Only two tsunamis in the historic record produced wave heights comparable to the 1998 event. In 1888, collapse of Ritter Volcano off New Britain Island produced 12 to 15 m waves and in 1930 a tsunami associated with a magnitude 5.8 earthquake produced 10.5 m waves in the Ninigo Island group (DAVIES, 1998a,b; HTDB/PAC,

Figure 4
Mangrove swamp on the back lagoon shore opposite the Arop villages, about 1 km from the open coast. Photo: L. Dengler.

2001). Six tsunamis are noted for the Aitape-Vanimo coastline in the past century, three of which may have caused damage (HTDB/PAC, 2001). The magnitude 7 December 1907 earthquake caused subsidence that destroyed the original Warupu village site and formed the present-day outlines of Sissano Lagoon (WELSCH, 1998). A tsunami caused by that earthquake may have reached 2 to 4 m in Aitape (EVERINGHAM, 1977). In September 1935, a magnitude 7.9 earthquake produced waves described by MCCARTHY (1963) as extending hundreds of yards inland in the Aitape region washing over much of the spit. A tsunami in February 1951 may have inundated the sand spit and caused one death (HTDB/PAC, 2001; WELSCH, personal communication). There is no question that the 1998 tsunami is unusual in the historic record. The relatively modest magnitude of the earthquake and the very large wave heights concentrated over a small area are difficult to explain and have been a matter of debate in the scientific community (TAPPIN *et al.*, 1999, 2001; GEIST, 2000, 2001; OKAL and SYNOLAKIS, 2001; SYNOLAKIS *et al.*, in press). For the people living in the Sissano Lagoon area, it was an unprecedented event.

Figure 5
Drawing by a child survivor of the tsunami. Students of Dr. Gillian Hills (University of Papua New Guinea) collected a series of drawings as part of post-tsunami trauma counseling. The straight lines are house stakes that were tossed about in the water. The circles show "whee-wheelum," a pidgin term for being caught in a large wave and spun around. The monsters in the lower right represent bad things in the deep. Photo: L. Dengler.

Direct Impacts

The tsunami affected 10 villages from the mouth of the Bliri River east of the Sierra Villages to the town of Aitape (Fig. 2). Approximately 12,000 people lived along this coastal stretch prior to the disaster. The tsunami occurred just after 7 PM local time on a Friday evening at the beginning of a four-day national holiday. Population in the area swelled with students home for holiday and workers from the cities or logging and mining operations returning to their home villages, and may have approached 15,000 at the time of the tsunami. The most severe devastation was centered on the east and west spits bounding Sissano Lagoon where the village of Warupu and the Arop villages (Arop 1 and 2) were completely destroyed. Nimas, the easternmost of the Sissano villages to the west of the lagoon was also destroyed. The other Sissano villages (Maindron, Mainya, Amsor) were heavily damaged, as were the Malol villages about 10 km to the east of the lagoon. Fatality estimates range from 2,182 (IFRC, 1999) to 2,202 (National Disaster and Emergency Services) and 2,324 (EM-DAT, 2001). It may have been even higher as the exact pre-event population is unknown. From the available death and survivor tallies (Table 1), it is estimated that about 40% of the population of the Arop and Warupu villages was killed. The casualty rate was less than 5% of the population in the other affected villages. At least 800 people received injuries requiring hospitalization (IFRC, 1998). Most casualties appear to have been caused when people were battered by debris (OCHA, 1998). A number of people also suffered pneumonia after salt-water inhalation. Shock and the long delay in receiving medical attention (see Response section below) exacerbated injuries.

No structures within 500 meters of the coast survived the tsunami in the coastal villages of Arop and Warupu. Houses were lightweight, constructed out of thatch and other traditional materials and elevated on posts. No trace of houses remained except, in a few cases, foundation posts (Fig. 6). The medical clinics, schools and subdistrict headquarters in Sissano were wood frame built on concrete slab foundations and were also damaged or destroyed (Tables 2, 3). All six structures making up the Sissano Subdistrict Headquarters were destroyed. Nine churches were destroyed and two partially destroyed. The most substantial building in the area was the Sissano Sacred Heart Mission built by German missionaries in 1926, capable of seating 500 people. The walls were composed of locally made cement breezeblocks. Only the foundation survived the tsunami (Davies, 1998).

Response

Response to all disasters moves through a three-phase set of actions, each with its own tools and objectives, and with foresight, a fourth phase to improve community

Table 1

Casualties and survival rate

Village	Known Deaths	Survivors	% Survival	School Population	School Deaths	% Survival
Malol	97	3816	97.5	400	18	95.5
Sissano (Maindron and Mainya)	24	1032	97.7	408	10	97.5
Sissano (Amsor and Nimas)	89	1153	92.8			
Arop 1 and 2	864	1404	61.9	314	117	62.7
Warupu	1099	1574	58.9	288	88	69.4
Total	2173	8979	80.5	1410	233	83.5

Source: National Disaster and Emergency Services, Dept. of Provincial and Local Level Government, Papua New Guinea, August 1998.

Figure 6
Foundation posts are all that remain of houses in the Arop 1 village site. Photo: L. Dengler.

Table 2

School losses

Name of School	1998 Enrollment	Number Class-rooms	Students Killed	Teachers Killed	Teachers Displaced	Status
St. Michael – Sissano Level 5 school	408	13	10[1]	2	14	Destroyed
Warapu Community School Level 4 school	288	11	88	1	11	Destroyed
Arop Community School Level 4 school	314	10	117	1	10	4 classrooms standing, abandoned
Malol Community School Level 5 school	400	14	18	0	15	Intact but abandoned
Yalingi Community School Level 3 school	122	5	0	0	5	Intact but abandoned
Aitape Vocational Inst				1[2]		
Total	1,532	53	233	5	55	

Source: National Disaster and Emergency Services, Dept. of Provincial and Local Level Government, Papua New Guinea, August 1998.

[1] *Nine (9) high school students killed while on school holidays.*

[2] *Teacher on holiday in the damaged area.*

resilience to future disasters. The phases vary in length and may overlap depending on the extent of damage, available resources, existing infrastructure, and the cultural and political setting.

Table 3

Damage to health care facilities

Name	Status
Sissano Health Centre	Half destroyed and abandoned
Warapu Aid Post	Completely destroyed
Arop Aid Post	Completely destroyed
Malol Aid Post	Intact but abandoned

Source: National Disaster and Emergency Services, Dept. of Provincial and Local Level Government, Papua New Guinea, August 1998.

- *Emergency Response*: Goal — to secure life and safety. Includes: situation assessment, search and rescue, securing operations areas, emergency medical response, evacuation, restoring or providing services vital to response such as transportation, communication and fire fighting, reducing secondary impacts such as disease prevention and contamination including collection and burial of bodies.
- *Relief*: Goal — to provide basic social necessities for survivors. Includes: providing food, shelter and sanitation, restoring infrastructure and lifelines including communications, and transportation, trauma counseling, education and rumor management to reduce anxiety among the population.
- *Recovery*: Goal — to reconstruct and rehabilitate impacted communities. Includes construction or restoration of permanent housing, sanitation and water systems, educational system and restoration of the economic framework. Long-term counseling, education and training may be required to restore economic viability to the community.
- *Reduction of vulnerability*: Goal — to reduce impacts of future disasters. Includes warning systems where appropriate, construction methods, land use decisions, education, and advance planning for future disasters.

Emergency Response

Residents of Aitape 20 km to the east of Sissano Lagoon were aware of some unusual ocean behavior soon after the earthquake as waves reached 3 m and caused some damage and a few injuries in Aitape Bay. Some injured arrived at the Raihu hospital in Aitape around 8:00 p.m., less than an hour after the event. Aitape officials were unaware of the much more significant damage in the Sissano Lagoon area because they received no communications from the impacted area and did not have the ability to carry out a damage assessment. Radio communication facilities at Arop and Sissano were destroyed in the tsunami. A Catholic missionary in Malol, Sister Margaret, repeatedly requested help via radio throughout the night but no one was monitoring transmissions until the morning (EMTV, 1998). Victims provided the initial response in the devastated area themselves, the survivors helping each other in the darkness. A government patrol officer in Sissano was able, with much difficulty, to reach Aitape by boat the following morning and was the first to give the outside world an idea of what had happened (WELSCH, personal communication). A mining company helicopter flew over the area in the morning and, noticing the damage, landed in the area with the crew providing the first outside response (DEPT. OF THE ARMY, 1998). As a result of the communications breakdown, rescue operations to the Sissano Lagoon communities did not begin until 16 hours after the event and were not fully mobilized until 2 days afterwards. The emergency response phase was at its peak the first week after the disaster. By the end of July it was winding down and declared over in early August; the Government-declared State of Emergency remained in place until October 1998 (IFRC, 1998).

1) Assessment

Once notified, the District Administrator of Aitape immediately returned from the regional capital and began a helicopter assessment. On July 18, the government requested assistance from Australia in planning the response. Australia sent logistics and disaster management experts who arrived within the next 2 days (AUSAID, 1998). A detachment of police and army was sent in from Wewak on the 19th (OCHA, 1998). The Prime Minister also visited the area on the 19th and declared a National State of Emergency on the 20th, backdating it to July 18 (OCHA, 1998). Damage and impact estimates changed rapidly over the next two weeks (Table 4) but it was quickly apparent that a major disaster had occurred. The United Nations Office for the Coordination of Humanitarian Affairs (OCHA) issued its first situation report

Table 4

Changing impact estimates

Date	Deaths	Hospitalized	Displaced	Missing	Source
18-Jul.	70	hundreds			IFRC
18-Jul.	71			100's	OCHA
19-Jul.	599				ABC
20-Jul.	725			2000–3000	OCHA
21-Jul.	750			3500	IFRC
21-Jul.	1,200	hundreds	3,000	1000's	OCHA
22-Jul.	1,600	more than 500			REUTERS
24-Jul.	1,500	700	5,200	6000	REUTERS
24-Jul.	1,500	700	5,163		OCHA
27-Jul.	2,000	1,131	10,068		REUTERS
28-Jul.	1,500	680	6,177	500+	OCHA
29-Jul.	1,638	550	6,100		AusAid
31-Jul.	2,173	800	8,500		IFRC
31-Jul.	2,100+		9,000	28	OCHA
7-Aug.	2,100+	668		500	OCHA
18-Aug.	2,182	473	9,199		IFRC
30-Oct.	2,182		10,483		IFRC

Source: Relief Web http://www.reliefweb.int/w/rwb.nsf.

and request for assistance on July 18 and international relief organizations quickly responded.

2) Medical response

The first trained medical personnel, a doctor and five nurses from the Raihu Clinic, arrived by helicopter at 11 AM on the 18th (DENGLER and PREUSS, 1999). During the initial 24 to 72 hours, survivors were brought to Raihu. The hospital, closed the previous day because of insufficient funds, reopened on the 18th to treat tsunami victims. Ironically, the timing of the closure was a benefit; the medical staff was still in the area but all the patients had been moved out, allowing room to treat tsunami victims. The most common injuries were lacerations, abrasions, soft tissue injuries and broken limbs caused by battering by debris. Several victims contracted pneumonia caused by salt-water inhalation (OCHA, 1998). Others received wounds from coral that were highly susceptible to septicaemia, blood poisoning caused by infection. Shock compounded the injuries of many and the delay in treatment, 35 degree Celsius tropical heat and high humidity caused wounds to become septic leading to gangrene in some cases.

The number of patients overwhelmed the Raihu clinic. The more seriously injured were airlifted to Vanimo, and later to Wewak. On July 19 an Australian Air Force Hercules already in Papua New Guinea was dispatched to the area with relief supplies. By Tuesday evening July 21, the Parachute Surgical Team of the Australian Defense Force sent two C-130's to establish a tent field hospital capable of providing Advance Trauma Life Support (ATLS) in Vanimo close to the Vanimo hospital. Over the next five days, the facility performed 12 to 20 surgical procedures a day. During the first three to four day period procedures included first look surgeries, first debridements (removing dead tissue from wounded areas), initial fracture management and initial amputations.

One week after the tsunami, medical teams from Australia, Japan, New Zealand, United States and the Netherlands were assisting in the hospitals in Aitape, Vanimo, Wewak and two field hospitals (JICA, 1998). Second level procedures including further debridement, fracture management, additional amputations and delayed primary closure, overlapped the end of the first phase of treatment. Beginning in the middle of the second, the first rounds of skin grafting and the first attempts at definitive fracture management were undertaken. Two weeks after the event, about 800 people were hospitalized and 24 amputations had been performed. All of the more seriously injured were transferred to the much better equipped Wewak hospital for further wound management.

3) Evacuation and security

Most of the survivors from the Arop villages and Warupu had been washed to the opposite side of the lagoon by the tsunami. Survivors, who were able, walked for several hours in the dark to inland villages with whom they had long-standing tribal relations—the Arop to Pou and the Warupu to Ramo (WELSCH, personal communication). The Sissano villagers headed to their inland garden sites near Olbrum and Rainbrum. Search and rescue operations continued to evacuate injured survivors over the next three days (DEPT. OF THE ARMY, 1998). Unlike most disasters where the damaged area remains populated, by the third day everyone was evacuated from the inundation zone. The injured had been air lifted out of the area and able-bodied survivors had self-evacuated, returning to the devastated villages only to look for missing relatives and to bury the dead. The government restricted access to the lagoon area on July 23 (IFRC, 1998). Government soldiers patrolled the area over the next two weeks, shooting dogs and pigs that were disturbing bodies and grave sites. The Government imposed a ban on alcohol in the area for security reasons.

4) Controlling contamination

Decomposition of bodies both onshore and within the lagoon was a contamination concern. Survivors began the search for relatives' bodies on the 18th and buried many in shallow graves on the sand spit using hand tools. By July 21, PNG Defense Forces and Police personnel were travelling into the area to burn and bury the unclaimed dead in mass graves (AUSAID, 1998). Recovering bodies from the lagoon was particularly difficult; relatives and government authorities used nets to collect many. Attempts to retrieve bodies in the lagoon were called off on July 23 due to safety concerns over disease and crocodiles that were feeding on the corpses. Government authorities officially sealed off a 40-km^2 zone containing the lagoon and restricted use of local food and water supplies (OCHA, 1998). The Government considered using explosives to widen the mouth of the lagoon to increase water circulation and reduce contamination. Explosives and detonation teams were moved into the Aitape area but officials shelved the idea on July 27 over concerns that the flushing could worsen the situation by contaminating offshore coastal waters. A fishing ban was imposed in the ocean along 100 km of coastline because of the contamination concerns. Coastal fishing remained off limits until late October and the lagoon remained closed to fishing until February 1999 (IFRC, 1999).

Relief

The Government of Papua New Guinea sent NDES staff to the area along with members of the PNG Defense Forces the day after the tsunami. They established care centers and, working with Australian advisors, set up emergency communications and airlifts for supplies. The Papua New Guinea Red Cross in cooperation with the International Federation of the Red Cross (IFRC) conducted a relief needs assessment over the first week of the disaster and the first supplies began arriving on the 18th. Other PNG relief efforts included two funds established by national newspapers, the Aitape Disaster Relief Fund and the Aitape Disaster Victims Fund and contributions from provincial governments. Response to international relief requests was quick and substantial. During the first three weeks after the disaster, 19 countries and 17 NGO's (Non-Governmental Organizations) donated money and supplies worth over 6 million US$ (OCHA, 1998). The Australian Government took the immediate lead due to their close proximity and historical relationship with Papua New Guinea. They provided four C-130 transport planes, a field hospital and medical team, advisors, relief supplies and over $600,000 (U.S.) in financial support. Additional contributions poured in through the Catholic Diocese. The Sissano tsunami was unusual in that the impacted area was small, the insult of short duration and the response oversubscribed; in contrast, the Bougainville relief effort that continued throughout the tsunami period was chronically underfunded (IFRC, 1998).

1) Coordination of relief efforts

The lack of regional infrastructure, the outpouring of national and international relief donations and the numerous governmental and NGO's operating in the area complicated the relief period. Most notably among the NGO's were the International Red Cross and the Diocese of Aitape District Disaster Committee (DADDC). Coordination and control difficulties emerged during the third week of the disaster. Under the State of Emergency, the central government was the controlling authority, however both district and provincial authorities were increasingly seeking roles (IFRC, 1998). The Aitape Restoration Committee (ARC) was established to coordinate the activities of all the organizations involved in the recovery. The ARC was composed of representatives of the central government, local governments, the army of Papua New Guinea, Australia, New Zealand, international relief NGOs, the police, and the DADDC (AusAid, 1998). After the first week, the majority of organizations and emergency teams left the area, leaving the following agencies with primary responsibilities:

Red Cross	Principal Relief Distributor
Salvation Army	Food, medicine, medical staff, counseling, vaccination
ADRA (Adventist Development and Relief Agency International)	Water and Sanitation
Catholic Diocese	Grief counseling, spiritual needs, education, health, physiotherapy
PNG Government	Food, clothing, transport, security, coordination

Although Aitape belongs to Saundaun Province, East Sepik Province played a leading role in relief activities, because of the road between Wewak and Aitape. The existing Red Cross supply base in Lae was the logistics base for Red Cross operations. After the State of Emergency was lifted in October, the Aitape/Lumi District Administrator assumed responsibility for relief and rehabilitation activities.

2) Care centers

Seven care centers were established for survivors (Fig. 2, Table 5). They were in two types of locations: 1) the garden lands owned by the villagers (Olbrum,

Table 5

Temporary shelter: Care centers

Care Center	Original Villages	Survivors in Center
Malol	Malol	3,816
Pou	Arop 1 and 2 Villages	2,046
Ramo	Warapu	1,644
Rowoi	Sissano (Amsor and Nimas)	1,153
Olbrum	Sissano (Maindron and Mainya)	1,032
Rainbrum	Sissano (Porono and Maintor)	89
Aitape	Yakoi	420
Total		10,200

Source: National Disaster and Emergency Services, Dept. of Provincial and Local Level Government, Papua New Guinea, August 1998.

Rainbrum) and 2) the square and the school sites of other villages (Pou, Ramo, Rowoi). The Malol center was first established near their original village site but was moved inland to Amut because of concerns about contamination from the bodies in the nearby lagoon. A center was also established in Aitape. Each care center accommodated victims from one or two villages. Elementary schools, churches, and medical clinics, locally called "Aid Posts" were established. Tarpaulins for temporary tent shelter, food, fuel for cooking, cooking supplies, clothing and water were provided at each center. A helicopter was used for the transportation of relief goods, and a heliport was set up at each center.

The system met the physical needs of most refugees (IFRC, 1998; WELSCH, personal communication). Friction developed among different kinship groups due to the large number of people forced into close proximity and old land ownership disputes. The situation was exacerbated by the influx of foreign foods and outside relief workers, disrupting the normal pattern of village life for both the hosts and the victims (WELSCH, personal communication). There was a dysentery scare at one care center that turned out to be false and no serious disease outbreaks were reported during the sheltering phase of the relief period. All of the care centers closed after three to four months, the last food distribution taking place in November 1998.

3) Transportation

PNG roadways fall into four classes: all-weather hard surface roads usable under wet and dry conditions, 2) dry weather roads with loose or light surfaces only accessible during the dry season, 3) vehicle tracks in the bush with no added surfacing materials accessible to four-wheel drive vehicles during the dry season only, and 4) foot tracks that are not passable by vehicles under any circumstances. At the time of the tsunami, a dry weather road extended from Aitape to Malol. Vehicle tracks and footpaths served the remainder of the area. Rowoi, Olbrum and Rainbrum were accessible only by canal or a walking trail through the bush. After the tsunami, boat traffic was banned on Sissano and Malol lagoons because of debris and sedimentation. The road to Pou and Ramo existed before the earthquake, but road surface conditions were bad and only four-wheel drive cars could pass when the rivers were low enough to ford. A new road to Pou and Ramo and the new village sites of Wipom, Barupu and Wauroin was constructed in October for relief and building supplies. Most supplies to the Pou and Ramo centers were transported by road after its construction. Helicopters remained the main supply access for the other care centers throughout the relief period until helicopter operations ceased in mid-December (IFRC 1998, 1999).

4) Trauma counseling

Post-traumatic stress was a significant problem for the population whether or not they had been physically injured. Many in hospital or in care centers were unresponsive during the first weeks afterwards (EMTV, 1998; O'Callaghan and Hamilton, 1998). The Catholic Diocese took the lead in counseling efforts, training grief counselors and developed plans for long-term counseling. A University of Papua New Guinea psychologist and a group of students spent two weeks assisting the Diocese's counseling efforts (Unicef, 1998; Hill, personal communication).

5) Rumors and public education

The tsunami was frightening and few survivors had any understanding of the physical processes that caused it. Rumors abounded in the weeks and months afterwards. The missionaries associated with the Catholic diocese in the region were careful to emphasize that the tsunami was a natural phenomenon; however a few missionaries from other denominations claimed the tsunami was a punishment from God and a larger event was imminent (Davies, 1998a). One particularly virulent rumor was that a foreign power (often identified as France) had detonated a bomb offshore and caused the tsunami (Davies, 1998a; Dengler and Preuss, 1999). Psychology students from the University of Papua New Guinea conducted a survey of 86 survivors in the three weeks immediately after the tsunami. Approximately 30% believed the tsunami to be caused by natural causes, 40% believed it a punishment from God, and 15% thought it caused by French bomb and the rest were unsure (Gillian Hill, personal communication).

Professor Hugh Davies of the University of Papua New Guinea was aware of the need to provide public information about the tsunami based on his experience during the volcanic eruption at Rabaul in 1994 (Davies *et al.*, 2001). Professor Davies authored "Earth Talk", a weekly column in The National, a Port Moresby newspaper that ran July 28 through November 17. He initiated a series of public information visits to hospitals and care centers and villages in August to help survivors understand that the tsunami was a natural process and to allay their fears that an even larger event was soon to come.

Recovery

The aim of the recovery phase is to restore self-sufficiency. In the Sissano disaster, the recovery effort was targeted at the establishment of new village sites. Reoccupying the original village sites was not an option. No structures remained at the Arop and Warupu village locations, much of the spit had eroded during the tsunami, the spits had become burial sites, and people were too frightened to return to the coast. Only in Mandroin did some villagers return to their homes after the government restrictions to the area were lifted (WELSCH, personal communication). Recovery activities coincided with and overlapped relief efforts and most of the organizations involved with relief were also responsible for recovery. Planning for new village sites began in the first weeks of the disaster; the recovery phase lasted two years and most activities were declared essentially complete by the Red Cross in mid-2000 (IFRC, 2000). The large number of agencies and organizations involved and different expectations among donors, aid providers and victims complicated recovery efforts (EHRLICH, personal communication; WELSCH, personal communication). Although international relief donations more than covered initial needs, many of these funds could not be used for recovery projects or only for particular types of projects, e.g., home construction not sanitation systems. More than 2 million kina (US$ 1.38 million) pledged by the national government to rebuild the villages never arrived (REUTERS, 1999). Uncertainties about land ownership also caused delays as donors did not wish to contribute to projects if land title was in doubt.

1) Coordination of recovery efforts

The Aitape District Disaster Committee (ADDC) was established to coordinate recovery in early September 1998. The Committee included: 1) the District administrator of Aitape (chair), 2) the administrators of the education, health care, fishery, industry, and social welfare departments, 3) two representatives from the Diocese of Aitape District Disaster Committee, 4) a representative from each care center, 5) the Red Cross, 6) representatives from other active NGOs, 7) two representatives from the state government; and 8) the representative of women (MAKI, personal communication). Subcommittees were established under the ADDC to address resettlement sites, health care, education, industry, and social welfare. The ADDC was intended to give the District government a coordinating role in planning and monitoring the recovery process. In practice, the ADDC had difficulties in getting participants to meet and acquiring support from the central government and was not able to effectively coordinate recovery efforts (EHRLICH, personal communication). The primary NGOs, the Red Cross and the Catholic Diocese, both developed their own rehabilitation plans that resulted in some competition between them and the government (WELSCH, personal communication).

2) Resettlement sites

The most important decision made in the first weeks of the disaster was the location of new settlement sites (Fig. 2). New village locations for the Arop and Malol peoples were on their traditional lands, selecting sites that were as far from the coast as their land boundaries allowed (WELSCH, personal communication). The Warupu chose a site halfway between the Ramo care center and the lagoon and called it Barupu. Most of the land was their traditional garden site; however the school and patrol post were on land claimed by the Ramo people (WELSCH, personal communication). The land negotiations for their resettlement were difficult and protracted (MAKI, personal communication). The Sissano peoples expanded the care centers on their inland garden sites to make Olbrum and Rainbrum permanent villages (IFRC, 1998). Housing was self-built collectively by villagers with building supplies and tools provided by relief organizations. All of the new village sites posed difficulties; the inland locations were hotter, damper and more mosquito-ridden than their original villages. Barupu was particularly problematic; the land was flat and very poorly drained marshland surrounded by sago swamp. The new roads built to access the village exacerbated the drainage problems.

3) Health, water and sanitation

A significant problem was securing an adequate supply of water and establishing sanitation systems. The water table is at or near the ground surface in the new village sites, making it difficult to build latrines. The new village locations initially used surface river water for cooking, washing, and waste. Coliform counts in the rivers had reached alarming levels by late November 1998 (AUSAID, 1998). Many villagers found the new conditions highly unsatisfactory, as fresh water had been readily available in their old villages and they had been supplied with bottled water at the care centers. The Red Cross developed a water and sanitation program, a community health project and nutrition assessment in August 1998 as part of the Aitape Post Disaster Programme. Seventeen rainwater catchment systems (tanks and corrugated metal roofs to collect rainfall) and twelve shallow water wells with hand pumps were installed. Heavy equipment excavated drainage ditches in Barupu. Five hundred and one pit latrines were constructed. The project was considered complete in October 1999 (IFRC, 2000).

No major health problems have been noted in the new villages since their establishment, although there were reports of anemia during the first year attributed to the lack of fish in diets (REUTERS, 1999) and an increased incidence of diarrhea in children (IFRC, 2000). The Red Cross initiated health education programs covering prevention of diarrhea, malaria and skin diseases, children's nutrition and HIV/AIDS awareness. They also distributed treated mosquito nets, health brochures and gave women from the new villages Community-Based Health training courses. Psychologically, many of the survivors exhibited symptoms of stress; increases in both domestic violence and alcohol-related crimes were noted a year after the tsunami (REUTERS, 1999). The new village sites far from the coast were difficult for most people to adapt to; they missed their relatives lost in the tsunami, their old villages next to the lagoon and their possessions acquired over many generations (WELSCH, personal communication).

The Father Antonia Rehabilitation Centre in Aitape, a former leprosy center specializing in artificial limbs, was renovated in October 1998 in Aitape to meet prosthetic needs and provide physiotherapy for amputees.

4) Transportation

Three new roads were planned to connect the resettlement sites: 1) Aitape – Malol, 2) Pou- New Arop 1– New Arop 2, and 3) Ramo – Barupu – Wipon- Rowoi-Olbrum (Fig. 2). The route from Aitape to Malol was planned because 1) a bridge on the road between Aitape and Malol was washed away, and 2) the Malol village was moved inland. The construction of this road was completed in December 1998 (MAKI, personal communication). The road from Pou to New Arop 1 (Wauroin) has also been completed. The route between Ramo to Olbrum has been completed as far as Barupu. The canal connecting the other resettlement sites such as Wipom, Rowoi, and Olbrum to the lagoon has also been restored.

5) Educational facilities

Approximately one month after the tsunami, schools for each displaced village were established at the care centers. The expectation was that return to their routines would be helpful for the traumatized children. However because of the high losses (approximately 15% of the 1998 school enrollment) the school setting accentuated the losses for both the children and the teachers (for example in one class only 7 children remained of 45 (DENGLER and PREUSS, 1999). Schools and housing for teachers were completed at the new settlements within two years of the tsunami.

6) Economic restoration

Restoring a viable, self-sufficient economic system was complicated by the scale of loss — all personal possessions including canoes, boats, tools, fishing supplies were destroyed by the tsunami. The closure of the lagoon and offshore fishing restrictions and the loss of inland garden land for new village sites also hindered recovery the first year. The coastal fishing closure was particularly difficult as it impacted many coastal villages and the Tumleo Islanders who had not been directly affected by the tsunami and were not recipients of international aid. Relief agencies supplied tools such as sago beaters, adzes and axes for making canoes, nylon string for making nets and other fishing supplies to the survivors. By 2001, fishing along the coast and in the lagoon had returned to near pre-tsunami levels. However, continued dissatisfaction with the new village sites, grief over losses, and frustration with relief agencies continues to hamper long-term recovery (WELSCH, personal communication).

Reduction of Vulnerability

The relocation of villages from the beach and sand spit has significantly reduced the exposure of the Sissano Lagoon population to the tsunami hazard. If the relocation remains permanent, it will reduce losses in future tsunamis in this area. The memory of the tsunami and on-going education programs should keep tsunami hazard awareness high in this generation of Sissano villagers. Outside of the tsunami-impacted area, it is unclear what long-term tsunami hazard reduction measures have been made. Tsunamis are now included in government earthquake awareness materials (Fig. 7), but there has been no widespread government effort to identify or develop education programs for other vulnerable areas.

On January 10, 2002, an earthquake ($M_w = 6.7$) occurred near Aitape. Hugh Davies made a reconnaissance visit to the area four days afterwards to collect geologic and damage information and document human response (DAVIES, personal communication). The earthquake damaged bush houses, concrete foundations, water tanks, and was strongly felt in much of the area affected by the tsunami. People who were living close to the coast at Sissano, Malol and Lemieng, and most likely at other sites he did not visit, voluntarily moved away from the coast at the time of the earthquake, as a precaution against possible tsunami. Some or all returned after it became obvious that no tsunami had been initiated.

The Red Cross planned a two-year program of community based self-reliance training in the Sissano villages that was scheduled to begin in February 2000 but it is unclear if the training has begun (IFRC, 2000). Red Cross was also able to procure a new Lae warehouse for warehousing relief supplies for future Papua New Guinea disasters.

Figure 7
Tsunami awareness information provided by the National Disaster and Emergency Services (NDES). The pidgin is loosely translated: An Earthquake's Worst Killer May Come From the Sea. In a low-lying coastal area, an earthquake may be your only warning that a tidal wave is about to strike. As soon as the shaking stops, start for high ground and stay there for one hour. If the sea goes out DO NOT walk onto the reefs because a tidal wave is sure to occur.

Impacts on Mitigation Outside of Papua New Guinea

The Papua New Guinea tsunami has had significant impact outside of the country in both increasing tsunami awareness and in approaches to assessing tsunami hazards.

Awareness

The 1998 tsunami was unusual in the attention given by international media. It played particularly large in the United States, Australia, New Zealand and Japan. Often news events that occur on Friday in remote countries get little play in the western media. The Papua New Guinea tsunami was different. The Associated Press listed over 300 postings about the tsunami between July 18 and July 23, making it the leading story of the week (Fig. 8). Members of the U.S. National Tsunami Hazard Mitigation Program and staff of the Tsunami Warning Centers gave 92 interviews to the media during the month following the tsunami including all major broadcasting companies (PAWG, 1998). Media reports covered not only the disaster itself, but also discussed tsunami hazards in their own regions. Several in-depth articles for the general public were the direct result of the Papua New Guinea tsunami (GONZALES, 1999; ROSENTHAL, 1999). EMTV, a Port Moresby television production company, produced a 50-minute documentary that was widely aired within the country and the

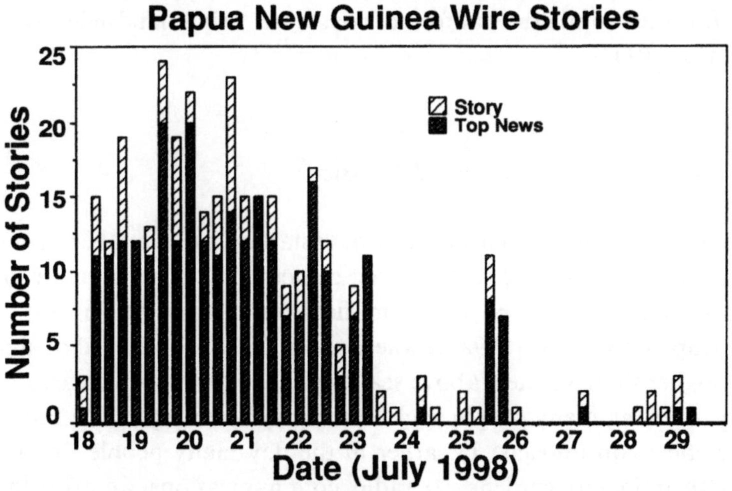

Figure 8
Papua New Guinea tsunami stories posted by the Associated Press wire service on the Washington Post web site (http://www.washingtonpost.com/wp-srv/searches/mainsrch.htm). Stories are listed by six-hour increments by Eastern Daylight Time; Top News refers to a short mention in the listing of leading news stories and Story refers to a separate article. The first stories were posted about 18 hours after the earthquake.

Australian Broadcasting Company ABC produced a Four Corners Documentary that aired in Australia, New Zealand and Papua New Guinea (EMTV, 1998; ABC, 1998).

Tsunami Hazard Assessment

Perhaps the greatest impact of the Sissano tsunami outside of Papua New Guinea has been recognition by the international tsunami community of the potential hazards posed by submarine slumps and landslides. Submarine mass movements have been recognized as potential tsunami sources for decades. Of the 81 historic California tsunami events listed by LANDER *et al.* (1993), 13 are designated as probable landslide sources. 95 of the 122 historic tsunami fatalities in Alaska are attributed to landslide tsunamis (LANDER, 1996). Yet there had been little attention in the scientific community to landslide sources prior to the Sissano tsunami. A 1993 conference to assess state-of-the-art tsunami modeling makes no mention of landslide sources (BERNARD and GONZALEZ, 1994) and a comprehensive volume on tsunami hazard reduction (HEBENSTREIT, 1997) has only one brief reference to landslides. Of the 250 tsunami publications listed by GeoRef between 1990 and 1998, only 11 include landslide in the title. In contrast GeoRef notes 20 articles (out of 213) between 1999 and 2001 that include landslides. The 2000 and 2001 Annual meetings of the American Geophysical Union and the 2001 International Tsunami Symposium featured tsunami landslide sessions and there have been several special landslide symposia in the past three years. Tsunami inundation projections currently being developed for the Southern California coast are using landslide source models (EISNER *et al.*, 2001).

Discussion

The Sissano tsunami was particularly devastating for a number of reasons. The most important was that peak wave impact coincided with maximum exposure. The wave heights estimated from post tsunami field teams show the highest waves from Arop to Warupu (KAWATA, 1999a,b) where about 7000 people lived on a narrow spit on land no higher than 3 meters above sea level. Exposure was increased by the four-day holiday, drawing many people back to their native villages. The time of day had two further effects on impacts: 1) earlier in the day many people would have been working at their inland gardens, 2) radio communications were no longer being monitored so that no one heard the requests for assistance until the following morning. In contrast, the June 23, 2001 Southern Peru tsunami that inundated a similar-sized area killed fewer than 100 in large part because it was mid-winter and the beach resort community that was most heavily damaged was virtually uninhabited (DENGLER, 2001).

Lack of awareness of the tsunami hazard also contributed to the high human losses. Few persons living in the area remembered the 1951 or 1935 tsunamis, both of which were much smaller and less devastating than the 1998 tsunami. Most residents had experienced strong earthquakes and had no reason to associate groundshaking with possible tsunami. About 20 minutes elapsed between the earthquake and the arrival of the first wave, sufficient time for many of the people to reach safety if they had immediately moved inland on feeling the ground shake. The matron at the Aitape hospital reported that some elderly survived the tsunami because they had either remembered previous tsunamis or had heard stories about them. After the earthquake, some got into boats and headed toward the back of the lagoon or moved to higher ground after the earthquake and others climbed trees (DENGLER and PREUSS, 1999).

Losses were further exacerbated by the lack of radio contact and inability to immediately communicate the magnitude of the disaster. Although a 24-hour emergency contact point had been established by the national government, it was not operational at the time of the tsunami. The communications system at the Sissano Sub-District office was destroyed by the tsunami and the Catholic Diocese radio network was not monitored in the evening. Members of provincial government were on vacation at the time of the event and the government had not the resources or the ability to conduct an adequate damage assessment immediately after the earthquake. No one expected a magnitude 7 earthquake to cause a major tsunami so there was no impetus to conduct an assessment. It is impossible to estimate the number of lives lost because of the delay in response, but the 16 hours or more that people were stuck in the water or trapped in mangroves clearly increased morbidity and contributed to the number of amputations.

Vegetation had a mixed effect on impacts. The heavily vegetated coastal area may have reduced the extent of inundation. On land, waves penetrated about 500 meters inland from the coast. In contrast, the tsunami caused by the June 2001 Peru earthquake penetrated over a kilometer inland over unvegetated beach sands and irrigated onion fields (DENGLER, 2001). The two tsunamis are not strictly comparable; peak wave heights in the Papua New Guinea tsunami were about twice as high as in Peru but the Peru wavelength was probably considerably greater. In the vicinity of the Arop school where the land was flat and heavily vegetated but no more than 3 meters above sea level, waves left debris about 500 to 600 meters from the coast. In the adjacent shallow lagoon, the waves swept over the spit and penetrated the far shore of the lagoon, damaging mangroves at the far shore of the lagoon, at distances of up to 1300 meters from the coast. The trees on the spit and along the coast increased the destructiveness of the tsunami. The spit and coastal area were covered with mature coconut palm trees, scattered casuarina trees and dense shrubbery. Many trees, particularly palms, were uprooted or snapped off at mid trunk level, swept into the lagoon or piled up against the back of the inundation area. The battering effects of the debris caused most injuries.

Some survivors were swept into the mangroves and impaled on broken limbs. Casuarina trees apparently withstood the tsunami impact better than other vegetation and some people survived by climbing these trees (DENGLER and PREUSS, 1999).

Air evacuation was the key to saving many lives. Although response was initially delayed, once the scope of the disaster became apparent, local companies and Australia and New Zealand provided rapid response. The Hevilift Company that provides airlift services for local oil companies dedicated several helicopters to initial response. Australian and New Zealand C-130 pilots capable of responding on short notice and trained to land on grass runways conducted an efficient operation. However, at one point the Hevilift Company ceased operations because of concerns regarding payment. Negotiating agreements or contracts in advance would expedite response in the future. The emergency medical operations were established quickly and cooperation among the local and international agencies involved with the medical response functioned very well during the emergency response period (DEPT. OF THE ARMY, 1998).

Relief efforts were well supported by national and international donors. Relief agencies were successful in communicating the magnitude of need to their supporters and the response was swift and substantial. However, coordination among relief organizations was difficult because of the different goals and organizational structures of the agencies involved. The recovery phase has been more difficult for several reasons. The magnitude of loss for the survivors was enormous and expectations that losses would be replaced could not be met (WELSCH, personal communication). Many donations were designated for relief only and could not be used for long-term recovery projects. All of the new village sites were less satisfactory than the original villages; the inland sites were foreign to peoples whose livelihood and culture was dependent on the coast, they were hotter and more insect-ridden, water and sanitation was difficult because of the high water table. The influx of outside relief workers and relief supplies disrupted the previously isolated peoples of the Sissano Lagoon area, created tensions and differing expectations among survivors and host communities.

The Sissano tsunami raised awareness of tsunami hazards nationally and in other tsunami-prone countries both within the scientific community and among the media. At least six scientific teams from four countries visited the impacted area within the first year. It has triggered documentaries, and numerous reports in national and international broadcasting and print media. In the scientific community, its most significant contribution has been to focus attention on landslide tsunami sources. It has influenced mitigation efforts through inclusion of landslide sources in hazard assessment, and in reiterating the importance of education as the most significant way to reduce casualties in a local tsunami disaster. However to have a significant long-term impact on reducing tsunami vulnerability, the lessons from Papua New Guinea and other recent tsunami disasters need to be institutionalized into national and international emergency response planning and the education infrastructure.

Lessons for Tsunami Mitigation

1) Narrow spits are extremely vulnerable to tsunami hazards. Some of the villages on the lagoon spit were too far from higher ground or inland sanctuaries for people to reach refuge in the time between the earthquake and the arrival of the first waves. Whenever possible, development on spits should be discouraged.

2) Tsunami education and awareness programs need to be established in all coastal communities. Many of the Sissano peoples could have reached safe haven if they had immediately responded to groundshaking; however most did not interpret the earthquake as a warning sign. Appropriate and consistent messages need to be developed through educational, government and NGO structures and disseminated using all available methods including unique cultural institutions and oral tradition. Education messages must emphasize groundshaking as the trigger for taking action.

3) 24-hour emergency contact points and a redundant communication system are essential to response. One of the tragedies of the Sissano tsunami was that 16 hours elapsed between the event and the arrival of first responders.

4) Inadequate assessment slows response. No one expected a damaging tsunami from an earthquake of this size. Aitape, the largest town in the area, had damage and injuries from the tsunami, but no one considered that other areas were hit harder. Assessment of the hardest hit areas was delayed, resulting in a belated response and additional fatalities. Areas vulnerable to tsunami had not been identified beforehand and there was no plan to assess such areas.

5) Rapid response medical teams and air evacuation save lives. The medical response to the Sissano tsunami worked well because of the training and readiness of the Australian and New Zealand teams and the rapid deployment of helicopters and transport planes. The long-standing relationship between Papua New Guinea and Australia aided coordination of the medical response effort. All tsunami vulnerable countries need to look at their response capabilities and develop agreements with other countries for rapid medical response if needed.

6) Vegetation as a tsunami abatement measure needs further research. Although trees may have reduced inundation, all of the affected villages were too close to the coast for this effect to be significant. Trees added to the damage by becoming battering rams in the fast moving water. Casuarina trees were somewhat more resistant than palm trees, although in areas of significant scour, their root systems were undermined. To use trees as a mitigation tool, understanding of the interaction of trees, roots and high flow regimes is needed.

7) Pre-event coordination planning for relief and recovery agencies could improve response. In the United States, adoption of the Standardized Emergency Management System (SEMS) has greatly improved emergency response during disasters by providing a framework for different organizations to work together. However relief and recovery organizations develop ad hoc working relationships with each other and with government agencies after each disaster. Developing a structural framework for relief and recovery agencies would improve coordination in both developed and developing countries.

8) Consider the cultural setting of the affected communities in developing relief and recovery programs. In the Sissano area, relief operations caused friction among groups that had traditionally been friends and recovery was hampered by differing expectations of what outside aid agencies could provide.

Acknowledgements

Thanks to Professor Hugh Davies, Department of Geology, University of Papua New Guinea who provided background information on PNG geology, geography, seismicity and tsunamis, and logistics assistance for the EERI PNG reconnaissance trip. Hugh has continued to provide information on the recovery process and answer our numerous questions. Michael Nongkas, an Honors Student at the Department of Geology UPNG, accompanied us in the Aitape region, translated, found vehicles and boats, and Anna Nasi, a second year student at UPNG from the Aitape region, provided a pre-tsunami vision of the Sissano Lagoon area. Malcolm Ehrlich and Kaigabu Kamnanaya of NDES gave perspectives on the governmental response. Diedre Christie and Veronica Gasu Yakuma of the PNG Red Cross provided additional information on Red Cross response. Gillian Hill, Psychology Department UPNG, shared the experiences of her post-tsunami field study. Thanks also to Costas Synolakis and Jose Borrero, USC, for their information from ITST trip #1 and Bruce Jaffee, USGS, on ITST trip #2. Dr. Rob Welsch, Dartmouth, generously shared his unpublished manuscript on the pre- and post-tsunami environment of the Sissano villages and answered many questions. Dr. Tom Cashman of the U.S. Dept. of the Army Medical Corps provided reports and information about the medical response. EERI through its NSF-funded Learning from Earthquakes Program financed the EERI field team investigation.

References

ABC (1998), 4 Corners: *The Survivors' Story* (video documentary), Australian Broadcasting Corporation.

AusAid (1998), *PNG Tsunami Relief Updates*, Australian Agency for International Development, Relief Web http://www.reliefweb.int/

BERNARD, E. N. and GONZALEZ, F. I. (1994), *Tsunami Inundation Modeling Workshop Report (November 16–18, 1993)*, National Oceanic and Atmospheric Administration (NOAA), Technical Memorandum ERL PMEL-100, 139 pp.

BORRERO, J. C. (2001), *Changing field data gives better model results: An example from Papua New Guinea*, Proc. International Tsunami Symposium 2001 (ITS 2001) (on CD-ROM), Pacific Marine Environmental Laboratory, NOAA, Session 2, No. 2–5, 397–405.

DENGLER, L. and PREUSS, J. (1999), *Reconnaissance Report on the Papua New Guinea Tsunami of 17 July 1998* EERI Newsletter, vol. 33 supplement

DEPT. OF THE ARMY (1998), *Memorandum for US CINCPAC/J07*, Tripler AMC, Hawaii 96859.

DAVIES, H. (1998a), *Tsunami PNG 1998 — Extracts from Earth Talk*, Waigani, Univ. Papua New Guinea, Port Moresby, PNG.

DAVIES, H. L. (1998b), *Sissano Tsunami: A Reconstruction of the Events of July 17, 1998, Based on Interviews with Survivors*, EOS, Transact., Am. Geophys. Union, vol. 79, Fall Meet. Suppl., F572.

DAVIES, H. L. (personal communication), Geology Department, University of Papua New Guinea

DAVIES, H. L., DAVIES, J. M., LUS, W. Y., PEREMBO, R. C. B., JOKU, N., GEDIKILE, H., and NONGKAS, M. (2001) *Learning from the Aitape Tsunami*, Proc. *International Tsunami Symposium 2001* (ITS 2001) (on CD-ROM), Pacific Marine Environmental Laboratory, NOAA, Session 2, No. 2–10, 415–424.

DENGLER, L. (2001), *Impacts of the 2001 Peru Tsunami in Camana*, Proc. *International Tsunami Symposium 2001* (ITS 2001) (on CD-ROM), Pacific Marine Environmental Laboratory, NOAA, Session 2, No. 2–7, 409.

DENGLER, L. and PREUSS, J. (1999), *Learning from Earthquakes: Reconnaissance Report on the Papua New Guinea Tsunami of July 17, 1998, EERI Special Earthquake Report, in Jan 1999 Newsletter of the Earthquake Engineering Research Institute*.

EHRLICH, M. (personal communication), National Disaster and Emergency Services, Port Moresby, PNG.

EISNER, R. K., BORRERO, J. C., and SYNOLAKIS, C. E. (2001), *Inundation Maps for the State of California*, Proc. *International Tsunami Symposium 2001* (ITS 2001) (on CD-ROM), Pacific Marine Environmental Laboratory, NOAA, NTHMP Review Session, R 4, 67–81.

EM-DAT (2001), *The OFDA/CRED International Disaster Database*, Universit Catholique de Louvain, Loouvain-la-Neuve, Belgium, http://www.cred.be/emdat/intro.html

EMTV (1998), *Wave: Aitape Tidalwave Disaster* (video documentary), EMTV, PNGFM Religious Television Association, Ango Drive, Boroko, PNG.

EVERINGHAM, I. B. (1977), *Preliminary Catalogue of Tsunamis for the New Guinea/Solomon Islands Region 1768–1972*, Australian Bureau of Mineral Resources, Report 180.

GELFENBAUM, G., JAFFE, B., NONGKASS, M., and DAVIES, H. L. (2001), *Sedimentary Deposits from the 17 July 1998 Papua New Guinea Tsunami*, Proc. *International Tsunami Symposium 2001* (ITS 2001) (on CD-ROM), Pacific Marine Environmental Laboratory, NOAA, Session 2, No. 2–13, 449–452.

GEIST, E. L. (2001), *Reply to Comment by E.A. Okal and C.E. Synolakis on "Origin of the 17 July 1998 Papua New Guinea Tsunami: Earthquake of Landslide?" by E.L. Geist*, Seismol. Res. Lett. *72*(3), 367–372.

GEIST, E. L. (2000), *Origin of the 17 July 1998 Papua New Guinea Tsunami: Earthquake or Landslide*? Seismol. Res. Lett. *71*(3), 344–351.

GOLDSMITH, P., BARNETT, A., GOFF, J., MCSAVENY, M., ELLIOT, S., and NONGKAS, M. (1999), *Report of the New Zealand Reconnaissance Team to the Area of the 17 July Tsunami at Sissano Lagoon, Papua New Guinea*, Bull. N.Z. Soc. Earthq. Engin. *32*, 102–118.

GONZALEZ, F. L. (1999), *Tsunami!*, Scientific American *280*(5), 56–65.

HEBENSTREIT, G. ed., *Perspectives on Tsunami Hazard Reduction: Observations Theory and Planning*, (Kluwer, Dordrecht, 1997) 218.

HILL, G. (personal communication), Univ. Papua New Guinea, Port Moresby, PNG.

HTDB/PAC (2001), *Historical Tsunami Database for the Pacific, 47 BC – 2000 AD*. CD-ROM, Tsunami Laboratory, ICMMG SD RAS, Novosibirsk, Russia.

IFRC (1998–2000), *Papua New Guinea: Tidal Wave Situation Reports*, International Federation of the Red Cross, ReliefWeb http://www.reliefweb/int

JICA (1998), *The Report on the Japan Disaster Relief Team for Papua New Guinea Tsunami Disaster*, JDR Report.

JENKINS, A. (1997), *Sissano Lagoon and Ali Island: Mangrove/Estuary and Coral Reef Ecosystems in the Path of Mainland Development*, Christensen Research Institute, PO Box 305, Madang 511, Papua New Guinea.

KAWATA, Y., TSUJI, Y., MATSUTOMI, H., IMAMURA, F., MATSUYAMA, M., TAKAHASHI, T., FUJIMA, K., BENSON, B., BORRERO, J., de LANGE, W., NOTT, J., OKAL, E., SYNOLAKIS, C., DAVIES, H., and LETZ, H. (1999a), *Field Survey on the 1998 Tsunami in the Northwestern Area of Papua New Guinea, Report for Grant-in-Aid for Scientific Research, Ministry of Education, Science, Sports and Culture*, Japan DPRI, Kyoto University, 81 pp.

KAWATA, Y., BENSON, B., BORRERO, J., DAVIES, H., de LANGE, W., IMAMURA, F., LETZ, H., NOTT, J., and SYNOLAKIS, C. (1999b), *Tsunami in Papua New Guinea Was as Intense as First Thought EOS*, Transac., Am. Geophys. Union *80*(9), 101, 104, 105.

LANDER, J. (1996), *Tsunamis Affecting Alaska 1737–1996*, NGDC Key to Geophys Res Documentation. *31*, U.S. Department of Commerce, NOAA, 195 pp.

LANDER, J. F., LOCKRIDGE, P., and KOZUCH, M. (1993), *Tsunamis Affecting the West Coast of the United States 1806–1992*, NGDC Key to Geophys. Records Documentation *29*, U.S. Dept. of Commerce, 241 pp.

MAKI, N. (personal communication), Earthquake Disaster Mitigation Research Center, RIKEN, 2465-1 Mikiyama, Miki, Hyogo, 673–0433, Japan.

MCCARTHY, J. K. (1963), *Patrol into Yesterday*, F. W. Cheshire, Canberra, 156 pp.

MCSAVENY, M., GOFF, J., DARBY, D. J., GOLDSMITH, P., BARNETT, A., ELLIOT, S., and NONGKAS, M. (2000), *The 17 July 1998 Tsunami, Papua New Guinea: Evidence and Initial Interpretation*, Marine Geology *70*, 81–92.

O'CALLAGHAN, M. L. and HAMILTON, P. (1998), *After the Wave, the Tsunami that Struck the Northwest Coast of Papua New Guinea Destroyed Several Thriving Communities*, The Australian Magazine, Sept 19–20, 20–24.

OCHA (1998), *UN Office for the Coordination of Humanitarian Affairs*, ReliefWeb http://www.reliefweb.int

OKAL, E. A. and SYNOLAKIS, C. E. (2001), *Comment on "Origin of the 17 July 1998 Papua New Guinea Tsunami: Earthquake or Landslide?*" by E. L. Geist, Seismol. Res. Lett., *72*(3), 362–366.

PAWG (1998), *National Tsunami Hazard Mitigation Program Tsunami Public Affairs Working Group Report* Anchorage, October 7–8, 1998, http://www.pmel.noaa.gov/tsunami-hazard/pawgoctober98report.html

RIPPER, I. D., LETZ, H., and MOIHOI, M. (1998), *Pre-tsunami Large Earthquakes of the Aitape Region North Coast Mainland of Papua New Guinea*, Papua New Guinea Department of Mining and Petroleum Geological Survey 98/7.

RIPPER, I. D., LETZ, H. and MOIHOI, M. (1999), *Felt Effects of the 17 July 1998 Sissano Lagoon Tsunami Earthquakes*, Geological Survey of Papua New Guinea Report 99/13.

REUTERS (1999), *PNG Tsunami Survivors Never Saw Government Aid*, Reuters News Service, July 18, 1999, http://www.reliefweb.int.

ROSENTHAL, A. M. (1999), *The Disaster in Papua New Guinea Has Awakened Fears of Tsunamis Occurring Elsewhere in the Pacific Including along North America's West Coast*, California Wild *52*(2) 24–32.

SYNOLAKIS, C. E., BARDET, J. P., BORRERO, J. C., DAVIES, H. L., OKAL, E. A., SILVER, E. A., SWEET, S., and TAPPIN, D. R. (in press), *The slump origin of the 1998 Papua New Guinea Tsunami*, Proc. Roy. Soc. (London).

TAPPIN, D. R., MATSUMOTO,T., WATTS, P., SATAKE, S., MCMURTY, G. M., MATSUYAMA, M., LAFOY, Y., TSUJI, Y., KANAMATSU, T., LUS, W., IWABUCHI, Y., YEH, H., MATSUMOTU, Y., NAKAMURA, M., MAHOI, M., HILL, P., CROOK, K., ANTON, L., and WALSH, J. P. (1999), *Sediment Slump Likely Caused 1998 Papua New Guinea Tsunami EOS*, Trans., Am. Geophys. Union *80*, 329–340.

TAPPIN, D. R., WATTS, P., MCMURTY, LAFOY, Y., G. M., and MATSUMOTU (2001), *The Sissano, Papua New Guinea Tsunami of July 1998 – Offshore Evidence on the Source Mechanism*, Marine Geology *175*, 21–23.

UNICEF, 1998, *Aitape Tidal Wave Disaster*, United Nations Children's Fund, 13 pp.

WELSCH, R. L. (1998), *An American Anthropologist in Melanesia: A.B. Lewis and the Joseph N. Field South Pacific Expedition 1909–1913*, Vol. I, *The Field Diaries* (University of Hawai'i Press, Honolulu) 632 pp.

WELSCH, R. L. (personal communication), Dartmouth College, Hanover NH, U.S.A 03755.

WELSCH, R. L., and Terrell, J. (1999), *Papua New Guinea Begins Rebuilding after Tidal Wave Disaster, In the Field:* The Field Museum's Membership Publication, *70*(1), p. 7.

WELSCH, R. L., and TERRELL, J., *Material culture, social fields, and social boundaries on the Sepik Coast of New Guinea.* In: *The Archaeology of Social Boundaries* (ed. Miriam Stark) (Washington: Smithsonian Institution Press, 1998) pp 50–77.

Pure appl. geophys. 160 (2003) 2033–2051
0033–4553/03/112033–19
DOI 10.1007/s00024-003-2418-9

Pure and Applied Geophysics

Large-scale Basement-involved Landslides, California Continental Borderland

M. R. LEGG[1] and M. J. KAMERLING[2]

Abstract—Large seafloor relief, shallow metamorphic basement, and seismic activity in the California Continental Borderland combine to produce major submarine slides capable of generating local tsunamis. The Catalina Schist basement complex of the Borderland contains melange units, faults between metamorphic facies, and foliation derived from a history of deep subduction underthrusting and subsequent tectonic exhumation along regional low-angle fault systems. Neogene volcanic and sedimentary rocks covered the exhumed metamorphic basement forming an interface with a strong contrast in material properties. Neogene transtensional block-faulting formed steep escarpments that cut through and elevated the low-angle detachment surface, providing slip surfaces with free faces susceptible to failure along the rheological boundary. Two basement-involved slope failure examples west of San Diego are described, including large block-glides and progressive rotational slumps along the steep northeast-facing Thirtymile Bank escarpment and along the southwest flank of Fortymile Bank.

Key words: Tsunami, submarine landslides, continental margin geology.

1. Introduction

The California Continental Borderland is a distinctive physiographic and structural province located offshore southern California, U.S.A., and northern Baja California, Mexico (SHEPARD and EMERY, 1941). The Borderland has numerous sets of parallel, generally northwest-trending ridges and islands separated by deep, elongate, fault-controlled basins (Fig. 1). This basin-and-ridge physiography presents high relief with numerous steep slopes that may be susceptible to failure. EMERY (1960) estimated that 63% of the Borderland is composed of slopes and deep irregular areas. Numerous submarine slope failures have been mapped on *sediment-covered* Borderland slopes (FIELD and EDWARDS, 1980, 1993; NARDIN *et al.*, 1979a,b; VEDDER *et al.*, 1986; CLARKE *et al.*, 1987). There are also many subaerial *basement-involved* landslides and landslide complexes 1–4 km across on Santa Catalina Island (BAILEY, 1941), and many large-scale, basement-involved, submarine landslides exist

[1] Legg Geophysical, Huntington Beach, California, U.S.A.
[2] Institute for Crustal Studies, University of California, Santa Barbara, California, U.S.A.

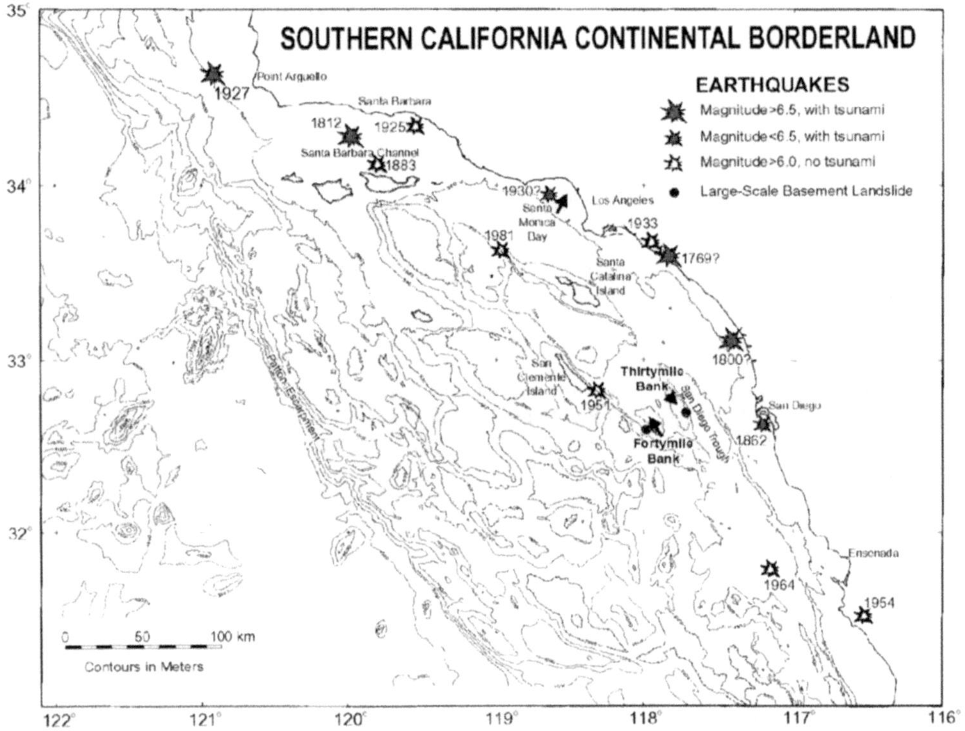

Figure 1
Map showing bathymetry of the California Continental Borderland, location of large historic earthquakes, and possible large-scale basement-involved landslides.

in the area. If slide movement is sufficiently rapid, such events could generate destructive tsunamis to the adjacent coastal areas (WATTS and RAICHLEN, 1994).

1.1 Tectonic Evolution of Borderland Slopes

Neogene oblique extension (LEGG, 1991; CROUCH and SUPPE, 1993; BOHANNON and GEIST, 1998) unroofed the regional Catalina Schist basement (KAMERLING and LUYENDYK, 1985). The Catalina Schist is considered a metamorphic core complex similar to the higher grade Franciscan subduction complex exposed within the California Coast Ranges onshore to the north (CROUCH and SUPPE, 1993; TEN BRINK *et al.*, 2000). Widespread extension-related volcanism filled Borderland basins with volcanic and volcaniclastic rocks and covered the exhumed Catalina Schist at the prominent detachment fault interface. The contact between the Neogene volcanic/volcaniclastic rocks and underlying Catalina Schist basement may be structural, due to continued detachment faulting, or depositional atop the exhumed basement surface. In either case, a prominent rheological discontinuity exists today where these

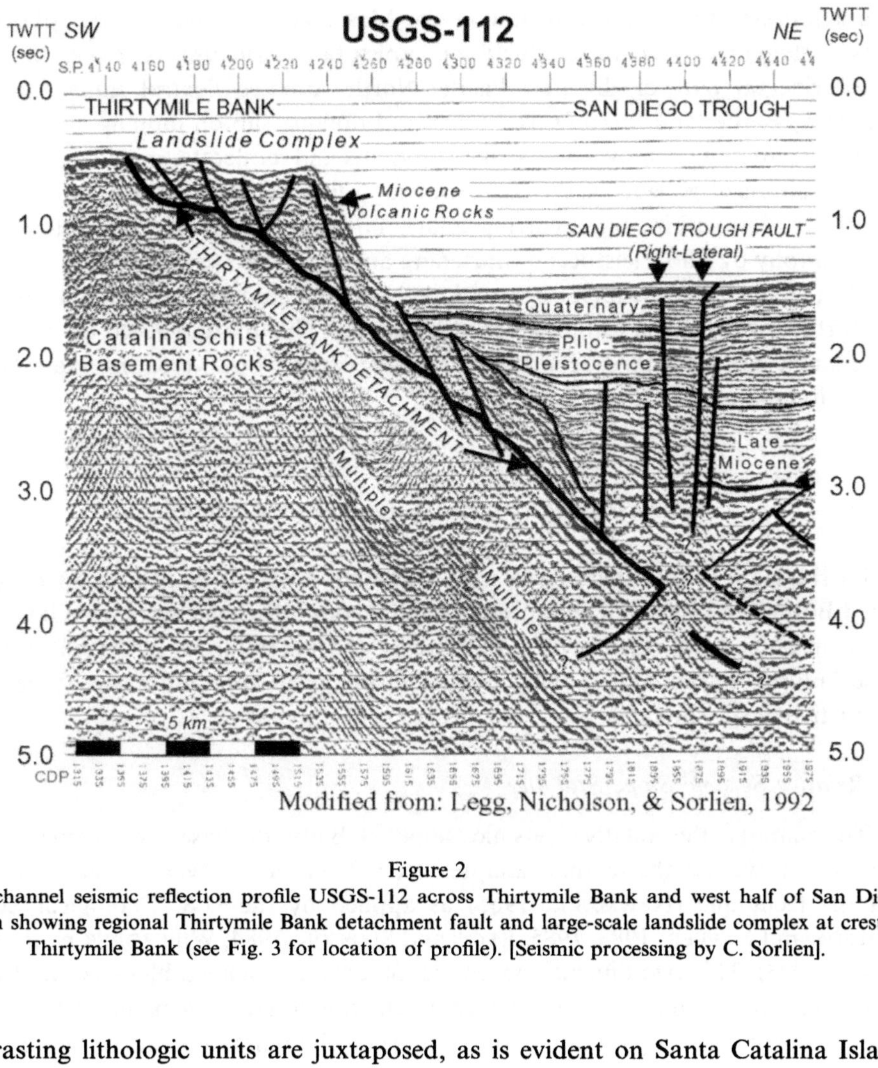

Figure 2
Multichannel seismic reflection profile USGS-112 across Thirtymile Bank and west half of San Diego trough showing regional Thirtymile Bank detachment fault and large-scale landslide complex at crest of Thirtymile Bank (see Fig. 3 for location of profile). [Seismic processing by C. Sorlien].

contrasting lithologic units are juxtaposed, as is evident on Santa Catalina Island (Cann, 1985) and in regional seismic reflection profiles (Fig. 2; Legg *et al.*, 1992; Bohannon and Geist, 1998).

Basin and ridge structure of the California Continental Borderland was established by Neogene block faulting and oblique extension (transtension). When the exhumed detachment fault systems, and subjacent ductile metamorphic basement rocks rose above the brittle-ductile transition level, moderate to high-angle normal faults broke through and dismembered the regional low-angle structure. Block faulting and tilting created the prominent Inner Borderland horst and graben structural style, uplifting ridges of both older basin fill and underlying basement rocks, and deepening new more localized sub-basins. In many areas the block faulting may have utilized pre-existing structures remnant from the ancient subduction structural fabric, including the Mesozoic to mid-Cenozoic subduction thrust system and overlying subduction complex fault structures (LEGG *et al.*, 1992). As a principal part of the new Pacific–North America dextral transform plate boundary, Borderland faulting likely involved oblique-slip on dipping structures or strain partitioning on subvertical strike-slip faults and moderate to low-angle normal faults. Continued growth of the San Andreas transform fault system and changing tectonic styles within the evolving plate boundary sustained the block fault physiography by Plio-Pleistocene shortening and strike-slip (transpression; WRIGHT, 1991). In many areas the high-angle and likely oblique faulting enhanced steep slopes and further elevated the Catalina Schist basement. The uplift exposed the low-angle detachment surface and overlying strata along steep slopes, creating gravitational instability and potential for slope failure.

2. *Earthquake-induced Slope Failure Potential*

Earthquakes may induce large-scale basement-involved landslides on oversteepened Borderland slopes. Topographic amplification of seismic shaking (GELI *et al.*, 1988) on submarine ridge tops may increase the potential for basement-involved slope failures. Large, fast moving, basement-involved slope failures may generate destructive local tsunamis.

2.1 Relative Seismic Slope Stability

To quantify the relative seismic slope stability of basement slopes in the Borderland, we use the method adapted from Newmark's dynamic slope stability analysis technique (NEWMARK, 1965) as applied by the U.S. Geological Survey (WILSON and KEEFER, 1983, 1985; LEGG and SLOSSON, 1984; WIECZOREK *et al.*, 1985; JIBSON, 1993). This two-dimensional model of a sliding inclined block is suitable for slumps and block slides which likely represent the initial movements of large-scale bedrock submarine slope failures. In equilibrium the block remains stationary, and the downslope forces are balanced by cohesion and basal friction. The downslope acceleration required to initiate block movement, called the critical acceleration, a_c is a measure of the seismic slope stability. Values of a_c are derived using the formula (modified from WILSON and KEEFER, 1985):

$$a_c/g = [\mathbf{C}'/(\gamma H)] + [(\rho - \rho_w)/\rho] \tan \phi' \cos \theta - \sin \theta \tag{1}$$

where g is the local gravitational acceleration, $\mathbf{C}$ is the cohesion of the materials, γ is the unit weight of the material, H is the thickness of the slide block, ϕ' is the effective angle of internal friction of the material, and θ is the slope angle. For submarine landslides, the additional effect of buoyancy forces must be included where ρ_w is the mass density of sea water, and ρ is the mass density of the slide material.

The values of cohesion, unit weight, and internal friction angle are related to site geology and water content or pore pressure. For submarine slides, the material is assumed to be saturated, and pore pressure may be assumed to be hydrostatic, including the overlying water column. Earthquake-induced pore pressure changes may be significant, but are ignored for simplicity in this analysis. Dynamic pore pressure effects would tend to reduce the critical acceleration values needed to initiate slope failure by weakening the slip surface. Reducing the effective angle of internal friction may account for this effect. Indeed, the natural slope angles may represent the real effective angle of internal friction.

Based upon subaerial landslide investigations, three general rock types, with associated shear strength parameters (Table 1) are used to determine the relative seismic slope stability values. Effective shear strength parameters are provided for saturated conditions as well as for slopes with bedding planes, joints, or other weak surfaces dipping parallel to or antiparallel to the downslope direction. For each rock type and slope gradient combination a range of critical acceleration, a_c, values is computed. These values are compared to the a_c limits for each relative seismic slope stability category and an appropriate stability level is assigned (Table 2).

For the large-scale submarine landslides considered in this study, the first term in equation (1) becomes insignificant as the block thickness increases to values in excess of 100 meters, and the cohesion term can be neglected. WIECZOREK *et al.* (1985) set the cohesion to zero for weakly cemented sandstones, shales and clays, which may be appropriate for the metamorphic basement (Catalina Schist) of the Inner Borderland. The relative seismic slope stability values shown in Table 2 are based upon the strength parameters shown below, which includes the small, but non-zero, cohesion factor. Therefore, the values used in this study are reasonable, and conservative. With regard to potential tsunami generation, large-scale and rapid slope movement is necessary and likely requires low to unstable values of relative seismic slope stability.

Table 1

Shear strength parameters for regional slope stability analysis (after LEGG and SLOSSON, 1984)*

Geologic Unit (Lithology)	Along Planes of Weakness**		Through Rock[!]	
	ϕ'	$C'/\gamma H$	ϕ'	$C'/\gamma H$
Serpentine – High talc content, usually sheared	7.5°	0.125	10°	0.5
Franciscan – Highly fractured, fault gouge and shattered rock	7.5°	0.25	15°	0.75
Franciscan – Relatively intact sandstone, graywacke, and well-consolidated shale, locally cemented	10°	0.5	20°	1.5

*Effective strength parameters, rock saturated but pore water pressure not considered in analysis (D. E. MORAN, pers. comm., 1981). Slide thickness, H, assumed to be 3 meters.

**Failure path along well-defined bedding planes, faults, or pre-existing slip surfaces.

[!]Failure path crossing bedding, through largely intact rock.

Table 2

Relative seismic slope stability (saturated, submarine case)

Slope		Lithology			
Degrees	Percent	Serpentine		Franciscan [Highly Fractured]	
		Dip	Anti-Dip	Dip	Anti-Dip
0–2.9	0–5	Moderate	Very stable	High	Very stable
2.9–8.5	5–15	Moderate	Stable	Moderate	Very stable
8.5–17	15–30	Low	High	Moderate	Very stable
17–27	30–50	Unstable	High	Low	Very stable
27–35	50–70	Unstable	Moderate	Unstable	Stable
> 35	> 70	Unstable	Moderate	Unstable	Stable-high

Very stable: Not likely to move under severe shaking, $a_c > 0.7\ g$.
Stable: May undergo slight movement under severe shaking, $0.5\ g \le a_c < 0.7\ g$.
High: May undergo moderate movement under severe shaking; Some slides related to steep slopes and adverse dips, $0.3\ g \le a_c < 0.5\ g$.
Moderate: May undergo major movement under severe shaking or moderate movement under moderate shaking; numerous slides, rock falls abundant, unconsolidated material undergoing deformation and failure, $0.1\ g \le a_c < 0.3\ g$.
Low: May undergo major movement under moderate shaking; abundant slides of all types, $0.01\ g \le a_c < 0.10\ g$.
Unstable: May undergo major movement under slight shaking; most of area and/or materials failing, e.g., oversteepened submarine canyon walls, $a_c < 0.01\ g$.
Adapted from LEGG and SLOSSON, 1984.

Earthquake-induced slope failure susceptibility can be derived using maps of bathymetry to measure slope gradient and seafloor geology to derive the relevant lithology group for each slope area. This technique has been used by the U.S. Geological Survey to predict subaerial seismic slope stability for microzonation studies (WIECZOREK *et al.*, 1981). Due to the limited scope of the present study, no regional mapping is attempted. Instead, a rough estimate of the relative slope stability for major Borderland escarpments is derived. The lithology group is estimated from seafloor geologic maps prepared by the U.S. Geological Survey (e.g., VEDDER *et al.*, 1974; CLARKE *et al.*, 1987). Seismic reflection profiles provide direct imaging of the suspected slide plane (detachment, Fig. 2), and the slope gradient of this surface is used to determine the relative seismic slope stability of the suspected slide block.

For regional slope stability analysis, most steep basement rock slopes of the Borderland have gradients within the 15–50 percent slope categories (middle rows in Table 2). For Borderland slope geometries where bedding or failure planes are inclined in the same direction as the submarine slope (Fig. 2; Table 2), the relative seismic slope stability is Moderate to Unstable. These low stability slopes may be prone to catastrophic failure and tsunami generation.

2.2 Slope Failure Potential

The slope failure susceptibility is independent of the earthquake activity. When expected shaking levels are considered, slope failure potential can be estimated. Using dynamic slope stability analysis, curves of predicted slope displacement can be computed for particular earthquake time histories by integrating strong-motion accelerograms for various values of critical acceleration (WIECZOREK *et al.*, 1985; WILSON and KEEFER, 1985). Alternatively, the predicted displacements may be correlated with other shaking measures such as Modified Mercalli Intensity (LEGG and SLOSSON, 1984) or Arias Intensity (WILSON and KEEFER, 1985; JIBSON, 1993). For potential tsunami generation, the submarine landslide must move rapidly over a substantial distance, perhaps kilometers. Based upon the predicted displacement from the simple dynamic analysis, the study by LEGG and SLOSSON (1984) presented five levels of slope failure intensity (Table 3). For large displacements, the simple assumptions for the dynamic analysis are invalid, and the predicted displacement may be underestimated. Nevertheless, the slope failure intensity levels are believed to provide useful indices of the relative slope failure severity expected in real earthquakes. It is suggested herein that only catastrophic slope failure would generate tsunamis, and a predicted slope movement in excess of about five meters is used as the threshold for catastrophic failure. More quantitative study of the dynamics of submarine slope movement is necessary to substantiate this assertion or derive better methods to estimate catastrophic failure potential (e.g., SYVITSKI and HUTTON, 2003; LOCAT *et al.*, 2003; MARTEL, 2003).

Table 3

Slope failure intensity matrix (probability of damage state in percent)

Slope Stability:Unstable: $a_c \leq 0.01\ g$					
Damage State	Modified Mercalli Intensity				
	VI	VII	VIII	IX	X
Light Disp < 0.5 cm	0	0	0	0	0
Moderate 0.5 cm < Disp < 5.0 cm	0	0	0	0	0
Heavy 5.0 cm < Disp < 50 cm	60	50	40	30	20
Severe 50 cm < Disp < 500 cm	30	40	45	50	55
Catastrophic 500 cm < Disp	10	10	15	20	25
$\sum$ Probabilities	100	100	100	100	100

Disp = Predicted slide displacement.
Adapted from LEGG and SLOSSON, 1984.

Numerous active faults within the Inner Borderland may produce strong earthquakes and shaking levels that exceed local critical accelerations and trigger slope movement. Most of the larger steep slopes in basement rocks are fault related, and consequently, adjacent to the strong earthquake source regions. Indeed, observations from submersible show that vertical and overhanging slopes are present along active Borderland faults (J. WARME, pers. comm., 1985; C. GOLDFINGER, pers. comm., 2000). Furthermore, the complex patterns of tectonic faulting break up the basement rock providing additional failure surfaces that tend to control the location and size of individual slope failures. Because major active faults lie subparallel to and along the large escarpments, the potential for large-scale, basement-involved, submarine landslides exists.

3. *Examples of Large-scale Basement-involved Landslides*

Two specific features are described herein that may represent large-scale, basement-involved, slope failures. Examples are based on detailed bathymetry from dense hydrographic soundings, multibeam swath bathymetry, and side-scan sonar systems that allow recognition of slide morphology. Both high-resolution, single-channel, and moderate resolution, multichannel, seismic reflection profiles help to define the internal structure and identify possible basal shear surfaces.

3.1 Thirtymile Bank

Thirtymile Bank is a northwest trending ridge of Catalina Schist mantled with Miocene volcanic and sedimentary rock. Seafloor samples from dredges and dart cores by the U.S. Geological Survey during the 1970s (Fig. 3; VEDDER, 1990; VEDDER *et al.*, 1974, 1976) consist of middle Miocene volcanic, volcaniclastic and sedimentary rocks in the area of the postulated submarine landslide complex. A sample from the crest of the triangular block was described as hornblende-hypersthene andesite with an altered glassy ground mass (VEDDER *et al.*, 1976); many of the samples in the landslide area include a tuffaceous matrix, altered, fractured and deformed. From a few kilometers northwest of the inferred slide complex to the northwest end of Thirtymile Bank, dredge and dart cores from the steep escarpment and crest of Thirtymile Bank recovered Catalina Schist basement rocks that included glaucophane schist and riebeckite schist (VEDDER *et al.*, 1974). The Thirtymile Bank detachment fault is exposed at the seafloor where these rocks were recovered, and correlates with the strong reflector imaged on USGS-112 (Fig. 2). The Catalina Schist and volcanic rocks, especially where ash is altered to bentonite or serpentine is present, provide well-defined slip surfaces for large-scale submarine landslides along the steep Thirtymile Bank escarpment.

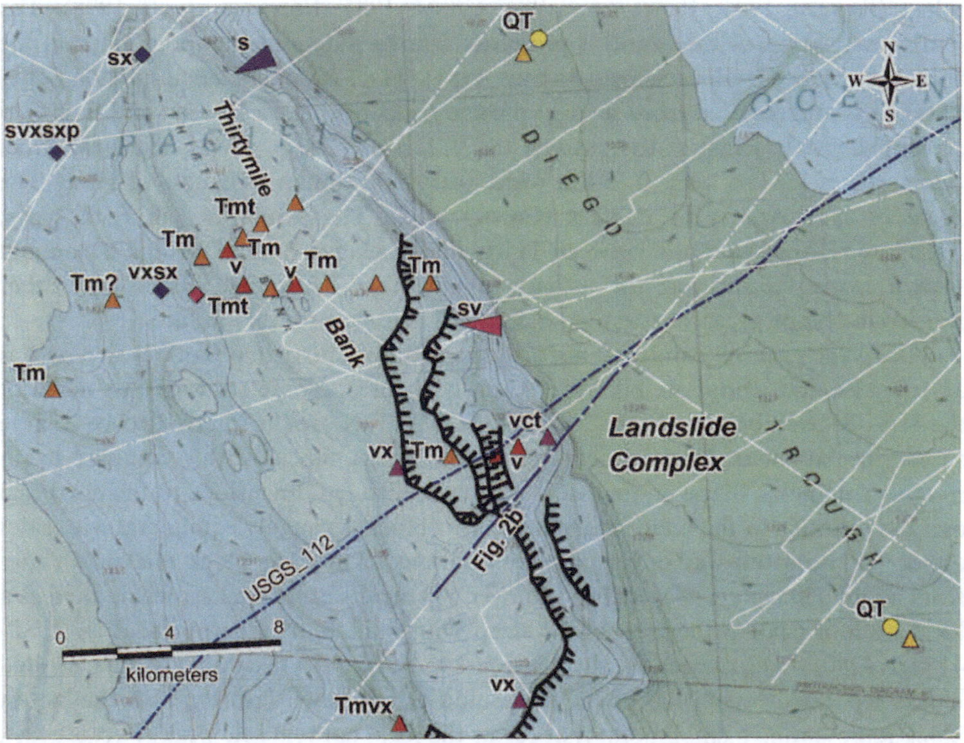

Figure 3

Bathymetric map showing the morphology of the Thirtymile Bank landslide complex, locations of tracklines for seismic profiles, and USGS bottom samples (VEDDER, 1990; VEDDER *et al.*, 1976). White lines are high-resolution, single-channel tracklines; dash-dot line is multichannel seismic trackline. Small symbols represent dart cores, large triangles are dredges. Triangles represent Tertiary (T), mostly middle Miocene, sample locations; diamonds represent Mesozoic(?) Catalina Schist basement rock samples; circles represent Plio-Pleistocene and younger (QT) sedimentary samples. Labels are: m = marine sedimentary, v = volcanic; vc = volcaniclastic; t = vitric tuff or bentonite; s = schistose rocks; x = detritus; p = phosphorite.

Comparison of the rocks and structure of Thirtymile Bank and Santa Catalina Island shows strong similarities. The island is composed of blueschist, greenschist, amphibolite, and sausserite gabbro intruded by Miocene igneous rocks (BAILEY, 1941; PLATT, 1975; SORENSEN *et al.*, 1991). Miocene sedimentary and volcanic rocks overlie the basement rocks. San Onofre Breccia occurs directly on the basement and above Miocene volcanic rocks. The San Onofre Breccia exposed on the mainland coast is composed of large blocks and debris from the Catalina Schist basement and represents talus or mass wasting debris off steep scarps. Modern large-scale submarine breccia deposition could generate local tsunamis. The rocks of the Catalina Schist basement complex are very weak and prone to failure as evidenced by the numerous subaerial landslides mapped on Santa Catalina Island in the Catalina

Schist (Bailey, 1941; Rowland, 1984; Cann, 1985; Francis *et al.*, 1998). The southwest margin of the island is primarily a large slide and slump complex in the blueschist facies Catalina Schist (Bailey, 1941). The Fisherman Cove slide on Santa Catalina Island resulted from strong volcanic rocks resting on sheared and fractured glaucophane-talc schist at the head of the slide area (Cann, 1985). The Fisherman Cove slide is 1.7-km long, 0.65-km wide, with a maximum thickness of 60 m. The angle of the slope is 11 degrees (Cann, 1985). The similarities in rock types, stratigraphy, and structure, between Thirtymile Bank and Santa Catalina Island and known slope failures on the Island suggest that Thirtymile Bank and other Borderland slope areas of Catalina Schist will also be prone to failure.

Along the generally straight, northwest-trending, Thirtymile Bank escarpment is an irregular area about six kilometers long and three kilometers wide just north of the U.S.–Mexico border (Figs. 2–3; USGS-112). This area is inferred to be a large submarine basement landslide complex. The largest block in this complex has a roughly triangular shape in map view with apex directed downslope to the northeast and base parallel to the main escarpment. This block is roughly equilateral with sides about 3.5 kilometers and covers an area of about 3 km^2. The block thickness varies from about 350 meters near its head to over 600 meters at its thickest where there is a local seafloor peak at the crest of the steep Thirtymile Bank escarpment (Fig. 2, SP-4140 to 4280). In cross section, there appears to be three to five individual back-tilted blocks separated by normal faults, in domino or bookshelf fashion, with the largest block representing a small horst that forms the local bathymetric peak. Farther east, within the subhorizontal sedimentary fill of the San Diego trough, there appears to be a large displaced block. This may be an older slide block, or alternatively, a remnant hanging wall block from the Neogene transtensional episode. An exploration industry seismic profile in this area shows possible sliding of the steep slope material out and over the flat-lying turbidites of the La Jolla submarine fan within the San Diego trough. The base of the tilted and displaced blocks is a strong reflector interpreted as the Thirtymile Bank detachment fault (Legg *et al.*, 1992; Bohannon and Geist, 1998). To the northwest of the triangular block along the escarpment, several smaller irregular-shaped blocks are apparent in the bathymetry, whereas a large rectangular block about three kilometers long, forms a bench along the steep escarpment to the southeast. The overall morphology of this area is similar to that of large subaerial basement landslide complexes mapped in the Peninsular Ranges of southern California (e.g., Tin Can Flat, Vallecito and Oak Mountain landslide complexes, Hart, 1991).

Like large-scale basement-involved subaerial landslides (Shreve, 1968; Hart, 1991), the failure mechanism of the Thirtymile Bank complex appears to be composite with slumping, as shown by the back-tilted block character, and block sliding. The buried rock mass in the San Diego trough appears to be displaced about two kilometers laterally and 800 meters vertically to the northeast of the crest of the escarpment where the horst block now lies. Such a transport distance, if by a slide mechanism, is similar to that of long run-out landslides, with a H/L ratio less than 0.5 (MELOSH, 1987). Long runout landslides typically occur rapidly, and therefore, if submarine, present potential sources for local tsunamis (see also VON HUENE *et al.*, 2003). The internal character of this block is difficult to determine, although there appears to be some irregular layering present in the seismic images (Fig. 2). We cannot say whether this is bedding with the available data, however, long runout landslides often preserve the relative stratification of the displaced rock mass (CAMPBELL *et al.*, 1995).

Burial of this block by La Jolla fan turbidites suggest that the feature is old. Seismic stratigraphic correlations and crude estimates of sedimentation rate in the San Diego trough (EMERY, 1960; LEGG, 1985; TENG, 1985) imply an age of late Miocene to Pliocene (>2 Million years ago). Other slide blocks still on the escarpment or overlying turbidites of the La Jolla fan may be Quaternary in age, perhaps within the past 100,000 years.

Individual slide blocks within the complex are located at all levels of the escarpment, from near the crest to the base. DENSMORE and HOVIUS (2000) suggest that earthquake triggered bedrock landslides tend to be more evenly distributed across the steep slopes; coseismic shaking induces slope movement at all slope levels, although some topographic amplification near the crest is possible. The nearby active San Diego trough fault zone may provide the necessary earthquake source region (Fig. 2). The remaining steep slope along the Thirtymile Bank escarpment and blocks of Cenozoic volcanic rocks atop the detachment surface suggests that future slope failures may occur due to this unstable configuration. Rapid movement of large submarine landslides toward San Diego, resulting from earthquake triggering or other instability, could generate a destructive tsunami along the adjacent mainland coast.

3.2 Fortymile Bank at North San Clemente Basin

Along the steep, southwest-facing escarpment that separates Fortymile Bank and North San Clemente Basin (relief exceeding 2000 meters), another large potential submarine landslide is identified. SeaBeam bathymetry (LEGG *et al.*, 1989) shows an irregular, north–northwest trending block about 7–8 km long by 4–5 km wide (Southwest Peak, Fig. 4) that appears displaced laterally about 5 km from the steep arcuate-shaped escarpment near the crest of Fortymile Bank. A high-resolution seismic profile that crosses this feature shows northeast-tilted reflectors within the offset block compared to subhorizontal reflectors within the main Fortymile Bank block (Fig. 5). Limited subbottom penetration and possible side-swipe artifacts preclude accurate delineation of the basal surface of the postulated slide block, although this surface may be exposed within the steep scarp along the active San Clemente fault. Earthquake-triggered slumping is considered the likely source of the 1998 Papua New Guinea tsunami (TAPPIN *et al.*, 2003), and the Fortymile Bank slide block is of similar scale (about 5–10 km^3).

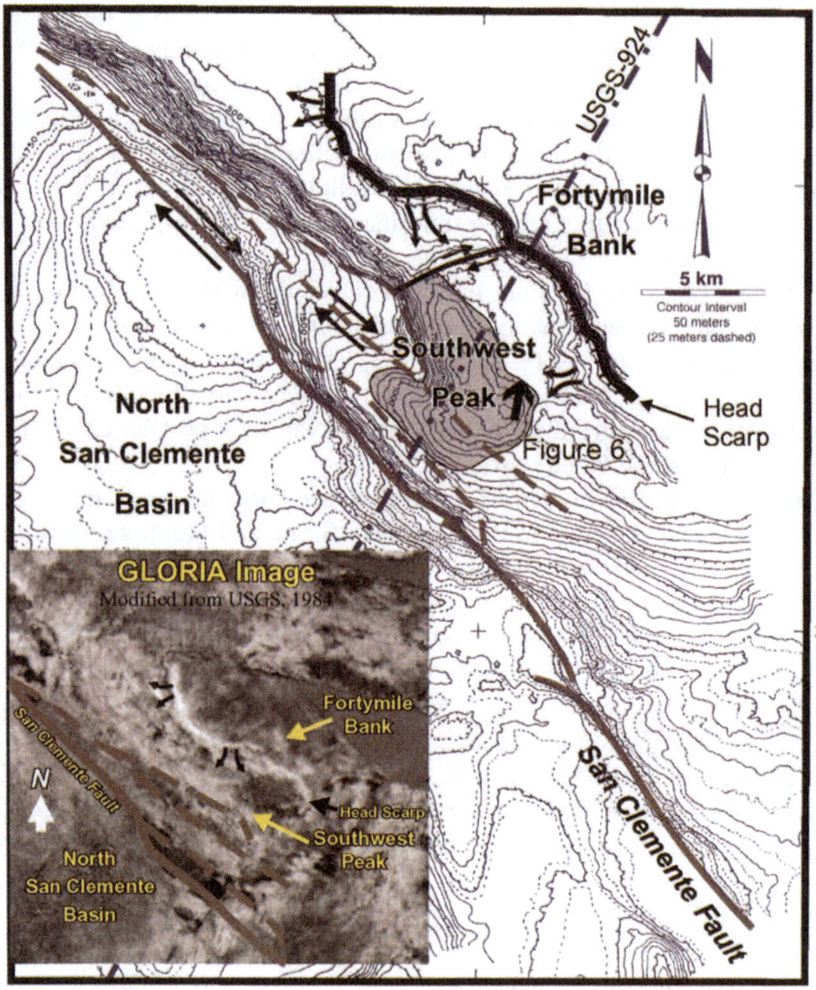

Figure 4
Map showing SeaBeam bathymetry (LEGG *et al.*, 1989) of the southwest flank of Fortymile Bank and GLORIA side-scan image (USGS, 1986) of proposed large-scale landslide blocks. Heavy line with ticks shows the major head scarp for the large-scale slope failure. Dashed lines are branch and secondary faults, and double arrows show postulated smaller landslide features. The principal displacement zone of the active San Clemente fault is shown by the solid line with arrows for sense-of-slip. Arrow labeled Figure 6 shows approximate 3-D view shown in subsequent figure.

Seafloor samples in the Fortymile Bank area are similar to those of Thirtymile Bank, with volcanic, volcaniclastic and sedimentary rocks of generally middle Miocene age reported (VEDDER *et al.*, 1974, 1976). Catalina Schist basement rock types are also reported from inferred outcrops on the steep, 1100-meter high escarpment to the northwest of Southwest Peak. Samples retrieved by submersible

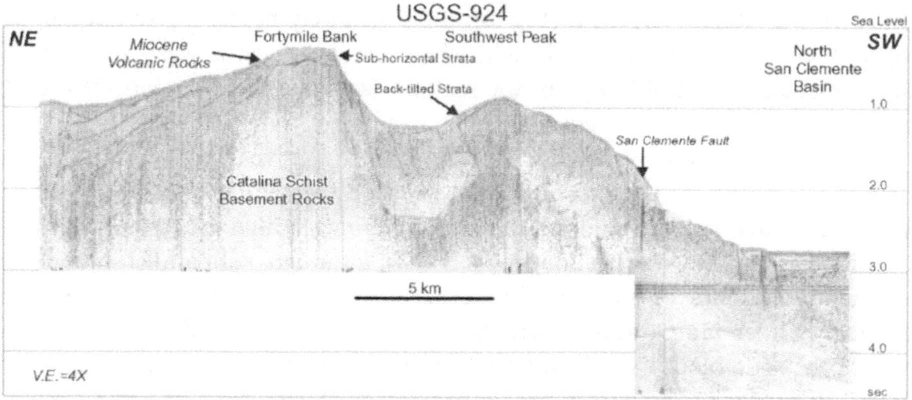

Figure 5
Sparker seismic reflection profile USGS-924 (VEDDER, 1975) across Fortymile Bank and San Clemente Basin showing back-tilted bedding in the inferred slide block.

from the upper Fortymile Bank escarpment, directly to the west and below the crest of Fortymile Bank, included tuffaceous and silty sandstones, as well as a one-half meter diameter "tectonic" rock covered with slickensides (M. LEGG, unpubl.). The cataclastic rock was retrieved from a small (100-m wide) mud-covered bench, located

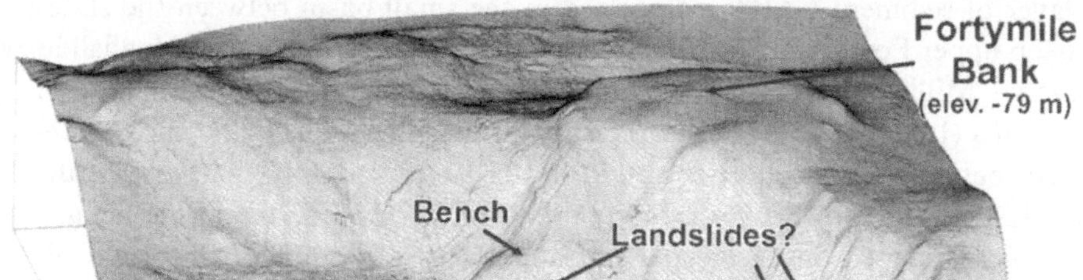

Figure 6
Three-dimensional perspective view from the south showing steep escarpment and possible landslide features near crest of Fortymile Bank.

midway up the steep escarpment (Fig. 6); the bench may be the top of another slide feature. About 1 km west of the bench, where the more gentle west-trending slope gradient increases to 30–40 degrees, the seafloor is covered by boulders (10s to 100s of centimeters in diameter) suggesting submarine mass transport.

The Fortymile Bank escarpment is related to active right-slip on the San Clemente fault zone (SHEPARD and EMERY, 1941; LEGG *et al.*, 1989). The principal displacement zone is located along the steep, 500-meter high escarpment located a few kilometers southwest of Southwest Peak, although subparallel branch and secondary fault zones lie along a hillside bench immediately below the postulated slide block. Other fault traces appear to disrupt and possibly offset the southwest corner of the slide block (Southwest Peak) to the northwest. The north trend of the elongate slide block, about twenty-five degrees clockwise from the San Clemente fault trend, may result from vertical axis rotation of an originally northwest-trending, fault-parallel, detached horst block in a zone of right-lateral shear.

Based on the tilted strata within the slide block, and the inferred displacement of about 5 kilometers, both slumping and block sliding may have occurred here. The basal surface may be the detachment fault that separates Cenozoic volcanic and sedimentary rocks from Catalina Schist basement rocks, as observed at Thirtymile Bank, but data are insufficient at this time to confirm this assertion. The estimated vertical drop of about 500–600 meters, compared to the lateral offset again is consistent with a long runout slide mechanism, but it is also possible that the feature is an older tectonic fault block remnant from the Neogene transtensional episode. A thin layer of sediment (< 100 m) ponded in the small basin between the slide block and steep upper Fortymile Bank escarpment suggests a more youthful landslide age – less than 1 Million years using the late Quaternary hemipelagic sedimentation rate for this area (LEGG, 1985). The steep escarpments and the strong lithologic/rheologic contrast, between Cenozoic rocks and schist basement, imply that large-scale slope failure is feasible for this unstable configuration. Indeed, numerous small slope failures were mapped on the steep escarpments based upon the seafloor morphology imaged by SeaBeam (LEGG *et al.*, 1989). It is possible that large-scale basement-involved slide blocks have overrun the active trace of the San Clemente fault, displacing the surface trace westward into the deep basin. Similar morphology occurs along the Elsinore fault where it marks the steep escarpment on the southwest flank of the Salton trough near the U.S.–Mexico border (PINAULT, 1984; LAMAR and ROCKWELL, 1986).

4. *Tsunami Potential*

All large earthquakes (Magnitude > 6.5) historically recorded in the southern California offshore region have generated local tsunamis (Fig. 1; December 21, 1812; November 4, 1927; MCCULLOCH, 1985). Two other large events (July 28, 1769 and

November 22, 1800) may have occurred in the coastal or offshore area based upon sparse intensity observations, but evidence of local tsunamis for these events is speculative at present (LEGG *et al.*, 1994; LISA GRANT, pers. comm., 2000). Some moderate offshore earthquakes (Magnitude = 5.5–6.5) apparently generated small local tsunamis (May 27, 1862; August 30, 1930, uncertain tsunami; March 10, 1933; LANDER *et al.*, 1993). To date, no local southern California tsunamis have been destructive. Repetition of tsunamis similar to earlier historical events may be damaging to the modern heavily developed coastal areas. Whether generated by tectonic displacements or submarine slope failure, the source mechanisms of historical Borderland tsunamis are unknown.

Using the equations demonstrated by VON HUENE *et al.* (2003) estimates of the maximum tsunami wave height generated by Borderland submarine landslides may be derived. The Fortymile Bank submarine landslide is best defined by the available data, and it is modeled in two ways: 1) as a rotational slump; and 2) as a translational slide. The slide dimensions were given above (7-km wide by 5-km initial length, and 650 meters mean water depth), and an initial failure thickness of 200 meters is assumed. For the rotational slump model, an angular displacement measured from the backtilted sediments is 0.1375 radians (about 7.9 degrees). For a mean slope of the failure scar of about 25 degrees, measured from the bathymetry, and an assumed radius of curvature of about 15 km, a maximum tsunami wave height of 7–9 meters is computed. For the translational slide model, the maximum tsunami wave height is predicted to be about 50 meters, similar to wave heights estimated for tsunami generation by the large Palos Verdes debris avalanche off Los Angeles (BOHANNON and GARDNER, 2003; LOCAT *et al.*, 2003). With nearly an order-of-magnitude variation in predicted wave heights from two different models, it is clear that the dimensions and mechanism of the submarine landslide be carefully determined before an accurate prediction of tsunami character can be made. Nevertheless, both models predict a wave height in the 10-meter range, which could be destructive to adjacent coastal populations.

5. Summary and Conclusions

Two large-scale basement-involved submarine landslide features have been described within the California Continental Borderland. Both appear to be multi-kilometer scale slide complexes with several individual blocks. The blocks are suggested to slide along well-defined shear surfaces, likely influenced by the Catalina Schist basement structure and lithology, and also appear to be back-rotated in seismic reflection profiles. The well-defined slide plane and rotation are characteristic of slumping, although some block sliding is also likely. Accumulations of incoherent reflection sequences, with hummocky surfaces, may represent break-up of the slide blocks into more plastic or fluidal slope failures on the steep (25 degrees) escarpments (e.g., NARDIN *et al.*, 1979b). Thus, the failures may initiate as slumps or slides, but evolve into plastic debris flows or possibly turbidity currents as movement progresses.

For regional prediction of potential slope failures, and slide-generated tsunami, simple estimates of seismic slope stability using the Newmark dynamic analysis technique may be appropriate offshore southern California. Application of this method to the Borderland suggests that earthquake triggering of mass movements on the unstable Catalina Schist basement slopes is feasible. Areas where pre-existing fractures and other weaknesses dip in the same direction as the steep slopes have the greatest displacement potential.

Timing or frequency of occurrence for large-scale basement-involved submarine landslides in the Borderland is presently unknown. The age of last movement for the two large features is undetermined at this time, although ponded sediments in the graben between the head scarp and large slide block at Fortymile Bank suggest a late Pleistocene age, within the past Million years. Stratigraphy from seismic profiles, and sedimentation rates inferred from bottom samples, may provide reasonable estimates of the age of some large Borderland submarine landslides, but such work remains to be completed. Combining slip rate data on offshore faults and historic seismicity may allow probabilitistic hazard assessments of the slope failure and tsunami potential.

The present study only begins to investigate the potential for large-scale basement-involved submarine landslides on the steep Borderland slopes. At present, only crude determinations of slide displacement, during single events or multiple, possibly progressive, failures has been made. This study predicts substantial initial tsunami wave heights, from 7–50 meters depending upon the slide model. High-resolution seismic profiling combined with high-resolution swath bathymetry are needed to more accurately describe existing submarine landslides. Theoretical studies on submarine landslide mechanics, analog modeling, and numerical simulations of the slope failure dynamics are needed to infer failure mechanisms and rates. These parameters can then be used to simulate observed failures and examine potential tsunami generation and risk to coastal populations.

Acknowledgements

We gratefully appreciate the discussions and comments provided by Chris Sorlien and Craig Nicholson and thoughtful reviews by Bob Bohannon and Phil Watts that helped to make this paper more accurate and readable. This paper is Contribution Number 371-111EQ from the Institute for Crustal Studies at the University of California, Santa Barbara.

References

BAILEY, E. H. (1941), *Mineralogy, Petrology, and Geology of Santa Catalina Island., California*, Ph.D. Dissertation, Stanford University, 193 pp.

BOHANNON, R. G. and GEIST, E. (1998), *Upper Crustal Structure and Neogene Tectonic Development of the California Continental Borderland*, Geolog. Soc. Am. Bull. *110*, 779–800.

BOHANNON, R. G. and GARDNER, J. V. (2003), *Submarine Landslides of San Pedro Sea Valley, Southwest Los Angeles basin*, Marine Geology, in press.

CAMPBELL, C. S., CLEARY, P. W., and HOPKINS, M. (1995), *Large-scale Landslide Simulations: Global Deformation, Velocities and Basal Friction*, J. Geophys. Res. *100*, 8267–8283.

CANN, L. (1985), *Slide-mudflow Fishermans Cove area, Santa Catalina Island.* In *Geology of Santa Catalina Island* (Gath, E. M. and Bottoms, M. M., eds) Santa Ana: South Coast Geological Society, pp. 73–81.

CLARKE, S. H., GREENE, H. G., KENNEDY, M. P., and VEDDER, J. G. (1987), *Geologic map of the inner-southern California continental margin Map No. 1a.* In *Geologic Map Series of the California Continental Margin* (Greene, H. G. and Kennedy, M. P., eds) California Division of Mines and Geology Area 1 of 7 (NOS 1206N-16) Scale 1:250,000.

CROUCH, J. K. and SUPPE, J. (1993), *Late Cenozoic Tectonic Evolution of the Los Angeles Basin and Inner California Borderland*, Geolog. Soc. Am. Bull. *105*, 1415–1534.

DENSMORE, A. L. and HOVIUS, N. (2000), *Topographic Fingerprints of Bedrock Landslides*, Geology *28*, 371–374.

EMERY, K. O., *The Sea off Southern California, a Modern Habitat for Oil* (New York, John Wiley and Sons, 1960)

FIELD, M. E. and EDWARDS, B. D. (1980), *Slopes of the southern California borderland: A regime of mass transport.* In Los Angeles: Pacific Section SEPM (M. E. Field, A. H. Bouma, I. P. Colburn, R. G. Douglas and J. C. Ingle, eds) Pacific Coast Paleogeography Symposium No. 4. pp. 169–184.

FIELD, M. E. and EDWARDS, B. D. (1993), *Submarine landslides in a basin and ridge setting, southern California.* In *Submarine Landslides; Selected Studies in the U.S. Exclusive Economic Zone* (W. C. Schwab, H. J. Lee and D. C. Twichell, eds) Denver, U.S. Geological Survey, Bulletin 2002, pp. 176–183.

FRANCIS, R. D., BOHANNON, R. G. *et al.*, (1998), *Geology and marine geophysics of Catalina Island and the California borderland.* In *Guidebook to Field Trip #7* (R. J. Behl, ed) 94th Annual Meeting, Long Beach, Cordilleran Section of the Geological Society of America, pp. 1–16.

GELI, L., BARD, P., and JULLIEN, B. (1988), *The Effect of Topography on Earthquake Ground Motion: A Review and New Results*, Bull. Seismol. Soc. Am. *78*, 42–63.

HART, M. W. (1991), *Landslides in the Peninsular Ranges, southern California.* In *Geological Excursions in Southern California and Mexico* (M. J. Walawender and B. B. Hanan, eds) San Diego, Geological Society of America Annual Meeting Guidebook, pp. 349–365.

JIBSON, R. W. (1993), *Predicting Earthquake-induced Landslide Displacements Using Newmark's Sliding Block Analysis*, Transportation Research Record *1411* (Washington, D.C., National Research Council Transportation Research Board), pp. 9–17.

KAMERLING, M. J. and LUYENDYK, B. P. (1985), *Paleomagnetism and Neogene Tectonics of the Northern Channel Islands, California*, J. Geophys. Res. *90*(B14), 12,485–12,502.

LAMAR, D. L. and ROCKWELL, T. K. (1986), *An overview of the tectonics of the Elsinore fault zone.* In *Neotectonics and Faulting in Southern California* (P. L. Ehlig, ed) Los Angeles, Cordilleran Section GSA Fieldtrip Guidebook, pp. 149–158.

LANDER, J. F., LOCKRIDGE, P. A., and KOZUCH, M. J., *Tsunamis affecting the West Coast of the United States, 1806–1992* (Boulder: U.S. Dept. of Commerce, NOAA, 1993).

LEGG, M. R. (1985), *Structural Geology and Tectonics of the Inner Continental Borderland Offshore Northern Baja California, Mexico* [unpubl. Ph.D. dissertation] Santa Barbara, University of California.

LEGG, M. R., LUYENDYK, B. P., MAMMERICKX, J., DE MOUSTIER, C., and TYCE, R. C. (1989), *Seabeam Survey of an Active Strike-slip Fault – The San Clemente Fault in the California Continental Borderland*, J. Geophys. Res. *94*, 1727–1744.

LEGG, M. R. (1991), *Developments in understanding the tectonic evolution of the California Continental Borderland.* In *From Shoreline to Abyss: Contributions in Marine Geology in Honor of Francis Parker

Shepard (R. H. Osborne, ed) Tulsa Soc. Econ. Paleontologists and Mineralogists Spec. Publ. No. 46. pp. 291–312.

LEGG, M. R., KUHN, G. G., and FROST, E. G. (1994), *Episodic Uplift of the Southern California Coast by Large Strike-slip Earthquakes* [abstract] in EOS Trans. Am. Geophys. Union Fall Meeting, San Francisco.

LEGG, M. R., NICHOLSON, C., and SORLIEN, C. (1992), *Active Faulting and Tectonics of the Inner California Continental Borderland* [abstract] in Trans. Am. Geophys. Union *73*, 588.

LEGG, M. R. and SLOSSON, J. E. (1984), *Probabilistic approach to earthquake-induced landslide hazard mapping.* In Proc. *Eighth World Conf. on Earthquake Engineering*, San Francisco, CA, pp. 445–452.

LOCAT, J., LOCAT, P., and LEE, H. J. (2003), *Numerical Modeling of the Mobility of the Palos Verdes Debris Avalanche, California, and its Implication for the Generation of Tsunamis*, Marine Geology, in press.

MARTEL, S. J. (2003), *Submarine Landslides as a Shear Fracture Phenomenon*, Marine Geology, in press.

MELOSH, H. J. (1987), *The mechanics of large rock avalanches.* In *Debris Flows/Avalanches: Process, Recognition, and Mitigation, Reviews in Engineering Geology* (J. E. Costa and G. F. Wieczorek, eds) VII, Geolog. Soc. Am. 41–49.

MCCULLOCH, D. S. (1985), *Evaluating tsunami potential.* In *Evaluating Earthquake Hazards in the Los Angeles Region* (J. I. Ziony, ed) U.S. Geolog. Survey Prof. Paper 1360, pp. 375–413.

NARDIN, T. R., EDWARDS, B. D., and GORSLINE, D. S. (1979a), *Santa Cruz Basin, California Borderland: Dominance of slope processes in basin sedimentation.* In *Geology of Continental Slopes* (L. J. Doyle and O. H. Pilkey, eds) Tulsa: SEPM Special Publication No. 27, pp. 209–221.

NARDIN, T. R., HEIN, F. J., GORSLINE, D. S., and EDWARDS, B. D. (1979b), *A review of mass movement processes, sediment and acoustic characteristics, and contrasts in slope and base-of-slope systems versus canyon-fan-basin floor systems.* In *Geology of Continental Slopes* (L. J. Doyle and O. H. Pilkey, eds). Tulsa: SEPM Special Publication No. 27, pp. 61–73.

NEWMARK, N. M. (1965), *Effects of Earthquakes on Dams and Embankments*, Geotechnique *15*, 139–160.

PINAULT, C. T. (1984), *Structure, Tectonic Geomorphology and Neotectonics of the Elsinore Fault Zone Between Banner Canyon and the Coyote Mountains, Southern California* [unpubl. M.S. thesis] San Diego State University.

PLATT, J. P. (1975), *Metamorphic and Deformational Processes in the Franciscan Complex, California: Some Insights from the Catalina Schist Terrane*, Geolog. Soc. Am. Bull. *86*, 1337–1347.

ROWLAND, S. M. (1984), *Geology of Santa Catalina Island: California Geology*, Calif. Dept. Conservation Div. Mines and Geology, pp. 239–252.

SHEPARD, F. P. and EMERY, K. O. (1941), *Submarine Topography off the Southern California Coast – Canyons and Tectonic Interpretation*, Geolog. Soc. Am. Special Paper 31.

SHREVE, R. L. (1968), *The Blackhawk landslide*, Geolog. Soc. Am. Special Paper 108.

SORENSEN, S. S., BEBOUT, G. E. and BARTON, M. D. (1991), *A field guide to the geology, petrology, and geochemistry of the Catalina Schist on Santa Catalina Island: Evidence for fluid-rock interaction, thermal evolution, and metasomatic alteration in a paleosubduction zone.* In *Geological Excursions in Southern California and Mexico* (M. J. Walawender and B. B. Hanan, eds) San Diego, Geological Society of America Annual Meeting Guidebook, pp. 272–296.

SYVITSKI, J. P. M. and HUTTON, E. W. H. (2003), *Failure of Marine Deposits and their Redistribution by Sediment Gravity Flows*, Pure appl. geophys. *160*, 2053–2069.

TAPPIN, D. R., WATTS, P., MCMURTRY, G. M., LAFOY, Y., and MATSUMOTO (2003), *Prediction of Slump Generated Tsunamis: The July 17, 1998 Papua New Guinea Tsunami*, Marine Geology, in press.

TEN BRINK, U. S., ZHANG, J., BROCHER, T. M., OKAYA, D. A., KLITGORD, K. D., and FUIS, G. S. (2000), *Geophysical Evidence for the Evolution of the California Inner Continental Borderland as a Metamorphic Core Complex*, J. Geophys. Res. *105*, 5835–5857.

TENG, L. (1985), *Seismic Stratigraphic Study of the California Continental Borderland – Structure, Stratigraphy, and Sedimentation* [unpubl. Ph.D. dissertation] Los Angeles: University of Southern California.

USGS, (1986), *Atlas of the Exclusive Economic Zone, western conterminous United States, U.S. Geological Survey*, Misc. Invest. I-1792, p. 15.

Vedder, J. G. (1975), *Acoustic Reflection Profiles, R/V KELEZ, May–June 1973, Leg 3, Offshore Southern California*, U.S. Geological Survey Open-File Report 75-265, scale 1:250,000.

Vedder, J. G. (1990), *Maps of California Continental Borderland Showing Compositions and Ages of Bottom Samples Acquired between 1968 and 1979*, U.S. Geological Survey Misc. Field Studies *Map MF-2122*, scale 1:250,000. 3 sheets.

Vedder, J. G., Beyer, L. A., Junger, Arne, Moore, G. W., Roberts, A. E., Taylor, J. C., and Wagner, H. C. (1974), *Preliminary Report on the Geology of the Continental Borderland of Southern California*, U.S. Geological Survey Misc. Field Studies *Map MF-624*, scale 1:500,000.

Vedder, J. G., Taylor, J. C., Arnal, R. E., and Bukry, D. (1976), *Map Showing Location of Selected Pre-Quaternary Rock Samples from California Continental Borderland*, U.S. Geological Survey Misc. Field Studies *Map MF-737*, scale 1:250,000. 3 sheets.

Vedder, J. G., Greene, H. G., Clarke, S. H., and Kennedy, M. P. (1986), *Geologic Map of the Mid-southern California Continental Margin Map No. 1A*. In (H. G. Greene and M. P. Kennedy, eds) Geologic Map Series of the California Continental Margin, Area 2 of 7. (NOS 1206N-15), Sacramento: Calif. Div. Mines and Geology, scale 1:250,000.

von Huene, R., Ranero, C. R., and Watts, P. (2002), *Tsunamigenic Slope Failure along the Middle America Trench in Two Tectonic Settings*, Marine Geology, in press.

Watts, P. and Raichlen, F. (1994), *Water Waves Generated by Underwater Landslides* [abstract] Seismol. Res. Lett. *65*, 25.

Wieczorek, G. F., Wilson, R. C. and Harp, E. L. (1985), *Map Showing Slope Stability during Earthquakes in San Mateo County, California*, U.S. Geological Survey Misc. Investigations *Map I-1257-E*, scale 1:62,500.

Wilson, R. C. and Keefer, D. K. (1983), *Dynamic Analysis of a Slope Failure from the 6 August 1979 Coyote Lake, California, Earthquake*, Bull. Seismol. Soc. Am. *73*, 863–877.

Wilson, R. C. and Keefer, D. K. (1985), *Predicting areal limits of earthquake-induced landsliding. In Evaluating Earthquake Hazards in the Los Angeles Region: An Earth-Science Perspective*, U.S. Geolog. Survey Prof. Paper 1360, pp. 316–345.

Wright, T. L. (1991), *Structural geology and tectonic evolution of the Los Angeles Basin, California*. In *Active Margin Basins* (Biddle, K. T., ed.) American Association of Petroleum Geologists, Memoir *52*, 35–134.

Pure appl. geophys. 160 (2003) 2053–2069
0033–4553/03/112053–17
DOI 10.1007/s00024-003-2419-8

Pure and Applied Geophysics

Failure of Marine Deposits and their Redistribution by Sediment Gravity Flows

J. P. M. SYVITSKI[1] and E. W. H. HUTTON[1]

Abstract—***SedFlux*** simulates the fill of sedimentary basins, and can be used to examine the location and attributes of sediment failure on continental margins and the runout of their associated sediment gravity flows. Numerical experiments show how the evolving boundary conditions of sea-level fluctuations, floods, storms, tectonic and other relevant processes control the rate and size of slope instabilities. By tracking deposit properties (pore pressures, grain size, bulk density, porosity), a finite-slope factor-of-safety analysis of marine deposits examines failure potential. A decider routine is used to determine whether the failed material will travel down slope as turbidity current or a debris flow. Examples provided insight into: (i) why fjords dominated by turbidity current deposition often contain debris flow deposits; (ii) how glaciated margins prograde seaward through shallow failures of low yield strength material; and, (iii) how large-scale basin subsidence can control the onset of canyon formation across continental slopes.

Key words: Sediment failure, gravity flows, numerical models.

1. Introduction

Stratigraphic simulation models (SSMs) are useful for predicting the time-varying impact of sedimentary processes, and the distribution of lithostratigraphic properties away from points of control (FRANSEEN *et al.*, 1991). The intent of this paper is to describe the general framework of ***SedFlux*** (SYVITSKI *et al.*, 1999), with particular reference to how sediment failure and sediment gravity flow routines may be used to explore the evolution of the seascapes of continental margins.

An early SSM to explore sediment failure was DELTA6 (SYVITSKI and ALCOTT, 1995). DELTA6 had delta foresets fail by oversteepening and subsequently move downslope as turbidity currents (using a non-dynamic highly parameterized scheme). Offshore slope failures were located by a finite-slope factor-of-safety (Janbu) technique. Failed sediment was transported offshore by SKRED, using a Lagrangian 1-D debris flow scheme using bilinear rheology (SYVITSKI and ALCOTT, 1995). More recently LOSETH (1999) described a SSM called DEMOSTRAT, that includes simple forms of marine sedimentation coupled to an infinite-slope stability analysis and

[1] Institute of Arctic and Alpine Research, University of Colorado at Boulder, 1560 30th Street, Campus Box 450, Boulder CO, 80309-0450, U.S.A.

rule-based gravity flow algorithms. SYVITSKI *et al.* (1999) described a beta version of 2-D ***SedFlux***-0.9F that incorporates advanced finite-slope stability analysis and fully dynamic sediment gravity flow modules. Here we describe the latest version of 2-D ***SedFlux***-1.1C with emphasis on slope stability and sediment failure experiments.

2-D ***SedFlux*** (SYVITSKI and HUTTON, 2001) simulates the lithologic character of basin stratigraphy (Fig. 1) by:

- spreading a river-delivered bedload of coarse material across the tidal range,
- dispersing the suspended sediment from the river through either surface (hypopycnal) or subsurface (hyperpycnal) plumes,
- dispersing and sorting the seafloor sediment by ocean storm events, failure of marginal deposits and their subsequent transport as sediment gravity flows (turbidity currents or debris flows),
- altering the accommodation space through thermal subsidence and tectonic displacements (formation of anticlines, synclines, growth faults, earthquake displacements, uplift), and
- compacting and preserving the sediment as a final deposit.

Recent unpublished additions include the determination of tsunami properties generated during sediment failure routines (HUTTON *et al.*, 2000). In addition, ***SedFlux*** offers the ability to include other seismic and failure modules discussed in this volume. The history and details of model development can be found elsewhere (SYVITSKI *et al.*, 1999; SYVITSKI and HUTTON, 2001).

2. Overview of 2-D ***SedFlux***

2-D ***SedFlux*** predicts basin stratigraphy in the horizontal and vertical direction. The initial basin geometry defines the initial bathymetry, at a user-specified horizontal resolution ($\sim$10 m). The vertical resolution is also user specified ($>$1 cm), within which deposited sediment have their characteristics averaged, except for grain size in which the entire size frequency distribution is tracked. Each of the modeled processes has its own unique resolution (temporal and spatial) independent of the ***SedFlux*** architecture.

Sediment is supplied to a basin through a single river, which has variable river mouth dynamics for each time step of the ***SedFlux*** simulation. The model river transports a multi-grain size suspended load and a single grain size bedload. Sediment enters the model domain solely through a river mouth located just upstream of a delta plain. The bedload component is deposited through stochastic processes across the delta plain and intertidal zone. The river's suspended sediment load is discharged as a surface plume, or as a hyperpycnal flow, depending upon the sediment-laden density of the river water.

Ocean energy is allowed to vary in intensity throughout the year, and can rework the seafloor, depending upon the strength of a particular storm, water depth, and

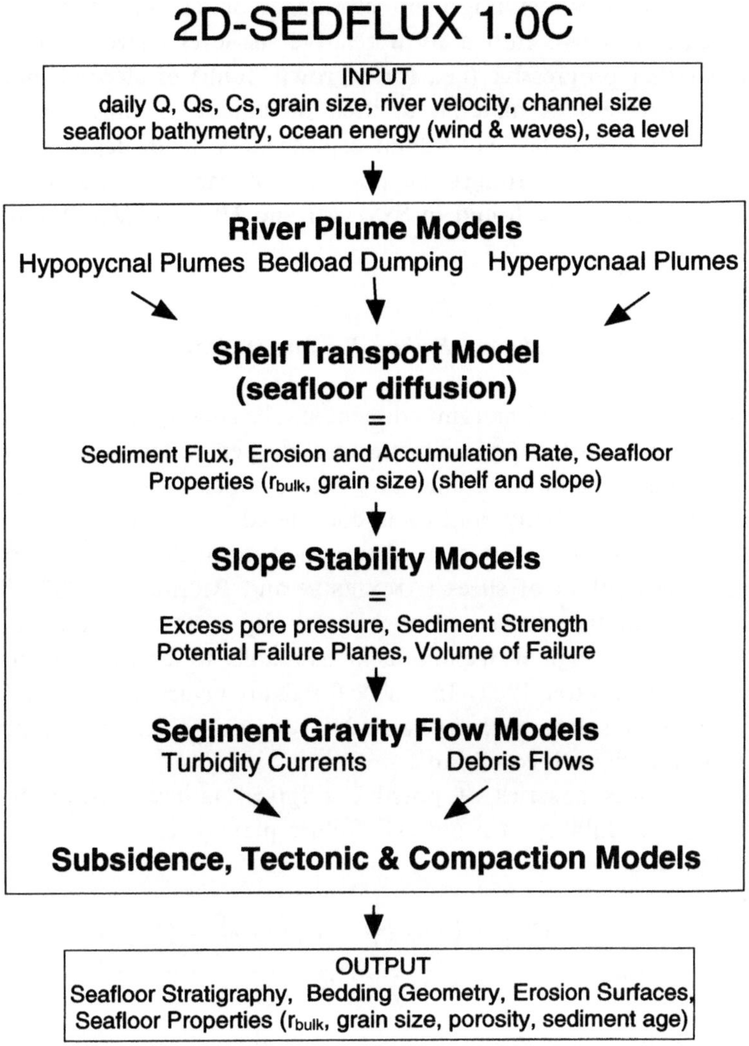

Figure 1
Flow chart of 2-D ***SedFlux*** version 1.1C, showing inputs, ***SedFlux*** model components, and outputs (modified from SYVITSKI and HUTTON, in press). Symbols include Q = water discharge, Qs = sediment discharge, and Cs = sediment concentration.

grain size properties of the seafloor. Sediment lags may locally develop and armor finer-grained material, preventing it from being winnowed away. Sediment failure occurs if potential failure planes are found to be unstable. Failed material is transported downslope as a turbidity current or as a debris flow.

Sea-level and base level are adjusted for each domain node (bin) at each time step. In this manner a eustatic sea-level curve can be used to control ocean level, while base level changes resulting from subsidence or tectonic causes can work to modify the eustatic curve into a local relative sea-level curve. Faulting is user-controlled as either progressive (i.e., for a growth fault) or stepped motion of the bedrock. The river erodes material uplifted on land over time, and the eroded material is transported either as bedload or suspended load, depending on its size characteristics. Details regardings the physics and the numerics of each of the individual processes can be found in SYVITSKI and HUTTON (2001) and references therein.

3. Sediment Failure Routine in ***SedFlux***

The failure of continental margin sediment and its subsequent movement plays an important role in transferring sediment into deeper water. Because ***SedFlux*** is designed to simulate the fill of sedimentary basins which have complex bathymetric shapes, a finite-slope stability routine is considered to be most appropriate. The geometry and location of failures are determined using the Janbu factor-of-safety analysis with the method of slices (ANDERSON and RICHARDS, 1987). The Janbu method is useful for the analysis of noncircular slip surfaces. The method ignores interslice forces, although a technique is available to correct for these forces (ANDERSON and RICHARDS, 1987). Interslice forces are invariably small, adjusting the calculated factor of safety by less than 10%, depending on the geometry of the problem as well as the soil conditions.

SedFlux examines a series of possible elliptical failure surfaces for stability (Fig. 2). The static stability of a possible failure plane is characterized through its factor of safety as,

$$F_{\text{total}} = \frac{\sum_{i=0}^{n} \left[b_i \left(c_i + \left(\frac{W_i}{b_i} - u_i \right) \tan \phi_i \right) \frac{\sec \alpha_i}{1 + \frac{\tan \alpha_i \tan \phi_i}{F_{\text{total}}}} \right]}{\sum_{i=0}^{n} W_i \sin \alpha_i} \tag{1}$$

where b is the width of a slice, c is the cohesion of the sediment, W is the weight per unit depth of the sediment, u is the excess pore pressure, ϕ is the sediment friction angle, α is the slope of the failure surface, and F_{total} is the factor of safety for the entire failure (with iterative convergence to a solution).

W is mass (M) of the sediment column times the acceleration due to gravity (g), therefore to include earthquake forces associated with ground acceleration (A_g), then $W = M\,(g + A_g)$. During an earthquake, a seismic wave exerts an additional body force on a deposit. Therefore, an earthquake probability frequency distribution (i.e., earthquake intensity as measured at the slip plane versus return interval) is used to explore dynamic slope stability. For such a scenario, associated ground acceleration (A_g) is acquired for each stability analysis performed in a simulation (typical values of A_g range from 0 to 0.3g). Thus an earthquake reduces the factor of safety,

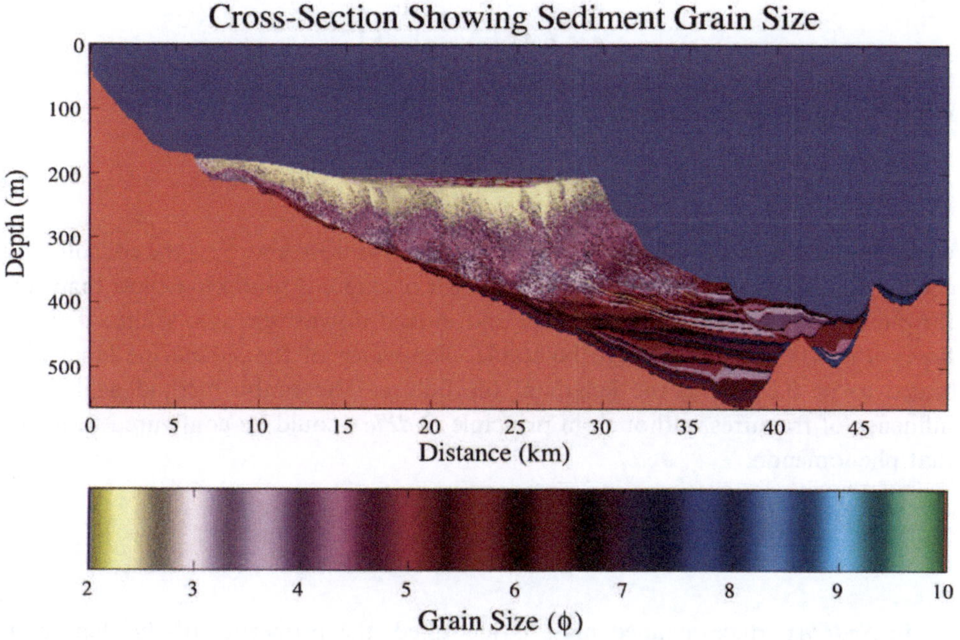

Figure 2
A 2-D ***SedFlux*** simulation of Knight Inlet, showing the sediment accumulation in terms of grain size. Distances and depths are from the beginning of the numerical model domain. Seafloor slopes were examined by FAIL subroutine and thirty-five mass failure events were generated in the 12,000-yr period; many more delta front failures due to foreset oversteepening shed low concentration turbidity currents. The 35-slope failures formed into debris flows (seen as layers of uniform grain size). The turbidites are found to be very thin bedded and graded sediment layers. Sea-level fell and rose during the simulation experiment (after MOREHEAD *et al.*, 2001).

increasing the likelihood of slope failure. WILSON and KEEFER (1983) provide another comparable method of introducing ground accelerations to a factor of safety analysis. Neither method, however, includes the contribution of water pressures generated by the earthquake itself, although in principle ***SedFlux*** could also be configured to include that phenomenon.

The excess pore pressure is obtained from consolidation theory under the assumption that these excess pressures are entirely the result of trapping pore water while compacting fine-grained sediment of low permeability. Using Gibson's graphical approximation (1958), we find

$$u_i = \frac{\gamma' z_i}{a_i} \quad , \qquad (2)$$

where γ' is the submerged specific weight of the sediment $(\gamma' = \gamma - \gamma_f)g$, and z is the depth of the failure plane with respect to the seafloor. The variable a is given through the relation,

$$a \equiv 6.4\left(1 - \frac{T}{16}\right)^{17} + 1 \ , \tag{3}$$

where T is a constant defined as,

$$T \equiv \frac{m^2 t}{C_v} \ , \tag{4}$$

where m is sedimentation rate, t is the time for deposition, and C_v is the consolidation coefficient for the sediment. If the static factor of safety is found to be less than some threshold, then the sediment is failed and moved downslope as a sediment gravity flow, otherwise it is deemed to be stable. The width of the submarine landslide is scaled to be 0.25 times the length of the failure. The Janbu method neglects the influence of fractures, although in principle ***SedFlux*** could be configured to include that phenomenon.

4. The Mass Flow Decider

In ***SedFlux***, once a failed mass is identified, the properties of the deposit are examined. If the failed material is clayey (user-defined, e.g., >10% clay), then the failed sediment mass is transported down-slope as a debris flow. An appropriate clay content is used as a proxy for ensuring low hydraulic conductivity and low permeability and thus the generation of a debris flow (MOHRIG *et al.*, 1999). This ensures that a viscoplastic (Bingham) rheology is achieved (ELVERHØI *et al.*, 2000). As outlined below, our modeled dynamics does not allow for the debris flow to erode the seafloor. Thus the grain size of the final deposit is equal to the homogenized grain size of the initial failed sediment mass.

If the material is sandy or silty, with little clay (user-defined, e.g., <10% clay) then the failed sediment mass is transported down-slope as a turbidity current. If the material moves as a turbidity current, flow accelerations may cause erosion of the seafloor and this entrained sediment may increase the clay content of the gravity flow compared to the initial failed sediment mass. Deposition of sand and silt along the flow path may result in the turbidity current transporting primarily clay in the distal reaches along the flow path (SKENE *et al.*, 1997).

5. Turbidity Currents

In 2-D ***SedFlux*** 1.1C, turbidity currents are initiated at the river mouth as a hyperpycnal flow, or from a slope failure located somewhere in the offshore, as allowed by the above decider routine. In the case of a slope failure, the failed mass does not transform directly into a turbidity current, rather the initial concentration of the gravity flow is set below some user-defined value less than the Bagnold limit (9% concentration by volume: MULDER *et al.*, 1997). Using this initialized concentration, flow height is established and gravity pulls the current down across the seafloor, while entraining water and sediment into the flow, depending on the dynamics associated with the evolving boundary conditions. Friction around and within the flow work to slow the flow, until all momentum is consumed.

Different turbidity current subroutines within ***SedFlux*** are available. INFLO is briefly reviewed here, but BANG is also available. BANG1D is 1-D, layer-averaged, and based on the Lagrangian form of the equations for the conservation of fluid, sediment, momentum and turbulent kinetic energy in a turbidity current (PRATSON *et al.*, 2000, 2001).

INFLO is a psuedo-2-D layer-averaged (specified but variable width), turbidity current model, configured in the Eulerian steady-state form, employing the conservation of fluid, sediment, and momentum. INFLO uses an internal friction parameterization and drag coefficient terms to close the solution (SKENE *et al.*, 1997; MULDER *et al.*, 1997). A turbidity current contains particles in suspension supported by fluid turbulence but maintaining enough of a sediment density that a gravitational driving force is produced that accelerates the flow downslope as a mixture of sediment and water. Energy is lost from friction of the flow across the seafloor, from friction beneath an overlying water mass, and from friction within the flow (grain-grain interaction). The balance of these terms is expressed as,

$$\frac{\partial U}{\partial t} = g_0 \sin(\beta C) - \frac{E + C_d}{h} U^2 - g_0 \left(\frac{eC - 1}{e - 1} \right) \cos(\beta C) \tan \phi \ , \qquad (5)$$

where, U is vertically integrated downslope velocity, g_0 is reduced gravity equal to $g(\rho_s - \rho)/\rho, \beta$ is slope, E is entrainment coefficient, C_d is drag coefficient, h is flow thickness, ρ is ambient fluid density, ρ_f is density of the flow, ρ_s is grain density, ϕ is angle of internal friction, C is bulk volume concentration of sediment calculated for each grain size, and g is acceleration due to gravity.

As the flow travels it entrains sea water and flow volume is increased. Fluid continuity is:

$$\frac{\partial Q}{\partial x} = EUW_F \ , \qquad (6)$$

where W_F is width of the flow (i.e., width of submarine canyon or basin floor), Q is volume discharge, x is distance in the downslope direction, and E is the entrainment coefficient that relates the amount of seawater that is entrained by flow.

Sediment can enter and exit the flow. As the flow falls below some critical velocity, sediment begins to be deposited. Above some critical shear stress, sediment from the seafloor is eroded and entrained into the flow. The continuity equation for suspended load is:

$$\frac{\partial J_i}{\partial x} = E_R - D_R = \left[\left(\frac{C_d \rho_f U^2 - \sigma_b}{\sigma_a}\right) \frac{\varphi_i W_F}{\text{day}}\right] - D_R \ , \tag{7}$$

where E_{Ri} is the rate of erosion averaged over one day of the i-th grain size of the seafloor, day is 86,400 s, σ_a is the gradient in the shear strength of the seafloor sediment, σ_b is the shear strength of the sediment at the seafloor, and D_{Ri} is the rate of deposition of the i-th grain size,

$$D_R = \begin{cases} 0 & \text{if } U > U_{cr} \\ \dfrac{\lambda_i J_i}{U}\left(1 - \dfrac{U^2}{U_{cr}^2}\right) & \text{if } U \leq U_{cr} \end{cases} \ , \tag{8}$$

and U_{cr} the critical velocity for deposition is given by,

$$U_{cr} = w_s \Big/ \sqrt{C_d} \ , \tag{9}$$

where w_s is the settling velocity of the i-th grain size.

INFLO uses a drag coefficient in three separate equations: (i) drag on the upper body of the turbidity current influencing the rate of entrainment (5); (ii) drag on the lower part of the flow linked to seafloor erosion (7); and, (iii) drag on settling particles, controlling the critical velocity of deposition (9).

6. *Debris Flows*

Debris flows are modeled after the properties of a Bingham plastic (viscoplastic) fluid, where deformation is driven by the excess of stress beyond the yield stress (JIANG and LEBLOND, 1992; PRATSON *et al.*, 2000; IMRAN *et al.*, 2001). The model neglects the tangential stress acting on the water-mud interface, because the viscosity of water is much smaller than that of the mud, and the basal shear of the mudflow is much greater than the interfacial shear (LIU and MEI, 1989). In addition, there is a no slip condition on the slide bottom.

The governing equations of the Lagrangian form of the depth-averaged debris flow equations, including both viscous and plug flow regions, are:

Continuity:

$$\frac{\partial D}{\partial t} + \frac{\partial}{\partial x}\left[U_p D_p + \frac{2}{3} U_p D_s\right] = 0 \ . \tag{10}$$

Momentum (shear layer):

$$\frac{2}{3}\frac{\partial(U_p D_s)}{\partial t} - U_p \frac{\partial D_s}{\partial t} + \frac{8}{15}\frac{\partial\left(U_p^2 D_s\right)}{\partial x} - \frac{2}{3} U_p \frac{\partial(U_p D_s)}{\partial x} = D_s g\left(1 - \frac{\rho_w}{\rho_m}\right) S - D_s g \frac{\partial D}{\partial x} - 2\frac{\mu}{\rho_m}\frac{U_p}{D_s} \ . \tag{11}$$

Momentum (plug flow layer):

$$\frac{\partial(U_pD_p)}{\partial t} + \frac{\partial\left(U_p^2D_p\right)}{\partial x} + U_p\frac{\partial D_s}{\partial t} + \frac{2}{3}U_p\frac{\partial(U_pD_s)}{\partial x}$$
$$= D_pg\left(1 - \frac{\rho_w}{\rho_m}\right)S - D_pg\frac{\partial D}{\partial x} - \frac{\tau_y}{\rho_m} , \tag{12}$$

where D is the total depth of the debris flow ($D_p + D_s$); D_p and U_p are the depth and layer-averaged velocity of the upper plug zone, respectively; D_s and U_s are the depth and layer-averaged velocity of the lower shear layer, respectively; g is acceleration due to gravity; S is slope; ρ_w is density of ocean water; ρ_m is density of the mud flow; τ_y is yield strength, and μ is kinematic viscosity. The yield strength controls how the deposit is stretched out. We typically use values of 100 Pa, but higher values will allow the debris flow to flow downslope more as a block then as a flow.

*7. Application of 2-D **SedFlux** 1.1C to Slope Stability Studies*

Our first example is from a 12,000-yr simulation of the sediment architecture forming Knight Inlet, a fjord in British Columbia, where the seafloor slopes are examined by the FAIL subroutine (Fig. 2; MOREHEAD *et al.*, 2001). Knight Inlet is a basin that is annually subjected to numerous gravity-flow events (MOREHEAD *et al.*, 2001). A river introduced a variable sediment load into the basin every day. An initial dropping relative sea-level (due to isostatic rebound as the glaciers retreated) and then a slow rise in relative sea-level (due to rising eustatic sea-level) was imposed on the simulated basin. Numerous small failures, near the delta foresets, generated turbidity currents that deposited their sandy load along the proximal part of the pro-delta slope (Fig. 2). Thirty-five large slope failures were initiated using the Janbu finite-slope factor-of-safety analysis (Fig. 2). Failure lengths ranged between 600 m and 3800 m, all were located in muddy facies on rather steep slopes (2° to 3.4°). The failure planes reached depths in the sediment column of 8 m to 30 m (Fig. 2). The failed sediment masses moved as debris flows, with run out distances of between 5 and 21 km. This situation of rarer debris flows mixed with numerous turbidity current events is common to fjord environments (HEIN and SYVITSKI, 1992; SYVITSKI and LEE, 1997).

The second simulation example is from a prograding margin fed with clayey sediment from an ice sheet and thus dominated by the deposition from debris flows (Fig. 3). Numerous slope failures occur near the shelf-slope break and the failed sediment masses moved as debris flows, depositing their sediment load along the length of the slope. Debris flow lobes accumulate near the base of the continental slope (Fig. 3B). Debris flow-dominated margins are common along glaciated and polar margins, where the sedimentary feed is via clayey basal till (VORREN *et al.*, 1998; ELVERHØI *et al.*, 1997). The ***SedFlux*** simulation can be compared with a

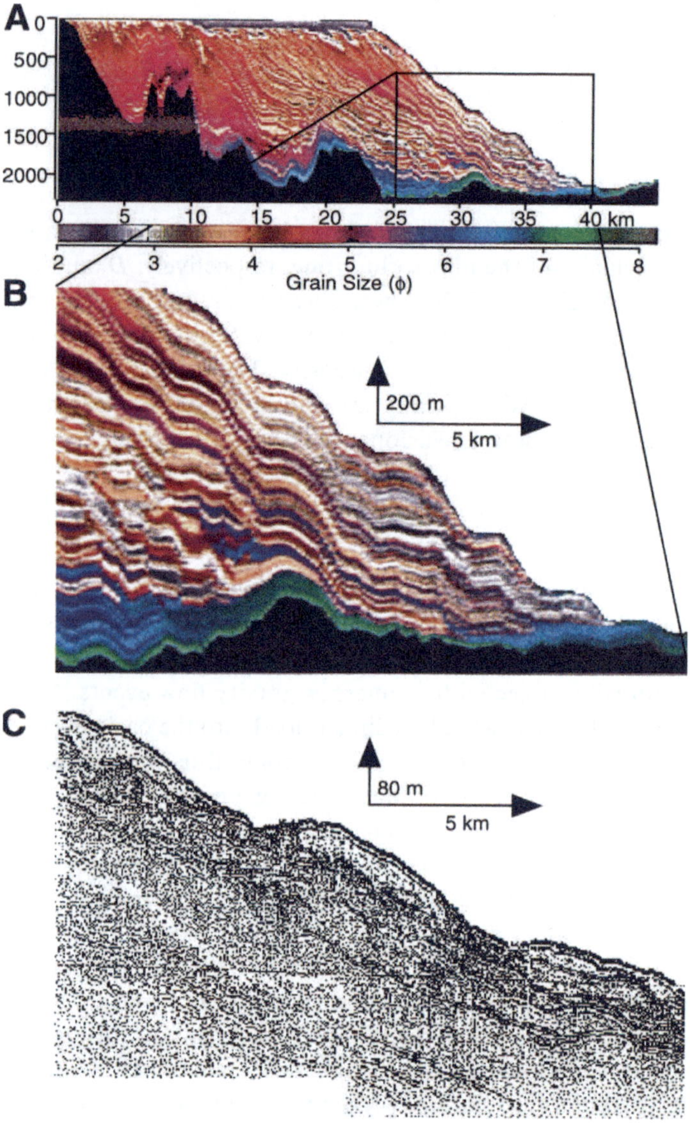

Figure 3

(A) A 2-D ***SedFlux*** run of a prograding debris flow-dominated margin. Numerous slope failures occur near the shelf-slope break and associated debris flows deposit their sediment load along the length of the slope. (B) Close-up of the amalgamation of debris flow lobes near the base of the continental slope. (C) 10-cubic inch Sleeve-Gun seismic record collected on HU93-030 off of the Kangerlussuaq Fan, East Greenland (for details on this record see STEIN and SYVITSKI, 1997). Debris flow lenses show characteristic transparent or chaotic internal reflections.

seismic record collected off of the Kangerlussuaq Fan, East Greenland (Fig. 3C, for details on this record see STEIN and SYVITSKI, 1997). Debris flow lenses show characteristic transparent or chaotic internal reflections.

A reasonable question arising from a comparison of our first and second examples would be why some debris flows produce thick deposits on the basin floor (i.e., Fig. 2), whereas other debris flows deposit their load along the entire length of a continental slope, stretched and thin (i.e., Fig. 3)? The answers lies among four parameters: failure size, steepness of the flow path (i.e., steepness of the continental slope), length of the runout path down a continental slope, and the yield strength of the failed material. If the failure size is small, the driving force (submerged failure weight) is concomitantly small and thus there is a tendency for sediment to deposit nearer the point of failure. If the flow path is steep, then there is a tendency for sediment to be deposited far from its point of failure, often near the slope-rise break. If the yield strength is low, there is a tendency for the deposit to stretch out. If the flow path is long, sediment will be more often retained on the continental slope. Thus the conditions for debris flows to deposit on the continental slopes of glaciated margins (Fig. 3) include numerous small failures of initial deposits possessing relatively low yield strengths, where debris flows travel across a relatively long and gentle slope. DIMAKIS *et al.* (2000) have used a similar analysis to explain the progradation of the Svalbard-Barents Sea margin.

Our third example (Fig. 4) is used to show how the *decider routine* works. Three 2000-year simulations are provided of a river delta prograding into an idealized margin. In each case, the exact sediment input to the numerical model (river system) is used. Sea-level is held constant. Sediment accumulates on a 0.75° slope over the first 25-km section of the basin, then the basin flattens outs at a water depth of 430 m. Less than 20% of the sediment discharged by the river is sand, the remainder is silt and clay. The river bedload is spread out over ~1 km distance from the river mouth, providing for steep foresets of muddy sand.

In the first realization (Fig. 4A), the decider is set to have failed sediment masses move seaward as a debris flow if the averaged grain size of the material is ≥10% clay. Many debris flow deposits can be seen and most of them have petered out at the base of the prodelta slope with little spillage onto the flat basin floor. Close to 50 debris flows were generated of a variety of size and grain size characteristics. Minimal turbidite deposits were generated (a small thin deposit can be seen at the 120 km location position). In the second realization (Fig. 4B), debris flows were only generated if the averaged grain size of the material is ≥20% clay. As a result there is a 66% decrease in debris flow generation and a concomitant increase in turbidity currents. The latter flowed out onto the basin floor, travelling tens of kilometers and depositing their load as thin bedded, vertically and laterally graded turbidites. In the third realization (Fig. 4C), no debris flows were generated as the criteria was set to ≥40% clay. Therefore all failed material was transported as turbidity currents. Turbidity current-dominated margins (Fig. 4C) give rise to steeper marginal slopes in

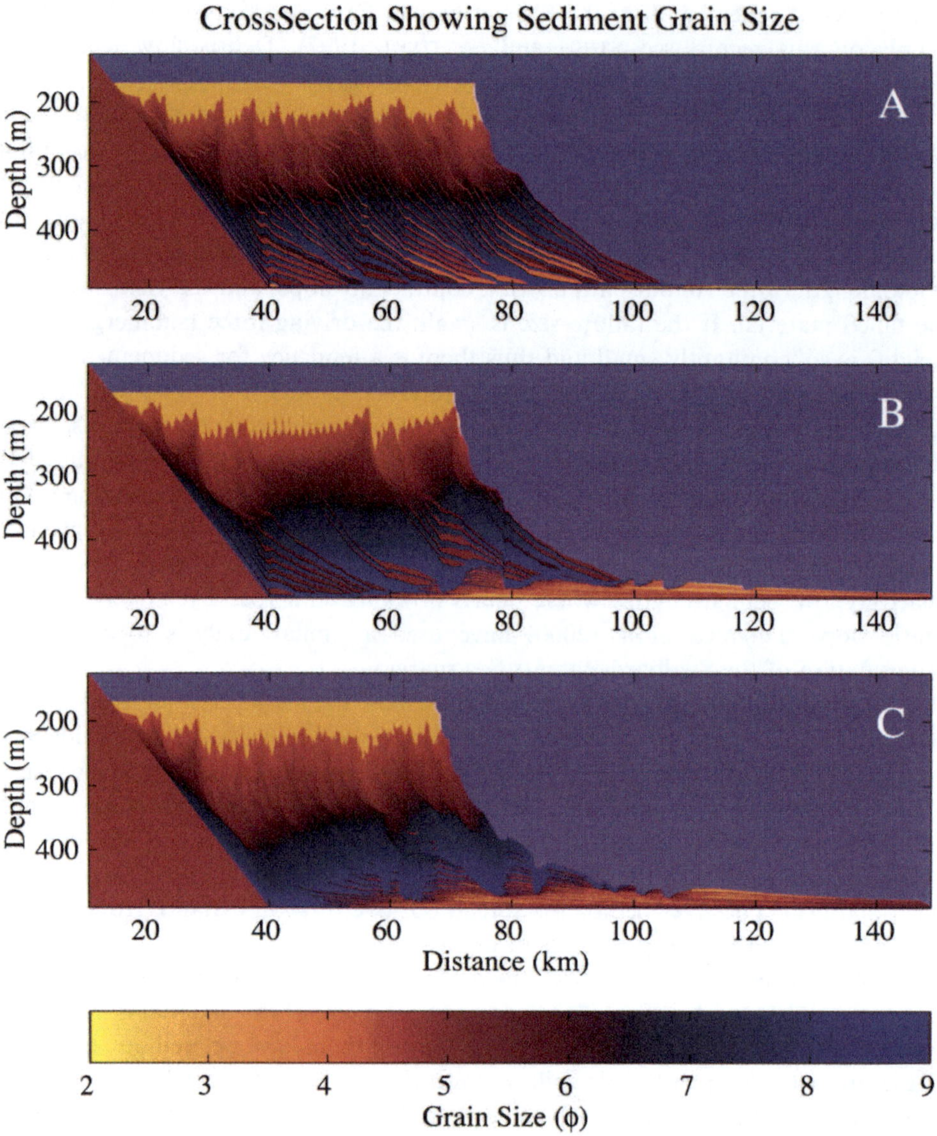

Figure 4

Three realizations of a deepwater deltaic margin prograding over a 3/4° slope onto a flat basin floor. The realizations show how ***SedFlux*** can use the grain size of the failed sediment mass to control the generation of debris flows and turbidity currents. In (A) the decider is set to have failed sediment masses move seaward as a debris flow if the averaged grain size of the material is ≥10% clay. The criterion changes to ≥20% clay in (B) and ≥40% clay in (C). As a consequence all failed sediment masses were translated as debris flows in (A) and as turbidity currents in (C), with the (B) realization falling some place in between. See text for details.

contrast to debris flow-dominated margins (Fig. 4A). The steeper slopes cause the turbidity currents to ignite and this in turn increases the rate older seafloor deposits are eroded, including older turbidite beds on the lower slope.

Our last example is using ***SedFlux*** to isolate the feedback of single process (crustal subsidence) on the formation of sediment failures near the shelf-slope break (Fig. 5). Subsidence is controlled by the sediment load (SYVITSKI and HUTTON, 2001). The model run simulates the growth of an unnamed NW Australian passive margin, using over 200,000 time steps, with the sediment input scaled to the Quaternary time period. Sea-level fluctuates 100 ± 20 m, through six separate episodes. The impact of subsidence includes development of accommodation space, thus providing the

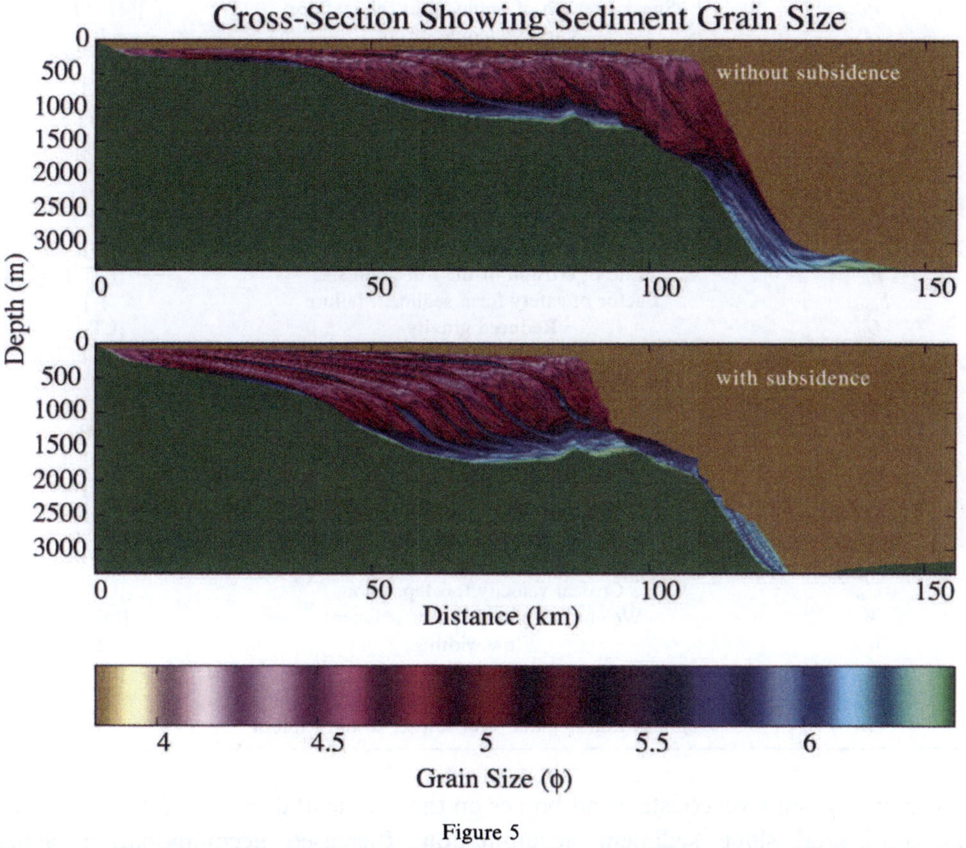

Figure 5
Formation of sediment failures near the shelf-slope break under different progradation-aggradation scenarios, as modeled by 2-D ***SedFlux***. The model run simulates a basin fill with over 200,000 time steps, and sediment input scaled to the Quaternary time period. Sea-level fluctuations were 100 m ± 20 m through six separate episodes. In simulation (A), subsidence was turned off. (B) provides a similar simulation except subsidence was turned on. Increased accommodation space associated with subsidence limits progradation of the coastline. Submarine canyons formed at the 120-km position in simulation (B) are exposed, yet in simulation (A) large canyons did not form and others are infilled (after SYVITSKI and HUTTON, 2001).

Table 1

Symbol list

Notation	Description	Unit
α	Slope of failure surface	[–]
β	Slope of the seafloor	[–]
ϕ	Angle of internal friction [sediment friction angle]	[–]
γ	Submerged density of sediment	[ML^{-3}]
φ	Volume concentration of *i*-th grain size in seafloor	[–]
ρ	Density of the ambient fluid	[ML^{-3}]
ρ_f	Density of the flow	[ML^{-3}]
ρ_s	Density of the sediment grain	[ML^{-3}]
σ_a	Shear strength vertical gradient in seafloor sediments	[ML^{-2}T^{-2}]
σ_b	Shear strength of sediment at the seafloor	[ML^{-1}T^{-2}]
A	Variable (equation 3)	[–]
A_g	Horizontal ground acceleration from an earthquake	[LT^{-2}]
B	Width of a slice in a failure	[L]
c	Sediment cohesion	[MT^{-2}]
C	Vertically averaged flow concentration	[–]
C_d	Drag coefficient	[–]
C_v	Consolidation coefficient	[L^{2}T]
D_{R_i}	Rate of deposition of the *i*-th grain size	[LT^{-1}]
E	Entrainment coefficient	[–]
E_{R_i}	Rate of erosion of the *i*-th grain size	[LT^{-1}]
F_{total}	Factor of safety for a sediment failure	[–]
G_0	Reduced gravity	[LT^{-2}]
H	Height of the flow	[L]
J	Flux of the *i*-th grain size between elements	[L^{3}T^{-1}]
M	Sedimentation rate	[LT^{-1}]
M	Mass of a column of sediment	[M]
Q	Volume discharge between flow elements	[L^{3}T^{-1}]
T	Time	[T]
T	Variable (equation 4)	[–]
U	Excess pore pressure	[ML^{-1}T^{-2}]
U	Vertically averaged flow velocity	[LT^{-1}]
U_{cr}	Critical velocity for deposition	[LT^{-1}]
W	Weight per unit depth of sediment	[MT^{-2}]
W_F	Flow width	[L]
w_s	Settling velocity	[LT^{-1}]
X	Horizontal position	[L]
Z	Depth of failure plane with respect to the seafloor	[L]

burial of low sea-level coastal sand bodies on the continental shelf, and the reduction in continental slope sediment accumulation. Increased accommodation space associated with the ongoing subsidence limits progradation of the coastline in the first model run (Fig. 5A). Submarine canyons formed at the 120 km position are exposed (Fig. 5A). Yet when modeled without the feedback of load-dependent subsidence, large canyons did not form and smaller canyons appear infilled (Fig. 5B). Similar scenarios have been used to explain the morphology of passive continental margins (O'Grady *et al.*, 2000).

8. Summary

2-D ***SedFlux*** version 1.1C provides the marine geotechnical community with a new tool in understanding the complex feedbacks between marine failure and erosion and deposition by sediment gravity flows. 2-D ***SedFlux*** contains many of the important processes that distribute sediment in a marine marginal setting to form a sedimentary architecture. 2-D ***SedFlux*** can be used to investigate new numerical expressions of slope stability or mass gravity subroutines. For example we have tested two versions of debris flow models to see how differences in numerical representations and theory can result in slightly different run out lengths and deposit shapes (SYVITSKI *et al.*, 1999). Various turbidity current models (1-D, 2-D; Eulerian; Lagrangian) are presently in the process of being tested.

Results from ***SedFlux*** simulations support field observation on why fjords contain many turbidites intermixed with rarer debris-flow deposits. The high sedimentation rate environment of the steep and clay-poor foresets gives rise to numerous small failures that result in turbidity current deposition. Infrequent but large failures of the clayey prodelta environment produce debris flows that deposit their load on the basin floor. ***SedFlux*** simulations also capture how glaciated margins prograde seaward through numerous shallow failures of low yield-strength deposits at the shelf-slope break. The clay-rich sediment (till) moves downslope as debris flows depositing their load along the long run out distance of the continental slope. Finally, ***SedFlux*** simulations show how large-scale basin subsidence can affect the onset of canyon formation across continental slopes.

New upgrades to the model include the prediction of the tsunami distributions derived from landslides (HUTTON *et al.*, 2000).

Acknowledgments

We thank our geotechnical-modeling contributors: Jane Alcott, Homa Lee, Gary Parker, Lincoln Pratson, Phil Watts, J. Locat, and Jeff Wong. We are grateful for the strong encouragement of the Office of Naval Research under the MG&G management of Dr. Joseph Kravitz. This manuscript forms a contribution to ONR's STRATAFORM project under the coordination of Dr. Charles Nittrouer.

REFERENCES

ANDERSON, M. G. and RICHARDS, K. S., *Slope Stability: Geotechnical Engineering and Geomorphology* (New York, John Wiley & Sons 1987).

BOOTH, J. S., SANGREY, D. A., and FUGATE, J. K. (1985), *A Nomogram for Interpreting Slope Stability of Fine-grained Deposits in Modern and Ancient Environments*, J. Sedimentary Petrology *55*, 29–36.

CARUCCIO, F. T., *Stratigraphic and geochemical controls on the occurrence of acidic mine waters and predictive technologies.* In *Quantitative Dynamic Stratigraphy* (T. A. Cross, ed.) (Prentice-Hall, N.Y. 1989) pp. 581–588.

DIMAKIS, P., ELVERHØI, A., HØEG, K., SOLHEIM, A., HARBITZ, C., LABERG, J. S., VORREN, T. O., and MARR, J. (2000), *Submarine Slope Stability on High-latitude Glaciated Svalbard-Barents Sea Margin*, Marine Geology *162*, 303–316.

ELVERHØI, A., NOREM, H., ANDERSEN, E. S., DOWDESWELL, J. A., FOSSEN, I., HAFLIDASON, H., KENYON, N. H., LABERG, J. S., KING, E. L., SEJURP, H. P., SOLHEIM, A., and VORREN, T. O. (1997), *On the Origin and Flow Behaviour of Submarine Slides on Deep-sea Fans along the Norwegian-Barents Sea Continental Margin*, Geo-Marine Lett. *17*, 119–125.

ELVERHØI, A., HARBITZ, C. B., DIMAKIS, P., MARR, J., MOHRIG, D., and PARKER, G. (2000), *On the Dynamics of Subaqueous Debris Flows*, Oceanography, *13*, 109–117.

FRANSEEN, E. K., WATNEY, W. L., KENDALL, S. G., and ROSS, W. (1991), *Sedimentary Modeling: Computer Simulation and Methods for Improved Parameter Definition*, Kansas Geolog. Survey Bull. *233*, 524 pp.

GIBSON, R. E. (1958), *The progress of consolidation in a clay layer increasing in thickness with time*. Géotechnique *8*, 171–182.

HEIN, F. J. and SYVITSKI, J. P. M. (1992), *Sedimentary Environments and Facies in an Arctic Basin, Itirbilung Fiord, Baffin Island, Canada*, Sedimentary Geology *81*, 1–29.

HUTTON, E. W. H., WATTS, P., and SYVITSKI, J. P. M. (2000), *Tsunami Generation during the Growth of a Continental Margin*, AGU 2000 Fall meeting EOS Supplement *81*(48), F649.

IMRAN, J., HARFF, P., and PARKER, G. (2001), *A Numerical Model of Submarine Debris Flow with Graphical User Interface*, Computers and Geoscience, *27*(6), 717–730.

JIANG, L. and LEBLONDE, P. H. (1992), *The Coupling of a Submarine Slide and the Surface Waves which it Generates*, J. Geophys. Res. *97*(C8), 12,731–12,744.

LIU, K. F. and MEI, C. C. (1989), *Slow Spreading of a Sheet of Bingham Fluid on an Inclined Plane*, J. Fluid Mechanics *207*, 505–529.

LOSETH, T. M. (1999), *Submarine Massflow Sedimentation: Computer Modelling and Basin Fill Stratigraphy*, Springer Lecture Notes in Earth Science *82*, 156 pp.

MARTINEZ, P. A. and HARBAUGH, J. W. (1993), *Simulating Nearshore Environments*, Computer Methods in Geosciences, Pergamon Press, *12*, 265 pp.

MOREHEAD, M. D., SYVITSKI, J. P. M., and HUTTON, E. W. H. (2001), *The Link between Abrupt Climate Change and Basin Stratigraphy: A Numerical Approach*, Global and Plantetary Change *28*, 115–135.

MORHIG, D., ELVERHØI, A., and PARKER, G. (1999), *Experiments on the Relative Mobility of Muddy Subaqueous and Subaerial Debris Flows, and their Capacity to Remobilize Antecedent Deposits*, Marine Geology *154*, 117–129.

MULDER, T., SYVITSKI, J. P. M., and SKENE, K. (1997), *Modelling of Erosion and Deposition by Sediment Gravity Flows Generated at River Mouths*, J. Sedimentary Res. *67*(3), 571–584.

NITTROUER, C. A. and KRAVITZ, J. H. (1996), *STRATAFORM: A Program to Study the Creation and Interpretation of Sedimentary Strata on Continental Margins*, Oceanography 9(3), 146–152.

NITTROUER, C. A. (1999), *STRATAFORM: Overview of its Design and Synthesis of its Results*, Marine Geology *154*(1–4), 3–12.

O'GRADY, D. B., SYVITSKI, J. P. M., PRATSON, L. F., and SARG, J. F. (2000), *Categorizing the Morphologic Variability of Siliciclastic Passive Continental Margins*, Geology *28*, 207–210.

PRATSON, L., IMRAN, J., PARKER, G., SYVITSKI, J. P. M., and HUTTON, E. W. H. (2000), D*ebris flow versus turbidity currents: A modeling comparison of their dynamics and deposits*. In A. Bouma (ed.), *AAPG Special Publication*. 57–71.

PRATSON, L., IMRAN, J., HUTTON, E. W. H., PARKER, G., and SYVITSKI, J. P. M. (2001), *BANG1D: A One-dimensional, Lagrangian Model of Subaqueous Turbid Clouds*, Computers and Geosciences *27*(6), 701–716.

ROBINSON, A. R., LERMUSIAUZ, P. F. J., and QUINCY SLOAN III, N. (1998), *Data Assimulation*. In *The Sea: Volume 10 – The Global Coastal Ocean: Processes and Methods* (K. H. Brink and A. R. Robinson, eds.) (John Wiley and Sons, New York 1998) pp. 541–594.

SCHÄFER-NETH, C. and STATTEGGER, K., *Icebergs in the North Atlantic: Modeling circulation changes and glacio-marine deposition*. In *Computerized Modeling of Sedimentary Systems* (J. Harff, W. Lemke and K. Stattegger, eds.) (Springer, New York 1999) pp. 63–78.

SHELTON, J. L. and CROSS, T. A., *The influence of stratigraphy in reservoir simulation*. In *Quantitative Dynamic Stratigraphy* (T. A. Cross, ed.) (Prentice-Hall, N.Y. 1989) pp. 589–600.

SKENE, K., MULDER, T., and SYVITSKI, J. P. M. (1997), *INFLO1: A Model Predicting the Behaviour of Turbidity Currents Generated at a River Mouth*, Computers and Geoscience *23*(9), 975–991.

STEIN, A. B. and SYVITSKI, J. P. M., *Glaciation-influenced debris flow deposits: East Greenland slope*. In *Glaciated Continental Margins: An Atlas of Acoustical Images* (T. W. Davies, T. Bell, A. Cooper, H. Josenhans, L. Polyak, A. Solheim, M. Stoker and J. Stravers, eds.) (Chapman and Hall, London 1997) pp. 134–135.

SYVITSKI, J. P. M. and ANDREWS, J. T. (1994), *Climate Change: Numerical Modelling of Sedimentation and Coastal Processes, Eastern Canadian Arctic*, Arctic and Alpine Research *26*(3), 199–212.

SYVITSKI, J. P. M. and ALCOTT, J. M. (1995), *DELTA6: Numerical simulation of basin sedimentation affected by slope failure and debris flow runout*. In *Proceedings of the "Pierre Beghin" International Workshop on Rapid Gravitational Mass Movements*, 305–312, CEMAGREF, Grenoble, France.

SYVITSKI, J. P. M. and HUTTON, E. W. H. *SEDFLUXIOC: An advanced process – response numerical model for the fill of marine sedimentary basins*. Computers and Geoscience *27*(6), 731–754.

SYVITSKI, J. P. and LEE, H. J. (1997), *Sequence Stratigraphy of Lake Melville, Labrador, during Ice-sheet Retreat Since 10,000 Years BP*, Marine Geology *143*, 55–80.

SYVITSKI, J. P., PRATSON, L., and MOREHEAD, M. (1997), *EARTHWORKS: A Large Spatial Scale Numerical Model to Study the Flux of Sediment to Ocean basins and Reworking of Deposits over Various Time Scales*, AGU 1997 Fall Meeting EOS Suppliment *78*(46), F258.

SYVITSKI, J. P. M., MOREHEAD, M., and NICHOLSON, M. (1998), *HYDROTREND: A Climate-driven Hydrologic-transport Model for Predicting Discharge and Sediment to Lakes or Oceans*, Computers and Geoscience *24*(1), 51–68.

SYVITSKI, J. P. M., PRATSON, L., and O'GRADY, D. (1999), *Stratigraphic predictions of continental margins for the navy parameterizations*. In *Numerical Experiments in Stratigraphy: Recent Advances in Stratigraphic and Computer Simulations* (J. W. Harbaugh, L. W. Whatney, E. Rankay, R. Slingerland, R. Goldstein and E. Franseen, eds.). SEPM Special Publication *62*, 219–236.

TETZLAFF, D. M. and HARBAUGH, J. W. (1989), *Simulating Clastic Sedimentation* (Van Nostrand Reinhold, New York) 202 pp.

VORREN, T. O., LABERG, J. S., BLAUME, F., DOWDESWELL, J. A., KENYON, N. H., MINERT, J., RUMOHR, J., and WERNER, F. (1998), *The Norwegian-Greenland Sea Continental Margins: Morphology and Late Quaternary Sedimentary Processes and Environment*, Quaternary Science Review *17*, 273–302.

WILSON, R. C. and KEEFER, D. K. (1983), *Dynamic Analysis of a Slope Failure from the 1979 Coyote Lake, California, Earthquake*, Bull. Seismol. Soc. Am. *23*, 863–877.

To access this journal online:
http://www.birkhauser.ch

Pure appl. geophys. 160 (2003) 2071–2086
0033–4553/03/112071–16
DOI 10.1007/s00024-003-2420-2

Pure and Applied Geophysics

Re-examination of the Source Mechanism of the 1998 Papua New Guinea Earthquake and Tsunami

FUMIHIKO IMAMURA[1] and KAZUMASA HASHI[1]

Abstract—Simulation of tsunami propagation and runup of the 1998 Papua New Guinea (PNG) earthquake tsunami using the detailed bathymetry measured by JAMSTEC and adding bathymetric data at depths less than 60 m is carried out, reproducing the tsunami energy focus into Warapu and Arop along the Sissano Lagoon. However, the computed runup heights in the lagoon are still lower than those measured. Even if the error in estimating the fault parameters is taken into consideration, computational results are similar. Analysis by the wave ray method using several scenarios of the source size of the tsunami and location by the wave ray method suggests that a source characterized by small size in water 1,000-m deep approximately 25 km offshore the lagoon, best fits the arrival determined from the interviews with eyewitnesses. A two-layer numerical model simulating the interaction of the tsunami with a landslide is employed to study the behavior of a landslide-generated tsunami with different size sand depths of the initial slide just outside the lagoon. A landslide model with a volume of 4–8 × 10^9 m^3 is selected as the best in order to reproduce the distribution of the measured tsunami runup in the lagoon. The simulation of a tsunami generated in two stages, fault and landslide, could show good agreement with the runup heights and distribution of the arrival time, but a time gap of around 10 minutes remains, suggesting that a tsunami generated by the mainshock at 6:49 PM local time is too small for people to notice, and the following tsunami triggered by landslide or mass movement near the lagoon about ten minutes after the mainshock attacked the coast and caused the huge damage.

Key words: Tsunami, Papua New Guinea, wave ray, landslide.

1. Introduction

In the evening of Friday, the 17th of July, 1998, a large tsunami swept across the sandbar that forms the outer margin of Sissano Lagoon, west Sepik, PNG. Initial news reports stated that the tsunami struck west of the town of Aitape in west Sepik province, hitting at least four villages. The tsunami occurred in the early evening at the start of the weekend. Many residents were preparing dinner or unwinding after a long work week. Within seconds, the scene changed to hell. An abnormal decline of the sea surface following the second earthquake was noticed, according to interviews with local residents. Immediately after, a black wall of water is said to have rushed shoreward. The damaged area is located on the sandbar of the Sissano Lagoon and

[1] Disaster Control Research Center, School of Engineering, Tohoku University, Aoba 06, Sendai 980-8579, Japan.

the tsunami easily overran the sandbar and washed away residents, livestock, and houses in a moment.

The first PNG tsunami investigation was carried out by the International Tsunami Survey Team (ITST) from 31 July to 7 August, 1998. The survey of the damaged area confirmed the 7–10 m wave reported and found one place where the waves were larger – up to 15 m with extreme overland flow velocities of 10 to 15m/s. The severe damage and significant wave heights were confined to a relatively short (30 km) stretch of coast between Aitape and Sissano Villages, and were especially large along the sandbar forming the outer margin of Sissano Lagoon (KAWATA *et al.*, 1999). Although several land-based field investigations and marine surveys were carried out to measure the detailed topography and find evidence for faults and submarine landslides, and despite several numerical analyses (TAPPIN *et al.*, 1999; TANIOKA, 1999; MATSUYAMA *et al.*, 1999) of the tsunami propagation, several questions remain unanswered regarding the distribution of runup heights in Sissano Lagoon and the arrival time of the tsunami.

Two earthquakes at 6:48 PM and 7:08 PM by local as well as global seismic records were noted, consistent with accounts of eyewitnesses in the area of west Sepik.

The narrow focus of the tsunami energy and tsunami heights larger than those estimated from the earthquake magnitude alone (KOSHIMURA *et al.*, 1998) suggests another or additional tsunami source closer to the shore than the epicenter beneath the New Guinea (NG) trough. Epicenter locations are not at the NG trough, but rather along the shore. Sissano Lagoon formed by subsidence during the 1907 earthquake, has important geographic characteristics. Several rivers along the coast supply substantial sediment to the coast during the rainy season. This sediment is deposited on a shallow sea region with a mild slope, suggesting that a marine landslide triggered by the earthquake might have taken place, in view of the high possibility of sizable sedimentation on a sea bottom of mild slope. This study focuses on the reconstruction of the tsunami source by carrying out a numerical analysis of wave propagation and applying the wave ray method.

2. Bathymetry Data and Condition for the Computation

Bathymetry and Tsunami Focusing

The propagation and evolution of a tsunami are strongly affected by the bottom bathymetry, which can influence the direction of the propagation, the energy concentration, and the distribution of tsunami heights along the shore. Tsunami analyses require integrated bathymetry and coastal topography for the study area. The first numerical analyses of the 1998 PNG tsunami used low-resolution bathymetry data, and the resulting simulations could not produce the distribution of measured tsunami runup heights along the Sissano Lagoon. JMSTEC (Japan

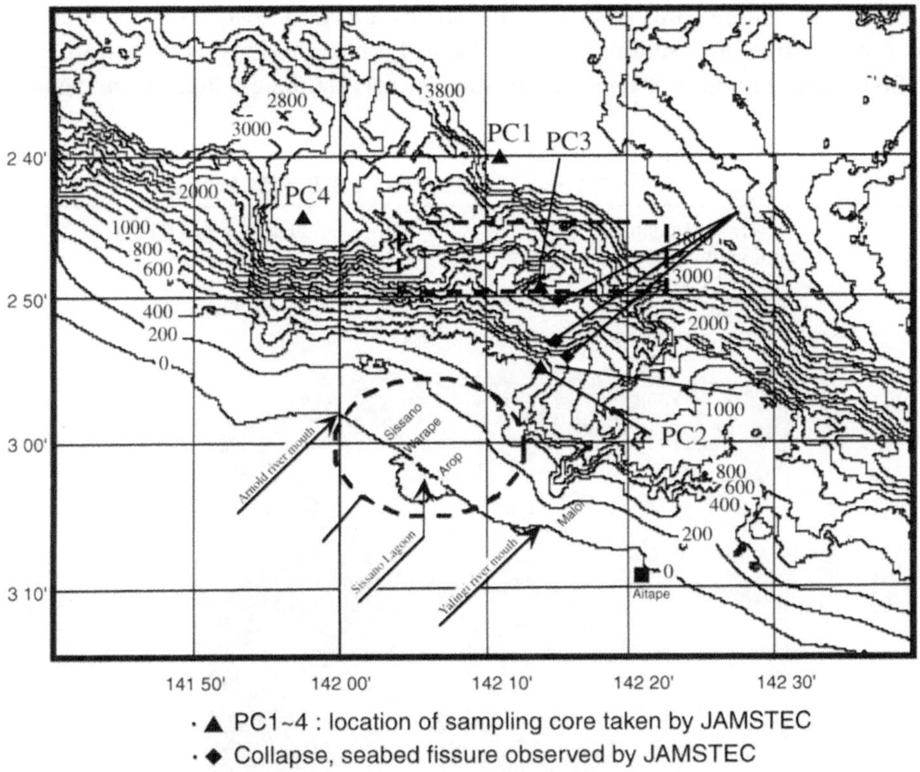

Figure 1

Newly obtained bathymetry off Sissano, Papua New Guinea. This bathymetry is limited to ocean more than 200 m deep. Rhombi indicate the landslide traces, collapse features, fractures and cracks observed during ROV dives.

Marine Science and Technology Center) carried out a marine survey and produced new bathymetric data, as shown in Figure 1. MATSUYAMA *et al.* (1999) simulated tsunami propagation assuming only fault motion to generate the tsunami with the new bathymetric data and demonstrated that tsunami energy is focused in the lagoon. However, the tsunami heights obtained by the simulation did not correspond well to those measured along the shore, the main causes being the lack of bathymetry data shallower than 200 m and the handling of runup by that model. For further numerical analyses, an appropriate runup condition with topographic data in the shallow sea and on land is required.

Present Numerical Condition

In the present study, a numerical simulation including full treatment of runup (rather than a purely reflective boundary) is compared to the measured runup heights in the lagoon. The new data are modified by editing the data measured at shallow sites (less than 100 m) by BRUCE (1999). Other conditions in the computation are summarized in Table 1. The modified sea bottom in the shallow area becomes steeper than the former estimate (Fig. 2), reducing the

Table 1

Numerical condition for the tsunami propagation and runup

Governing Equation	Shallow water theory
Condition along the shore	Runup
Spatial Grid size	200 meter
Time step	0.25 second

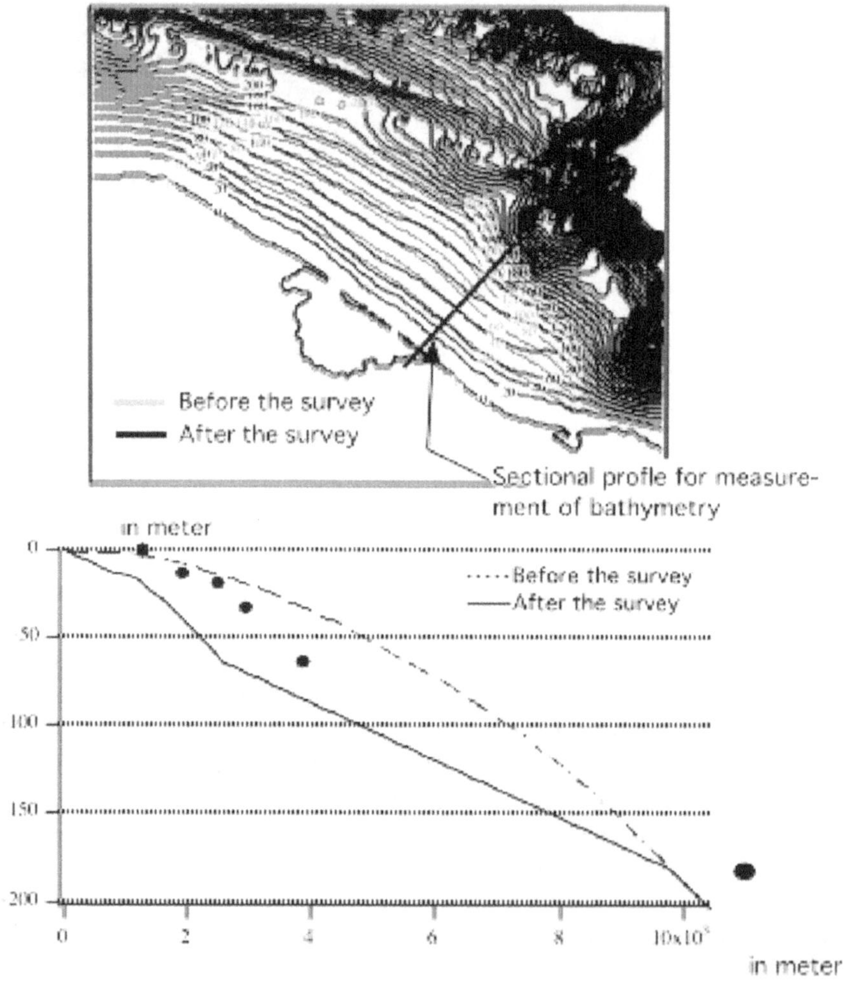

Figure 2

Contour maps of water depth before and after the addition of bathymetric data from the area less than 60 m deep (upper) and the sectional profile of the sea bottom in a shallow region (lower).

numerical dissipation in the computation due to larger steepness of the wave front, on a gently sloping sea bottom. Figure 3 demonstrates the energy concentration of the tsunami into the lagoon by using the ray method from a

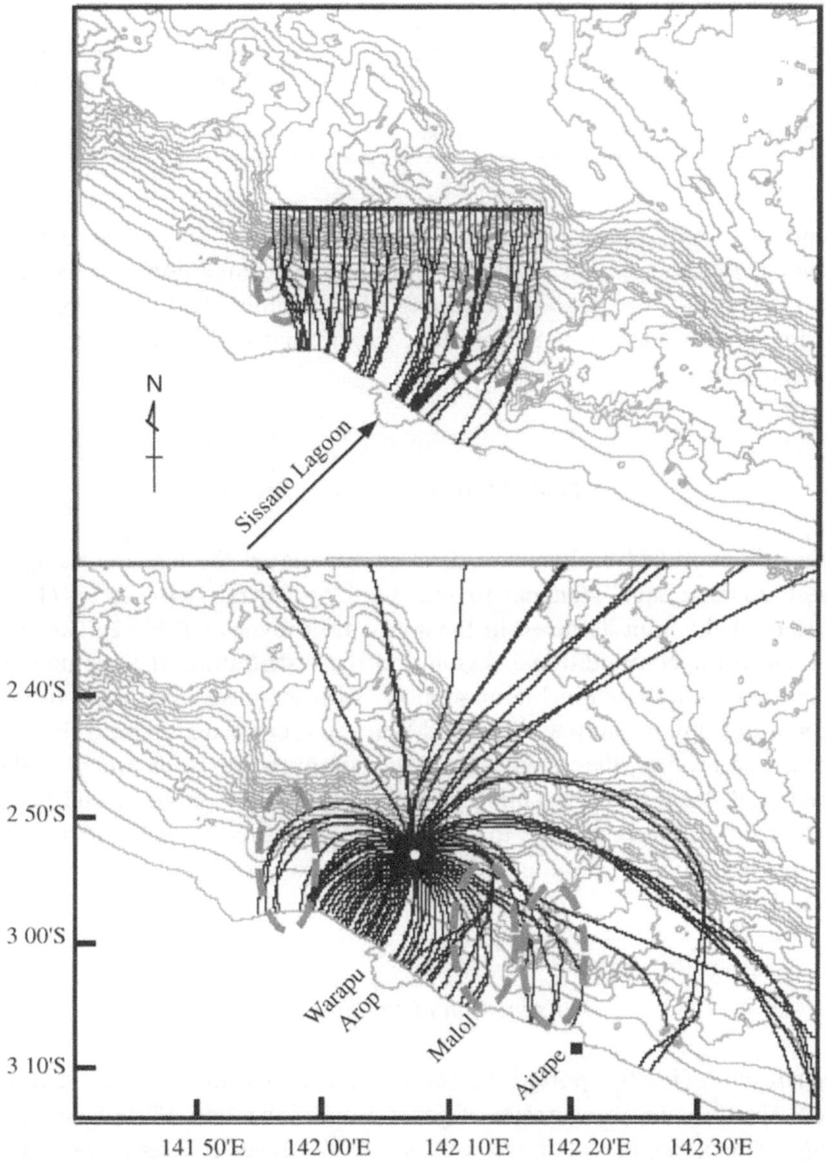

Figure 3
Pattern of energy radiation from both line and point sources using the wave-ray method, which indicates energy concentration into the lagoon due to refraction by the Sissano shelf and canyon at the northeast edge of the shelf.

Table 2

Assumed volume and location of the initial landslide

	Location in Fig. 5	Diameter (km)	Averaged Thickness (m)	Total Volume ($\times 10^9$ km^3)
Case-1	Point 8	10 km	30 m	2.3
Case-2	Point 6	10 km	30 m	2.3
Case-3	Point 8	20 km	30 m	9.2
Case-4	Point 8	10 km	60 m	4.6
Case-5	Point 8	15 km	45 m	7.9

line and a point source, with the modified new bathymetric data. Being independent of the type of the tsunami source, the simulation reveals the energy concentration into the lagoon, meaning the Sissano shelf and concave ridge submarine canyons act as an important factor in focusing the tsunami toward Sissano Lagoon.

3. Fault Motion Generated Tsunami

Using the modified bathymetric data, we simulated the tsunami using only an east-west trending fault identical to that used by MATSUYAMA *et al.* (1999). The parameters of the fault assumed in the study are shown in Table 2. The maximum water level in the computational region and the distribution of the tsunami runup along the coast are shown in Figure 4, along with the measured data. Although the location of maximum runup at Arop and Warapu is consistent in both the computed and measured profiles, the magnitude of the computed runup is much smaller than observed. A dislocation on the fault more than three times larger would be required to produce the measured data in the simulation. This simulation shows that tsunami generation by the fault alone could not reproduce the observed runup heights of more than 10 m.

4. Estimation of the Source Area

TAPPIN *et al.* (1999) reported the results of a marine investigation, suggesting a sediment slump as the mechanism of generation of the 1998 PNG tsunami. Visual imaging using a remote operated vehicle (ROV) revealed fresh brecciated talus deposits, vertical and near-vertical cliffs, seabed fissures, and headwall collapse, indicating that this area recently has been subject to sea bottom movement. However, neither large-scaled debris flows nor significant landslides were reported, partially

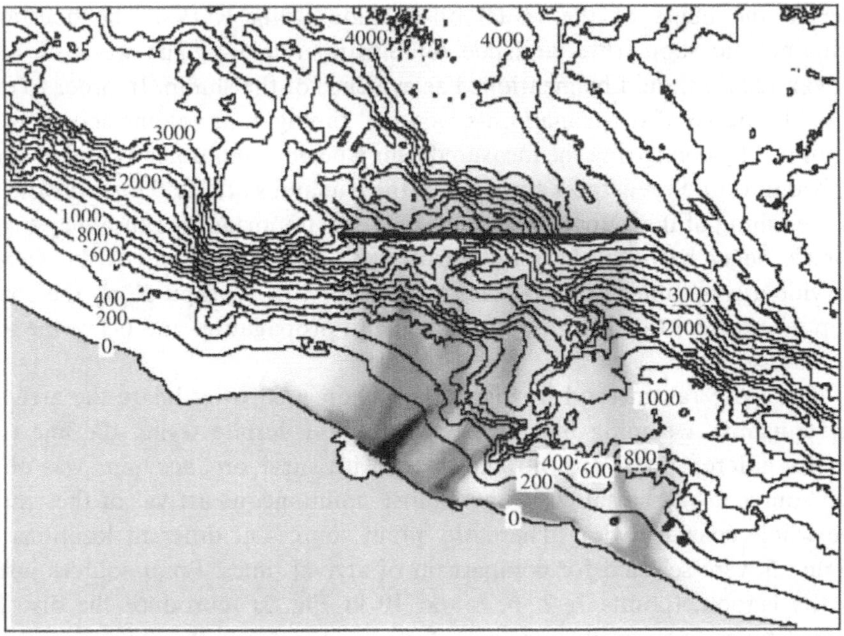

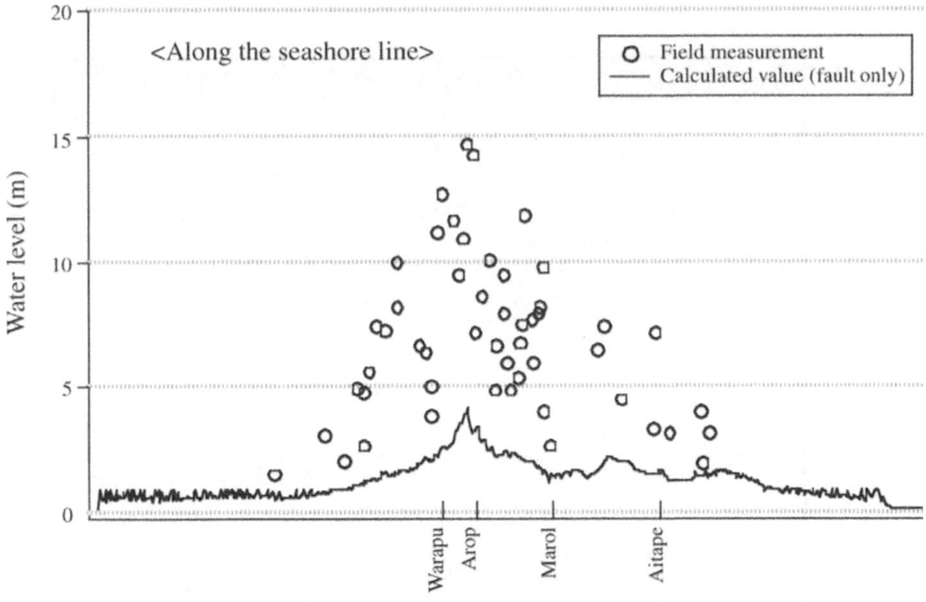

Figure 4
Spatial distribution of the model results of the maximum water level for a fault-generated tsunami (upper) and the profile of field-measured runup heights along the shore, indicating that the computed result is less than the measured.

due to the limits of the visual observation using ROVs. The question of the tsunamigenic submarine landslide off Sissano remains. The work of SWEET and SILVER (2003) should be mentioned as evidence of the slump. In order to discuss the possible source of the tsunami, its location, timing, and volume should be carefully examined by comparing the measured runup heights and other available data such as arrival time and eyewitness accounts of the tsunami's attack. The survivors' accounts of the timing of the tsunami (DAVIES, 1998) are important because they suggest a 20-minute delay between the main earthquake and the arrival of the tsunami, and provide arrival times at each place from Aitape to Warapu along the coast, which help us interpret the direction of the tsunami propagation and hence the location of the source.

The wave ray method in Figure 3 is again used to estimate the arrival time of the tsunami. Assuming first a line source, and despite trying the line tsunami at several different locations, no agreement with survivors' accounts was obtained. A line source always resulted in the almost simultaneous arrival of the tsunami over the whole coastal region. Therefore, point sources at different locations shown in Figure 5 were selected for comparison of arrival times. Point sources just offshore in the lagoon, (points 1, 2, 6, 8 and 10 in Fig. 5) reproduce the distribution of arrival times fairly well. For those farther away from the lagoon, such as points

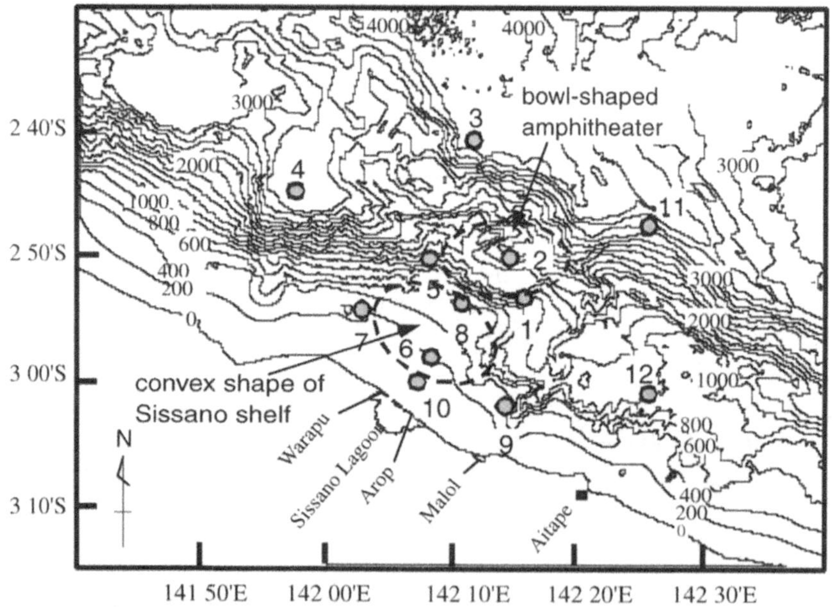

Figure 5
Locations of the model landslide sources to estimate the location of the tsunami source by the comparison with the estimated arrival time.

1 and 2 near the bowl-shaped amphitheater, the arrival times approximate the account of the eyewitness . On the other hand, point sources at points 6 and 8 located on the convex part matching the reported time of arrival at the Sissano shelf show a distinctive U-shaped profile, which can be selected among all shown in Figure 6 as best. Thus we conclude that survivors' accounts of the timing suggest a point source for the tsunami on the convex part of the Sissano shelf, around 25 km offshore.

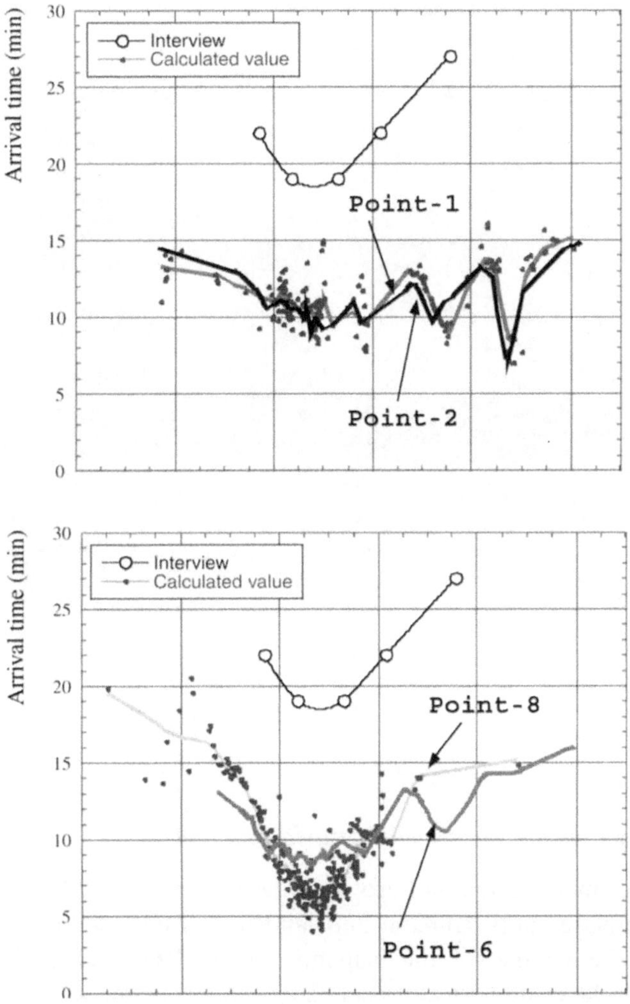

Figure 6
Comparison of the arrival time between the estimated and computed results at different locations. Models using points 6 and 8 show the best agreement with profile of estimates based on eyewitness accounts.

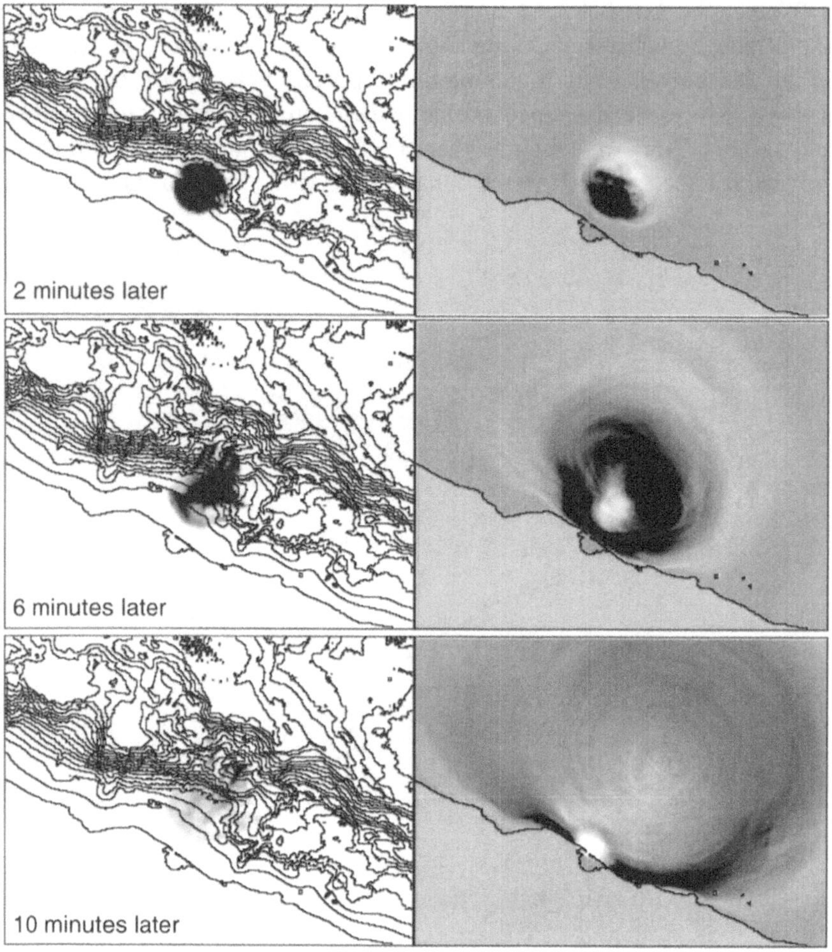

Figure 7
Landslide (left) and tsunami (right) propagation pattern 2, 6, and 10 minutes after the landslide generation. The tsunami caused by the landslide (model 1) propagates circularly.

5. *Landslide-generated Tsunami*

Two-layer Model for the Landslide-generated Tsunami

The wave ray method does not provide wave runup heights; therefore a numerical model that simulates both tsunami and landslide is necessary in order to discuss quantitatively the volume of the tsunami source. IMAMURA and IMTEAZ (1995) developed a two-layer numerical model using shallow water wave theory integrated over each layer with a non-horizontal bottom/interface. Hydrostatic pressure distribution, uniform density and velocity distributions are assumed in each layer. First, tsunamis caused only by landsliding at specific locations, such as points 6 and

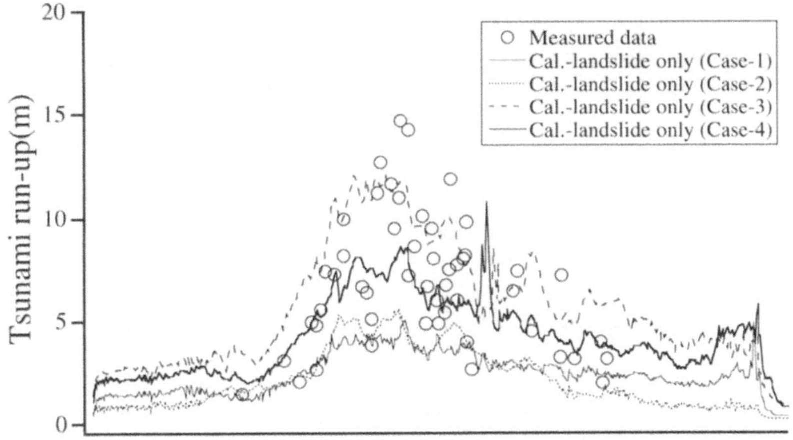

Figure 8
Comparison of the runup heights between the measured and computed results for models 1 to 4 with various volumes for the landslide.

8, are simulated. Figure 7 shows an example of the propagation of both the water surface and the landslide, which are simulated as the upper and lower layers in the model. The movement of the landslide is started when sudden ground motion or some other external disturbance alters the balance of internal forces, the resisting force and the driving force. The landslide flowing down to the PNG trench decreases the water surface toward the shore at its back and increases it toward the open ocean at the front, generating a typical wave propagation pattern as shown in Figure 7. Eyewitnesses to the PNG tsunami reported the initial motion of the tsunami as a recess of the ocean. The point source area assumed here, with an area smaller than that of the faulting source, results in a circular propagation pattern. Most of the wave energy is directed towards the coastal area, as shown in the lower part of Figure 3, which contributes to the concentration in the lagoon of the tsunami.

Volume of the Source

Simulations assuming various sizes of the initial landslide area were carried out to determine the best-fitting model by comparing the model results with runup measured in the field. Figure 8 shows an example of the results of model runup distribution along the shore for Cases 1 to 4. The computed profiles all agree well with the field observations, however simulations with sources of more than 4.0×10^9 m^3 in Cases 3 and 4 show the best agreement with the measured runup. After assuming various shapes for the source, we concluded that a landslide having a diameter (d) of 15 km at a depth (p) of 30–60 m was the best; these parameters result in volumes ranging from 4 to 8×10^9 m^3. Figure 9 shows the computed runup distribution for a source having $d = 15$ km and $p = 45$ m (Case 5, volume is

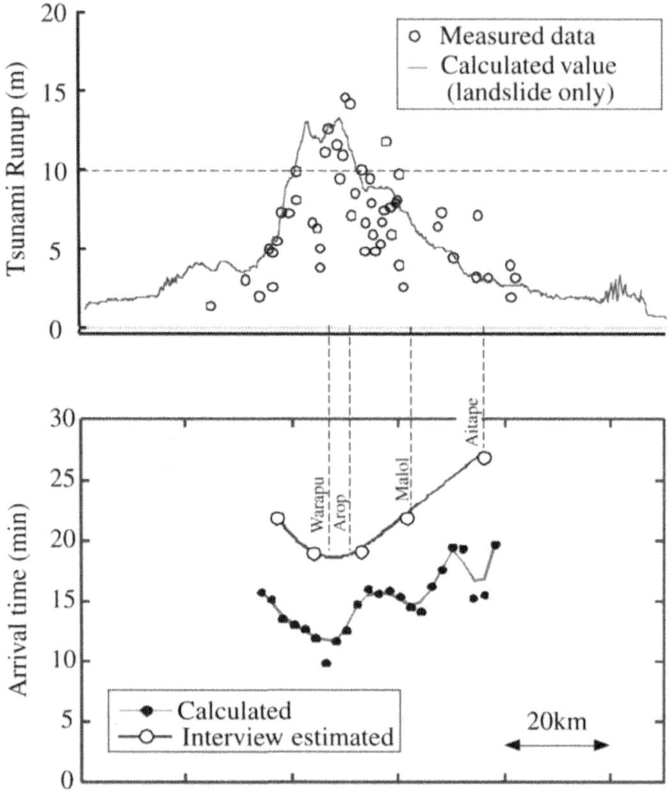

Figure 9
Comparison of the runup heights (upper) and arrival times (lower) along the shore between the estimated and computed results for model 5.

7.9×10^9 m^3). These results also reproduce well both the measured runup and the estimated arrival time.

Estimated Arrival Time and Measured Runup

Here we deal with two items that limit the location and size of the tsunami source; timing of the tsunami's arrival and its measured runup. Although the profile of the arrival time computed with the best-fit model corresponds to that determined from the interview, a gap of about 10 minutes between them remains. As long as the bathymetry is correct, we can obtain an accurate travel time from source to shore using numerical simulation. DAVIES (1998) carried out interviews with the local people to estimate the arrival time relative to the main and following shocks as shown in Figure 10. Unless a clock or TV is available during the event, eyewitness recollections of the arrival time relative to the shocks are more reliable than their remembrance of absolute arrival time. If both computed and estimated time are

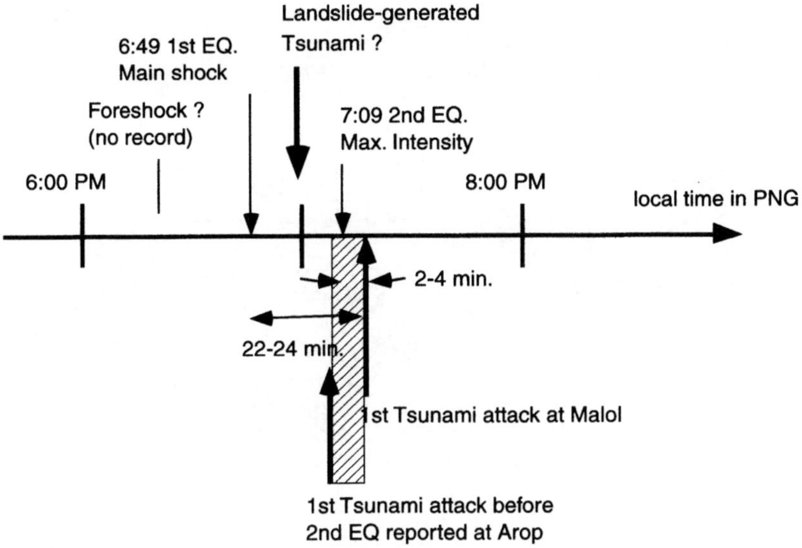

Figure 10
Timetable for the earthquake and tsunami estimated from interviews with survivors. The time gap of the arrival time in Figure 9 suggests the generation of the tsunami due to the landslide around 7:00 PM local time.

reliable, the time lapse in Figure 9 indicates that the landslide-generated tsunami originated 10 minutes after the main shock (6:49 PM local time). Therefore, we interpret that the landslide-generated tsunami took place around 7:00 PM local time.

The ITST tried to measure the heights of tsunami traces in as many areas as possible. Several different values are plotted for the same area so that the data of the measured runup in Figure 9 are scattered around the lagoon. Since the computed results to be compared with those measured are selected and plotted as the maximum value along transects running across the shore line, we should select the measured data in the same way, we should use the envelope of scattered runup data. However, local effects such as a splash and wave reflection from trees or houses can increase tsunami heights locally. It seems that the runup heights of 15 m at Arop might be the result of such local effects. Without information on the situation of the tsunami traces and surrounding area, we could not judge which data had been so amplified and therefore were too large to be compared with the computed data. In light of this, it is suggested that the numerical result for Case-5 is good enough to reproduce the main features of the tsunami along Sissano Lagoon, except for locally high runups.

Multi-generated Tsunami

We propose a tsunami generated by both fault and landslide with a time gap to explain the 1998 PNG tsunami. Figure 11 shows the present model of source areas

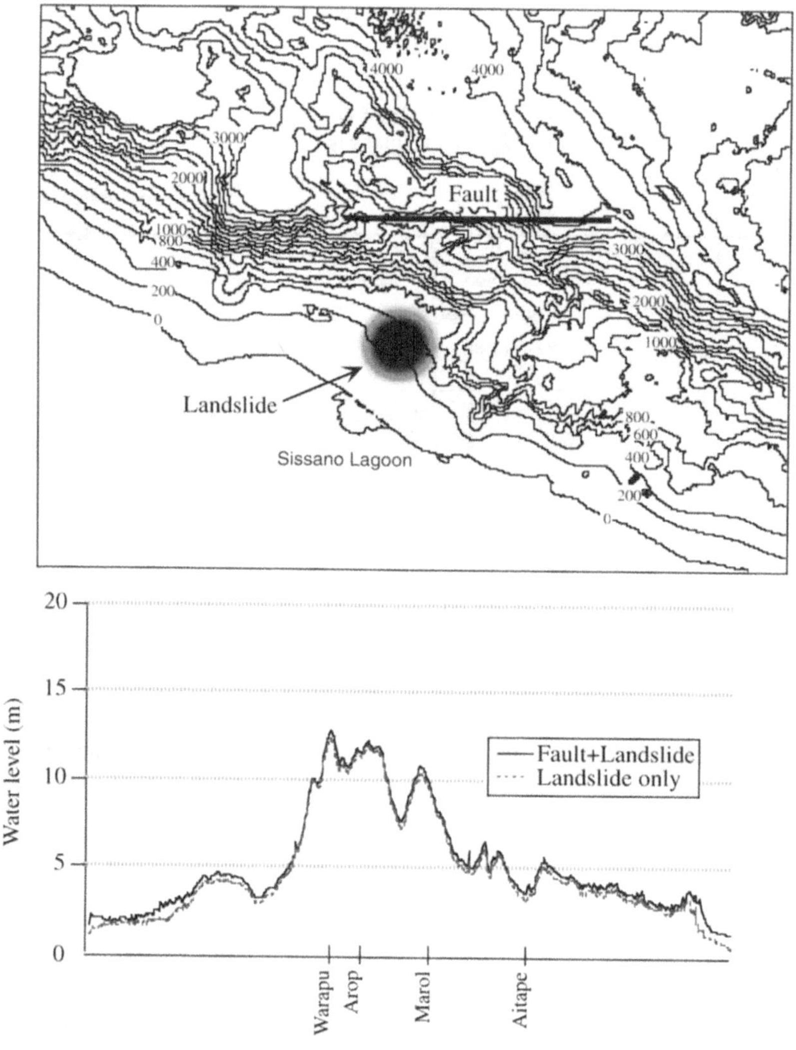

Figure 11
Location of the sources for combined fault and landslide generated tsunami (upper) and resulting runup heights compared with the model for landslide only.

and their locations. The distribution of runup heights from this model along the lagoon is also plotted in the figure and compared with the result from only the landslide, indicating that the addition of the fault-generated tsunami does not increase the maximum runup. This is probably because the fault-generated component of the tsunami is small and the time gap between fault motion and the generation of the landslide can shift the phase of the wave crest for both tsunamis, possibly causing destructive interference of the waves. The location of the fault off the convex part of the Sissano shelf and of the landslide on the shelf closer to the coast leads to radiation of the wave toward the Pacific Ocean for the fault component and into Sissano Lagoon for the landslide component, because of trapped and leaky waves on the shelf (KAJIURA, 1972). This is consistent with the result of TANIOKA (1999) that the far-field tsunami observed in Japan can be explained by the faulting alone.

6. Conclusions

The measured runup and estimated arrival time make it possible to estimate the source mechanism of the tsunami observed along Sissano Lagoon. The wave ray method and two-layer model for the landslide-generated tsunami suggest the tsunamis by both fault and landslide, with the slide located around 25 km offshore of the lagoon and attaining volume of 4 to 8×10^9 m^3. The time gap, around 10 minutes, of generation of both tsunamis explains the arrival time well. New questions arise however; the volume of 4 to 8×10^9 m^3 is large enough to be observed by the visual image investigation by JAMSTEC. Actually, no significant traces except for cracks and headwall collapse (SWEET and SILVER, 2003) were reported.

Acknowledgements

This research was partially supported by a Grant-in-Aid from the Ministry of Education, Science and Culture, Japan.

REFERENCES

BRUCE, J. (1999), Private communication.

DAVIES, H. L. (1998), *Sissano Tsunami: A Reconstruction of the Events of July 17 Based upon Interviews with Survivors*, EOS, Trans. Am. Geophys. Union *79*(45), F572–F749, [abstract].

KOSHIMURA, S., HAMZAH, L., and IMAMURA, F. (1998), *Preliminary Report on Earthquake and Tsunami of the 1998 Papua New Guinea*, Disaster Control Research Center, Tohoku University, 18 pp.

IMAMURA, F. and IMTEAZ, M. M. A. (1995), *Long Waves in Two Layers: Governing Equations and Numerical Model*, J. Sci. Tsunami Hazards *13*(1), 3–24.

KAJIURA, K. (1972), *The Directivity of Energy Radiation of the Tsunami Generated in the Vicinity of a Continental Shelf*, J. Oceanog. Soc. Japan *28*, 260–277.

KAWATA, Y., BENSON, B. C., BORRERO, J. C., BORRERO, J. L., DAVIES, H. L., DE LANGE, W. P., IMAMURA, F., LETZ, H., NOTT, J., and SYNOLAKIS, C. (1999), *Tsunami in Papua New Guinea was as Intense as First Thought*, EOS, Trans. Am. Geophys. Union *80*, 101 and 104–105.

MATSUYAMA, M., WALSH, J. P., and YEH, H. (1999), *The Effect of Bathymetry on Tsunami Characteristics at Sissano Lagoon, Papua New Guinea*, Geophys. Res. Lett. *26*, 3513–3516.

SWEET, S. and SILVER, E. A. (2003), *Tectonics and Slumping in the Source Region of the 1998 Papua New Guinea Tsunami from Seismic Reflection Images*, Pure Appl. Geophys. *160*, 1945–1968.

Tanioka, Y. (1999), *Analysis of the Far-field Tsunamis Generated by the 1998 Papua New Guinea Earthquake*, Geophys. Res. Lett. *26*, 3393–3396.

Tappin D. R. *et al.* (1999), *The Sediment Slump Likely Caused New Guinea Tsunami*, EOS, Trans. Am. Geophys. Union *80*, 329, 334 and 340.

Pure appl. geophys. 160 (2003) 2087–2118
0033 – 4553/03/112087 – 32
DOI 10.1007/s00024-003-2421-1

Pure and Applied Geophysics

The July 1998 Papua New Guinea Earthquake: Mechanism and Quantification of Unusual Tsunami Generation

KENJI SATAKE[1] and YUICHIRO TANIOKA[2]

Abstract—The unusual tsunami generated by the July 17, 1998 Papua New Guinea earthquake was investigated on the basis of various geophysical observations, including seismological data, tsunami waveform records, and on-land and submarine surveys. The tsunami source models were constructed for seismological high-angle and low-angle faults, splay fault, and submarine slumps. Far-field and near-field tsunamis computed from these models were compared with the recorded waveforms in and around Japan and the measured heights along the coast around Sissano Lagoon, respectively. In order to reproduce the far-field tsunami waveforms, small sources such as splay fault or submarine slump alone were not enough, and a seismological fault model was required. Relocated aftershock distribution and observed coastal subsidence were preferable for the low-angle fault, but the low-angle fault alone could not reproduce the large near-field tsunamis. The low-angle fault with additional source, possibly a submarine slump, is the most likely source of the 1998 tsunami, although other possibilities cannot be excluded. Computations from different source models showed that the far-field tsunami amplitudes are proportional to the displaced water volume at the source, and the comparison with the observed tsunami amplitudes indicated that the displaced water volume at the 1998 tsunami source was $\sim$0.6 km^3. The near-filed tsunami heights, on the other hand, are determined by the potential energy of displaced water, and the comparison with the observed heights showed that the potential energy was $\sim 2 \times 10^{12}$ J.

Key words: Tsunami, Papua New Guinea, Sissano Lagoon, submarine slump.

1. Introduction

The July 17, 1998 Papua New Guinea earthquake (M_s 7.1) caused unusual tsunami. Around Sissano Lagoon on the northern coast of New Guinea Island, very large tsunami, more than 10 m height, caused significant damage including over 2000 casualties (KAWATA *et al.*, 1999; MCSAVENEY *et al.*, 2000). The large tsunami heights and damage, however, were limited in an approximately 40 km section of the coast. The tsunami was also recorded on several tide and ocean bottom pressure gauges in and around Japan with amplitudes of up to 10 cm. The tsunami magnitude, M_t, of this

[1] Active Fault Research Center, National Institute of Advanced Industrial Science and Technology, Tsukuba, 305-8567 Japan. E-mail: kenji.satake@aist.go.jp.
[2] Institute of Seismology and Volcanology, Hokkaido University, Sapporo 060–0843 Japan. E-mail: tanioka@eos.hokudai.ac.jp.

earthquake was estimated as 7.5 (KIKUCHI *et al.*, 1999). The difference in surface wave and tsunami magnitudes, $M_t - M_s$, is 0.4, marginal for this event to be a "tsunami earthquake" (ABE, 1989). "Tsunami earthquakes" are unusual earthquakes that excite much larger tsunamis than expected from seismic waves (KANAMORI, 1972).

The earthquake occurred along the New Guinea (or Wewak) trench (Fig. 1), where the North Bismark (or Pacific) plate obliquely subducts beneath the Australian plate (TREGONING *et al.*, 1998). Typical interplate earthquakes along subduction zones are tsunamigenic and affect extended coastal regions. Some subduction zones generate unusual "tsunami earthquakes" whose source regions are located offshore near trench axis (SATAKE and TANIOKA, 1999). For the purpose of investigating the generation mechanism of this unusual tsunami, several submarine surveys of the tsunami source area were conducted (TAPPIN *et al.*, 1999). Swath bathymetric surveys provided detailed digital bathymetric data, which can be used for tsunami numerical computations. Sidescan sonar surveys and visual observations by using Remotely Operated Vehicles and manned submersibles explored the ocean bottom features. Single- and multi-channel seismic surveys imaged subsurface features of the tsunami-generating ocean bottom.

Despite such extensive measurements and surveys made during several years after the tsunami, no conclusive evidence of the tsunami source has been found and the tsunami generation mechanism is still in debate (GEIST, 2001; OKAL and SYNOLAKIS, 2001). Various modeling results have been published on this tsunami. These source models can be classified into seismological fault models (MATSUYAMA *et al.*, 1999; TANIOKA, 1999; GEIST, 2000) and submarine landslide models (HEINRICH *et al.*, 2000,

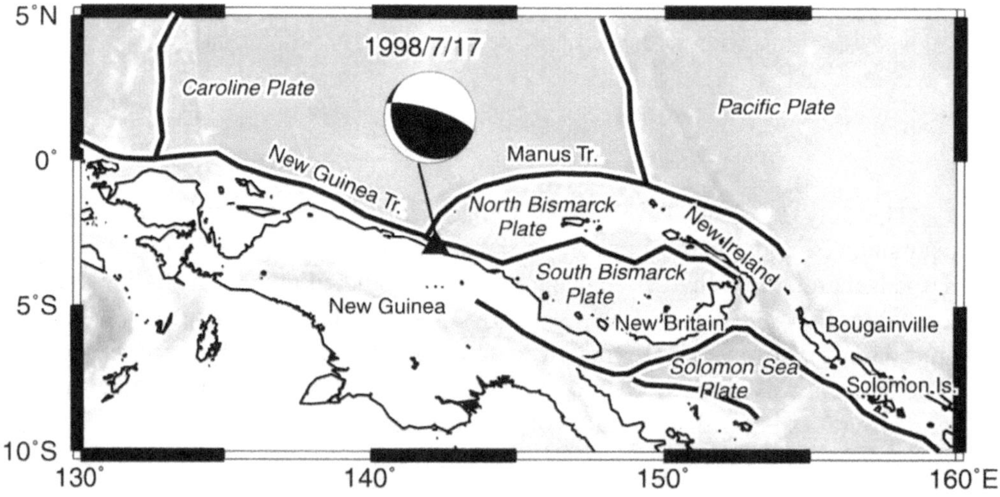

Figure 1
Tectonic map around New Guinea Island. The epicenter (solid triangle) and the Harvard CMT solution of the 1998 Papua New Guinea earthquake are shown.

2001; SYNOLAKIS *et al.*, 2002; TAPPIN *et al.*, 2001). Each study claimed that their model could satisfactorily explain the observed tsunami heights, but no quantitative comparison of different models has been made.

In this paper we first review and examine various observations made after the earthquake and tsunami regarding the source mechanism. On the basis of these observations, we construct several tsunami generation models, compute tsunami generation and propagation, and compare the observed and computed tsunami heights. We will show that large-scale seismological fault motion is required to explain the far-field tsunami waveforms. We will also show that the low-angle fault motion alone cannot reproduce the large near-field tsunami, hence additional source, either splay fault or submarine slump, is required. Several source models can explain the observed tsunami data; it is difficult to obtain the unique tsunami source model from the tsunami data alone. However, the comparisons with the observation can constrain the important tsunami source quantities: the displaced water volume and potential energy.

2. *Observations*

2.1 Seismological Observations

The source parameters of the 17 July 1998 Papua New Guinea earthquake provided by NEIC (USGS) are as follows. The origin time: 8 h 49 m 13.2 s (GMT), epicenter: 2.961°S, 141.926°E. The relocated epicenters by SYNOLAKIS *et al.* (2002) and HURUKAWA *et al.* (2003) are essentially the same location.

The Harvard CMT solution (DZIEWONSKI *et al.*, 1999) shows that the two nodal planes are: strike $\phi = 146°$, dip $\delta = 19°$, rake $\lambda = 127°$, and $\phi = 287°$, $\delta = 75°$, $\lambda = 78°$ (Figs. 1 and 2). The seismic moment is estimated as 3.67×10^{19} Nm, which corresponds to $M_w = 7.0$. Strikes of both planes, roughly NW-SE, are parallel to the coastline of New Guinea Island and the trench axis. The first nodal plane would indicate a low-angle thrust mechanism of interplate earthquakes at the New Guinea trench, while the second plane would indicate a high-angle fault parallel to the trench axis. GEIST (2000) showed that the Harvard CMT solutions of past earthquakes in the region have various focal mechanisms, and the July 1998 event was not a typical event reflecting the regional seismotectonics.

The aftershocks are distributed in an area with approximately 50 km × 40 km north of Sissano Lagoon (HURUKAWA *et al.*, 2003). Figure 2 shows the distribution of aftershocks that occurred within 24 hours after the mainshock. HURUKAWA *et al.* (2003) used teleseismic and regional data but with the station corrections estimated from locally recorded aftershocks. The absolute location errors are estimated to be less than 5 km. All the aftershocks are located to the east of the mainshock epicenter, indicating an eastward propagation of faulting.

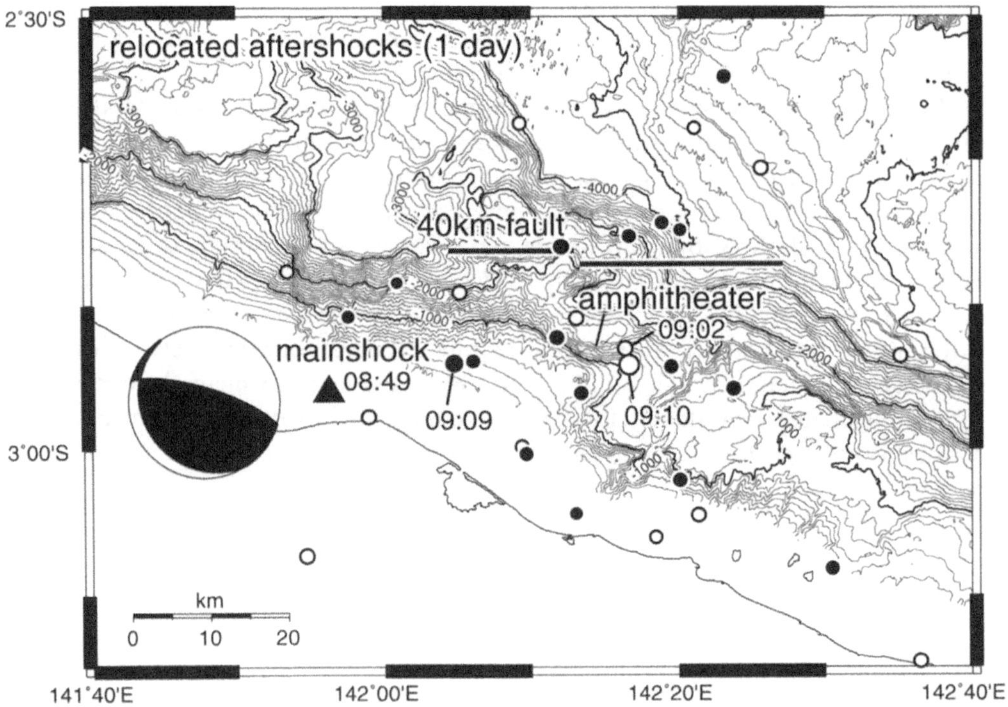

Figure 2
Epicenter distribution of aftershocks within 1 day of the 1998 mainshock. Solid circles are epicenters relocated with smaller errors (<0.1° in longitude and latitude and <20 km in depth), while open circles have larger depth errors (HURUKAWA *et al.*, 2003). Detailed bathymetry mapped by SEABEAM 2100 system on JAMSTEC's *Kairei* is also plotted (TAPPIN *et al.*, 1999). The Harvard CMT solution and locations of 40-km fault and amphitheater are also shown.

The mainshock rupture process was analyzed from teleseismic body waveforms by KIKUCHI *et al.* (1999). The estimated source duration is 19 s, typical for this size of earthquakes. In other words, the mainshock was not a slow earthquake. They also showed that the mainshock cannot be explained by a single-force mechanism which would be expected from a submarine landslide.

The choice of actual fault plane is controversial. HURUKAWA (1999), from distribution of aftershocks relocated by teleseismic data only, preferred the high-angle fault as the actual fault plane. KIKUCHI *et al.* (1999) performed the body wave inversions for both low-angle and high-angle faults and found that variance reduction was slightly better for the high-angle, though the difference was insignificant. These studies favor the high-angle fault as the actual fault plane, indicating that the earthquake was an intraplate event. The high-angle fault is more efficient in producing large tsunamis (KIKUCHI *et al.*, 1999; GEIST, 2000; GEIST, 2001). OKAL and SYNOLAKIS (2001) strongly argued against this choice of fault plane.

HURUKAWA *et al.* (2003), from more accurately relocated aftershocks, concluded that the low-angle fault was the actual fault plane. Available seismological data do not yield conclusive results for the choice of fault plane, hence we will consider both planes in our modeling.

Hydroacoustic T phase of the mainshock and aftershocks was examined by OKAL (OKAL and SYNOLAKIS, 2001; SYNOLAKIS *et al.*, 2002). He indicated that one of the aftershocks, which occurred at 9:02, generated unusual T phase; strong amplitude for an m_b 4.4 event, duration comparable to that of the mainshock, and significant high-frequency contents. He interpreted this unusual event as a submarine slump. From the fact that T phase from this event was not recorded in Taiwan, he estimated the source at the amphitheater, because the raypath would be blocked by the upraised block.

2.2 Field Survey Results

The tsunami heights show unusual distribution. Figure 3 shows the maximum tsunami heights measured by International Tsunami Survey Team (ITST) (KAWATA *et al.*, 1999). The maximum runup height, 15 m, was measured at Arop, near the bottom of the sand spit separating the lagoon and outer ocean. On both sides of this peak the tsunami height decays quickly with distance. The tsunami heights exceeding 10 m were observed in a 20-km section of the coast around Sissano Lagoon, and a damaging tsunami exceeding 5 m was limited within 40 km of the coast. Such a local concentration of tsunami heights is very unusual for either tsunamigenic or tsunami earthquakes, and implies local effects such as secondary tsunami source or refraction due to local bathymetry.

Lack of local instrumental records such as tide gauges makes it difficult to accurately estimate the timing of tsunami arrival. Field survey teams interviewed survivors regarding ground shakings and tsunami arrival times (KAWATA *et al.*, 1999). Some coastal residents felt ground shaking only once while others felt it twice or three times. Some reported that the first event was the strongest while others said the second one was stronger. Such inconsistencies in eyewitness accounts are not surprising, considering the unexpected hazard and possible panic, but make it difficult to estimate the tsunami arrival time from eyewitness accounts. Those who felt more than one shaking reported that the tsunami arrived after the last one. From this information, SYNOLAKIS *et al.* (2002) estimated that the tsunami arrived at 09:09–09:10 and ruled out the possibility that the tsunami was generated by the mainshock. They assumed that the last shaking before the tsunami arrival was from the aftershock which occurred at 09:09. While it is a possible interpretation, they also needed to assume that the first shaking out of three was from an unrecorded foreshock, and failed to explain the account that the shaking from the aftershock was stronger. It seems dangerous to over rely on the eyewitness accounts and assumptions, hence we will not constrain the tsunami arrival time in this paper.

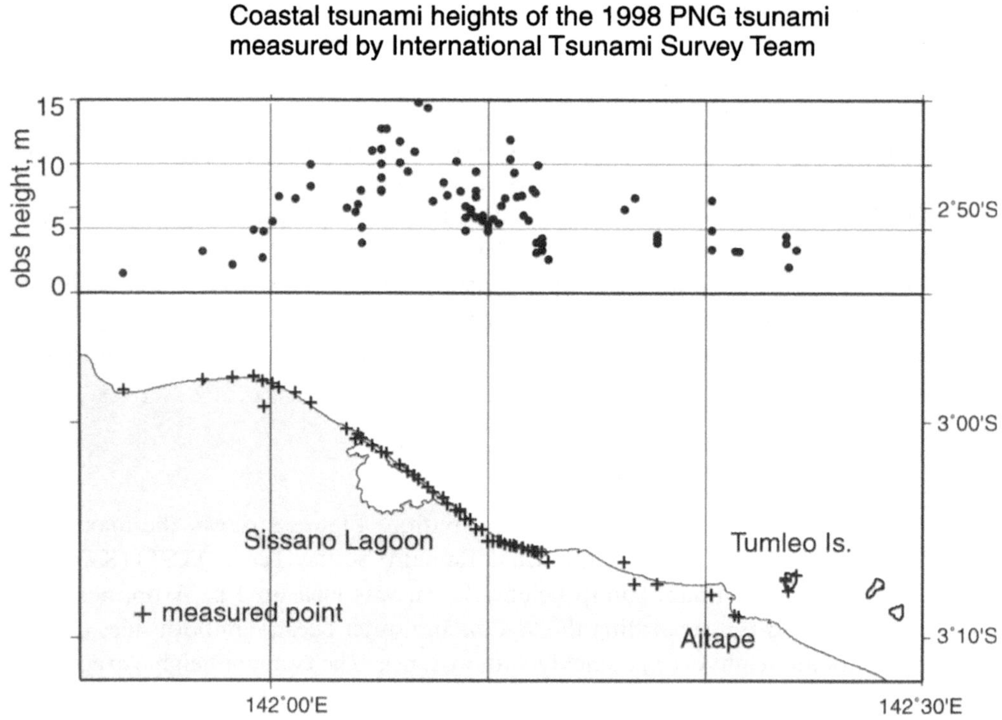

Figure 3
Tsunami heights along the northern coast of New Guinea Island measured by International Tsunami Survey Team (KAWATA *et al.*, 1999). The map shows the location of measurement points (plus signs), while the measured water heights are shown in the upper diagram as a function of longitude.

Another piece of information from a field survey team is evidence of subsidence. While KAWATA *et al.* (1999) could find no evidence or eyewitness accounts of permanent coastal movement, MCSAVENEY *et al.* (2000) reported field evidence of coseismic subsidence by 30–40 cm on the spit separating Sissano Lagoon and the outer ocean, which was recently confirmed by DAVIES *et al.* (2001).

2.3 Far-field Tsunami Observations

The tsunami was also recorded on six tide gauges (Naze, Kushimoto, Mera, Yaene, Muruto and Chihijima) and two ocean bottom (OB) pressure gauges (Boso-OB and Muroto-OB) around Japan (Fig. 4). Their amplitudes reached 10 cm (Fig. 5). From these records, the tsunami magnitude M_t was determined as 7.5 (KIKUCHI *et al.*, 1999). Because of small signal-to-noise ratio, the tsunami first arrivals are not clear enough on these records to help estimate the tsunami origin time (Fig. 5).

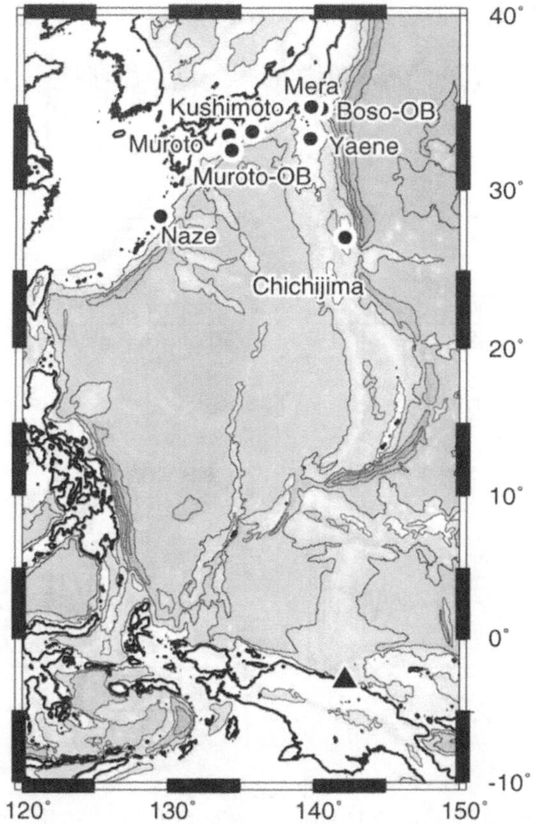

Figure 4
Location of two ocean bottom pressure gauges and six tide gauge stations in and around Japan, where the 1998 tsunami was recorded. The map also shows the area and bathymetry used in numerical computation of far-field tsunami. The epicenter is indicated by a solid triangle.

TANIOKA (1999) modeled these far-field tsunami data from three different fault models, including high-angle and low-angle faults, and concluded that the high-angle fault model best explains the observations. While details of the source process such as the choice of fault plane may be difficult to resolve from far-field data (OKAL and SYNOLAKIS, 2001), the far-field data should constrain the overall size of the tsunami.

2.4 Submarine Surveys

Between 1999 and 2001, several research cruises have been conducted in the tsunami source area offshore Sissano Lagoon. The first cruise in January 1999 with R/V *Kairei* of Japan Marine Science and Technology Center (JAMSTEC) performed swath bathymetry and side-scan surveys, high-frequency (4 kHz) sub-bottom profiling, gravity and magnetic survey and piston core sampling at four sites (TAPPIN

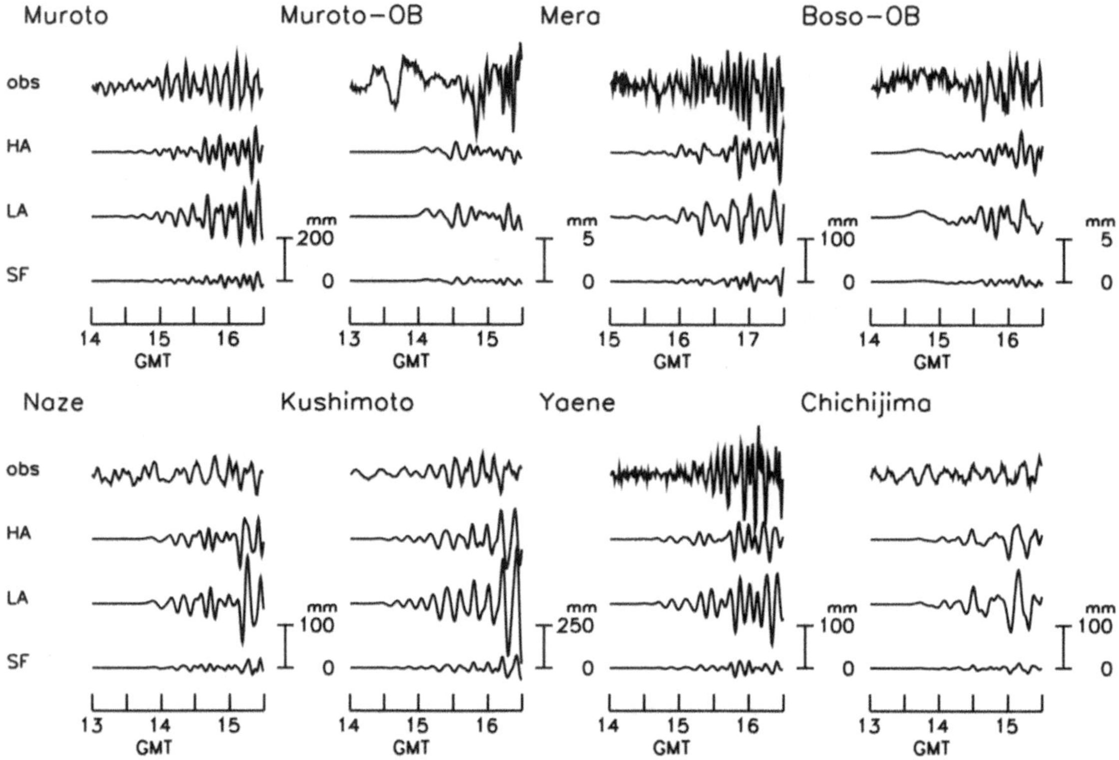

Figure 5
Observed (top) and computed (three traces) tsunami waveforms at the two OB pressure gauges and six tide gauge stations (see Fig. 4 for the locations).

et al., 1999). Detailed bathymetric mapping revealed interesting submarine features, including arcuate amphitheater and fault linearments. The digital bathymetric data were obtained for depth deeper than 200 m and to be used for tsunami numerical computations (Fig. 2).

The amphitheater is located at approximately 2.83°S, 142.26°E or about 30-km northeast of Sissano Lagoon (Fig. 6). The scarp of the amphitheater is approximately 10-km wide and 1-km high. To the north of the amphitheater, there is an upraised block 600 m high. While the arcuate geomorphology of the amphitheater would indicate that large-scale submarine landsliding was involved in the formation process, detailed features show that it was not a recent event. Many submarine galleys on the scarp show that erosion has continued for some time. Reflectivity on the amphitheater wall was not very strong, indicating that the slope is covered with sediments (Fig. 6). The piston cores obtained below the amphitheater contain no turbidite sediments in the top 8 m of sea bottom (TAPPIN *et al.*, 1999).

Two major fault linearments are identified; one is approximately 40-km long in the E-W direction on the northern slope of the upraised block and the other is 15-km

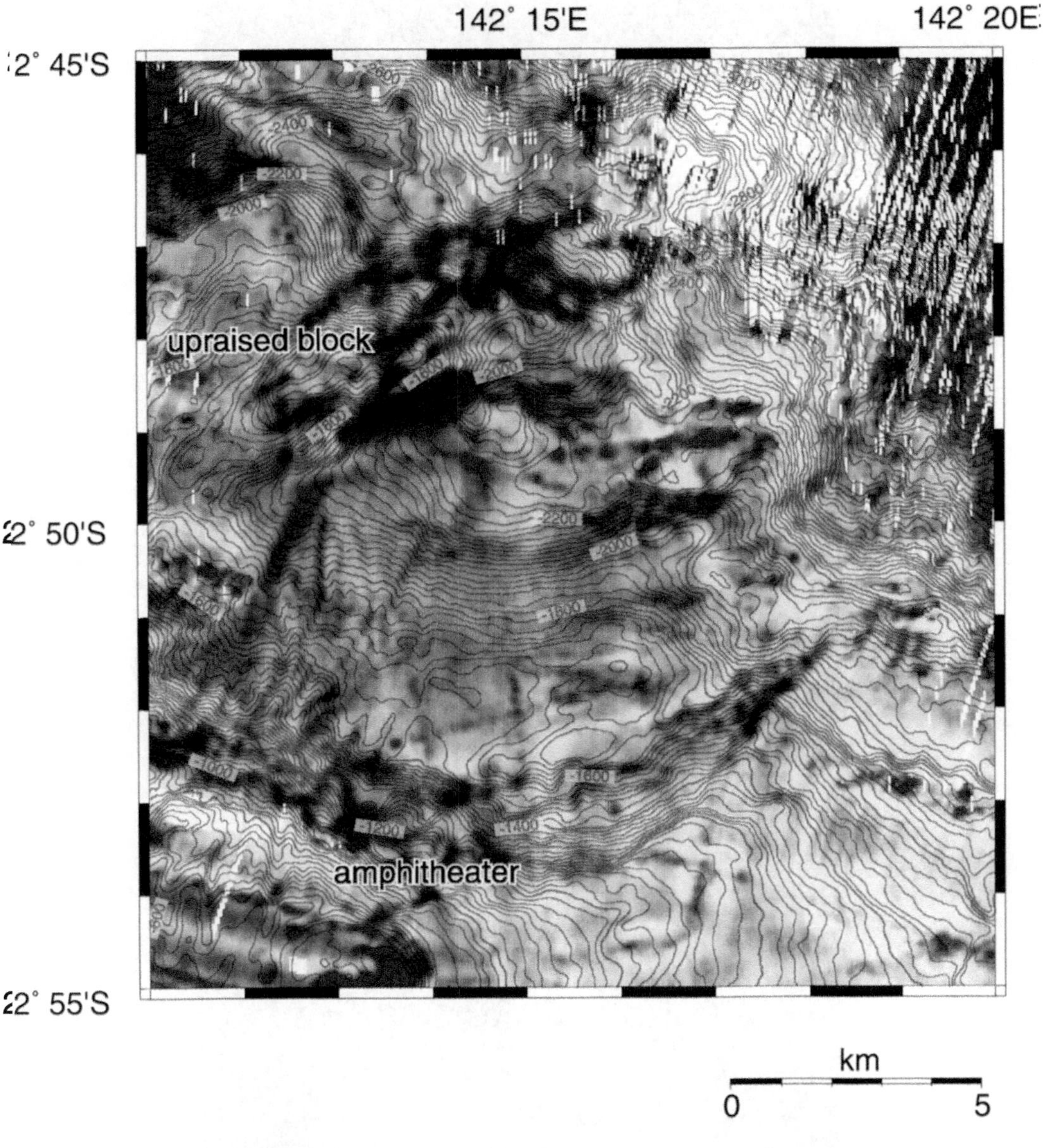

Figure 6
Detailed bathymetry and backscatter image around the amphitheater and upraised block. SEABEAM 2100 data obtained during the JAMSTEC's cruise are processed into 50-m grids. The bathymetry contour is 25 m. The darker color indicates stronger reflection of backscatter data.

long in the ENE-WSW direction just south of the block. Another important bathymetric feature is convex-shaped slope, probably submerged delta, just off Sissano Lagoon. This peculiar topography acts as a refractor and concentrates tsunami energy, as we will see.

95 08 02
13 13 26

Two other cruises in 1999 by JAMSTEC's R/V *Natsushima* provided visual observations of ocean bottom; one with Remotely Operated Vehicle (ROV) *Dolphin* and the other with manned submersible *Shinkai 2000*. Many features indicating recent seafloor disturbance, such as fresh fissures (Fig. 7), talus deposits and scarp, were observed around the amphitheater. Along the fault linearments, few features of such recent movement were recognized. These indicate that strong shaking was centered at the amphitheater, and the linearments are not surface expressions of the earthquake fault. TAPPIN *et al.* (2001) attributed the above bottom features to evidence of submarine slump, however these features alone do not prove the existence of submarine slump. Similar features were also reported from other tsunami sources where no submarine slumps were inferred (TAKUECHI *et al.*, 1998). Fig. 7 compares fresh cracks observed in Papua New Guinea, Japan Trench and near-by Okushiri Island in Japan Sea.

Another cruise in 1999 with R/V *Maurice Ewing* performed a multi-channel seismic reflection survey (SWEET *et al.*, 1999). One of the profiles shows a depression of 100-m deep and 750-m long, and SWEET *et al.* (1999) interpreted it as the rotational slump with a maximum depth of 600 m below the seafloor with the center-of-mass drop of 300 m.

The most recent cruise in February 2001 with JAMSTEC's R/V *Natsushima* performed dense single-channel seismic surveys in the amphitheater and surrounding region. The surveys confirmed the slump feature found by R/V *Maurice Ewing*, but also found several other slump features in and around the amphitheater. Slumps are typical features on the continental slope of this active margin. The timing of the slumps cannot be estimated from these seismic survey data.

3. Tsunami Source Models

On the basis of various observations, we construct tsunami source models to compute and test near-field and far-field tsunamis.

3.1 Initial Water Heights from Fault and Slump Models

The tsunami sources are disturbances on the water surface. When the wavelength is much longer than the water depth, long-wave or shallow water approximation

◄

Figure 7
Photographs of open cracks observed by Remotely Operated Vehicle or manned submersible. Top is on the amphitheater off Sissano, Papua New Guinea. The middle is at Japan Trench where the 1933 Sanriku earthquake occurred and the bottom is near Okushiri Island within the source region of the 1993 SW Hokkaido earthquake. The horizontal extent of each photograph is a few to several meters. Downloaded from JAMSTEC's website (www.jamstec.go.jp).

holds. Then the water surface deformation can be assumed to be the same as the bottom deformation caused by geological processes. In the case of fault source, ocean bottom displacement η can be computed once the fault parameters are provided (e.g., OKADA, 1985). Fault parameters required for the computations are: location (coordinate and depth), size (length L and width W), geometry (strike ϕ, dip δ, rake λ) and slip amount u (Fig. 8).

For the submarine slumps, we use a static water-surface deformation estimated by an empirical formula derived by GRILLI and WATTS (1999). SYNOLAKIS *et al.* (2002), TAPPIN *et al.* (2001) and MATSUYAMA and YEH (2003) also use this formulation in modeling the submarine slumps. These studies modeled tsunami by sliding a rigid body, whose shape is a semi-ellipse, down on a uniformly sloping seabed. The slide speed is expressed by a hyperbolic-tangent function of time. The empirical prediction for the maximum free-surface depression η_{max} caused by a landslide was introduced as

$$|\eta_{max}| = 0.218\,T\,(\sin\theta)^{1.38}\left(\frac{L}{d}\right)^{1.25}, \quad (1)$$

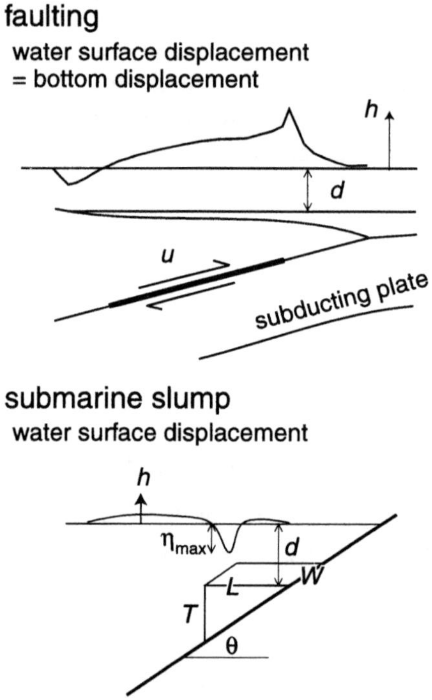

Figure 8
Schematic cross sections of fault (top) and submarine slump (bottom) sources and the produced displacements of water surface. Note that the different scale (tens of km for the top, while up to a few km in the bottom).

where T is the thickness and L is the length of the semi-ellipse sliding rigid body, d is the initial water depth at the middle of the sliding body, and θ is the seabed slope angle. Because equation (1) only gives the maximum water height above the slide source, we use the following functional form to estimate the two-dimensional distribution

$$\eta(x,y) = \eta_{\max}\left[\exp\left\{-12\left(\frac{x-L/2}{L}\right)^2\right\} - \exp\left\{-108\left(\frac{x}{L}\right)^2\right\}\right] \times \exp\left\{-\left(\frac{y}{W}\right)^2\right\} . \quad (2)$$

This expression yields no volume change when integrated over the source area. The parameters characterizing the slump source are summarized as follows: the slide thickness T, slide length L, slide width W, the slope angle θ, and water depth d. The slide volume Q is computed as a product of T, L, and W (Fig. 8).

3.2 Quantification of Source Models

In order to make quantitative comparisons among the tsunami source models, we introduce two parameters: the volume of displaced surface water V and the potential energy E. The displaced water becomes the initial condition for tsunami computations, and it is computed by integrating the initial surface displacement distribution η (x,y),

$$V = \int |\eta| dS . \quad (3)$$

Note that the surface integral of water height η becomes zero for the slump source, hence we needed to use the absolute value $|\eta|$. For the fault source, the amount of water displacement is assumed to be the same as the ocean bottom deformation caused by faulting, because of the long wavelength compared to the water depth (the long-wave, or shallow-water approximation). For the submarine slump source, on the other hand, such approximation does not hold, so the displaced water volume is different, and much smaller, than the estimated slump volume Q in the previous section.

The potential energy of displaced water can be computed as (e.g., KAJIURA, 1981)

$$E = \frac{1}{2}\rho g \int |\eta|^2 dS , \quad (4)$$

where ρ is density of water. The above integrations are made numerically over the computational area.

3.3 Seismological Fault Models

Most tsunamis are generated by ocean-bottom deformation due to earthquake faulting. Hence it is natural to first consider a fault motion as the tsunami source for the July 1998 PNG tsunami. The CMT solution indicates that the fault motion was either high-angle fault dipping NE or low-angle fault dipping SW. We will consider both planes.

Model HA is a seismological high-angle fault (Figs. 9 and 10b). Considering the CMT solution, the geometry is assumed as $\phi = 290°$, dip $\delta = 70°$, and $\lambda = 90°$. From the aftershock distribution and body wave analysis, we estimate that the fault length $L = 40$ km, width $W = 15$ km, and average slip $u = 2$ m. The seismic moment becomes 3.6×10^{19} Nm, the same as the CMT solution, assuming that the rigidity $\mu = 3 \times 10^{10}$ N/m^2. The top edge of the fault is assumed to reach the ocean bottom. The computed ocean bottom deformation (Fig. 10b) shows that the NE

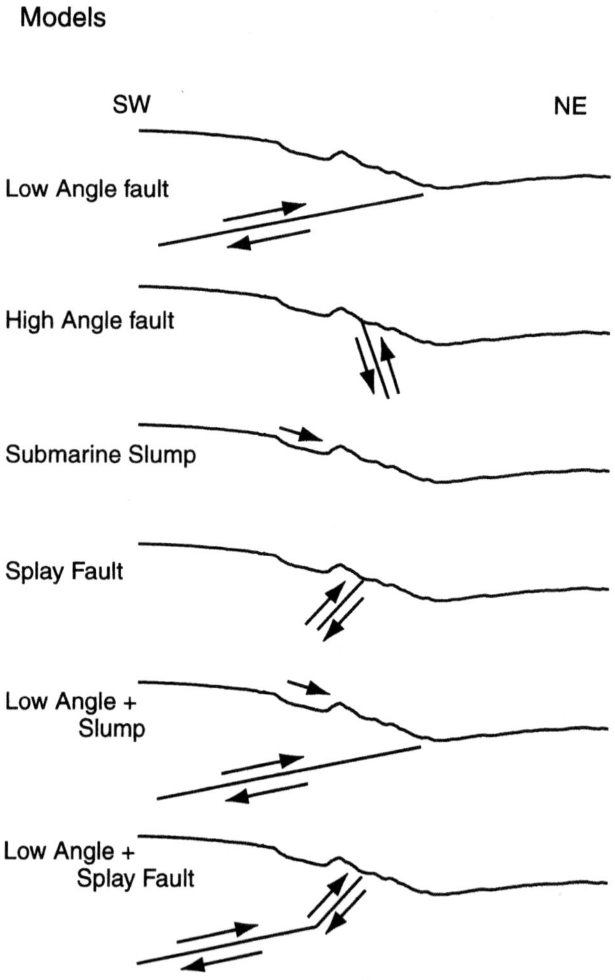

Figure 9
Schematic cross sections of several models examined in this paper. The profile of ocean bottom is in SW-NE direction through the amphitheater and upraised block.

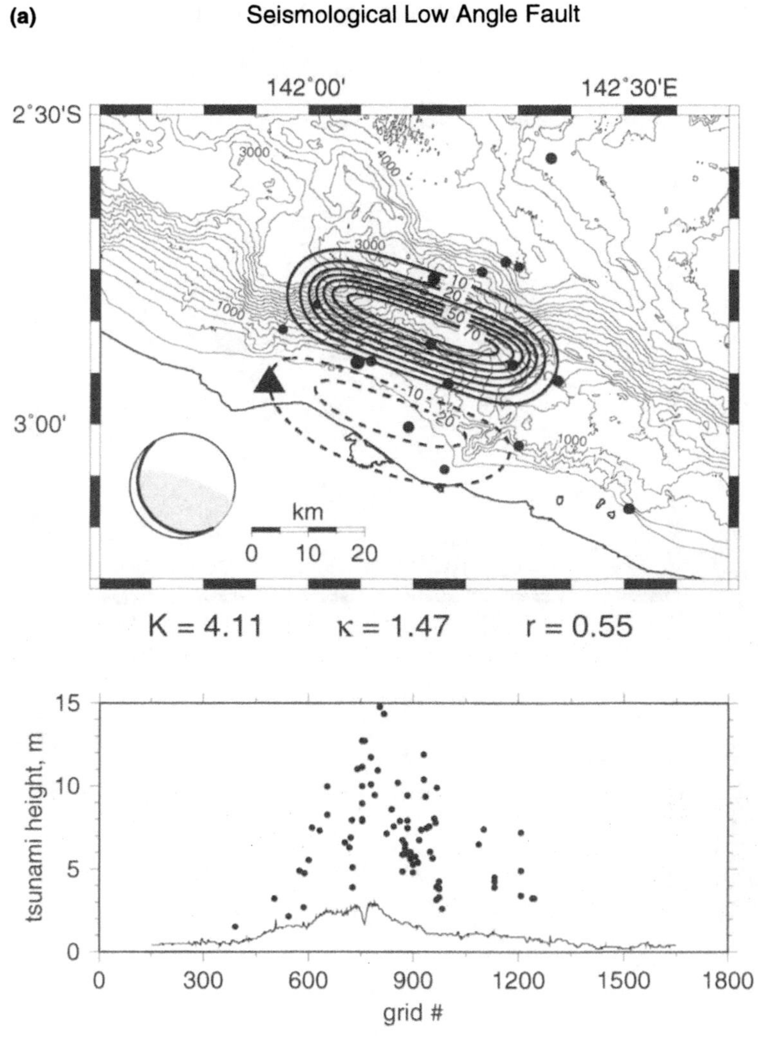

Figure 10
Results of tsunami numerical computations. The top figure shows displacement of water surface overlaid on bathymetry and aftershock distribution (HURUKAWA *et al.*, 2003). The bottom diagram shows comparison of the observed and computed tsunami heights along the coast shown above. Statistical parameters (see text for definitions) of the comparison are also shown. (a) Seismological low-angle fault (LA) model. (b) Seismological high-angle fault (HA) model. (c) Splay fault (SF) model.

(offshore) side is uplifted as large as 1.2 m and the SW (onshore) side is subsided as large as 0.7 m. This pattern is opposite to the geomorphological linearment of the "40-km fault" identified in the bathymetric survey.

Model LA is a seismological low-angle fault (Figs. 9 and 10a). From the CMT solution and local topography, the fault geometry is assumed as $\phi = 110°$, dip $\delta = 20°$, and $\lambda = 90°$. The fault size and the slip are the same as model HA. The fault

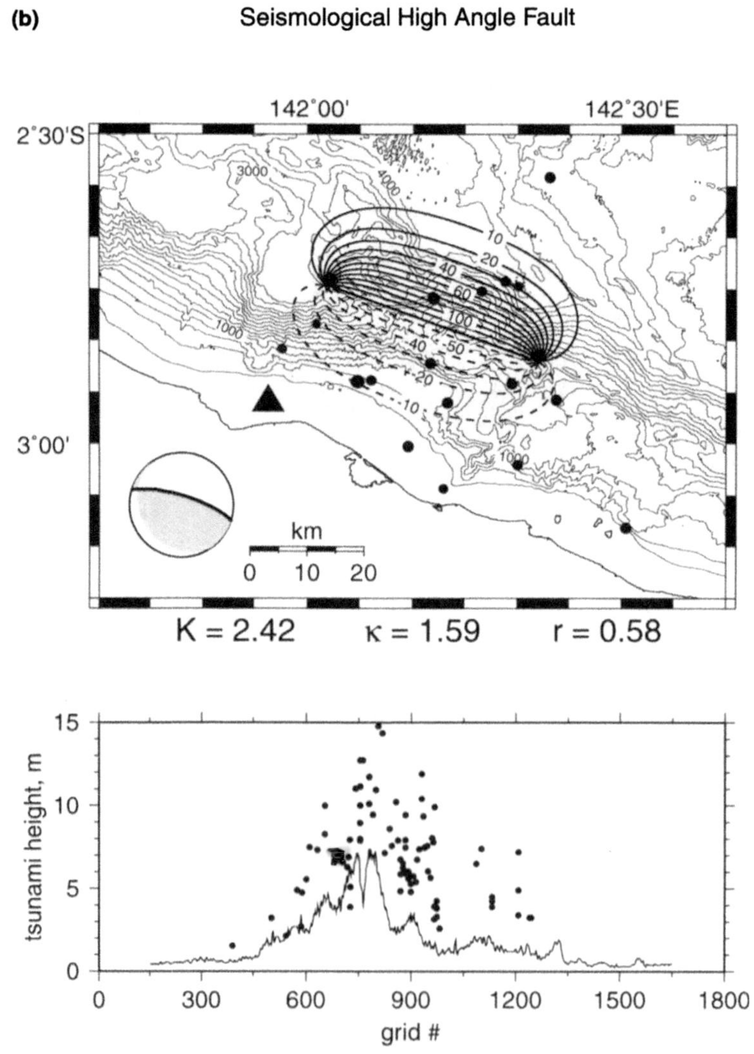

Figure 10
(Contd.)

is expected to lie on the subducting plate, so that the top edge is assumed to be at 5-km depth. The faulting produces uplift (max 0.8 m) offshore and subsidence (< 0.2 m) on the coast near Sissano Lagoon (Fig. 10a). The coastal subsidence is consistent with the field observation (McSaveney *et al.*, 2000).

The geomorphology of the "40-km fault" indicates that the NE (offshore) side has subsided relative to the SW (onshore) side. Such a fault motion can be modeled as a splay fault branching from the megathrust fault between the subducting and overlying plates (Fig. 9). We therefore model such a splay fault, model SF, in

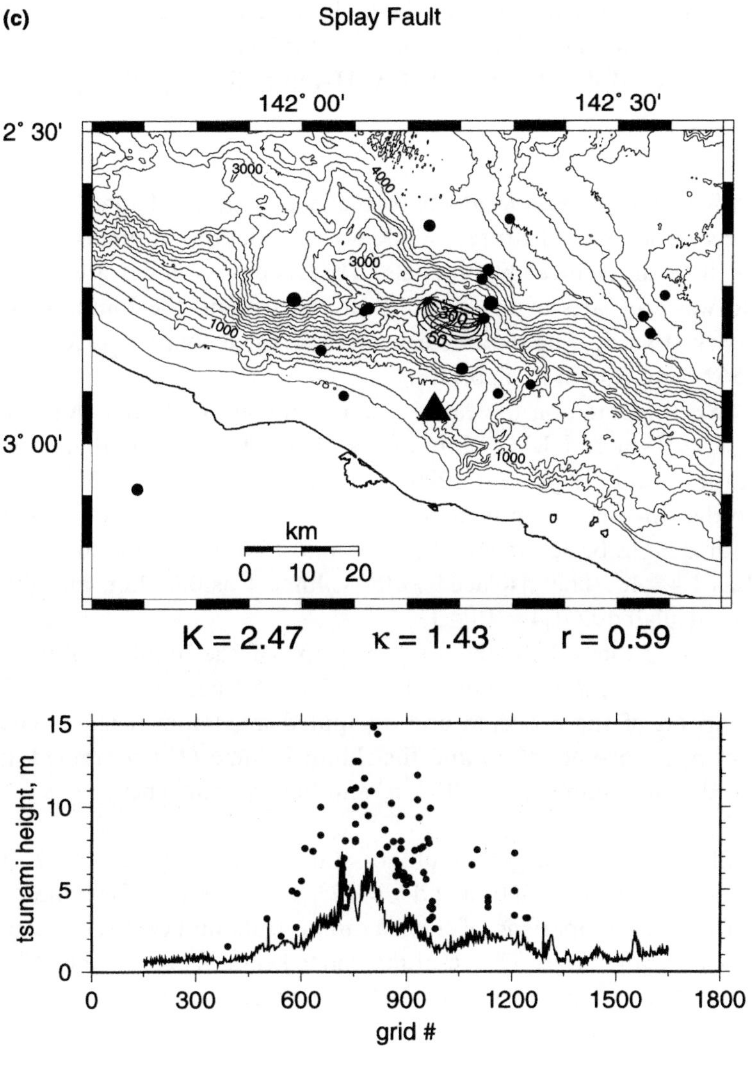

Figure 10
(Contd.)

addition to the above seismological fault models. The fault geometry is assumed as $\phi = 110°$, dip $\delta = 45°$, and $\lambda = 90°$; the strike is parallel to the LA fault, but the dip angle is larger than the CMT solution. Such a fault motion does not represent the mainshock, rather additional motion with insignificant moment release or aseismic motion. The fault size is assumed as $L = 10$ km, width $W = 7$ km. The slip on the fault is assumed to be 6 m. This fault model yields the seismic moment of 1.3×10^{19} Nm, a third of the CMT solution, assuming the same rigidity as used for the deeper fault.

The displaced water volume V for the fault models is 0.52, 0.65 and 0.16 km^3 for low-angle (LA), high-angle (HA) and splay fault (SF) models. The potential energy E is 0.9, 1.6 and 1.4×10^{12} J for models LA, HA and SF, respectively.

3.4 Submarine Slump Models

The marine surveys, both swath bathymetry and single-channel seismic surveys, identified several slumping features. While no information on timing is available, we will model three different slump models. Three slump sources are located on the eastern and western parts of the amphitheater scarp and the southern slope of the upraised block (Fig. 6). The parameters for these three slump sources, models SSE, SSW and SSN, are shown in Table 1.

Model SSE is located on the eastern half of the amphitheater slope. The size of the slump, $L = 4$ km and $W = 2$ km, are estimated from the bathymetry (Fig. 11). At this location, water depth $d = 1200$ m and the slope $\theta = 12°$. The thickness T is adjusted so that the maximum tsunami height becomes similar to the observed. The final T is estimated to be 350 m and η_{max} becomes 16.5 m, and the slump volume Q is computed as 1.4 km^3. The displaced water volume V is 0.05 km^3 and the potential energy E is calculated as 1.4×10^{12} J.

Model SSW is located on the western slope of the amphitheater. The size is $L = 4$ km, $W = 2$ km, $d = 1500$ m and $\theta = 27°$. Thickness T is estimated as 100 m from comparison of the observed and computed maximum heights. The resulting water height η_{max} becomes 10 m and the slump volume Q is computed as 0.8 km^3. The displaced water volume V is 0.05 km^3 and the potential energy E is calculated as 0.8×10^{12} J.

Model SSN is located on the southern slope of the upraised block. The size is $L = 1$ km, $W = 2$ km, $d = 2000$ m and $\theta = 27°$. The thickness T is estimated as large as 1500 m from the comparison of the maximum tsunami heights, though this value may be unrealistically large. The resulting water height η_{max} becomes 45 m and the slump volume Q is computed as 3 km^3. The displaced water volume V is 0.05 km^3 and the potential energy E is calculated as 4.8×10^{12} J. The volume and potential energy are computed for each model and shown in Table 2.

4. Tsunami Numerical Computations

4.1 Near-field Tsunami Computation

4.1.1 Bathymetry data and grid size

Swath bathymetric survey using the SEABEAM 2100 system on R/V *Kairei* provides basic bathymetric data for tsunami numerical computations. Because of safety reasons, the survey was limited to the offshore region where the depth is more than 200 m, or approximately 10–20 km offshore the coastline (see Fig. 1b). For

Table 1

Parameters for tsunami computations

Model		Displacement	M_o or Q	τ, s	Max, m	Min, m	Mean, cm	rms, cm	V, km^3	E.10^{12} J	K	κ	r
LA	Low Angle	U = 2 m	M_o = 3.6E+19	30	0.78	−0.24	5.6	13.9	0.52	0.9	4.11	1.47	0.553
HA	High Angle	U= 2 m	M_o = 3.6E+19	30	1.22	−0.68	7.0	18.8	0.65	1.6	2.42	1.59	0.580
SF	Splay Fault	U = 6 m	M_o = 1.3E+19	5	3.70	−0.70	1.8	17.6	0.16	1.4	2.47	1.43	0.592
SSE	Slump	h_{max} = 17 m	Q = 1.4 km^3	1	5.33	−15.70	0.6	17.7	0.05	1.4	1.87	1.66	0.191
SSW	Slump	h_{max} = 10 m	Q = 0.8 km^3	1	3.33	−9.77	0.6	13.7	0.05	0.8	2.25	1.83	0.359
SSN	Slump	h_{max} = 45 m	Q = 3.0 km^3	1	19.41	−54.80	0.5	32.5	0.05	4.8	2.30	1.70	0.523
LA+SF	Low Angle + Splay Fault			–	4.00	−0.58	6.9	24.0	0.64	2.6	2.21	1.41	0.670
LA+SSE	Low Angle + Slump			–	6.02	−14.92	6.1	22.5	0.57	2.3	1.60	1.68	0.234
LA+SSW	Low Angle + Slump			–	4.09	−9.13	6.1	19.6	0.57	1.7	1.74	1.70	0.407
LA+SSN	Low Angle + Slump			–	19.80	−54.70	6.2	35.4	0.57	5.7	2.17	1.55	0.461

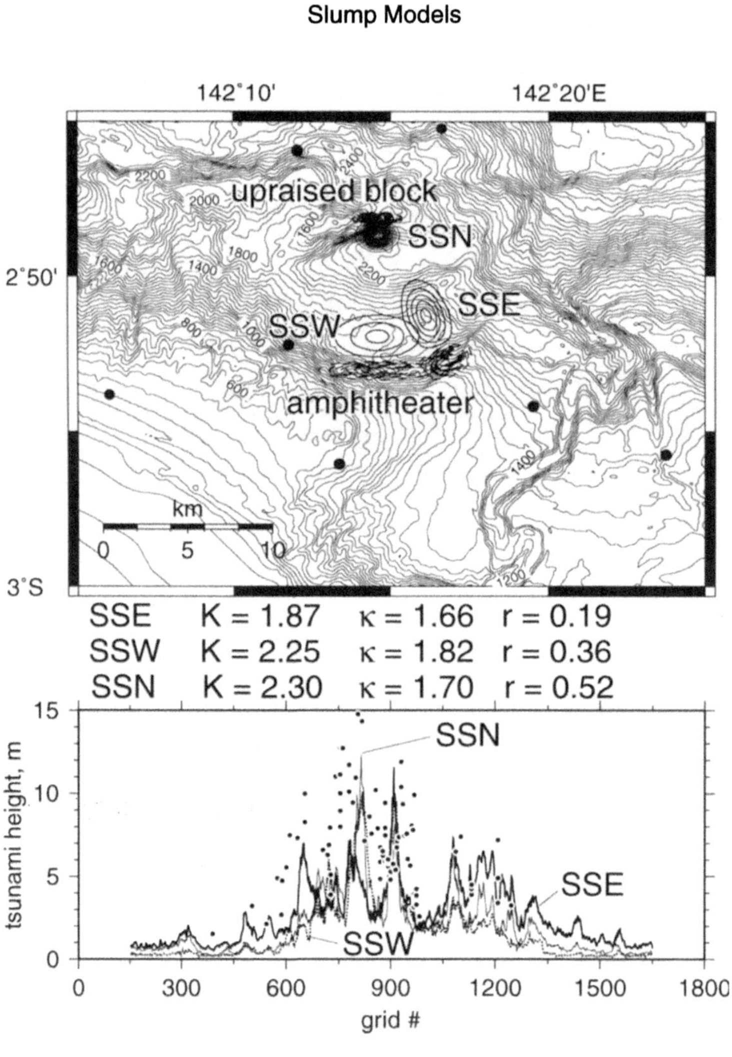

Figure 11
Results of tsunami computations for submarine slump models. The top map shows the water surface displacement associated with three slump locations (SSE, SSW and SSN). Aftershocks located by HURUKAWA *et al.* (2003) are also shown. The bottom diagram compares the observed and computed tsunami heights along the coast around Sissano Lagoon (see Fig. 10). Statistical parameters of the comparison are also shown.

tsunami computation, the bathymetric data between 200 m depth and the coastline are important and essential. We assumed that the water depth is 10 m on the coastline and estimated the water depth by interpolation. Digitized contour lines from an Australian Hydrographic chart (No. 389, published in 1968) are used to constrain the interpolation. The area for the computation ranges 141°40′–142°40′E and 3°15′–2°30′S.

Table 2

Fault and slump parameters

Fault models	Name	L (km)	W (km)	Strike (°)	Dip (°)	Ud (m)	Depth (km)	M_o (Nm)	M_w
LA	Low Angle	40	15	110	20	2	5	3.6E+19	7.0
HA	High Angle	40	15	290	70	2	0	3.6E+19	7.0
SF	Splay Fault	10	7	110	45	6	0	1.26E+19	6.7
Slump models		T (m)	b (m)	d (m)	θ (°)	Ly (m)		η_{max} (m)	Q (km^3)
SSE	South east	350	2000	1200	12	2000		16.5	1.4
SSW	South west	100	2000	1500	26.6	4000		10.3	0.8
SSN	North	1500	1000	2000	26.6	2000		45.4	3

We gridded the bathymetric data into two different intervals, 5 sec and 2 sec of the arc, or approximately 150 m and 60 m, in order to examine the effects on the tsunami computations. Results of numerical computations, maximum water heights along the coast, are compared for the different grid sizes. Water heights computed from model LA are very similar for the 2 sec and 5 sec grids. Submarine slump models, on the other hand, produce very different results. The coastal tsunami heights computed on the 2 sec grid are more than twice larger than those computed on the 5 sec grid. Such a difference may come from the relatively small size of the slump source. We therefore use the 2 sec (60 m) grid in this study, which is comparable to that used by MATSUYAMA *et al.* (1999) and much finer than other studies (HEINRICH *et al.*, 2000; SYNOLAKIS *et al.*, 2002; TAPPIN *et al.*, 2001).

4.1.2 Governing equations

For the tsunami propagation, shallow water or long-wave theory is usually used. The equation of motion, or conservation of momentum, can be given as (e.g., SATAKE, 1995)

$$\frac{\partial \mathbf{V}}{\partial t} + (\mathbf{V} \cdot \nabla)\mathbf{V} = -g\nabla h + C_f \frac{\mathbf{V}|\mathbf{V}|}{d+h} , \tag{5}$$

where **V** is the depth-averaged horizontal velocity vector, h is the water elevation or tsunami amplitude, d is the water depth and g is gravitational acceleration. The second term in the right-hand side represents bottom friction where C_f is the non-dimensional frictional coefficient. The equation of continuity, or conservation of mass, can be written as

$$\frac{\partial (d+h)}{\partial t} = -\nabla\{(d+h)\mathbf{V}\} . \tag{6}$$

Equations (5) and (6) can be linearized as

$$\frac{\partial \mathbf{V}}{\partial t} = -g\nabla h \ , \tag{7}$$

$$\frac{\partial h}{\partial t} = -\nabla \cdot (d\mathbf{V}) \ , \tag{8}$$

when the tsunami amplitude h is small compared to water depth d and the bottom friction can be neglected. While such assumptions may not hold for the present case, the major advantage of the linear theory is the superposition principle; the computational results can be easily scaled to estimate with different initial water heights.

We compared the effects of governing equations for the fault and slump models. For model LA, the computed tsunami heights along the coast from nonlinear and linear equations are very similar, indicating that the result is insensitive to the choice of governing equations. On the other hand, for model SSE, the computed tsunami heights are very different for linear and nonlinear cases; the coastal tsunami are much larger, about three times on the average, for the linear case than the nonlinear case. Part of this discrepancy is due to the assumption of total reflection on the coastal line. The maximum water heights appear somewhat offshore, rather than on the coastline, in the nonlinear computation. Because such a phenomenon does not reproduce the observed tsunami, and considering the merit of the superposition principle, we decide to use linear equations in this study. SATAKE (1995) also showed that use of finer grids is more effective than to use the nonlinear equation to reproduce the observed tide gauge records for the 1992 Nicaragua earthquake tsunami.

4.1.3 Finite-difference computation

Equations (7) and (8) are numerically solved with the finite-difference method. The staggered grid is used on the spherical coordinate system with the grid interval of 2 sec of the arc. The explicit equations are given in SATAKE (1995). The time-step is set at 0.2 s in order to satisfy the stability condition of the finite-difference method. The tsunami computation for 30 min after the generation is computed and the maximum water heights on the coastal points are output for comparison.

4.2 Far-field Tsunami Computations

4.2.1 Bathymetry data and grid size

The computed region extends 120°–150°E and 10°S–40°N. The predicted bathymetry data from satellite altimetry (SMITH and SANDWELL, 1997) are used. The predicted data are generally reliable for deep ocean (> 1,000 m), but less accurate in shallow water. For the Japanese coast where tide gauges are located, we use actually measured bathymetry data. The grid size is basically 1 min of the arc, but a finer grid (20 sec) system is used near the coastal tide gauge stations.

4.2.2 Governing equations

For a tsunami traveling a long distance, the effects of dispersion and the Coriolis force become significant. Because of the relatively small source size of the 1998 PNG tsunami, the dispersion effect cannot be ignored (TANIOKA, 1999). The governing equation including the first-order dispersion effect is the linear Boussinesq equation. Including the Coriolis force, it can be written as

$$\frac{\partial \mathbf{V}}{\partial t} = -g\nabla h - 2\Omega \times \mathbf{V} + \frac{d^2}{3}\nabla\nabla \cdot \frac{\partial \mathbf{V}}{\partial t}, \tag{9}$$

where Ω is the rotation vector of the Earth. The second term in right-hand side represents the Coriolis force and the third term represents physical dispersion.

4.2.3 Finite-difference computation

Equations of motion (9) and continuity (8) on the spherical coordinate system are solved by using a staggered finite-difference method. The Alternating Direction Implicit (ADI) scheme with double-sweep algorithm (KABILING and SATO, 1993) is employed. Specific expressions of the equations and details of the method are given in TANIOKA (2000).

4.3 Parameterization of Comparisons

For the comparison between the observed and computed heights, we introduce three statistical parameters. The first parameter is geometric mean K of the amplitude ratio O_i / C_i where O_i and C_i are the i-th observed and computed heights. The second parameter is an index of scatter κ as defined by AIDA (1978) as

$$\log \kappa = \left[\frac{1}{n}\sum_i \left(\log \frac{O_i}{C_i}\right)^2 - (\log K)^2\right]^{1/2} . \tag{10}$$

If the logarithmic amplitude ratios obey the normal distribution N(log K, log κ), then parameter κ can be considered as an error factor, because its logarithm shows the standard deviation. More practically, 68% of the amplitude ratio data is expected to lie within the range of factor κ. The third parameter is the linear correlation coefficient

$$r = \frac{\sum_i (O_i - \overline{O})(C_i - \overline{C})}{\sqrt{\sum_i (O_i - \overline{O})^2}\sqrt{\sum_i (C_i - \overline{C})^2}} , \tag{11}$$

where $\bar{O}$ and $\bar{C}$ are the means of O_i's and C_i's, respectively.

The first parameter, K, indicates the relative size of the observed and computed tsunami models, whereas the latter two indicate the goodness of the model. The smaller κ and the larger r are the measures of the better model.

5. *Comparison of Observed and Computed Tsunamis*

5.1 Comparisons at Far-field

Computed tsunami waveforms are compared with the observed waveforms at the two ocean-bottom and six tide gauge stations in Figure 5. As stated before, the tsunami first arrivals of the observed waveforms are not clear, hence we attempt to compare the overall pattern of waveforms and amplitudes. The large amplitude wave recorded before 14:00 GMT at Muroto-OB is background noise unrelated with the tsunami.

The tsunami waveforms and amplitudes computed for models HA and LA are basically similar at all the stations. This is not surprising, considering the long travel distance (OKAL and SYNOLAKIS, 2001). The amplitudes of later phase are somewhat larger for LA than HA at some coastal stations, although the difference is not significant compared with the difference between the observation and computations. Hence it is difficult to conclude which model is better.

Far-field tsunami waveforms are also computed from model SF, splay fault model. As shown in Figure 5, the computed waveforms from this small fault are much smaller than the observation or computed waveforms from seismological fault (HA and LA) models. The small source produces a shorter wavelength tsunami which is easily dispersed during the propagation. The slumping sources (models SSE, SSW and SSN) have even smaller source dimension, hence the computed waveforms would be even smaller and unable to explain the observations. An accurate computation of far-field tsunami waveforms from such small sources would require enormous computer memory and currently is impractical.

From the comparisons at far-field, we conclude that seismological fault motion is required to explain the observed tsunami waveforms. Short-wavelength source, such as splay fault or submarine slump, alone cannot reproduce the far-field tsunami.

5.2 Comparisons at Near-field

Comparisons of the observed and computed tsunami heights along the northern coast of New Guinea Island are shown in Figures 10–12.

Figure 10a shows the comparison for model LA. The computed tsunami heights are less than 3 m, much smaller than the observed heights. The geometrical ratio K is about 4, indicating that the computed heights are on average about a quarter of the observed. The computed tsunami does not reproduce the unusual height distribution centered at Sissano Lagoon. The tsunami source may be too close to the submerged delta, a possible refractor, for the tsunami energy to concentrate toward Sissano Lagoon.

Figure 10b is for model HA. The computed tsunami heights reproduce the unusual tsunami heights around Sissano Lagoon. The offshore location of displacement peak may be favorable for an incoming tsunami to refract at the submerged delta, as indicated by MATSUYAMA *et al.* (1999). The computed tsunami heights are still smaller than the observed. The geometrical ratio, $K = 2.42$, indicates that the computed heights are less than a half of the observed heights on the average.

Computed tsunami heights from model SF, splay fault, are shown in Figure 10c. The computed results reproduce large tsunami peaks around Sissano Lagoon, nonetheless amplitudes are still smaller than the observation. The geometrical average, $K = 2.47$, is similar to that of model HA. The computed tsunami heights become larger if we increase the slip amount on the splay fault. However, such a fault would yield larger seismic moment, and would be incompatible to the seismic observations.

Tsunamis computed from submarine slump models are shown in Figure 11. All three models, SSE, SSW and SSN, show similar tsunami heights along the coast. All of them have several narrow peaks around Sissano Lagoon, and reproduce the observed tsunami heights. The maximum amplitudes are also similar to the field measurements, while the average computed heights are still smaller as indicated by the geometrical ratio K (about 2). From these comparisons, it is difficult to judge which of the slump models best explains the tsunami observations. We note that any tsunami source around the amphitheater reproduces tsunami concentration toward Sissano Lagoon.

5.3 Combination of Source Models

We now consider combinations of fault and other sources, because small sources such as splay fault or submarine slump alone cannot reproduce the far-field tsunamis while low-angle fault alone cannot reproduce the near-field tsunamis.

We first examine a combination of models LA and SF, as the splay fault is considered as branching from the megathrust fault. The computed near-field tsunami heights are compared with the observation in Figure 12a. The tsunami heights are large, about 8 m, around Sissano Lagoon. The maximum height is still somewhat smaller than the observation, and the geometrical ratio $K = 2.2$. On the other hand, the error factor, $\kappa = 1.41$, is the smallest, and the correlation coefficient $r = 0.67$ is the largest among all the models. The statistical parameters indicate that this combined model best explains the observation among the models considered in this study.

We next combine model LA and submarine slump sources. Results from a combination of models LA and SSE are shown in Figure 12b. The computed tsunami heights show a few peaks, with the largest located east of Sissano Lagoon. The maximum height is more than 12 m, close to the observed, but the location is somewhat different. The geometric ratio $K = 1.6$ indicates that the computed heights are on the average still smaller than the observed. The error factor $\kappa = 1.68$ and the correlation coefficient $r = 0.23$ show that the fit is not very good compared with the other models.

Results from a combination of models LA and SSW are shown in Figure 12c. The computed tsunami height distribution shows two narrow peaks near Sissano

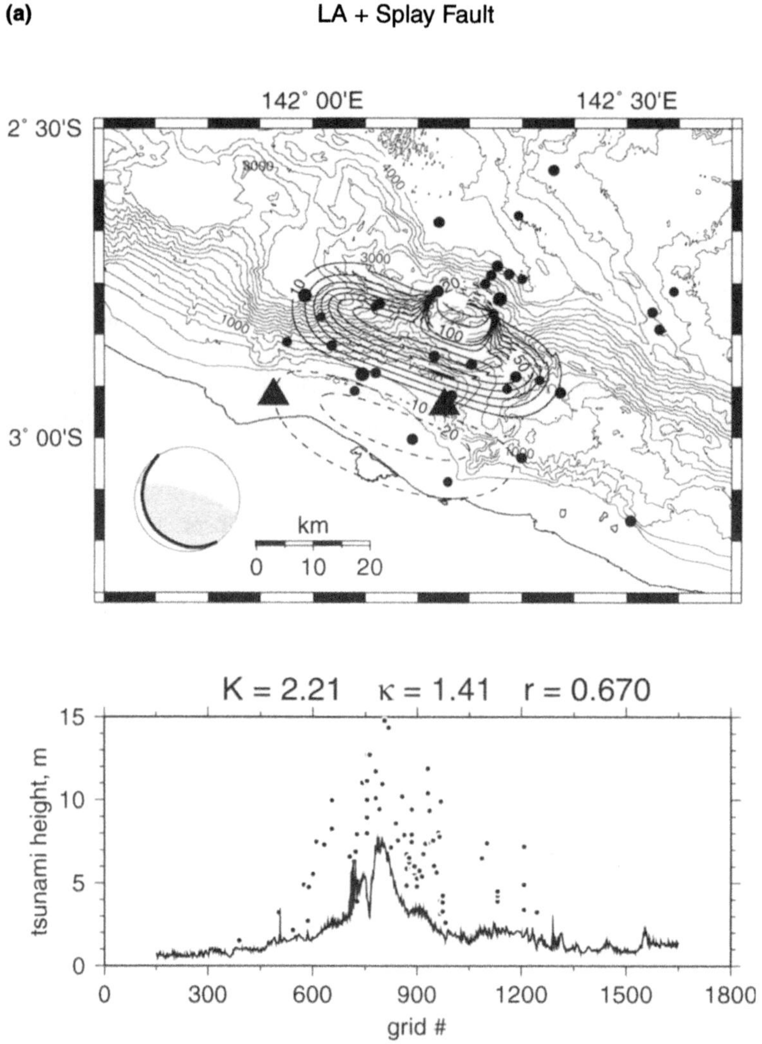

Figure 12
Results of tsunami computations for a combined source of low-angle (LA) fault and additional source. The top figure shows displacement of water surface overlaid on bathymetry and aftershock distribution (HURUKAWA *et al.*, 2003). The bottom diagram shows comparison of the observed and computed tsunami heights along the coast shown above. Statistical parameters of the comparison are also shown. (a) Combination of LA and SF (Splay fault) models (b) Combination of LA and submarine slump (model SSE). (c) Combination of LA and submarine slump (model SSW). (d) Combination of LA and submarine slump (model SSN).

Lagoon with the heights of 12 m, consistent with the observation. The geometric ratio $K = 1.74$ indicates the average computed height is still smaller than the observed. The error factor $\kappa = 1.70$ is similar to the previous case, but the correlation coefficient $r = 0.41$ is better.

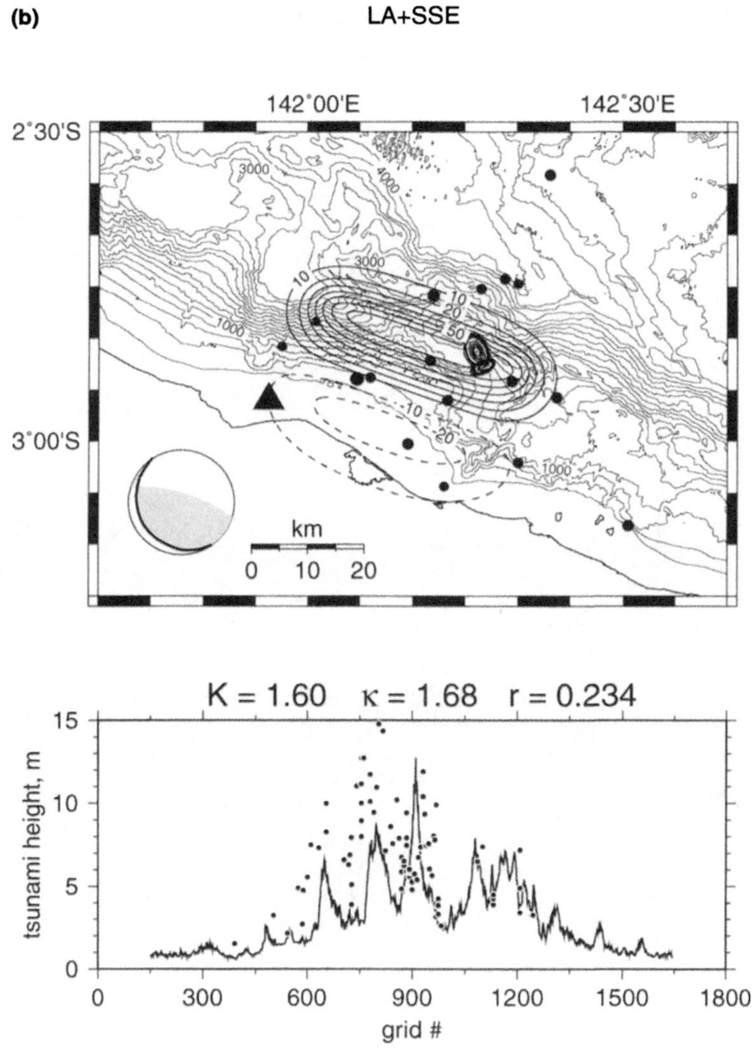

Figure 12
(Contd.)

Results from a combination of models LA and SSN are shown in Figure 12d. The overall shape of the computed tsunami height distribution is similar to case LA + SSE, although the peaks are narrower. The maximum height is 11.5 m, similar to the observation. The geometric ratio $K = 2.17$. The error factor $\kappa = 1.55$ is the smallest and the correlation coefficient $r = 0.46$ is the largest among the three combination models of LA and submarine slump.

Visual inspection of Figures 12 indicates that combination models of low-angle fault and submarine slump explain the observation equally well. The statistical

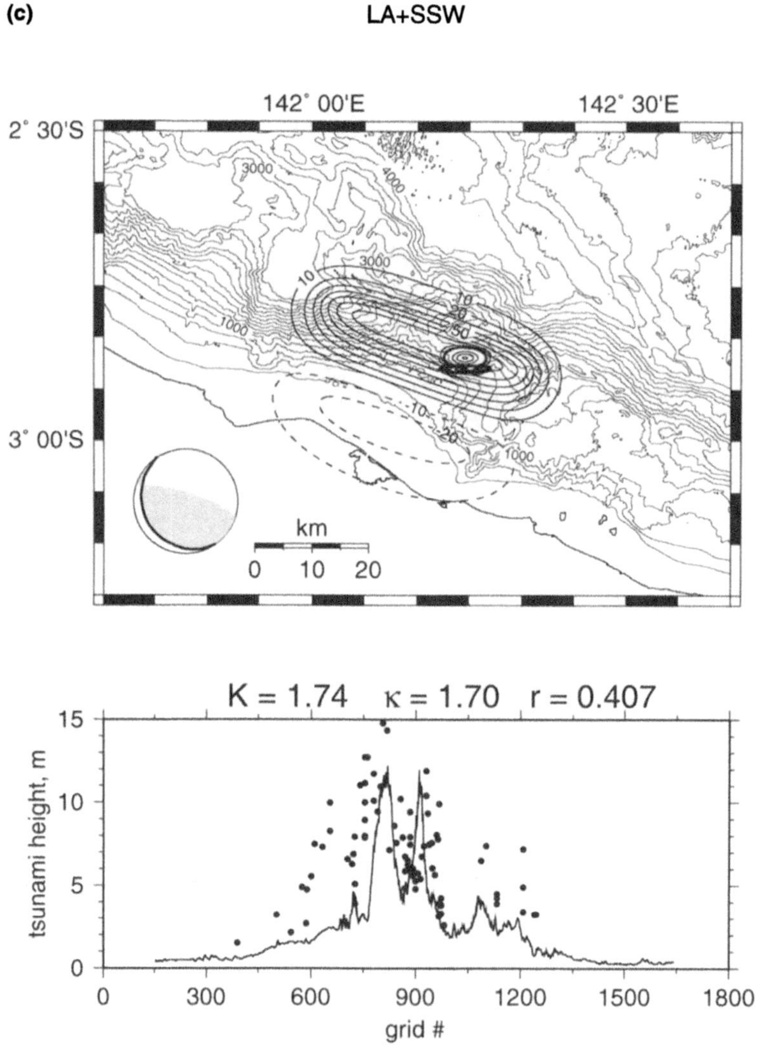

Figure 12
(Contd.)

parameters indicate that the combination of LA and SF is the best among these models, but the differences are in significant. From these comparisons, we conclude that additional tsunami source, in addition to the low-angle fault, is required to reproduce the near-field tsunamis, however it is difficult to identify and locate the additional source from the observed tsunamis. The slump parameters such as size or volume cannot be uniquely estimated; the three different slump volumes ranging from 0.8 to 3.0 km^3 produced similar tsunami heights. Are there any parameters to quantify the tsunami source? We will examine other parameters in the next section.

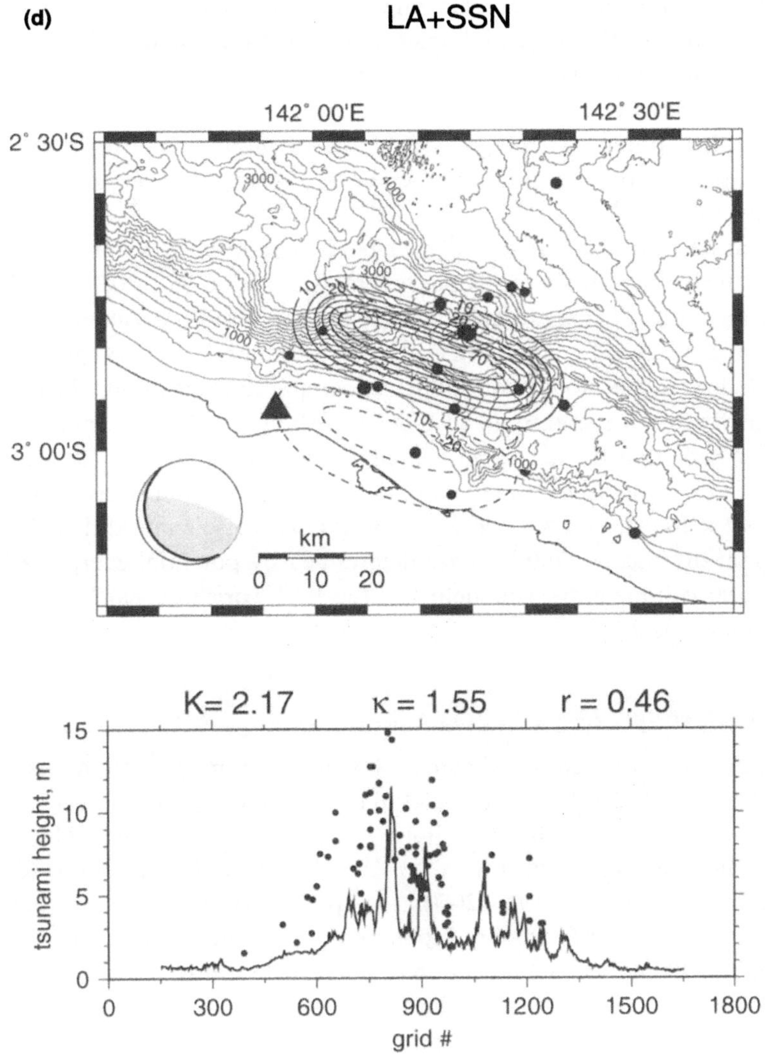

Figure 12
(Contd.)

6. Discussion and Conclusions

6.1 Scaling of Far-field and Near-field Tsunamis

We have seen that the tsunami heights at far-field are similar from models LA and HA, but much smaller from model SF (Fig. 5). The displaced water volume V are 0.6, 0.7 and 0.2 km^3, for these three models (Table 1). This indicates that the far-field tsunami amplitudes are roughly proportional to the source volume. On the other hand, the potential energy E for these models is 0.9, 1.6 and 1.4×10^{12} J, more or less similar values. These comparisons show that the far-field tsunami amplitudes are controlled by the displaced water volume at the source, rather than the potential energy. From the comparison with the observed amplitudes, the source volume of the 1998 Papua New Guinea tsunami is estimated as $\sim$0.6 km^3.

Computations and comparison of near-field tsunami yield different results. The displaced water volume for various models ranges from 0.05 to 0.7 km^3 (Table 1), more than an order of magnitude different, but the resulting tsunami heights are basically similar (except for model LA). The potential energy, E, ranges from 0.8 to 4.8×10^{12} J, but those models reproducing comparable maximum heights to the observations (except for slump model SSN) are about 2×10^{12} J (Table 1). We can therefore conclude that the near-field tsunami heights are controlled by the potential energy rather than displaced water volume, and the potential energy for the 1998 tsunami is $\sim 2 \times 10^{12}$ J. SATAKE and KANAMORI (1991) also showed that the tsunami amplitudes (at distances <700 km) are determined by potential energy rather than displaced water volume in their modeling of the 1984 Torishima earthquake tsunami on the Japanese coast.

6.2 Tsunami Generation Process

Low-angle fault and additional source, possibly a submarine slump, is the likely source of the 1998 Papua New Guinea tsunami. Seismological data do not provide conclusive evidence regarding the choice of actual fault plane. The relocated aftershock distribution (HURUKAWA *et al.*, 2003) and the observed coastal subsidence (MCSAVENEY *et al.*, 2000) are preferable for the low-angle fault. Tsunamis recorded in and around Japan require seismological fault, either low-angle or high-angle, as the tsunami source; small-scale source such as splay fault or submarine slump alone cannot reproduce the far-field tsunami observation. On the other hand, low-angle fault alone cannot reproduce the large tsunamis observed near Sissano Lagoon, and additional source is required. Comparisons of the observed tsunami heights with those computed from different models indicate that several models can reproduce the observations, indicating the difficulty of determining a tsunami source from the observed tsunami data alone. A combination of low-angle and splay fault (model LA + SF) and another combination of low-angle fault plus slump source (model LA + SSE) equally explain the observations. Lack of convincing field evidence makes the controversy inconclusive. Abundance of small-scale submarine slump features mapped by the single-channel seismic survey supports that one of such small-scale slumps is a possibility of additional tsunami source.

References

Abe, K. (1989), *Quantification of Tsunamigenic Earthquakes by the Mt. Scale*, Tectonophysics *166*, 27–34.

Aida, I. (1978), *Reliability of a Tsunami Source Model Derived from Fault Parameters*, J. Phys. Earth *26*, 57–73.

Davies, H. L., Davies, J. M., Lus, W. L., Perembo, R. C. B., Joku, N., Gedikile, H., and Nongkas, M. (2001), *Leaning from the Aitape Tsunami*, ITS 2001 Proceedings 415–424.

Dziewonski, A. M., Ekström, G. and Maternovakaya, N. N. (1999), *Centraoid-moment Tensor Solutions for July-September, 1988*, Phys. Earth Planet. Inter. *114*, 99–107.

Geist, E. (2000), *Origin of the 17 July 1998 Papua New Guinea Tsunami; Earthquake or Landslide*?, Seism. Res. Lett. *71*, 344–351.

Geist, E. L. (2001), *Reply to Comment on "Origin of the 17 July 1998 Papua New Guinea Tsunami; Earthquake or Landslide*?" *by E. L. Geist*, Seism. Res. Lett. *72*, 367–372.

Grilli, S. T. and Watts, P. (1999), *Modeling of Waves Generated by a Moving Submerged Body. Applications to Underwater Landslides*, Engin. Analysis with Boundary Elements *23*, 645–656.

Heinrich, P., Piatanesi, A., and Hebert, H. (2001), *Numerical Modelling of Tsunami Generation and Propagation from Submarine Slumps: the 1998 Papua New Guinea Event*, Geophys. J. Int. *145*, 87–111.

Heinrich, P., Piatanesi, A., Okal, E., and Hebert, H. (2000), *Near-field Modeling of the July 19, 1998 Tsunami in Papua New Guinea*, Geophys. Res. Lett. *27*, 3037–3040.

Hurukawa, N. (1999), *A Fault Plane of the 1998 Papua New Guinea Earthquake Estimated from Relocated Aftershocks Using Data of the International Data Center of CTBT (in Japanese)*, Zisin, J. Seism. Soc. Jpn *52*, 95–99.

Hurukawa, N., Tsuji, Y., and Waluyo, B. (2003), *The 1998 Papua New Guinea Earthquake and its Fault Plane Estimated from Relocated Aftershocks*, Pure Appl. Geophys., *160*, 1829–1841.

Kabiling, M. B., and Sato, S. (1993), *Two-dimensional Nonlinear Dispersive Wave-current and Three-dimensional Beach Deformation Model*, Coastal Eng. Japan *35*, 195–212.

Kajiura, K. (1981), *Tsunami Energy in Relation to Parameters of the Earthquake Fault Model*, Bull. Earthq. Res. Inst., Univ. Tokyo *56*, 415–440.

Kanamori, H. (1972), *Mechanism of Tsunami Earthquakes*, Phys. Earth Planet. Inter. *6*, 246–259.

Kawata, Y., Benson, B. C., Borrero, J. C., Borrero, J. L., Davies, H. L., Lange, W. P. D., Imamura, F., Letz, H., Nott, J., and Synolakis, C. E. (1999), *Tsunami in Papua New Guinea was as Intense as First Thought*, EOS, Trans. Am. Geophys. Union *80*, 101, 105.

Kikuchi, M., Yamanaka, Y., Abe, K., and Morita, Y. (1999), *Source Rupture Process of the Papua New Guinea Earthquake of July 17, 1998 Inferred from Teleseismic Body Waves*, Earth Planets Space *51*, 1319–1324.

Matsuyama, M., Walsh, J. P., and Yeh, H. (1999), *The Effect of Bathymetry on Tsunami Characteristics at Sissano Lagoon, Papua New Guinea*, Geophys. Res. Lett. *26*, 3513–3516.

Matsuyama, M. and Yeh, H. (2003), *Effects of Tsunami at Sissano Lagoon, Papua New Guinea, due to Submarine—Landslide and Tectonics Origins.*

McSaveney, M. J., Goff, J. R., Darby, D. J., Goldsmith, P., Barnett, A., Elliott, S., and Nongkas, M. (2000), *The 17 July 1998 Tsunami, Papua New Guinea: Evidence and Initial Interpretation*, Marine Geology *170*, 81–92.

Okada, Y. (1985), *Surface Deformation due to Shear and Tensile Faults in a Half-space*, Bull. Seism. Soc. Am. *75*, 1135–1154.

Okal, E. and Synolakis, C. E. (2001), *Comment on "Origin of the 17 July 1998 Papua New Guinea Tsunami; Earthquake or Landslide*?" *by E. L. Geist*, Seismol. Res. Lett. *72*, 362–366.

Satake, K. (1995), *Linear and Nonlinear Computations of the 1992 Nicaragua Earthquake Tsunami*, Pure Appl. Geophys. *144*, 455–470.

Satake, K. and Kanamori, H. (1991), *Abnormal Tsunamis Caused by the June 13, 1984, Torishima, Japan, Earthquake*, J. Geophys. Res. *96*, 19,933–19,939.

Satake, K. and Tanioka, Y. (1999), *Sources of Tsunami and Tsunamigenic Earthquakes in Subduction Zones*, Pure Appl. Geophys. *154*, 467–483.

Smith, W. H. and Sandwell, D. T. (1997), *Global Seafloor Topography from Satellite Altimetry and the Depth Soundings*, Science *277*, 1956–1962.

SWEET, S., SILVER, E. A., DAVIES, H., MATSUMOTO, T., WATTS, P., and SYNOLAKIS, C. E. (1999), *Seismic Reflection Images of the Source Region of the Papua New Guinea Tsunami of July 17, 1998 (abstract)*, EOS, Trans. Am. Geophys. Union *80*, suppl., F750.

SYNOLAKIS, C. E., BARDET, J.-P., BORRERO, J. C., DAVIES, H. L. G., OKAL, E. A., SILVER, E. A., SWEET, S., and TAPPIN, D. R. (2002), *The Slump Origin of the 1998 Papua New Guinea Tsunami*, Proc. Roy. Soc. London, ser A, *458*, 763–789.

TAKUECHI, A., and SHIPBOARD SCIENTIFIC PARTY of R/V YOKOSUKA (1998), *Bottom Response to a Tsunami Earthquake: Submersible Observations in the Epicenter Area of the 1993 Earthquake off Southwestern Hokkaido, Sea of Japan*, J. Geophys. Res. *103*, 24,109–24,125.

TANIOKA, Y. (1999), *Analysis of the Far-field Tsunamis Generated by the 1998 Papua New Guinea Earthquake*, Geophys. Res. Lett. *26*, 3392–3396.

TANIOKA, Y. (2000), *Numerical Simulation of Far-field Tsunamis Using the Linear Boussinesq Equation—The 1998 Papua New Guinea Tsunami*, Meteorol. Geophys. *51*, 17–25.

TAPPIN, D. R., WATTS, T. M. P., SATAKE, K., MCMURTY, G. M., MATSUYAMA, M., LAFOY, Y. TSUJI, Y., KANAMATSU, T., LUS, W., IWABUCHI, Y., YEH, H., MATSUMOTO, Y., NAKAMURA, M., MAHOI, M., HILL, P., CROOK, K., ANTON, L., and WALSH, J. P. (1999), *Sediment Slump Likely Caused 1998 Papua New Guinea Tsunami*, EOS, Trans. Am. Geophys. Union. *80*, pp. 329, 334, 340.

TAPPIN, D. R., WATTS, P., MCMURTY, G. M., LAFOY, Y., and MATSUMOTO, T. (2001), *The Sissano, Papua New Guinea Tsunami of July 1998 — Offshore Evidence on the Source Mechanism*, Marine Geology *175*, 1–23.

TREGONING, P., LUMBECK, K., STOLZ, A., MORGAN, P., MCCLUSKY, S. C., VAN DER BEEK, P., MCQUEEN, H., JACKSON, R. J., LITTLE, R. P., LAING, A., and MURPHY, B. (1998), *Estimation of Current Plate Motions in Papua New Guinea from Global Positioning System Observations*, J. Geophys. Res. *103*, 12,181–12,203.

Pure appl. geophys. 160 (2003) 2119–2146
0033–4553/03/112119–28
DOI 10.1007/s00024-003-2422-0

Pure and Applied Geophysics

Field Survey and Numerical Simulations: A Review of the 1998 Papua New Guinea Tsunami

PATRICK J. LYNETT,[1] JOSE C. BORRERO,[2] PHILIP L.-F. LIU[3] and COSTAS E. SYNOLAKIS[2]

Abstract—The Papua New Guinea (PNG) tsunami of 1998 is re-examined through a detailed review of the field survey as well as numerous numerical computations. The discussion of the field survey explores a number of possible misinterpretations of the recorded data. The survey data are then employed by a numerical model as a validation tool. A Boussinesq model and a nonlinear shallow water wave (NLSW) model are compared in order to quantify the effect of frequency dispersion on the landslide-generated tsunami. The numerical comparisons indicate that the NLSW model is a poor estimator of offshore wave heights. However, due to what appears to be depth-limited breaking seaward of Sissano spit, both numerical models are in agreement in the prediction of maximum water elevations at the overtopped spit. By comparing three different hot-start initial profiles of the tsunami wave, it is shown that the initial shape and orientation of the tsunami wave is secondary to the initial displaced water mass in regard to prediction of water elevations on the spit. These numerical results indicate that agreement between numerical prediction of runup values with field recorded values at PNG cannot be used to validate either a NLSW tsunami propagation model or a specific landslide tsunami hot-start initial condition. Finally, with the use of traditional tsunami codes, a new interpretation of the PNG runup measurements is presented.

Key words: Submarine landslide, runup, Boussinesq.

1. Introduction

On July 17, 1998 at 08:49 GMT (18:49 local), an earthquake of $M \approx 7$ occurred near the Pacific coast of western Papua New Guinea. Shortly after the earthquake, a destructive tsunami caused extensive damage along the coast from the town of Aitape west to the region around Sissano Lagoon (see Fig. 1). In fact, the death toll was the worst from a tsunami in the past 50 years, with over 1000 persons killed by the tsunami waves. The exact causative mechanism of the tsunami has been the subject of considerable debate, although recently published works (i.e. SYNOLAKIS *et al.*, 2002) strongly indicate a slump source.

[1] Department of Civil Engineering, Texas A&M University, College Station, TX 77843–3136, U.S.A.
[2] School of Engineering, University of Southern California, Los Angeles, CA 90089-2531, U.S.A.
[3] School of Civil and Environmental Engineering, Cornell University, Ithaca, NY 14853, U.S.A.

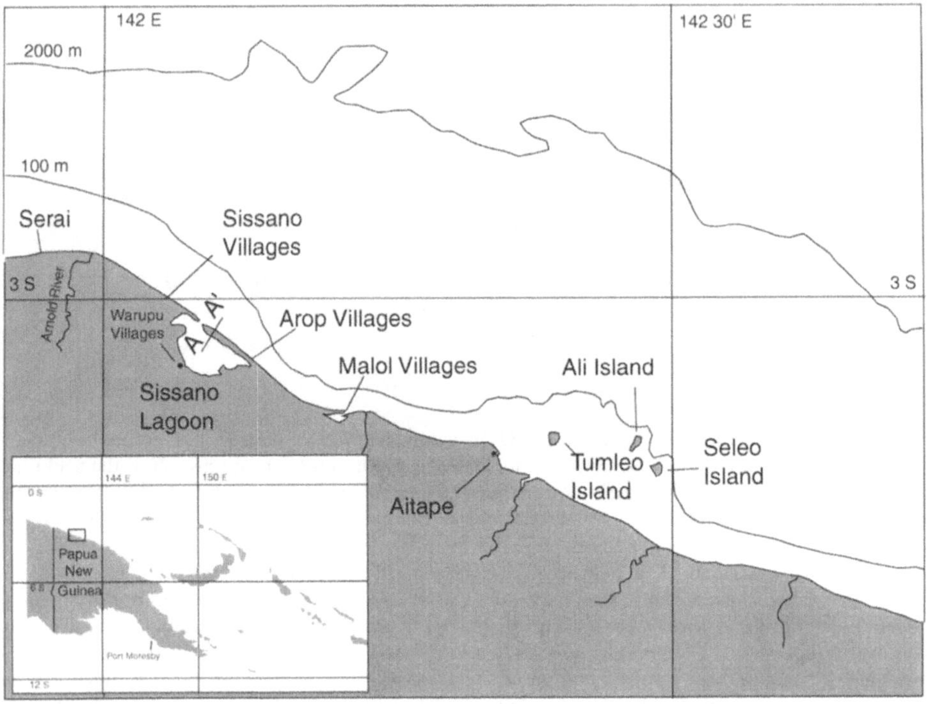

Figure 1
A map of Papua New Guinea and the affected region.

Landslide and slump sources, while poor generators of farfield tsunamis (OKAL, 2003), can generate locally extreme wave heights. Local events are, by definition, caused by tsunamis that cannot transmit energy across large distances, whether it be due to properties of the tsunami wave or a manner of physical boundary. In the case of landslide and slump tsunamis, the deterring factor is a function of the generated wave properties. Typically, a landslide source region encompasses a much smaller area than a dislocation source. Consequently, significantly less total energy is imparted into the water; this energy spreads rapidly and inhibits the possibility of farfield tsunamis. An additional effect of the smaller source region is the generation of a tsunami composed of relatively shorter wavelengths. As the tsunami propagates in the open ocean, the different components will travel at different speeds, opposed to the very long tsunamis of dislocation sources. This dispersion of different frequencies will also act to lessen the amplitude of the tsunami, spreading out energy in the direction of propagation.

Depending on the wavelength of the tsunami, frequency dispersion effects can be important in the nearfield as well, over the distance of a few wavelengths. It will be shown in the numerical section of the paper that this is the case for the PNG tsunami. Additionally, the effect of the initial shape and orientation of the tsunami on offshore wave heights and wave runup will be analyzed. Lastly, the interpretation of the field data when used in conjunction with a well-studied tsunami code is discussed. First, however, a detailed review of the field survey is presented.

2. *Review of Field Survey*

2.1 Background

While it is customary during a post tsunami survey to measure the maximum runup and the maximum inundation, in Papua New Guinea the worst damage was concentrated in the region around Sissano Lagoon where measurements of this type were impossible. Sissano Lagoon has the unique feature of a long narrow sand spit between the lagoon and the sea, which has no elevation higher than 3 m and is as narrow as 150 m in some locations. Because of the low relief of the topography and the extremely large wave heights, the sand spit was completely overtopped and water was able to flow over the spit and into the lagoon. Hundreds of villagers lived along the sand bar, and all of their dwellings, canoes and other signs of human habitation were uniformly stripped from the beach.

This fact warrants special care when dealing with measurements taken along the Sissano spit. The data are neither runup nor inundation distances. Since the spit was overtopped and the flow was allowed to propagate across the waters of the lagoon, no true value of "runup" is available. Rather, what we have for this region are measurements of the maximum water elevation, or flow depths when topography is factored in, with the value inferred from local clues such as debris in trees or broken branches up to a certain level.

Despite the trained eyes of the post event field survey team, there is always the possibility of misidentification of a water mark. In terms of broken branches, the surveyor does not know whether it was the tsunami, a big windstorm the week before, or just a dead branch falling from above. Debris in trees also presents problems, as it could have fallen from a taller tree above, or been blown in by the wind or intentionally put there by a former resident (in the case of rags, or buckets).

Within two weeks of the event, the ITST (International Tsunami Survey Team) was mobilized to document the effects along the shore of Papua New Guinea. The team was comprised of participants from Papua New Guinea, Japan, Australia, New Zealand and the United States (KAWATA *et al.*, 1999).

The group split in two with part of the team going by boat along the northern coast from Wewak towards Sissano, and the other half making the journey over land. The boat-based group was able to survey several offshore islands in addition to points along shore (see Fig. 2). The teams met up in Aitape where a relief command center was set up to aid the thousands of victims of the tsunami.

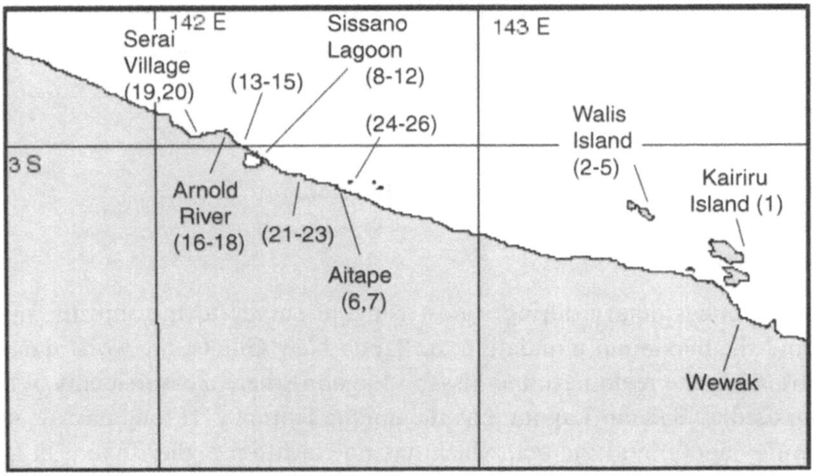

Figure 2
The north coast of Papua New Guinea showing the route taken by the boat-based survey team.

Again the team was split and continued by land and sea towards Sissano. The boat-based team was able to survey as far west as the lumber mill at Serai. The land-based team could only get as far as Malol. However, helicopters were used to ferry the land-based team in the most affected area around Sissano Lagoon.

2.2 Recorded Field Data

The boat-based group was able to obtain twenty-seven measurements on their passage from Wewak to Sissano. Data were taken using standard field surveying procedures which included measuring horizontal inundation, vertical runup and noting possible flow depths through identification of water marks, such as broken branches or debris left in trees.

The following lists all 27 of the points taken by the boat-based team, which included the second and fourth authors. Figure 2 should be referenced to determine the specific location of each of the measurement points.

Site 1 – Purak Village, Kairiru Island

Residents reported feeling the earthquake and noticed a small surge. The runup was up a gentle slope. The uncorrected runup was 1.4 m with an inland penetration of 9.8 m.

Site 2 – Mohu Island, near Walis Island

An uninhabited island. A small log was found above the high tide line along with some other small debris. The log was measured at 1.3 m elevation and 14 m inland penetration along a linear beach slope.

Site 3 – Kameilam Village, Walis Island

Shown this village by locals in a canoe after stopping at Mohu Island. The locals reported having felt the earthquake and hearing the sound of water rushing over the offshore reef. It was probably after dark when the waves affected this area. Maximum runup of 0.3 m with 3.5 m inland penetration.

Site 4,5 – Big Walis Village, Walis Island

Local residents reported feeling the earthquake and hearing the sound of water rushing over an offshore reef. Similar situation as in the previous village. Maximum runup of 0.5 m with 5.5 and 5.2 m inland penetration.

Site 6 – Aitape

Aitape was heavily affected by the earthquake and tsunami. The town was being used as a staging area for the relief effort. This transect had a maximum runup of 3.4 m at a distance of 70 m from shore. The maximum inland penetration was 157 m at an elevation of 2.34 m.

Site 7 – Aitape

This transect had a maximum runup of 3.5 m at a distance of 70 m from shore. The maximum inland penetration was 154 m at an elevation of 1.55 m. This transect crossed over the coast road to a school building. There were erosional features at the base of the school building.

Site 8 – Eastern Spit, Sissano Lagoon

The first transect of the boat team in the devastated area. Flow depths were inferred from broken branches and elevated debris, and measured with a laser distance finder. Water marks were found as high a 9.6 and 10.8 m at distances of 73 and 117 m from shore. This entire area was previously the site of a village. The only remains are the stilts upon which the houses stood. A profile of transect 8 is shown in Figure 3.

Site 9 – Eastern Spit, Sissano Lagoon

Similar to line 8, a profile is shown in Figure 3. This transect is the location of the famous "bucket in the tree" photograph. This is the image where a bucket is shown hanging in the small branches of a tree. The branches extend out over the water of the lagoon. The height of this bucket was assumed to be the height of the flow and documented as such in initial reports. The bucket in the tree is show in Figure 4.

Upon further inspection, the authors are not so confident in the bucket as a reliable flow depth indicator. The fact that so many branches remain on the tree is one indicator that such a deep sustained flow did not occur at this location. How the bucket actually got there is still open to debate. Some have suggested that the tree was bent over by the force of the wave and able to pick up debris.

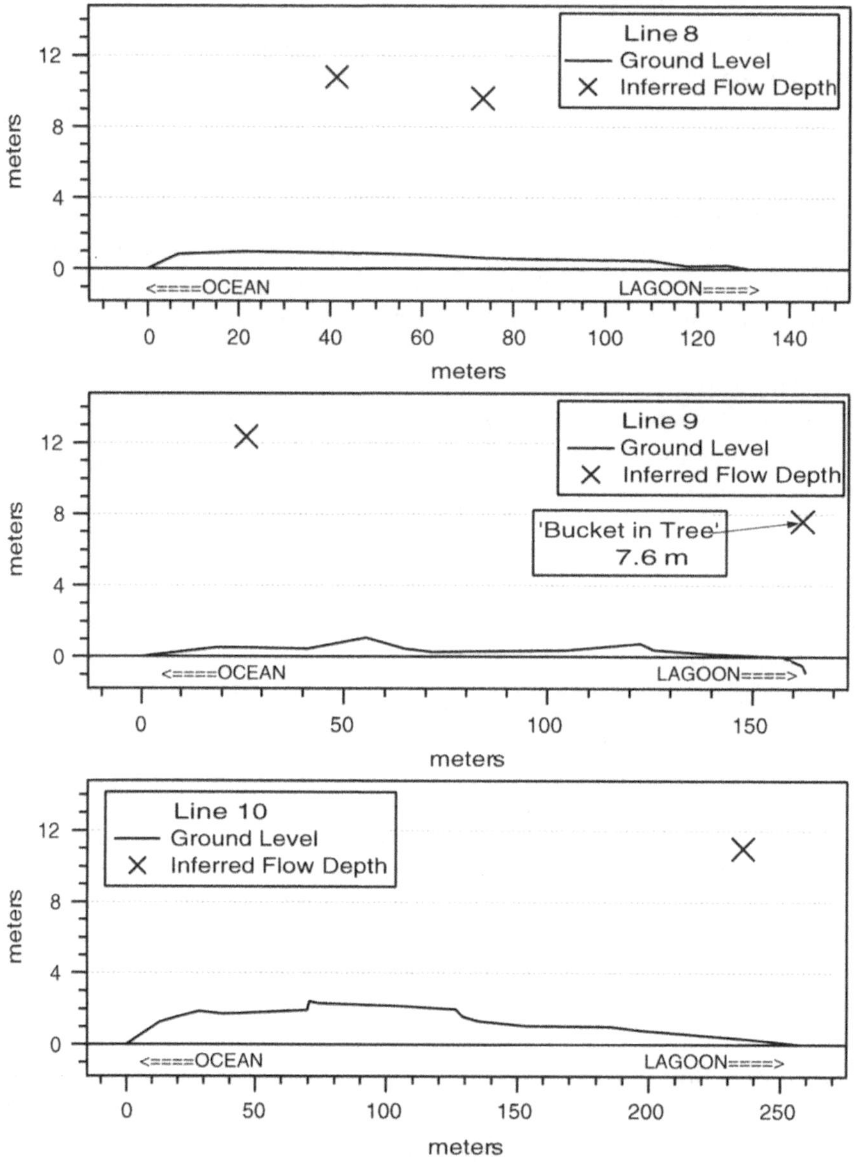

Figure 3
Profiles of line 8, 9, and 10 taken during the 1998 Papua New Guinea tsunami field survey.

Site 10 – Western Spit, Sissano Lagoon

The first transect on the western side of the spit. Numerous houses existed here and were washed away. One potential flow marker was found on this line, a palm frond at 11 m elevation on the back side of the spit near the lagoon.

Figure 4
Photo showing house stilts bent back by the force of the wave. A bucket is seen to be hanging from the branches of the tree in the distance. How the bucket got there is still subject to debate.

Site 11 – West of Sissano Village

Near the main village of Sissano. The flow penetrated across a low flat area, and penetrated to the jungle behind the village. We were able to measure up to the edge of the jungle and no further. Debris was piled up at the landward extent of our transect to a height of 4.6 m above the ground (6.9 m above sea level). The transect of this location is shown in Figure 5.

Site 12 – East of Sissano Village

Near the main village of Sissano, but east of line 11. The tsunami traveled across a low flat area, and penetrated the jungle behind the village. We were able to measure up to the edge of the jungle and no further. Debris was piled up at the landward extent of our transect to a height of 4.5 m above the ground (6.4 m above sea level). This site was the location of a governmental building. The building was a sturdy wooden frame structure built on a concrete foundation. At the time of our survey the building was completely wiped from its foundation. The foundation itself was left with a sand layer up to 9 cm thick. The transect of this location is shown in Figure 5.

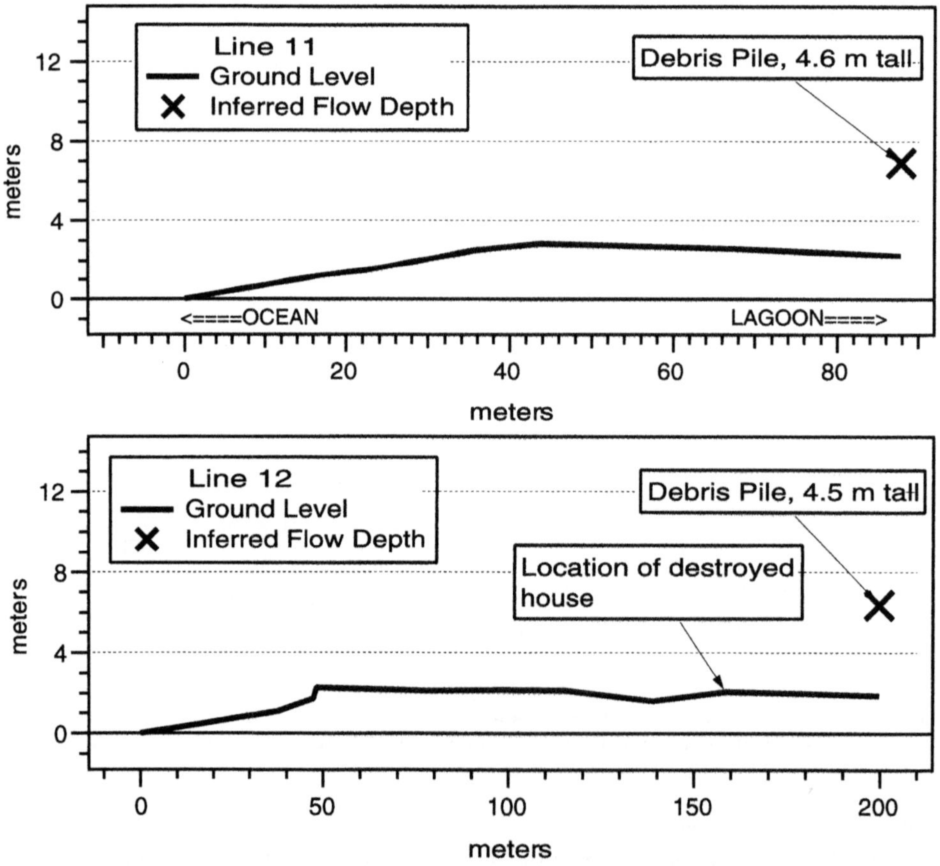

Figure 5

Profiles of lines 11 and 12 taken during the 1998 Papua New Guinea tsunami field survey.

Site 13,14 – Between Sissano and the Arnold River

Measured inland from shore until the jungle became too thick. A debris pile was located 42 m inland. Site 14 was nearly identical, measured 100 m to the west.

Site 15 – East of Arnold River

This transect featured a flow indicator at 3.2 m above ground level a distance of 62 m inland and an elevation of 2.13 m. The maximum inland penetration was 111 m inland at an elevation of 0.5 m above sea level.

Site 16 – Arnold River, East Bank

On the east bank of the Arnold River, a small abandoned village was found. There was a structure near the bank of the river that was partially destroyed by the wave. The structure was built on the river flood plain some 20 m from the river but nearly 100 m from the ocean. The structure was probably damaged by the wave coming in from the river as well as over the beach from the ocean. A suspicious palm frond was noticed hanging from a tree trunk, initially it was believed to be a flow mark, however a local resident informed the team that it was actually a marker placed there by residents to mark private property. Details of the flow characteristics at the mouth of the Arnold River are given in Figure 6.

Site 17 – Arnold River, West Bank

The west bank of the Arnold River was another difficult site to interpret. There were marks of wave uprush, and debris on the floor of the dense forest beyond the tree line. The difficulty in understanding the flow marks was compounded by the location, a river mouth, where unusual focusing and channeling of the flow may have

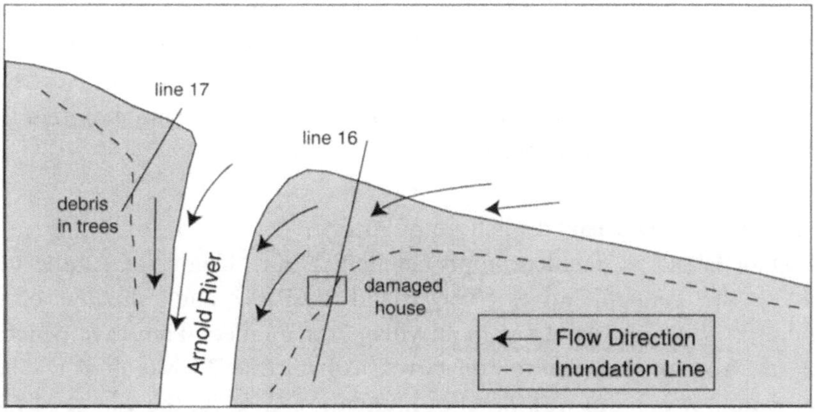

Figure 6

Detail of the inundation measurements at the mouth of the Arnold River.

taken place. A suspicious stick was found in a tree at 5.2 m elevation, but no other similar flow marks were observed. The maximum inland penetration was 86 m to an elevation of 2 m.

Site 18 – West of Arnold River

Continuing west of the Arnold River was a long, relatively straight beach. An unused structure was found 115 m inland beyond the tree line. It was undamaged, but there was a debris line at the foot of the structure. The wave had to overtop a 2.5 m beach crest at 75 m inland before depositing the debris at the structure – 115 m inland, 2.3 m elevation.

Site 19 – Serai Village

Serai Village is approximately 5 km west of the heavily damaged Sissano Villages. However at Serai, there was no damage. The wave did not reach the houses. The water overtopped the berm crest at 3.8 m elevation and 38 m inland. Its maximum excursion was 57 m at an elevation of 3.4 m. It should be noted that the beach in this area is much steeper than at Sissano.

Site 20 – Serai Village

Just west of Serai is the Erkan Log Yard. This is a location where large ships come in to have logs loaded for export. The beach is steep and backed by high steep cliffs. The local residents noted that there were several large rockslides on these cliffs during the earthquake. These rockslides confirmed survey team members in their speculations, developed earlier in the field, that underwater landslides may have triggered the tsunami.

Sites 21,22,23 – Returning to Aitape

These three points were taken along the beach between the Arnold River and Serai. They each gave consistent values of debris in the trees fronting the beach. The heights of the debris above sea level were 7.2, 5.1, and 6.9 m at distances of 29 m, 26 m and 27 m, respectively. Because of a lack of time, the maximum inundation was not found for each of these profiles.

Sites 24,25,26 – Tumleo Island offshore of Aitape

The small Island of Tumleo, approximately 1 km offshore of Aitape was also affected by the tsunami waves. The island has a broad, shallow coral reef surrounding it. Site 23 was at the main village on Tumleo. The wave penetrated a distance of 70 m and maximum runup was 2.4 m. Site 25 was at Ali Village on Tumleo. Runup was 3.7 m with 88 m of inundation. Site 26, the village of Nowom, faced towards the direction of the incoming wave. Runup was 4.2 m and inundation was 96 m.

3. *Numerical Modeling of the Tsunami*

The extensive field work described in the previous section provides a unique opportunity to achieve an understanding of the atypical source mechanism of this devastating tsunami. Through comparing the numerical runup/flow depth values with those that were recorded by the field team, numerical models can shed light onto the characteristics of the tsunami, which might then provide information of the dynamics of the underwater landslide. Additionally, the field data allows for validation of the numerical models, making them increasingly trustworthy in future simulations.

The tsunami generated by this slump is not the typical tsunami that has been modeled for decades using the nonlinear shallow water (NLSW) equations. These equations are accurate for large scale tsunamis, such as those that impact the entire Pacific rim, whose wavelengths are very large compared to the ocean depth. Local tsunamis, such as at PNG, have wavelengths that are still large compared to the water depth, on the order of 10 times the depth. For waves of this length, the NLSW equations have large, leading order errors. Therefore, it is suspect to use NLSW models to draw conclusions about local tsunami events. On the other hand, the Boussinesq equations are well-known to be accurate for waves of this length.

The Boussinesq equations are derived under the assumption that nonlinearity and frequency dispersion are weak, i.e.,

$$O(\varepsilon) = O(\mu^2) \ll 1, \tag{1}$$

where $\varepsilon =$ wave amplitude/water depth and $\mu =$ water depth/wavelength. The resulting approximate continuity equation, in dimensional form, is

$$\frac{\partial(h+\zeta)}{\partial t} + \nabla\cdot[(h+\zeta)\mathbf{u}_\alpha] - \nabla\cdot\left\{h\left[\left(\frac{h^2}{6}-\frac{1}{2}z_\alpha^2\right)\nabla(\nabla\cdot\mathbf{u}_\alpha) - \left(\frac{h}{2}+z_\alpha\right)\nabla\left(\nabla\cdot(h\mathbf{u}_\alpha)+\frac{\partial h}{\partial t}\right)\right]\right\} = 0. \tag{2}$$

Equation (2) is one of three governing equations for ζ and $\mathbf{u}_\alpha$, the free surface displacement and the depth-integrated horizontal velocity vector. The other two equations come from the horizontal momentum equation, which is given in vector form as

$$\frac{\partial\mathbf{u}_\alpha}{\partial t} + \mathbf{u}_\alpha\cdot\nabla\mathbf{u}_\alpha + g\nabla\zeta + \frac{\partial}{\partial t}\left\{\frac{1}{2}z_\alpha^2\nabla(\nabla\cdot\mathbf{u}_\alpha) + z_\alpha\nabla\left[\nabla\cdot(h\mathbf{u}_\alpha)+\frac{\partial h}{\partial t}\right]\right\} = 0. \tag{3}$$

Equations (2) and (3) are the coupled governing equations, written in terms of $\mathbf{u}_\alpha$ and ζ, for weakly nonlinear, weakly dispersive waves generated by a seafloor movement. This set of equations is most commonly known as the "extended" Boussinesq equations, and will be referred to as "BOUS" in this paper. We note here that $\mathbf{u}_\alpha$ is evaluated at $z = z_\alpha(x, y, t)$, which is a function of time. The choice of z_α is made based on the linear dispersion characteristics of the governing equations (e.g., NWOGU 1993; CHEN and LIU, 1995). Assuming a fixed seafloor, in order to extend the applicability of the governing equations to relatively deep water (or a short wave), z_α is recommended to be evaluated as $z_\alpha = -0.53\,\mathrm{h}$. In the following analysis, the same relationship is employed.

Also, a set of model equations can be derived where it is assumed that frequency dispersion is unimportant, i.e.,

$$O(\mu) \ll 1 \tag{4}$$

which yields the well-studied NLSW model:

$$\frac{\partial(h+\zeta)}{\partial t} + \nabla \cdot [(h+\zeta)\mathbf{u}_\alpha] = 0, \tag{5}$$

$$\frac{\partial \mathbf{u}_\alpha}{\partial t} + \mathbf{u}_\alpha \cdot \nabla \mathbf{u}_\alpha + g\nabla\zeta = 0. \tag{6}$$

3.1 Numerical Model

The structure of the current numerical model is described in detail in LYNETT and LIU (2002), where it was employed to model landslide generated waves and runup. The numerical scheme is a high-order predictor-corrector method, employing a third-order in time explicit Adams-Bashforth predictor step, and a fourth-order in time Adams-Moulton implicit corrector step (PRESS *et al.*, 1989). The implicit corrector step must be iterated until a convergence criterion is satisfied. All spatial derivatives are differenced to fourth-order accuracy, yielding a model which is numerically accurate to $(\Delta x)^4, (\Delta y)^4$ in space and $(\Delta t)^4$ in time. Runup, inundation, and overtopping by the tsunami can be examined. The moving boundary scheme employed here is the technique developed by LYNETT *et al.* (2002). Founded around the restrictions of the high-order numerical wave propagation model, the moving boundary scheme utilizes linear extrapolation of free surface and velocity through the shoreline, into the dry region. This approach allows for the five-point finite difference formulas to be applied at all points, even those neighboring dry points, and thus eliminates the need of conditional statements. To simulate the effects of wave breaking, the eddy viscosity model (ZELT, 1991; KENNEDY *et al.*, 2000) is used here. Readers are directed to KENNEDY *et al.* (2000) for a thorough description and validation of the breaking model, and the coefficients and thresholds given therein are used for all the simulations presented in this paper. Both BOUS and NLSW simulations utilize the same numerical algorithm and initial conditions for the simulations to be presented, the only difference being the dispersive, $O(\mu^2)$, terms are not calculated for the NLSW runs.

3.2 Dispersion Effects on Two PNG Initial Conditions

In this section, two of the reported initial conditions for the PNG event, used previously in NLSW models, will be employed in both NLSW and BOUS simulations. These "hot start" initial conditions are meant to yield a reasonable facsimile of the free surface disturbance caused by the slump. Thus, the actual movement of the slump is not being recreated here; rather, a very simplified approach has been taken. The initial free surface condition, with zero velocity, is placed above the slump region and evolves after the numerical simulation is started. These initial conditions are based on a number of observations of the post-slump seafloor profile, as well as assumptions of the failure motion (SYNOLAKIS *et al.*, 2002). Figures 7 and 8 depict the two initial conditions to be examined in this section.

The initial condition in Figure 7, called in this paper the short-fat initial condition, represents the initial condition founded on recent observations of the offshore slump. This condition can then be thought of as more up-to-date than the long-narrow initial condition shown in Figure 8, which had been developed before offshore studies of the slump area had been undertaken. It must be noted that there is a large degree of uncertainty in even the short-fat initial condition, as these initial conditions attempt to approximate a dynamic situation (tsunami generation and seafloor movement occurring simultaneously) with a static, "hot start" initial condition. It is, however, difficult to quantify this error, as little can be inferred of the exact motion of the slump.

For simplicity, the short-fat initial condition will be referred to as SFIC, and the long-narrow initial condition as LNIC. The bathymetry used in these simulations is derived from the bathymetry collected during a post tsunami offshore cruise by the Japanese research vessel Kairei (TAPPIN *et al.*, 1999), and is limited to depths greater than 200 m. For the region between the shoreline and 200 m depths, both linear interpolation and local bathymetric charts were used. Before looking at the numerical output, the expectation of the importance of frequency dispersion effects can be roughly inferred. For both initial conditions, the wavelength is on the order of 10 km in a water depth of about 1.5 km. From linear analysis, it should be expected that initially this wave will behave as an intermediate water depth wave, not a shallow water depth, or long, wave. For reference, the ratio of the wave speed of a 10 km wave to the wave speed of a long wave in 1.5 km of water is 0.9. Thus, initially, frequency dispersion will play an important role. However, as the wave propagates into shallower water, frequency dispersion effects become increasingly minor. The question to be answered is: Will the dispersion effects near the source, in the deeper water, play a significant enough role to alter the waveform in the shallow water? In an attempt to resolve this question, for each of the two initial conditions, simulations were run using both the BOUS (including weak frequency dispersion effects) and NLSW (non-dispersive) models.

First, we examine the LNIC. Figure 9 shows a summary of the numerical results. In the top row of the figure, snapshots of the free surface are taken from both

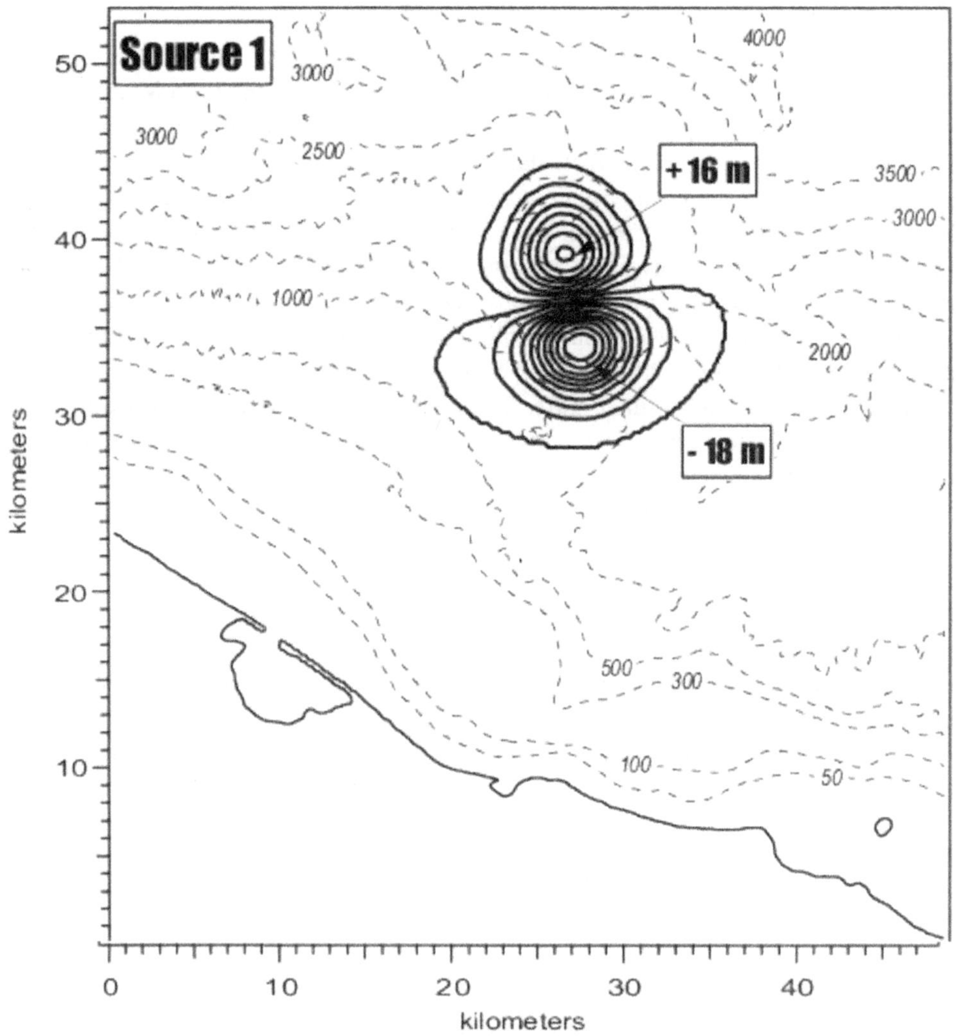

Figure 7
The short-fat initial condition (SFIC) for the Papua New Guinea event. The island is located on the bottom of the figure; Sissano Lagoon is shown.

simulations at a time of 3 minutes after the hot start. Both images show a leading depression wave traveling towards the spit, while a leading elevation wave encroaches upon the shore to the west of the spit. Differences between the two are obvious, with the BOUS simulation exhibiting a well-defined secondary depression wave moving towards the spit. Also, the wave predicted by NLSW at this time has a larger height than that predicted by BOUS. The middle plot in the figure shows the difference in the predicted maximum free surface elevation between the models, where positive values indicate the NLSW predicted a larger free surface elevation at that point. In

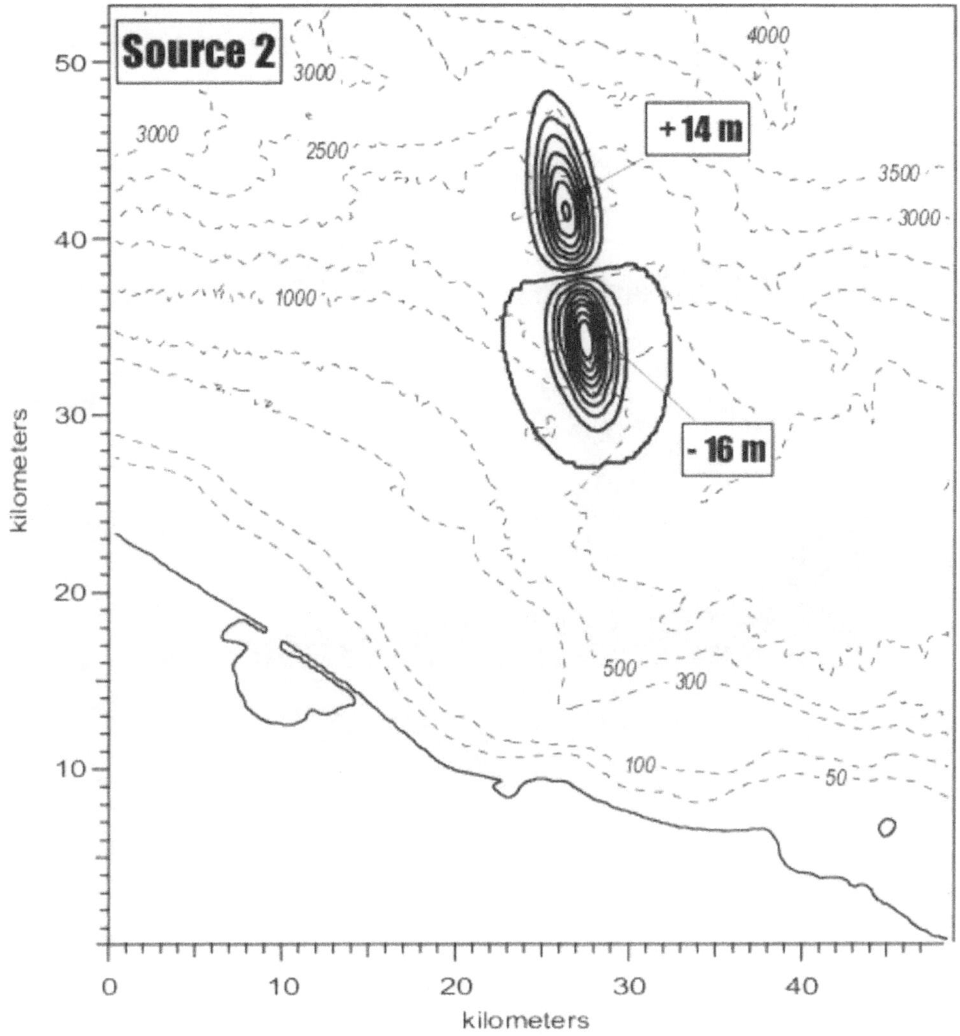

Figure 8
The long-narrow initial condition (LNIC) for the Papua New Guinea event. The island is located on the bottom of the figure; Sissano Lagoon is shown.

the deeper water, near the source, there is virtually no difference between the models. As the wave approaches the spit, the scenario changes, and NLSW predicts a considerably larger wave, nearly 10 m higher. Interestingly, there is a sharp contrast between the large difference between the models and a much smaller difference as the wave travels shoreward. This contrast is attributed to wave breaking, in fact, it occurs in the same region where breaking is initiated in both the models. The nearshore cancellation of the large differences in the models just seaward of the

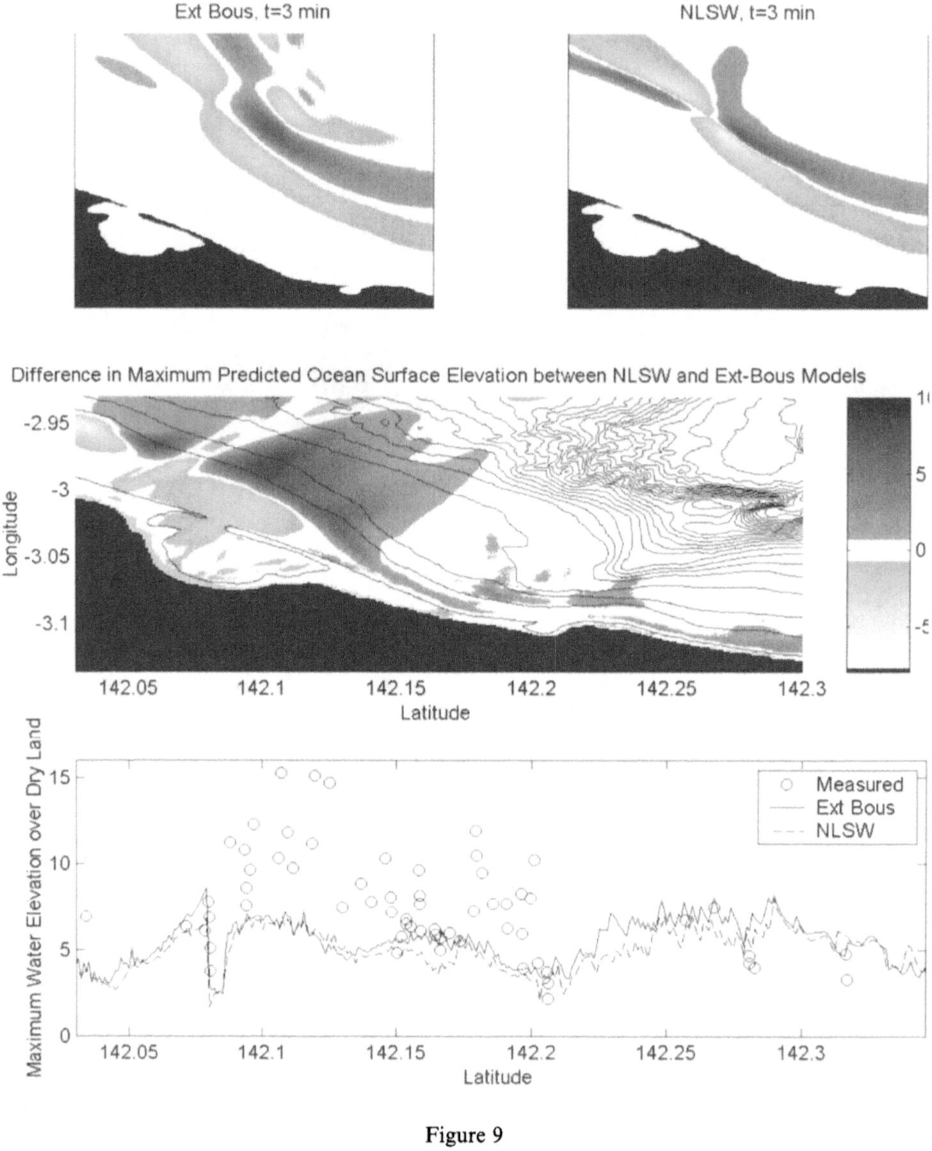

Figure 9
Dispersion effects with the long-narrow initial condition. The top plots (in the top row) show the instantaneous free surface elevation predicted by the NLSW and BOUS models. The middle plot shows the difference in maximum free surface elevation between the models. Positive values indicate that the shallow water model predicts a higher free surface at that location. The lower plot compares the predicted inland water elevations with field data.

breaking line is a strong indication that breaking in this area is depth-limited. The lower plot in the figure shows the maximum free surface elevation predicted over dry land, including over the inundated spit. Also plotted are the recorded field data. The BOUS and NLSW predict very similar flooding patterns, although there exist local differences on the order of 40%, particularly east of latitude 142.15. It may be possible that east of this latitude, breaking is not governed by depth, whereas seaward of the spit, breaking is depth-limited.

Numerical results for the second initial condition, SFIC, are shown in Figure 10. All of the conclusions stated in the above paragraph, for LNIC, are directly applicable to initial condition SPIC as well. Some differences do exist however. Examining the middle plot in Figure 10, we see that the region of largest difference between the NLSW and BOUS models is shifted to the east, as compared to the same plot in Figure 9. This is due to the fact that with SFIC, wave energy is focused slightly more to the east than with LNIC, and where the largest waves are located will also be the location of the largest difference between the NLSW and BOUS models. The numerical results using SFIC appear to give a better agreement than LNIC with the field data for water elevation over dry land.

Thus far, only the spatial differences due to dispersive effects have been examined. It would likewise be interesting to look at the time-dependent aspects, such as differences in tsunami arrival time, and number of crests. To do this, time series of the numerical free surface elevation are recorded at three different locations. The three locations are located near the shoreline, and are given in Figure 11. The time series comparisons at these three locations are shown in Figure 12 for the SFIC. Clearly, BOUS and NLSW predict very different waveforms. In all of the plots in Figure 12, the BOUS model gives a first crest arrival time roughly 30 seconds later than the NLSW model. This time corresponds to the arrival time of the second trough in the NLSW results. The number of distinct wave crests contained in the waveform is the same with both models. While it is immediately evident that frequency dispersion has a leading order impact on the wave form, it is equally clear that this impact would be difficult to extract from eyewitness accounts. An eyewitness arrival time record with an accuracy of seconds would be a tough find, especially in the PNG area.

It would seem that for the PNG case, in the region of primary interest (near the lagoon), frequency dispersion effects are unimportant to the prediction of runup and inundation. This is not because dispersion effects are unimportant offshore, it is only because of what appears to be depth-limited breaking negating a large difference in offshore wave height. While use of NLSW might lead to good runup prediction, other properties of the tsunami, such as wave shape and arrival time, show larger errors. It should be noted, however, that the large uncertainty associated with the tsunami generation dynamics in the source region represents an error much larger than that due to neglect of frequency dispersion.

3.2.1 Sensitivity of predicted PNG runup to initial condition

In the previous section, two initial conditions, LNIC and SFIC, were discussed. In this section, the relation between these two initial conditions will be more closely scrutinized, along with a third initial condition. This third initial condition, meant to

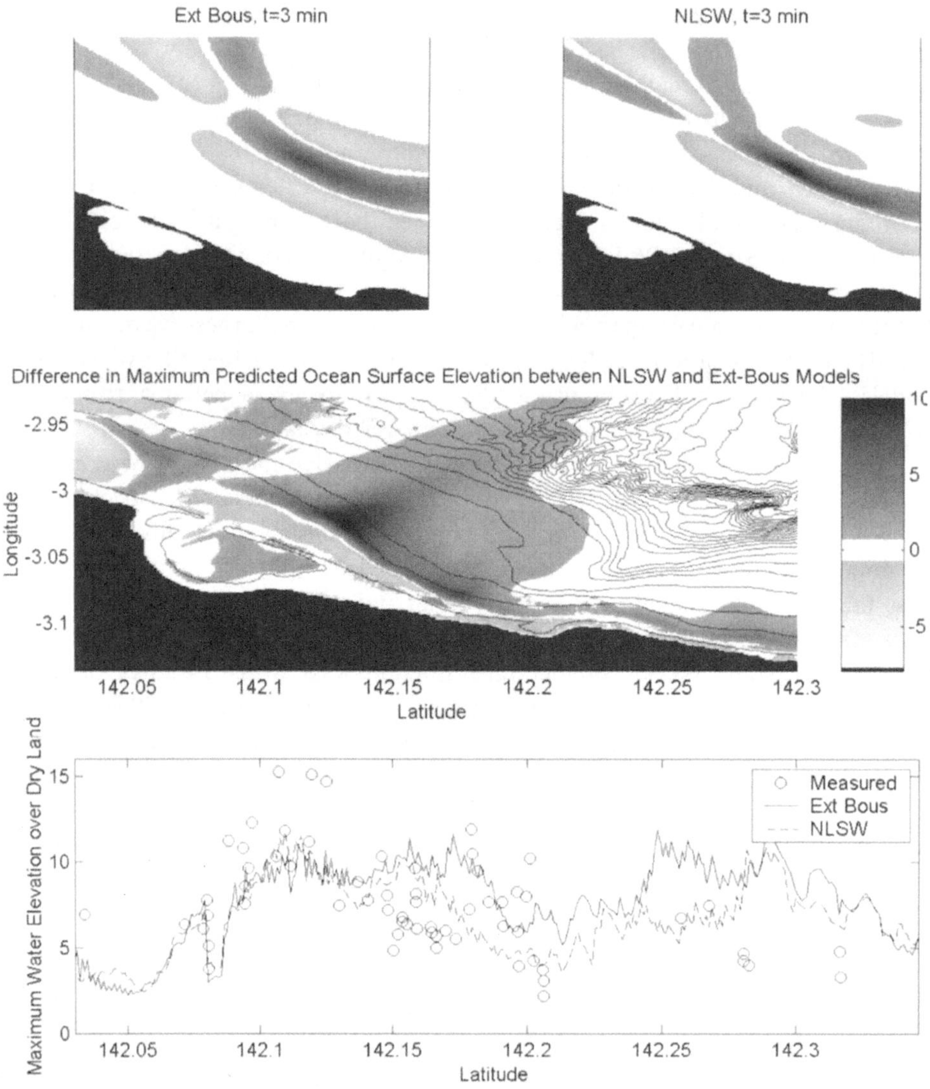

Figure 10
Dispersion effects with the short-fat initial condition. The top plots (in the top row) show the instantaneous free surface elevation predicted by the NLSW and BOUS models. The middle plot shows the difference in maximum free surface elevation between the models. Positive values indicate that the shallow water model predicts a higher free surface at that location. The lower plot compares the predicted inland water elevations with field data.

be close to the simplest reduction to the complex hot start problem, is a single sine wave fit with a Gaussian distribution in the transverse direction. The sine wave has a length of 10 km and a width of 5 km. The sine wave initial condition, or SineIC, is centered at the same location as the other two initial conditions.

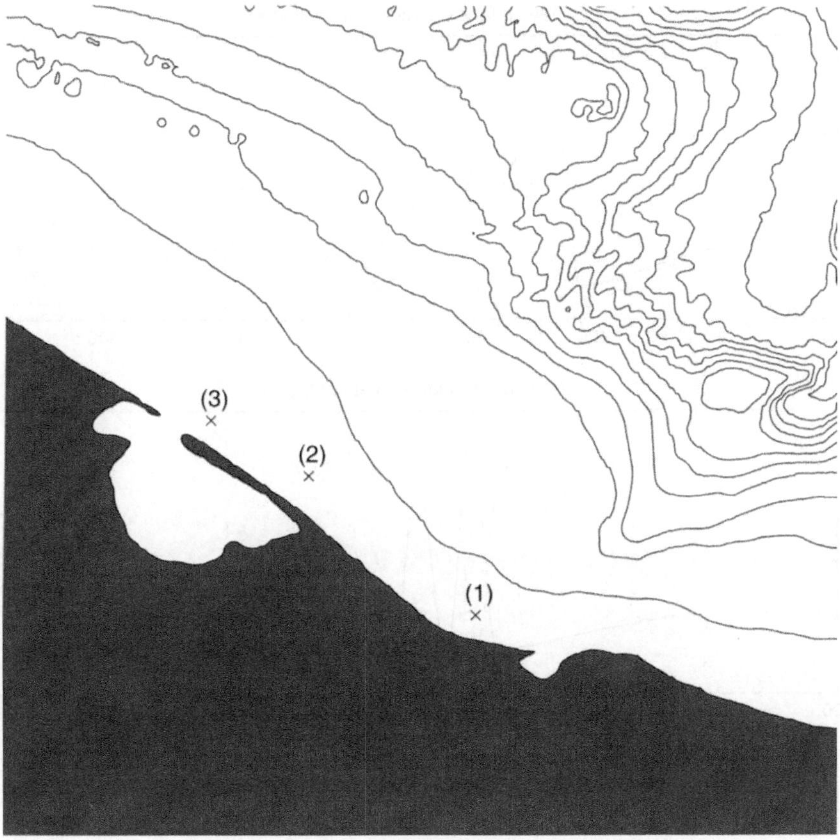

Figure 11
Locations of the PNG time series comparisons. Depth contours are every 100 m.

Figure 13 shows a number of snapshots of the free surface for all three initial conditions. In the first column are the numerical results, using BOUS, with SFIC, the second column with LNIC, and the third column SineIC. The first row are the initial conditions, at time = 0. The second row, at time = 1 minute, shows the three initial conditions spreading out radially. At time = 3 minutes, shown in the third row, it is clear how the different initial conditions each produce very different offshore wave heights and different number of waves. In the last row, at time = 7 minutes, breaking has initiated in all of the simulations, creating similar wave heights offshore of the lagoon for all three initial conditions, but still different number of waves.

A summary of the numerical results from the three initial conditions is given in Figure 14. In the top row are shown the maximum free surface elevations predicted by the three initial conditions. SFIC gives the largest maximum free surface elevation, and also predicts the highest flooding of dry land, as shown by the lower plot. LNIC and SineIC are in close agreement overall.

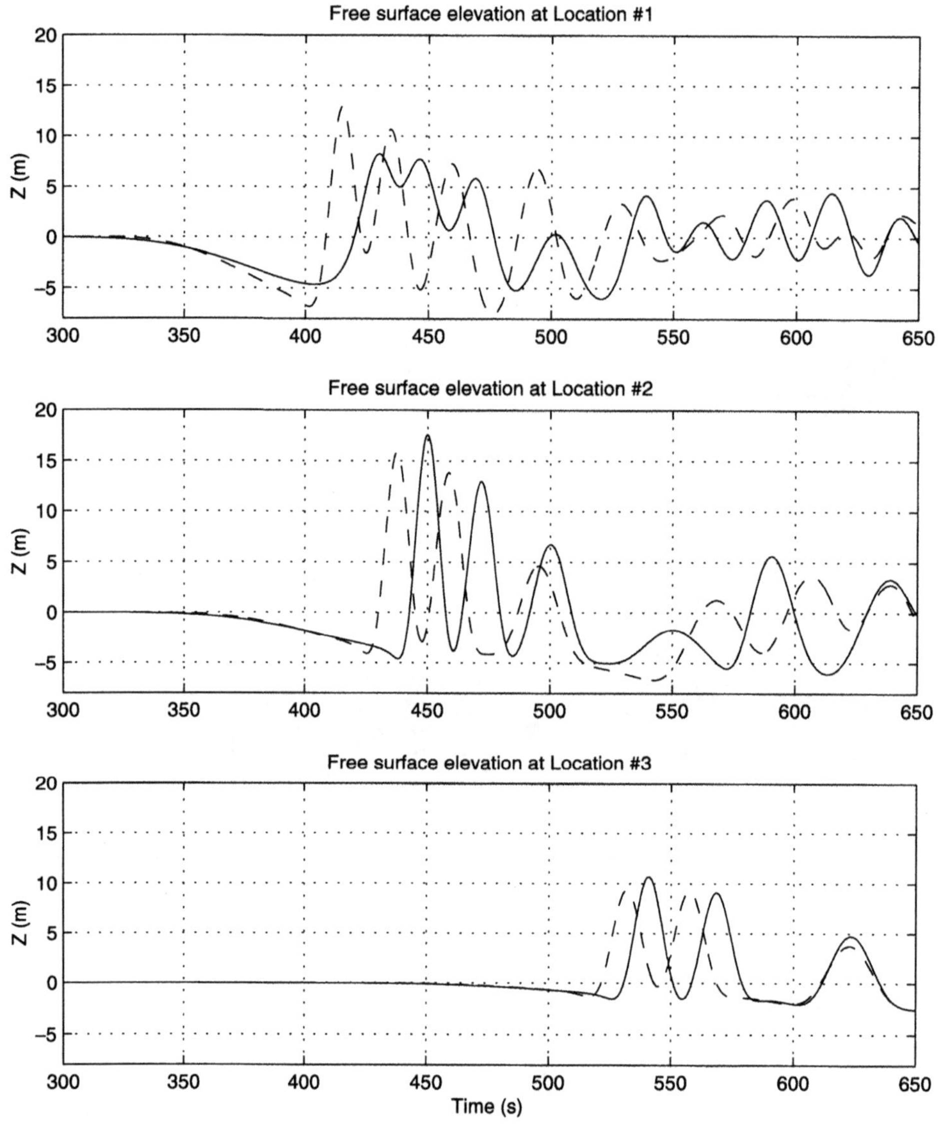

Figure 12
PNG time series comparisons of the free surface for the three locations shown in Figure 11, for the SFIC. BOUS results are given by the solid line, NLSW by the dashed line.

Let us now examine the maximum predicted free surface elevation averaged over the spit (roughly 142.09 to 142.2 degrees). Over this distance, the SFIC predicts heights in the range of 10 meters. The LNIC and the SineIC predict nearly the exact inundation patterns, with an average height near 6 meters. It is interesting to note that despite the large difference in shape between LNIC and SineIC, the

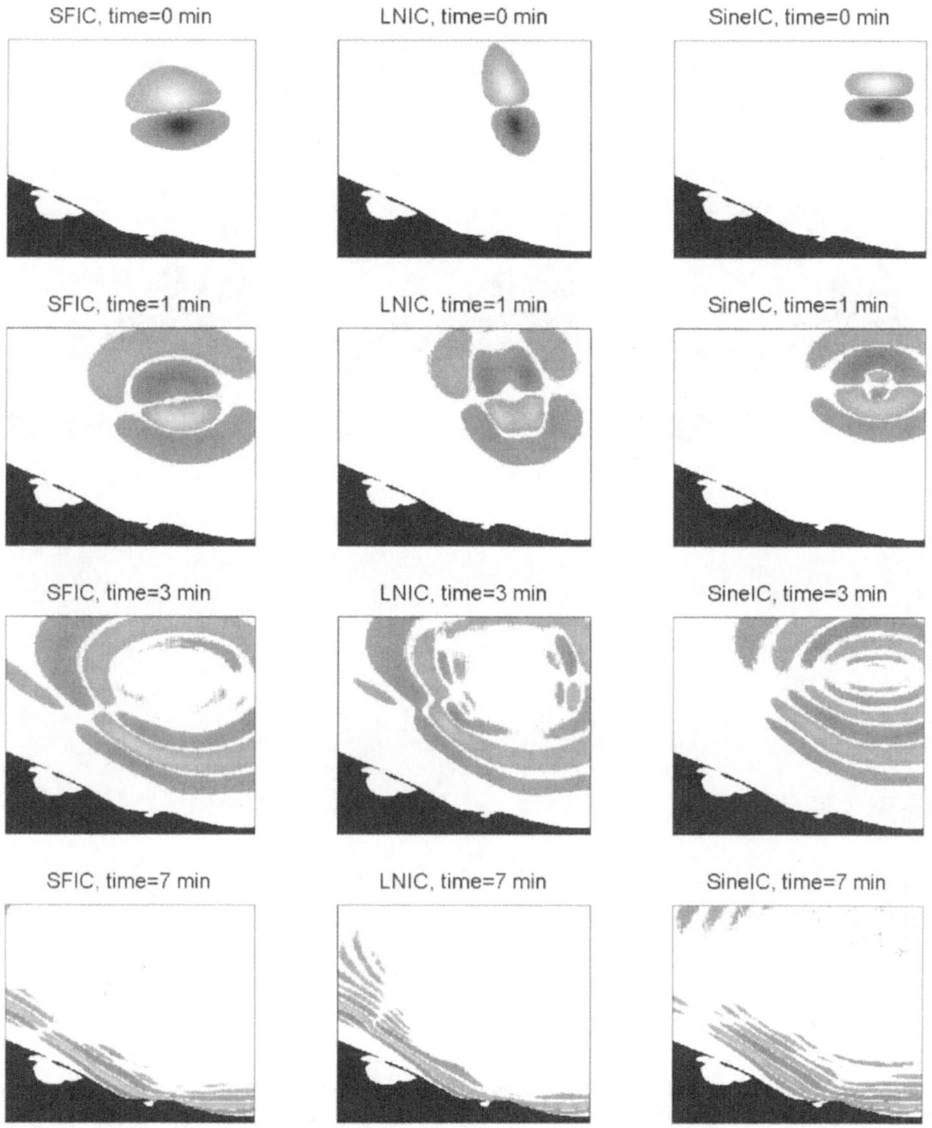

Figure 13
The evolution in time of three different initial conditions. In the first column are the results from the short-fat initial condition, the middle column the long-narrow initial condition, and the last column the sine-wave initial condition. The first row is for the initial time, the second row for $t = 1$ min, the third for $t = 3$ min, and the fourth for $t = 7$ min.

predicted runup are in good agreement. One possible explanation for this agreement is that the displaced water mass of the initial condition is the dominant factor in predicting runup for this case, whereas the shape and orientation will play

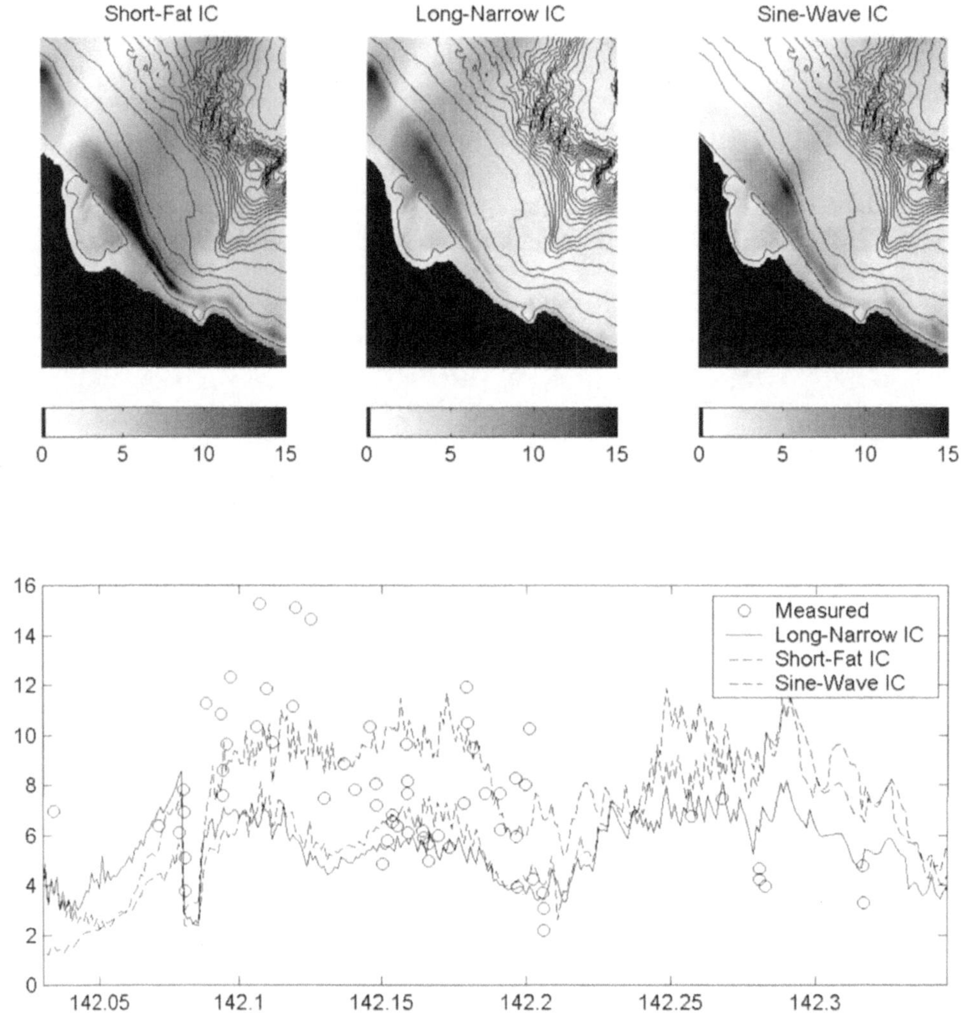

Figure 14

Comparisons from the three different initial conditions. The top row shows the maximum ocean surface predicted by the different initial conditions, and the bottom plot compares the inland free surface elevations.

secondary roles. The initial displaced mass, or the displaced mass of the "hot start" condition, is given by:

$$M = \rho \int_0^L \int_0^W |\zeta(x, y, t = 0)| \, dx \, dy \tag{7}$$

where L and W are the domain widths in the y and x directions and ρ is the density of water. The following ratios are calculated:

$$\frac{M(\text{LNIC})}{M(\text{SFIC})} = \frac{M(\text{SineIC})}{M(\text{SFIC})} = 0.58. \tag{8}$$

Note that the maximum depression of SineIC was chosen so that its initial M was equal to that of LNIC. Not coincidentally, the ratios of M are very close to the ratios of maximum inundation heights over the spit (6 m/10 m). Thus, it would seem that the finer detail of the initial condition plays a small role in maximum runup for the PNG case. However, these details will play a significant role in the prediction of wave arrival time, the maximum runup away from the spit, and the number of distinct waves hitting the coastline.

3.2.2 Interpretation of numerical output with respect to the measured field data

While the numerical model employed in the previous section is capable of simulating the tsunami overtopping of the Sissano spit, many other frequently used tsunami models are not. Presently, most published and validated tsunami inundation models compute either the largest inundation height over dry land, whether runup or inundation depth (MOST) or a water surface elevation at the first offshore wet point (TSUNAMI-N2) or water elevation at an offshore threshold depth (e.g., YEH *et al.*, 1996). In the case of a low relief topography where overtopping is possible, unless care is taken in the computation, models may output as "runup" what is really computed flow depth over the existing topography (see Fig. 15). In order to actually compute runup in Sissano, the model must compute evolution of the breaking wave over the spit, into and across the lagoon, and finally into mangroves and palm forest. Breaking wave computations are chancy at best with the tried shallow water wave models, and the propagation of a broken wave over extreme shallow water can be unstable as well.

The tsunami propagation and inundation model used in this section solves the NLSW equations in characteristic form. The runup code uses a moving boundary

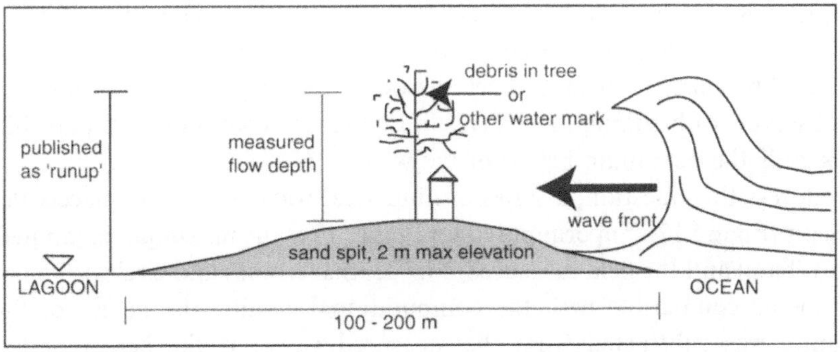

Figure 15
A cartoon showing the difference between measured flow depths and published runup in the case of the Papua New Guinea tsunami.

algorithm that extends the computational domain as needed for the flow conditions. The particular version used is known by the acronym MOST (TITOV and GONZALEZ, 1997), and is currently used at the University of Southern California and at NOAA's Pacific Marine Environmental Laboratory. It has been validated repeatedly through comparison with laboratory data and comparisons with field measured tsunami runup data (TITOV and SYNOLAKIS, 1995, 1997, 1998).

To test the effect of the sand spit at Sissano Lagoon on the model results, two different simulations were run. One with the existing topography and bathymetry, and another with a 1:30 slope along the entire shore, including the lagoon. A 1:30 slope is used because it is close to the existing slope. An initial grid spacing of approximately 140 m was used. Note that the numerical model used (MOST) develops a variable spaced grid from the input bathymetry. In these cases, the near shore and overland grid spacing was approximately 80 m.

The initial wave used in these simulations is the SFIC. The results from the two cases are shown in Figure 16. The shoreline shown in the runup figures is that of the undisturbed real shoreline. In the upper panel, the concentric black contour lines are the contours of the initial surface displacement. In the lower panel, the solid lines show the computed runup using the existing topography. MOST calculates the largest inundation height over dry land, whether runup or flow depth. In the case of a uniformly sloping beach, maximum runup is the highest inundation height. In the case of the model with the Sissano spit the highest inundation height is over the spit. Note how the runup values drop off in the region of the narrow sand spit. The dashed lines show the computed runup in the case of a uniform plane beach. The computed runup values along the sand spit are considerably higher and closer to what was reportedly measured in the field. They are also consistent with laboratory data on solitary waves, which suggest that the maximum runup is up to twice larger than the flow depth at the initial shoreline.

As discussed earlier, the field measurements taken along the Sissano sand spit are subject to considerable uncertainty. Are the measured values representative of the flow depths, or are they local maxima? Were the watermarks really caused by this event? Is it possible that the flow conditions were so extreme that local extrema could be generated by splash that cannot yet be calculated adequately by the models? Even casual observers of breaking long waves in real beaches would notice that the splash spray exceeds the maximum height of the wave.

To address this question, a series of numerical wave gages was placed along the eastern spit (Figure 17). Superimposed on Figure 17 is the maximum computed water elevation (bars) and the field measured flow depths (crosses and circles). To make the measurements compatible with the computational results, the height of the local topography was subtracted from the measured value. If the height of the local topography was not measured, a nominal 2 m, the maximum recorded elevation over the spit, was subtracted. Adjusting these values and comparing it to a computed flow depth brings the results of the numerical simulation in better agreement with the

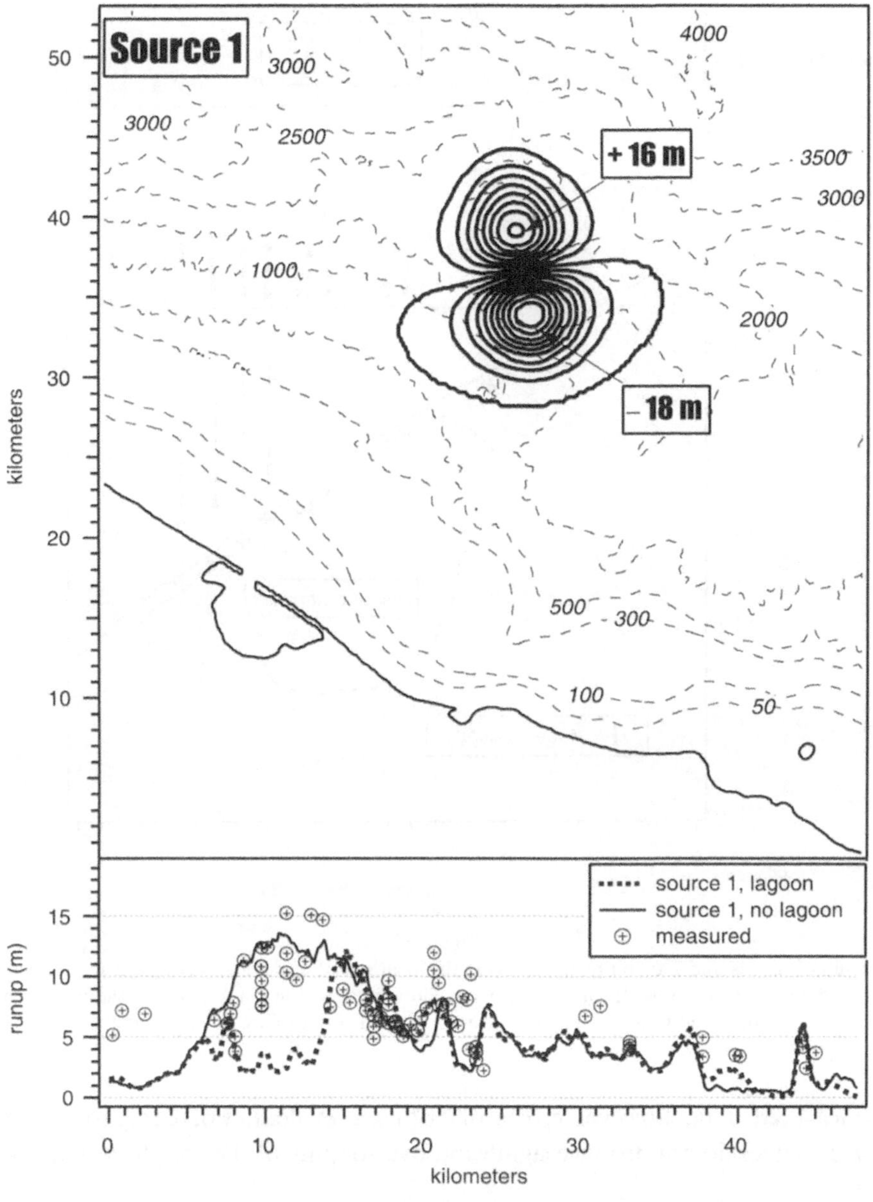

Figure 16
Upper panel – the offshore bathymetry and contours of the initial wave shape (thick black contours). Lower panel, computed runup for cases with existing topography including the lagoon and with a 1:30 plane beach along the entire shore. The spike on the right side of the figure refers to runup on Tumleo Island off Aitape. Measured runup there was 4.5 m.

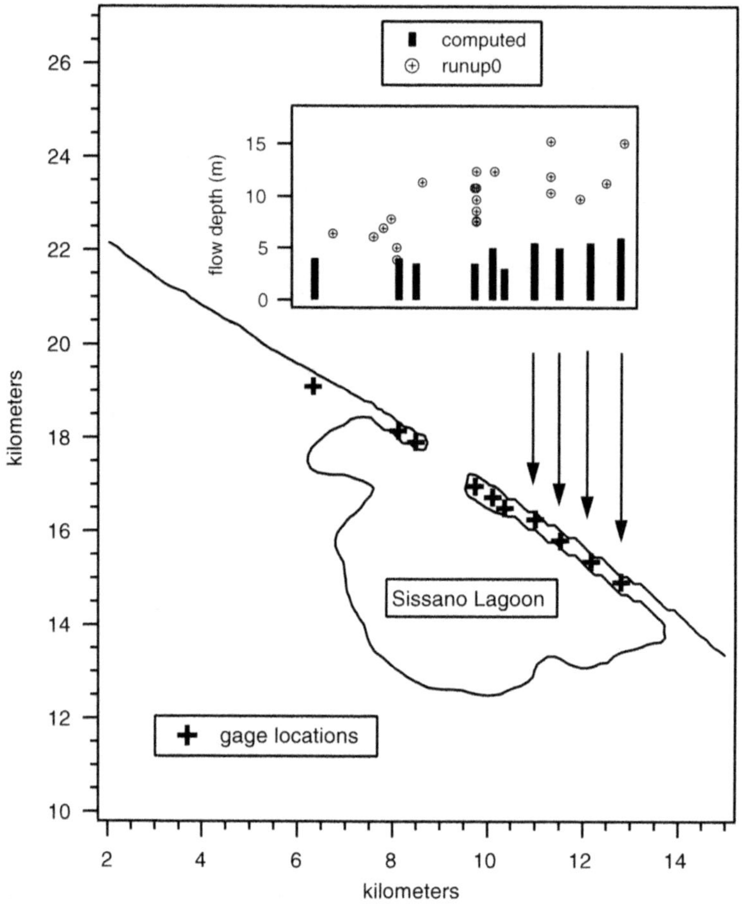

Figure 17
A close-up of the Sissano Lagoon region showing the locations of numerical wave gages (crosses) and the corresponding computed flow depths (black bars). Plotted above the black bars are the measured data converted to a flow depth above the ground surface (circles with cross).

field-measured data, however errors of 50% are evident. For comparison, purely tectonic sources do not produce significant overtopping of the spit (SYNOLAKIS *et al.*, 2002).

Due to the problems with the field data recorded after the 1998 tsunami survey, it is proposed here that a new runup plot be considered when discussing the Papua New Guinea tsunami. In this new runup plot, the points that are "true" runup values will be kept the same while points collected along the sand spit or in other areas where a true runup reading could not be attained will be differentiated and noted as "measured flow depths."

An example of this revised runup plot is shown in Figure 18.

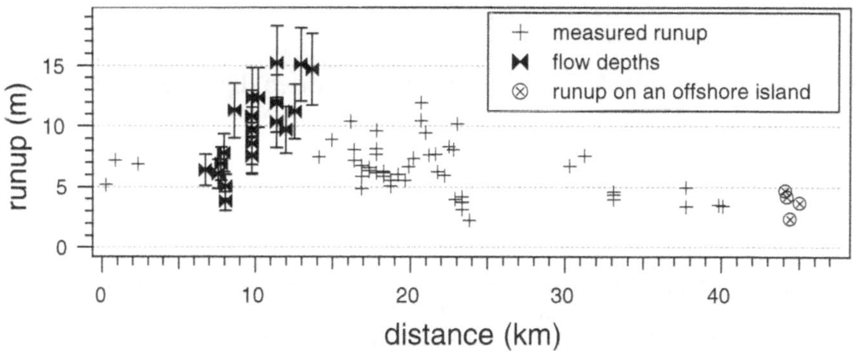

Figure 18
A proposed alternate version of the Papua New Guinea runup plot, with the values around Sissano Lagoon differentiated as 'flow depths' and plotted with 20% error bars.

4. Conclusion

The Papua New Guinea tsunami was a brutal display of the power of nature and a catalyst for a paradigm shift within the tsunami research community. As this shift occurs, the applicability of using traditional dislocation source tsunami propagation codes for landslide tsunamis must be checked. Landslides represent a different type of source, creating waves with different general characteristics. Landslides tsunamis, as demonstrated by the PNG wave, can have lengthscales too small to be accurately modeled with nondispersive models. Offshore wave elevation predictions of the Boussinesq and NLSW models differ by as much as 10 meters where the Boussinesq model predicts a height of 5 meters. However, due to bathymetry effects, namely depth-limited breaking, the inundation due to the PNG tsunami can be accurately simulated with a NLSW model near the spit. This effect is site-specific to PNG, and therefore the good runup prediction of NLSW models for this event cannot be used as justification to apply a NLSW model to a future landslide tsunami occurrence.

The breaking effects offshore of the spit must also be considered when justifying an initial condition for the PNG tsunami. As shown in this paper, unique initial conditions do not lead to unique maximum sea-surface elevation patterns over the spit. In fact, it was shown that the water elevations over the spit are far more closely related to the magnitude of the displaced mass contained in the initial condition. This is not true away from the spit, though, where the breaking appears not to be depth-limited. Finally, given the behavior of numerical bores in well-used tsunami codes, it is recommended that computations with current inundation tools which model evolution over narrow spits be bracketed between two runup distributions, one over a uniform beach and one over the spit.

Acknowledgements

The research reported here is partially supported by Grants from National Science Foundation (CMS 9908392 and CMS 0217744).

References

Chen, Y. and Liu, P. L.-F. (1995), *Modified Boussinesq Equations and Associated Parabolic Model for Water Wave Propagation*, J. Fluid Mech. *228*, 351–381.

Kawata, Y., Benson, B., Borrero, J. C., Davies, H. L., deLange, W. P., Imamura, F., Letz, H., Nott, J., and Synolakis, C. E. (1999), *Tsunami in Papua New Guinea was as Intense as First Thought*, EOS, Transact. Am. Geophys. Union *80*(9), 101, 104–105.

Kennedy, A. B., Chen, Q., Kirby, J. T., and Dalrymple, R. A. (2000), *Boussinesq Modeling of Wave Transformation, Breaking, and Runup. Part I: 1D*, J. Waterway, Port, Coastal and Ocean Eng. 126(1), 39–47.

Lynett, P., Wu, T.-R., and Liu, P. L.-F. (2002), *Modeling Wave Runup with Depth-Integrated Equations*, Coast. Eng. 46(2), 89–107.

Lynett, P. and Liu, P. L.-F. (2002), *A Numerical Study of Submarine Landslide Generated Waves and Runup*, Roy. Soc. London A, *485* (2028), 2885–2910.

Nwogu, O. (1993), *Alternative Form of Boussinesq Equations for Nearshore Wave Propagation*, J. Waterway, Port, Coastal and Ocean Eng. *119*(6), 618–638.

Okal, E. (2003), *Normal Mode Energetics for Far-field Tsunamis Generated by Dislocations and Landslides*, Pure Appl. Geophys. *160*, 2189–2221.

Press, W. H., Flannery, B. P., and Teukolsky, S. A., *Numerical Recipes*. (Cambridge University Press 1989), pp. 569–572.

Synolakis, C. E., Bardet, J.-P., Borrero, J. C., Davies, H. L., Okal, E. A., Solver, E. A., Sweet, S., and Tappin, D. R. (2002), *The Slump Origin of the 1998 Papua New Guinea Tsunami*, Roy. Soc. London A *458*, 763–789.

Tappin, D. R. *et al.* (1999), Sediment Slump Likely Caused 1998 Papua New Guinea Tsunami. EOS *80*(30), 329.

Titov, V. and Gonzalez, F. (1997), *Implementation and Testing of the Method of Splitting Tsunami (MOST) Model*, NOAA Technical Memorandum ERL PMEL – 112.

Titov, V. V. and Synolakis, C. E. (1995), *Modeling of Breaking and Nonbreaking Long Wave Evolution and Runup Using VTCS-2,* J. Waterway, Port, Coastal and Ocean Eng. *121*, 308–316.

Titov, V. V. and Synolakis, C. E. (1997), *Extreme Inundation Flows during the Hokkaido-Nansei-Oki Tsunami*, Geophys. Res. Lett. *24*, 1315–1318.

Titov, V. V. and Synolakis, C. E. (1998), *Numerical Modeling of Tidal Wave Runup*, J. Waterway, Port, Coastal and Ocean Eng. *124*(4), 157–171.

Yeh, H., Liu, P. L.-F., and Synolakis, C. eds (1996), *Long-wave Runup Models*, Proc. 2nd Int. Workshop on Long-wave Runup Models, World Scientific Publishing Co., Singapore.

Zelt, J. A. (1991), *The Runup of Nonbreaking and Breaking Solitary Waves*, Coast. Eng. *15*, 205–246.

Pure appl. geophys. 160 (2003) 2147–2153
0033–4553/03/112147–07
DOI 10.1007/s00024-003-2423-z

Pure and Applied Geophysics

Tsunami Wave Height Dependence on Landslide Volume

T. S. Murty[1]

Abstract—Tsunami generation from submarine landslides depends mainly on the volume of the slide material and also on other factors which include: angle of the slide, water depth, density of the slide material, the speed with which the material moves, duration of the slide, etc. Based on an incomplete data set of volume V of slide versus maximum amplitude H of the resulting tsunami waves, gleaned through available literature, a simple linear regression relationship was developed. Another partial data set was developed also from published literature, on V versus H values, based on numerical models. It was found that the agreement between the results of the numerical simulations and the observations is rather poor. It is not clear why this is so, and which data set is of questionable relevance. This is not to cast doubt on numerical models that do not use volume of the slide in an explicit manner.

Key words: Landslide, volume, tsunami, amplitude.

1. Introduction

Intuitively one would expect the amplitude of a tsunami generated by a submarine landslide to be directly proportional to the total volume of the slide material (Harbitz *et al.*, 1993) even though several other parameters also play important roles. These factors, in random order, are the following:

a) Depth at which the slide occurs, or rather, depth of water above slide.
b) Angle of the slide from the horizontal (or vertical) direction.
c) Total distance moved by the slide.
d) Duration of the slide.
e) Density of the slide material.
f) Coherent nature of the slide.
g) Grain size and spectrum.
h) Characteristic speed with which the slide moves.

It is quite possible that there are several other important factors not included in the above list. The aim of this study is modest and simple; to obtain a relationship between the volume V of the slide (expressed in millions of m^3) and the maximum amplitude H (in meters) of the resulting tsunami waves, based on observational data available in the published literature and to compare this relationship with the results

[1] W.F. Baird & Associates Coastal Engineers Ltd., Ottawa, Canada K1V 0Y3.

of some numerical models. The observational data and numerical model results that are included in this study have been selected using only the following criterion, namely whether values for V and H are provided explicitly by the authors. If these are not provided, those events and models are not included here. Even though one can examine the accuracy of this data, it is beyond the scope of this short communication, and no attempt was made to verify the data.

If different numbers are provided by different authors for the same event, all the data have been used as separate entries, rather than using an average value. Even though, ideally a multivariate analysis could be performed using data on the parameters listed above, it is not attempted here and will form the basis for a future study.

The basic argument used here is that no matter how important all the other parameters may be, if there is no volume, there is no tsunami. It is that simple. Hence, the slide volume must remain the most important and basic parameter. The limited goal then is to obtain an order of magnitude relationship between V and H.

It should be noted that the relationship between the volume of the slide and the height of the resulting tsunami is phenomenological. Volume-based correlations are also a classical and reasonably successful approach in the field of geology, such as correlations for runout distances, etc. (BLYTH and FRETIAS, 1974; MCLEAN and GRIBBLE, 1979; and WHITTEN and BROOKS 1981).

2. *State of the Art in the 1970s*

Until the 1970s, there was no vast literature on numerical models for tsunamis from submarine landslides. However, there was considerable work on analytical models and laboratory studies. These approaches could not deal with realistic geometries and mostly dealt with very simplified situations.

MURTY and BROWN (1979) studied the tsunami of April 27th, 1975 from the submarine landslide in Kitimat Inlet on the coast of British Columbia, Canada, and MURTY (1979) simulated this tsunami using two different models. The results of an analytical theory and also an asymptotic theory agreed reasonably well with observations.

3. *Observational Data*

Table 1 lists the volume of the Slide V, the maximum height H of the tsunami wave, a brief description of the event, and the source of the data. Uncertainties in the data are clearly pointed out. At best, this table is only a partial list of landslides generated tsunami events worldwide and several important events are probably missing from the list.

Table 1

Observational data on landslide generated tsunamis worldwide

Event	V	H	Source
Vaiont Valley, Italy Oct. 1993	250	100	BOLT *et al.* (1975)
Kitimat, B.C., Canada April 27, 1975	55	7.6	BORNHOLD and HARPER (1998)
French Rivera near Nice Oct. 16, 1979	35 Average $V_{min} \approx 10$ (RZADKIWICZ *et al.*) $V_{max} \approx 65$ (George P.C.)	10 Anon, 1980	RZADKIWICZ *et al.* (1998) G. PARARAS-CARAYANNIS (1999) Personal Communication ANON (1980)
Lituya Bay, Alaska, U.S.A. July 9, 1958	31	30	HARBITZ *et al.* (1993). They refer to PUGH and HARRIS (1982)
Kitimat, B.C. Canada April 27, 1975	26	4.3	MURTY (1979)
Lomblen Island, Indonesia July 19, 1979	20 This is estimated based on discussions with colleagues, taking the following dimensions for the Slide Length = 2000 m. Width = 1000 m. Thickness = 10 m	11.1	TSUJI and HUSNI (1998)
Skagway, Alaska, U.S.A. Nov. 3, 1994	16.5	8.4	MADER (1997)
Skagway, Alaska, U.S.A. Nov. 3, 1994	15.3	10.7	CAMPBELL and NOTTINGHAM (1999)
Tafjord, Norway April 7, 1934	1.25	3 After the disturbance spread out	HARBITZ *et al.* (1993) Their Table 1 is confusing to me and hence the data of Table 1 is not used, Loen, 1905, $V = 5 \times 10^4$, $H = 40$ Tafjord, 1934, $V = 1$ to 1.5×10^6 LOEN, 1936, $V = 10^6$, $H = 74$
Skagway, Alaska Oct. 29, 1966	0.0306	0.4	NOTTINGHAM (1999)
Lake URI, Switzerland, April, 1992 artificially induces rockfall	0.016	0.3 Field measurements and comparison with physical model results	MULLER (1994)

V is Volume of the Slide (in millions of m^3). H is the Maximum Height of the Resulting Tsunami (in m).

Since the Lituya Bay tsunami could have been due to a rock fall and not due to a submarine slide, regression was made with and without its inclusion. The difference in the results was not significant.

4. Model Results

Table 2 summarizes the results of some of the numerical simulations reported in the literature. Again it should be pointed out that this is an incomplete list.

Some of the above models are based on pioneering work by JIANG and LEBLOND (1992, 1993 and 1994). Even though the mathematics in these models is quite satisfactory, because of the somewhat idealistic geology that was inherent in such models, the results could sometimes be confusing and cannot be considered predictive for real events.

In recent years geotechnical engineers made good progress in more reliable mass failure prediction than was possible before. In the coming decade or so when there is more confidence in such predictions, undoubtedly the accuracy of tsunami simulations will also improve.

5. Discussion of Results

Figure 1 summarizes the results. A regression line was fitted making use of observational data alone. The results of the numerical simulations are plotted on the same diagrams. Based on this regression the following relationships are obtained.

$$H = 0.3945\,V \tag{1}$$

or

Table 2

Results of numerical simulations of landslide generated Tsunamis

V	H	Source and Comments
750	10.5	KULIKOV *et al.* (1999)
600	10.0	STREIM and MILOH (1976) Analytical Model
230	4.5	KULIKOV *et al.* (1999)
20	0.5	RUBINO *et al.* (1994) Their diagrams and numbers are hard to read, even with a magnifying glass. Our reading of the value for H could be wrong. They refer to slides with $V = 300$ as small.
1.25	4.0	KULIKOV *et al.* (1999)
0.8	6.5	RABINOVICH *et al.* (1999)

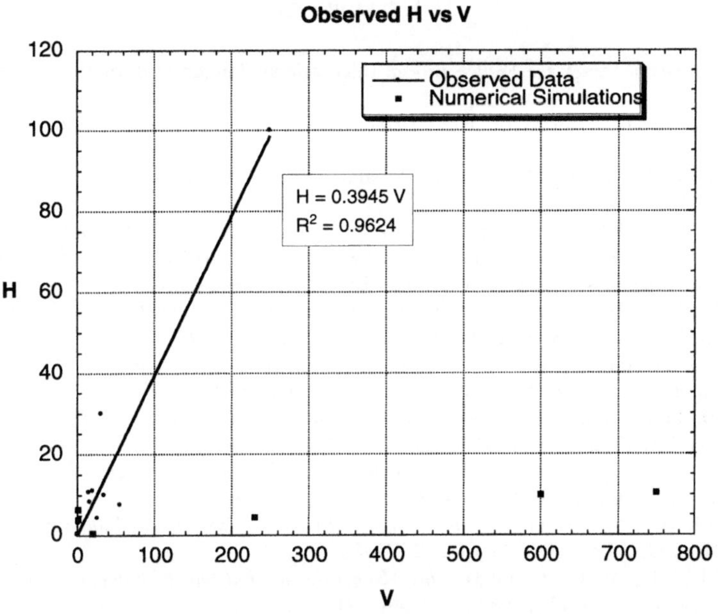

Figure 1
Observed maximum height of the tsunami (H in m) versus volume of the slide (V in millions of m^3).

$$V = 2.3994H, \tag{2}$$

where V is the volume of the submarine landslide in millions of cubic meters and H is the maximum amplitude of the resulting tsunami waves in meters. For example: $V = 10$ gives $H \approx 4$.

It is not clear why there is such poor agreement between observations and the results of some numerical simulations. However, it should be pointed out that all numerical simulations do not explicitly use the volume of the slide as one of the input parameters and quite accurate tsunami simulations have been made (WATTS, 1998; RAICHLEN *et al.*, 1996; and HEINRICH 1992). Since the aim of the present study is limited to finding a relationship between the volume of the slide and height of the resulting tsunami, no comparison was made with the results of the above models.

In particular, the work of Watts makes use of scaling which is deterministic and somewhat novel in tsunami research. In a sense, these models offer more detail in exchange for more parameters, some of which have limited availability at this time. However, with the increased interest in recent years in submarine landslides-generated tsunamis, there are good prospects for obtaining better information on the various parameters identified in the above studies.

References

Anon (1980), *Tsunami Struck the French Rivera*, International Tsunami Information Centre, Newsletter, *13*(1), 10.

Blyth, F. G. H. and De Fretias, M. H., *A Geology for Engineers* (Edward Arnold, Pitman Press, Bath 1974).

Bolt, B. A., Horn, W. L., MacDonald, G. A., and Scott, R. F. *Geological Hazards* (Springer-Verlag, Berlin, 1975) 328 pp.

Bornhold, B. D. and Harper, J. R., *Engineering Geology of the Coastal and Nearshore Canadian Cardillera, Proc. 8th Int. Congress, Int. Assoc. for Engineering Geology and the Environment*, Sept 98, Vancouver, Canada (A.A. Balkema/Rotterdam/Brookfiled, 1998) pp. 63–75.

Campbell, P. A. and Nottingham, D. (1999), *Anatomy of a Landslide-Created Tsunami at Skagway, Alaska, November 3, 1994*, Science of Tsunami Hazards *17*(1), 19–48.

Cornforth, D. H. and Lowell, J. A., *The 1994 submarine slope failure at Skagway, Alaska*. In: *Landslides*, Vol. 1 (K. Senneset ed.) (Balkema, Rotterdam, 1996) pp. 527–531.

Harbitz, C. B., Pedesen, G., and Gjevik, B. (1993), *Numerical Simulation of Large Water Waves Due to Landslides*, J. of Hydraulic Engin. *19*(12), 1325–1342.

Heinrich, P. (1992), *Nonlinear Water Waves Generated by Submarine and Aerial Landslides*, J. of Water Ways, Port Coastal and Ocean Engin. ASCE, *118*(3), 249–266.

Jiang, L. and LeBlond, P. H. (1992), *The Coupling of a Submarine Slide and the Surface Waves which it Generates*, J. of Geo. Phys. Res. *97*(C8), 12,731–12,744.

Jiang, L. and LeBlond, P. H. (1993), *Three-Dimensional Modeling of Tsunami Generation Due to a Submarine Mudslide*, J. of Phys. Oceanography. *24*(3), 559–572.

Kulikov, E. A., Fine, I. V., Rabinovich, A. B., Bornhold, B. D., and Thomson, R. E. (1999), *Numerical Simulation of Submarine Landslides and Tsunami in the Strait of Georgia*, Proc. of the 1999 Canadian Coastal Conf. May 1999, Victoria, BC, Canada, *2*, 845–862.

Kulikov, E. A., Rabinovich, A. B., Thomson, R. E., and Bornhold, B. D. (1996), *The Landslide Tsunami of November 3, 1994, Skagway Harbor, Alaska*. J. of Geo. Phys. Res. *101*(C3), 6609–6615.

Mader, C. O. (1997) *Modelling the Skagway Tsunami*, Science of Tsunami Hazards *15*(1), 41–48.

McLean, A. C. and Gribble, C. D., *Geology for Civil Engineers* (George Allen and UNWIN, London, 1979), 310 pp.

Muller, D. (1994), *Physical Modelling and Field Measurements of Impulse Waves*, Proc. of the Int. Symp: *Waves-Physical and Numerical Modelling*, Univ. of British Columbia, Vancouver, Canada, Aug. 1994, *1*, 307–315.

Murty T. S. (1979), *Submarine Slide-generated Water Waves in Kitimat Inlet, British Columbia*, J. Geo. Phys. Res. *84*(C12), 7777–7779.

Nottingham, D. (1999), *October 166 Tsunami*, personal communication. 4 pp.

Rabinovich, A. B., Thomson, R. E., Kulikov, E. A., Borhnold, B. D., and Fine, I.V. (1999), *The Landslide Generated Tsunami of November 3, 1994 in Skagway Harbor, Alaska: A Case Study*, Geo. Phys. Res. Lett. *26*(19), 3009–3012.

Raichlen, F., Lee, J. J., Petroff, C., and Watts, P. (1996), *The Generation of Waves by a Landslide:Skagway, Alaska-A Case Study*, Chapter 101 in *Coastal Engineering 1996*, (B.L. Edge ed.), Proc. of 25th Int. Conf., ASCE 2, 1293–1306.

Rubino A., Backhaus, J. O., and Pierini, S. (1994), *Tsunamis Generated by Mud Slides*, Int. Symp: *Waves-Physical and Numerical Modelling*, Univ. of B.C. Vancouver, Canada, 466–473.

Rzadkiewicz, R., Heinrich, P., Savoye, B., and Bourillet, J. F. (1998), *Numerical Modelling of a Land Slide Tsunami: The 1979 Nice event (French Riviera)*, Abstract, *Int. Conf. on Tsunamis, Paris, France*, May 1998, 26–28.

Streim, H. L. and Miloh, T. (1976), *Tsunamis Induced by Submarine Slumpings off the Coast of Israel*, Int. Hydrographic Rev. *LIII*(2), 41–55.

Tsiji, Y. and Husni, M. (1998), *The landslide tsunami of Waiteba Bay of July 19, 1979, Lomblen Island, Indonesia, Prof. of Int. Workshop on Tsunami Disaster Mitigation*, January 1998, Tokyo, Japan, Japan Meteorological Agency and Science and Technology Agency, 87.

WATTS, P. (1998), *Wavemaker Curves for Tsunamis Generated by Underwater Landslides*. J. of Waterway, Port, Coastal and Ocean Engin. ASCE *124*(3), 127–137.

WHITTEN, D. G. A. and BROOKS, J. R. V., *The Penguin Dictionary of Geology* (Penguin Books. New York, 1981), 493 pp.

Pure appl. geophys. 160 (2003) 2155–2176
0033–4553/03/112155–22
DOI 10.1007/s00024-003-2424-y

Pure and Applied Geophysics

Some Aspects of Energy Balance and Tsunami Generation by Earthquakes and Landslides

L. J. Ruff[1]

Abstract—Tsunamis are generated by displacement or motion of large volumes of water. While there are several documented cases of tsunami generation by volcanic eruptions and landslides, most observed tsunamis are attributed to earthquakes. Kinematic models of tsunami generation by earthquakes — where specified fault size and slip determine seafloor and sea-surface vertical motion — quantitatively explain far-field tsunami wave records. On the other hand, submarine landslides in subduction zones and other tectonic settings can generate large tsunamis that are hazardous along near-source coasts. Furthermore, the ongoing exploration of the oceans has found evidence for large paleo-landslides in many places, not just subduction zones. Thus, we want to know the relative contribution of faulting and landslides to tsunami generation. For earthquakes, only a small fraction of the minimum earthquake energy (less than 1% for typical parameter choices for shallow underthrusting earthquakes) can be converted into tsunami wave energy; yet, this is enough energy to generate terrible tsunamis. For submarine landslides, tsunami wave generation and landslide motion interact in a dynamic coupling. The dynamic problem of a 2-D translational slider block on a constant-angle slope can be solved using a Green's function approach for the wave transients. The key result is that the largest waves are generated when the ratio of initial water depth above the block to downslope vertical drop of the block $H_0/W \sin \delta$ is less than 1. The conversion factor of gravitational energy into tsunami wave energy varies from 0% for a slow-velocity slide in deep water, to about 50% for a fast-velocity slide in shallow water and a motion abruptly truncated. To compare maximum tsunami wave amplitudes in the source region, great earthquakes produce amplitudes of a few meters at a wavelength fixed by the fault width of 100 km or so. For submarine landslides, tsunami wave heights — as measured by b, block height — are small for most of the parameter regime. However, for low initial dynamic friction and values of $H_0/W \sin \delta$ less than 1, tsunami wave heights in the downslope and upslope directions reach b and $b/4$, respectively.Wavelengths of these large waves scale with block width. For significant submarine slides, the value of b can range from meters up to the kilometer scale. Thus, the extreme case of efficient tsunami generation by landslides produces dramatic hazards scenarios.

1. Introduction

Tsunamis are caused by a rapid displacement of a large water volume. Earthquakes are the most significant source of large tsunamis. Great underthrusting earthquakes in subduction zones uplift the ocean above the fault area by a meter or more, the disturbed ocean surface then propagates away as gravity waves. The largest underthrust earthquakes generate tsunami waves that can cause devastation on both

[1] Department of Geological Sciences, University of Michigan, Ann Arbor, MI, U.S.A.

local and remote coasts. Other sources of tsunamis are impacts, volcanic eruptions, and landslides.

For a violent volcanic eruption to cause a tsunami, it must either explode underwater or cause debris to fall into the water. This happenstance is rather rare, but one well-known example is the Krakatau eruption in 1883 (NOMANBHOY and SATAKE, 1995). Fortunately, large impacts are even rarer, though potentially very devastating (e.g., WARD, 2000). Landslides are acknowledged to be a ubiquitous and costly natural hazard, but their dynamic energy release is usually small compared to large earthquakes. There are a few documented cases of large subaerial landslides that fell into water and caused large localized waves. One of the most spectacular examples was the Lituya Bay landslide triggered by shaking from the large strike-slip earthquake in southeastern Alaska on July 10, 1958 (MILLER, 1960). The landslide fell into a narrow fjord and caused a "slosh" wave that reached a height of 520 m, then continued out the fjord where it was at a height of 10 meters as it entered the ocean. The wave amplitude quickly diminished as it spread out in the open ocean. This indicates that localized tsunamis can be extremely hazardous along a limited coastal extent, yet the total displaced water volume is too small to cause destructive trans-Pacific tsunami waves.

Submarine landslides are mostly hidden from our view. Studies of marine sedimentation show that submarine landslides are a common process and the style ranges from translational block sliding to fluid-like flows (HAMPTON *et al.*, 1996). Hereafter, this wide range of mass wasting is referred to as "landslides"; and, the later quantitative modeling will be restricted to translational block sliding. Despite the overall geologic fact that submarine slides must be common, it is rare to directly observe any slide event. The most famous submarine slide was the Grand Banks slide of November 18, 1929, which was documented by the breaking of trans-Atlantic cables that crossed the slide area. Later studies showed that the peak slide velocity was about 3 m/s, and that the volume of material is about 1 to 5×10^{11} m^3 (PIPER *et al.*, 1985; HASEGAWA and KANAMORI, 1987). This submarine slide raised a regional tsunami that was destructive along the Burin Peninsula of Canada.

Some earthquakes cause tsunamis that are unusually large relative to the seismic waves (KANAMORI, 1972; ABE, 1979). Several explanations have been proposed for these "tsunami earthquakes," including an unusual earthquake rupture process and triggered submarine landslides. One recent case of triggered subaerial/submarine slides is the 1992 Flores Island event where it appears that the landslides caused a local "enhancement" of tsunami heights (HIDAYAT *et al.*, 1995; IMAMURA *et al.*, 1995). Another recent example where a submarine slide may have caused a large localized tsunami is the 1998 Papua New Guinea event (KAWATA *et al.*, 1999; TANIOKA, 1999; TAPPIN *et al.*, 2003; SWEET and SILVER, 2003). As more detailed teleseismic and regional quantitative tsunami wave modeling is combined with marine geophysical surveys, more cases of triggered-landslide tsunamis may be exposed. These examples now compel us to develop both generalized and detailed models of tsunami wave generation by landslides.

To determine whether large localized tsunami waves are caused by earthquake faulting or by a submarine slide, we need to develop a simple yet useful way to compare the generation of tsunami waves by earthquakes and landslides (MURTY, 2003). To zeroth order, larger tsunamis are generated by larger earthquakes and larger landslides. Of course, wave propagation effects away from the source region are also important for final runup heights (see WARD, 2000 for review and references). We know that seismic moment is the best overall measure of earthquake size, and seismic moment combined with fault size and orientation quickly give us the displaced water volume. For landslides, it would seem that total material volume, or total distance traveled, or some combination of these two parameters would be the best measures of landslide size. But how does this size translate into a predictable tsunami wave height? An experimental approach can determine some empirical rules for specific parameter choices of landslides (e.g., WATTS, 1998), but it would be difficult to test the parameter range relevant to the geologic spectrum of slides (these issues are discussed in WATTS and BORRERO, 2003). The main contribution of this paper is to answer the above question with the simplest possible dynamic model of a submarine landslide.

For earthquake generation of tsunami waves, purely kinematic models specify that ocean surface vertical deformation mirrors the solid surface vertical deformation due to static earthquake faulting (Fig. 1). Some seismologists have extended this kinematic approach to landslide generation, but the fundamental problem is

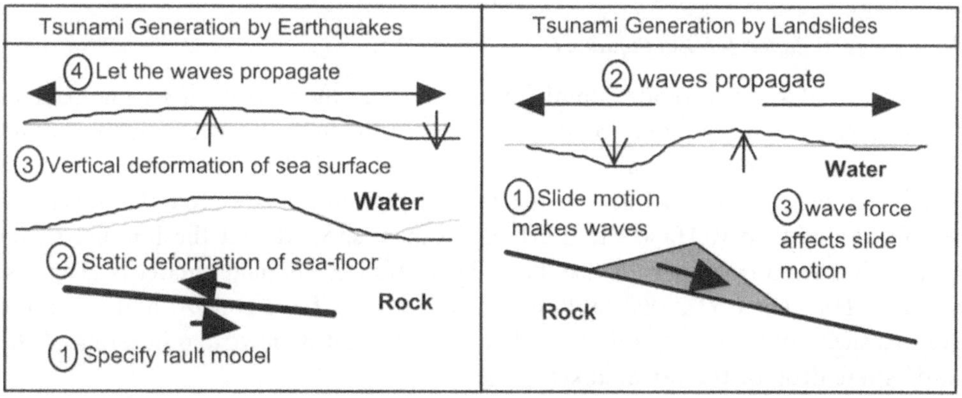

Figure 1
Schematic introduction to tsunami generation by earthquakes and submarine landslides. On the left, the kinematic modeling scheme for earthquake faulting is depicted. On the right, the interaction between wave generation and wave resistance force on the slider is depicted. In general, we must solve the full dynamic problem to generate waves from a block sliding due to its gravitational force; though, strong interaction only occurs in a limited part of the parameter regime.

that an *ad hoc* specification of slide velocity will be dynamically incorrect. At the other extreme of the modeling spectrum are fully coupled and dynamic nonlinear numerical models of slide and wave interaction and propagation (e.g., SYNOLAKIS and TADEPALLI, 1993; TITOV and SYNOLAKIS, 1993; JOHNSON and MADER, 1994; GRILLI and WATTS, 1999). Here, we find a useful intermediate approach where the slide velocity is determined by the landslide dynamics and the tsunami wave generation interacts with landslide dynamics, yet the approximations allow for an "almost-analytic" solution that yields general statements on the basic principles of tsunami generation by landslides. Another key question is how to compare the size of earthquakes and landslides. Energy release is probably the best relative measure of earthquake and landslide size. The efficiency of energy conversion into tsunami waves is then a key parameter. We will also estimate maximum wave amplitude at the source for both earthquakes and landslides; this alternative measure of tsunami size may be more useful for modeling and hazards assessment.

2. *Energy Balance and Tsunami Generation by Earthquakes*

Kinematic models for tsunami generation by earthquakes are well developed and accepted (OKAL, 1988; SATAKE, 1995). Consequently, the conversion of energy from earthquake faulting into tsunami waves has received little attention. In order to compare earthquakes and landslides based on energy, we will calculate the efficiency of tsunami wave generation by earthquakes. First we review energy basics and then show that only a small fraction of earthquake energy goes into tsunami waves.

2.1 Energy Balance for Earthquakes

Energy balance is a fundamental consideration in the physics of seismic sources and waves. Since earthquakes reduce the shear stress in some rock volume around the fault, stored elastic strain energy is released and deposited into two main reservoirs: waves and "friction" in the faulted region. Shortly after the advent of seismic source theory, HASKELL (1964) and others showed that the internal strain energy change associated with the shear dislocation earthquake model is given as: $\Delta E = \tau_{av} DA$, where D is the fault slip averaged over fault area A, and τ_{av} is the average shear stress as the fault slips. This formula can be rewritten in terms of the static stress drop of the earthquake:

$$\Delta E = A[\Delta\sigma D/2 + \tau_{\text{final}} D], \tag{2.1}$$

where $\Delta\sigma$ is the stress drop, and τ_{final} is the shear stress on the fault after the earthquake. Note that gravitational energy changes do not appear in Equation (2.1). As shown by DAHLEN (1977), the earthquake-induced changes in rotational, gravitational, and hydrostatic energies can be much larger than the earthquake energy in Equation (2.1), nonetheless all of these large first-order energy changes are perfectly balanced so that the net energy available for fault friction and seismic waves is precisely the same as given by ignoring gravitational forces and the hydrostatic stress state.

Although seismologists can estimate $\Delta\sigma$, D and A from various observations, the level of absolute stress (i.e., τ_{final}) is still a matter of great speculation and opinions. Indirect estimates based on the accessory observations of heat flow suggest that τ_{final} is not too much larger than $\Delta\sigma$ in strike-slip settings, but may be ten times larger than $\Delta\sigma$ in subduction zones (e.g., RUFF, 1999). Given this uncertainty in absolute stress level, it is reasonable to use the lower bound on earthquake energy, which is based on observed quantities:

$$\Delta E_{\min} = \Delta\sigma DA/2. \tag{2.2}$$

The next problem is how the shear energy release in Equation (2.2) is partitioned between fault friction and seismic waves. The partitioning is determined by the space-time history of dynamic friction during faulting, and careful seismic wave studies are required to quantify this partitioning (e.g., KIKUCHI and FUKAO, 1988; WINSLOW and RUFF, 1999). Here, we ignore this partitioning problem and use $\Delta E_{\min}$ as a measure of overall earthquake energy. Recall that seismic moment is defined as: $M_o = \mu DA$, where μ is source region shear modulus. Hence we can rewrite $\Delta E_{\min}$ as:

$$\Delta E_{\min} = \Delta\sigma DA/2 = \Delta\sigma M_o/2\mu. \tag{2.3}$$

Seismic moment is an earthquake parameter that can be reliably and systematically determined. Let us accept one of the key maxims of earthquake systematics that all large shallow earthquakes have the same stress drop of about 3 MPa. Since a crustal shear modulus is about 30 GPa for all shallow tectonic environments around the world, we can evaluate Equation (2.3) into the following universal formula that immediately gives an energy estimate once seismic moment is known (KANAMORI, 1977; VASSILIOU and KANAMORI, 1982):

$$\Delta E_{\min} = 5 \times 10^{-5} M_o. \tag{2.4}$$

To give some numbers for great underthrust earthquakes, let a fault zone extend for 100 km in the downdip direction and 200 km along strike, and let the slip be 10 meters. The seismic moment is 6×10^{21} Nm, which translates into a moment magnitude M_w of 8.5. We can estimate the earthquake energy by multiplying this moment by the universal factor in Equation (2.4) to obtain the value of 3×10^{17} Joules. The largest earthquake in the 20th century (1960, Chile) has an energy estimate of about 10^{19} J. There were about ten underthrusting earthquakes with $M_w \geq 8.5$ in the 20th century.

2.2 Tsunami Generation by Earthquakes

Large shallow earthquakes that occur beneath the ocean can cause tsunamis that bring destruction to local coastlines, or across the ocean in some cases. Tsunamis are caused by the fault-induced seafloor deformation displacing large volumes of water (Fig. 1). While this basic principle has been understood for a long time, quantitative modeling of tsunami waveforms is a rather recent advance.

If the ocean is viewed as a uniform global layer, then seismic source theory and normal mode formalism can be extended to include tsunami water waves, which are dominated by wave periods much longer than earthquake duration times (OKAL, 1988; WARD, 2000). However, the strong variations in bathymetry play a dominant role in the propagation of tsunami waves. Thus, an alternative approach is to place a finite fault within an elastic half-space, calculate the static vertical displacements, displace the water up and down to disrupt the ocean surface, then numerical methods propagate the tsunami waves across the detailed bathymetry of the region (e.g., SATAKE, 1995). This methodology provides results consistent with other constraints. The method works because earthquake duration is usually less than one minute, even for great earthquakes, while the fault-induced tsunami waves have significant energy at wave periods more than one minute. This separation of time scales between the seismic source duration and tsunami wave period allows us to complete the seismic source process before we activate the tsunami wave propagation dynamics. While this kinematic approach is now well-accepted, the implications for earthquake energy balance have not been fully considered.

2.3 Tsunami Energy Balance for Earthquakes

At first glance, we might think that an ocean layer on top of the solid surface significantly alters the earthquake energy balance because the earthquake must "lift" the entire ocean. However, a more careful look at the energy balance shows that the first-order energy required to "lift the ocean" is balanced by the first-order change in the hydrostatic energy of the ocean-rock column. The details of this energy balance are not detailed here; they follow the same arguments of DAHLEN (1977) for the whole earth problem.

Thus, we only need to be concerned about the second-order energy that is carried away from the source region by the tsunami waves. In the standard kinematic modeling of tsunami waves, the wave energy that propagates away is the potential energy given up by the "hills and holes" in the water surface returning to the reference level state. Since the "hills and holes" are only a few meters high, even for great earthquakes, this amount of potential energy is much less than the energy required to lift the entire ocean layer by a meter. But is this second-order tsunami wave energy comparable to the minimum earthquake energy? We can derive a simple relationship that connects seismic moment to tsunami wave energy for the typical geometry of a great underthrust earthquake. Place the up-dip edge of the 100-km wide fault in an elastic half-space at a depth of 10 km below the traction-free surface, and the dip angle is 20°. Given the shear dislocation of 10 m, the maximum vertical uplift at the surface is 4.7 m just above the up-dip edge, and the maximum surface subsidence is 1.3 m about 50 km behind the down-dip edge of the fault. To give our generic earthquake a finite size, let the along-strike length be 200 km, then we have the same moment as before (6×10^{21} Nm), and recall that the value for $\Delta E_{\min}$ is 3×10^{17} J. After displacing the water surface by the same amount as the vertical ground motion, the gravitational potential energy available for the tsunami waves is 7.1×10^{14} J. As the tsunami waves begin to propagate away from the fault region, this energy becomes equally partitioned between the potential and kinetic energy of the waves. In particular, let the waves propagate in a uniform 4-km thick ocean layer, then we find that the equi-partitioning was achieved to within 1% after just 11 minutes of wave computation. The total tsunami wave energy is a small fraction of the minimum earthquake energy of 3×10^{17} J — only about 0.002 of the available energy need be diverted into tsunami wave energy. Given the huge errors in the actual measurement of earthquake energy, loss of this tiny fraction would hardly be noticed. On the other hand, the tsunami waves associated with this small energy fraction can be large and destructive.

Figure 2 shows a simplified diagram of the vertical deformation that can be used to develop a simple algebraic connection from earthquake parameters to tsunami

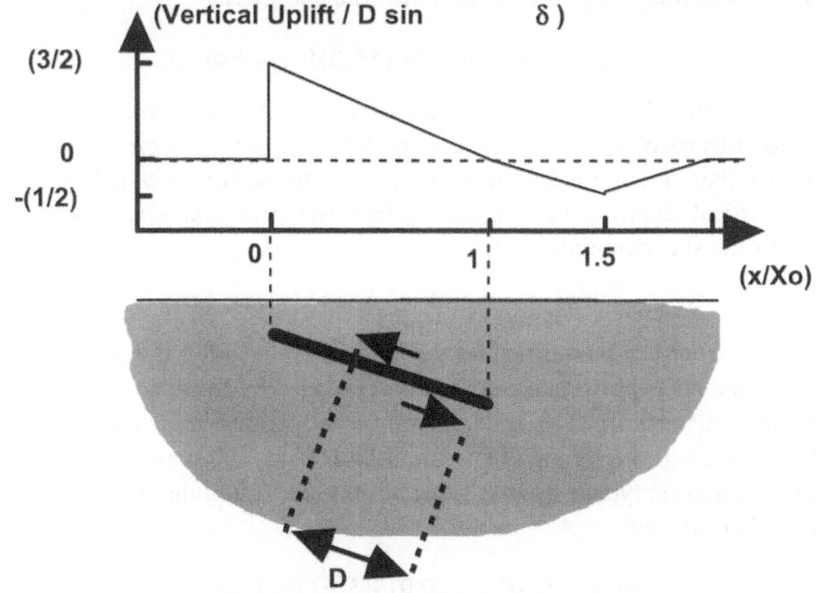

Figure 2
Stylized faulting model and consequent free-surface vertical deformation are shown in this 2-D cross-section view. The simplified surface deformation closely matches the exact results, yet allows for a simple analytical formula for the tsunami wave energy.

wave energy. We estimate the maximum uplift above the updip edge as: ΔV_{max} $(x/X_o = 0)$ = 1.5 D sin δ; and the maximum subsidence behind the fault as ΔV_{min} $(x/X_o = 1.5) = -D/2$ sin δ, where X_o is the horizontal projection of the fault width, i.e., $X_o = W$ cos δ, δ is fault dip angle, and D is the fault shear dislocation. Note that application of the above simplistic description to our above example great earthquake fault yields ΔV_{max} of 5.1 m and ΔV_{min} of −1.7 m, close to the values from the full numerical calculation. The simplistic triangular functions for vertical motion can be integrated along x and fault strike to obtain the following formula for tsunami wave energy: $\Delta E_{tsunami} = g\rho_w WL$ (5/12)cos δ(D sin $\delta)^2$, where ρ_w is water density and g is gravitational acceleration. Note that this formula contains the product of the fault area and displacement, hence we multiply and divide by the shear modulus μ, to extract the seismic moment of the earthquake:

$$\Delta E_{\text{tsunami}} = M_o g\rho_w/\mu(5/12)\cos\delta\, D\sin^2\delta. \tag{2.5}$$

To demonstrate that the above simple expression is useful, substitute the particular values for M_o, D, and μ for our example great earthquake to obtain: $\Delta E_{tsunami} = 9.1 \times 10^{14}$ J, just 28% larger than the value calculated for the exact 3-D faulting model. Accepting the above algebraic formula as a good estimator of tsunami energy, we can rewrite it to look quite similar to the earthquake energy formula. First, let us define the approximate static stress drop as: $\Delta\sigma = \mu\ D/W$, or $\Delta\sigma/\mu = D/W$, and finally $D = W\ \Delta\sigma/\mu$. Thus, the energy is:

$$\Delta E_{\text{tsunami}} = M_o\Delta\sigma g\rho_w/\mu^2(5/12)W\cos\delta\sin^2\delta. \tag{2.6}$$

Recognizing that $M_o(\Delta\sigma/2\mu)$ is the minimum earthquake energy ΔE_{min}, we have for the ratio of tsunami to earthquake energy: $\Delta E_{tsunami}/\Delta E_{min} = g\rho_w/\mu$ (5/6) W cos δ sin^2 δ. Notice that W sin δ is the depth range of fault width W, which provides the "natural" vertical depth scale for the hydrostatic pressure: $g\rho_w W$ sin δ, that is compared to the shear modulus as:

$$\Delta E_{\text{tsunami}}/\Delta E_{\text{min}} = g\rho_w W\sin\delta/\mu\ (5/6)\ \sin\delta\cos\delta. \tag{2.7}$$

Thus, aside from the geometric factor of 5/6 sin δ cos δ that is always less than 1, the ratio of tsunami to earthquake energy is controlled by the ratio of $g\rho_w W$ sin δ to μ. Given that the maximum depth extent of the seismogenic plate interface in subduction zones is about 40 km (TICHELAAR and RUFF, 1993), we can let W sin δ be 40 km, choose μ as its typical crustal value of 30 GPa, to obtain an upper numerical bound on this ratio as:

$$\Delta E_{\text{tsunami}}/\Delta E_{\text{min}} = 0.013(5/6)\sin\delta\cos\delta. \tag{2.8}$$

Therefore, this approximate algebraic formulation shows that the tsunami wave energy shall always be less than 1% of the minimum earthquake energy. This result shows that the interaction of tsunami waves back on the earthquake rupture process can be ignored, hence kinematic models of tsunami wave generation by earthquakes are fully justified from an energy perspective.

3. Energy Balance and Tsunami Generation by Landslides

Landslides move due to the force of gravity. The energy release of a landslide is thus the difference between net gravitational potential energy before and after the event. The change in strain energy of the solid earth can be ignored to second-order. For submarine slides, it is the difference between rock and water density that gives net displaced mass. As a surficial process, we can construct good estimates of the total mass of material and the center of mass displacement for many cases. Thus, in contrast to earthquake energy, the overall static energy release of landslides might be reliably estimated for both recent and ancient landslides.

There have been many spectacular and hazardous subaerial landslides, although it now seems that many of the largest slides occur under water. Marine geophysical surveys are capable of shedding light on these submarine slides (see reviews by MOORE *et al.*, 1994; HAMPTON *et al.*, 1996). For reference, the energy release of the 1929 Grand Banks slide is about 3×10^{17} J. However MOORE *et al.* (1994) discuss ancient landslides off many margins of the world, and the Nuuanu slide off the Hawaiian Islands has an energy release of about 2×10^{20} J. When we compare this gravitational energy release to the earthquake energy of the 1960 Great Chilean earthquake (about 10^{19} J), we see that these ancient landslides are some of the most energetic events on earth.

Where does the landslide energy go? It is partitioned into the two main reservoirs of the ocean and the solid earth (Fig. 3). As for earthquakes, the partitioning within the solid earth reservoir between friction and seismic waves is controlled by dynamic friction during the slide. The ocean reservoir is further divided into "friction" (i.e., hydrodynamic drag and related phenomenon such as acoustic waves) and surface gravity waves — tsunamis. This partitioning is determined by the details of block motion, in particular whether the fastest slide velocity is close to the wave propagation velocity. Unfortunately, we cannot just look at the "before and after" of a submarine landslide and guess how large the tsunami waves were; we must first solve the dynamic problem of landslide motion.

3.1 Kinematic Versus Dynamic Landslide Models

Submarine slides are a complex geologic process, but are idealized hereafter as two-dimensional translational slides down a constant-angle slope. A kinematic description simply specifies the time history of block displacement; then, the tsunami waves that are generated by this history have no influence back on the block (e.g., IWASAKI *et al.*, 1996; MA *et al.*, 1999). The extreme specification is to make the slider

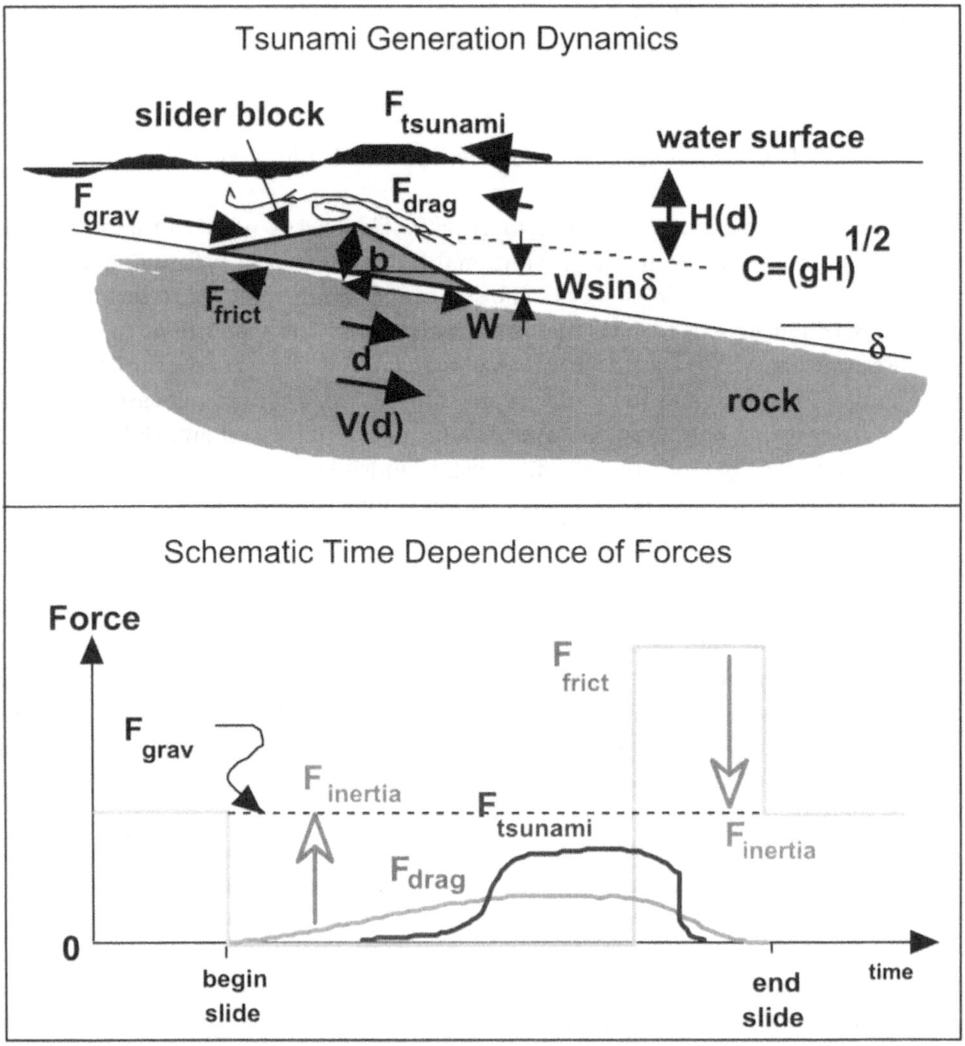

Figure 3
Force balance for the submarine slider block problem. See text for description of all variables. A symmetric triangular slider block was used to produce a simple yet reasonable wave impulse response. The downslope gravitational force is constant throughout the motion, the three resistive forces are interface friction, hydrodynamic drag, and wave resistance. The difference between gravitational force and the sum of resistive forces is the inertia force which thence gives block motion history. A schematic history of a particular slide event is depicted below to suggest the interplay of these force terms.

velocity so fast that it emulates the earthquake generation problem where the source starts and stops before any gravity waves can propagate. While this instantaneous kinematic model is acceptable for the earthquake problem, we cannot accept it for the landslide problem because this *ad hoc* specification may not conserve energy. To experiment with physically plausible kinematic models, one could calculate hundreds or thousands of possible histories — but which one is correct? Perhaps the best approach is to solve the dynamic problem where the velocity of the block and the time-dependent tsunami wave heights are determined in a self-consistent fashion. Since landslides are idealized simply, the mathematical problem reduces down to a single-particle integro-differential equation with nonlinear terms. Although the necessity of solving the dynamic problem greatly complicates tsunami wave generation — it is quite beneficial in the final analysis because the dynamically-consistent solutions only traverse a small part of the total kinematic parameter regime. Hence the dynamic solution actually simplifies the scientific problem of tsunami generation by landslides.

3.2 Dynamics of the 2-D Slider Block

Figure 3 shows the physical elements in the force balance. The key is that the net gravitational energy given up by the block as it moves down the slope is:

$$\Delta E_{grav} = \Delta\rho WbL\, g\, d \sin\delta, \tag{3.1}$$

where $\Delta\rho$ is the difference between rock and water density, W is half the downslope block width, b is the block height at the center of the symmetric triangular cross-section, L is the along-strike block length, δ is the slope dip angle, and d is the current block displacement along the slope. The gravitational force acting along the downslope direction is the derivative of Equation (3.1) with respect to d:

$$F_{grav} = \Delta\rho\; WbL\, g \sin\delta \tag{3.2}$$

and it is constant before, during, and after the slide event. The density difference $\Delta\rho$ in Equations (3.1) and (3.2) can be viewed as either the net mass that slides down (good for the static view) or as the net effect of differential hydrostatic pressure acting on the block (better for the dynamic view).

In general, after the slider has stopped with final displacement d, the net gravitational energy release in Equation (3.1) is partitioned into a total of four reservoirs: friction along the sliding interface, seismic waves radiated into the solid, frictional drag within the water, and water gravity waves. During the slide, the force balance consists of the following three force types: (i) the constant gravitational force in the downslope direction; (ii) the three resistive forces of sliding interface friction, hydrodynamic drag, and gravity wave generation; and (iii) the inertia force which translates the mismatch between gravitational and resistive forces into block acceleration and deceleration. This force balance along the downslope direction is written as: $F_{inertia} = F_{grav} - F_{frict} - F_{drag} - F_{tsunami}$. After dividing the entire equation by the constant F_{grav}, we obtain the following non-dimensionalized force terms:

$$F_{inertia}/F_{grav} = 1 - F_{frict}/F_{grav} - F_{drag}/F_{grav} - F_{tsunami}/F_{grav}, \tag{3.3}$$

where the resistive forces are positive when directed upslope, and F_{inertia} is positive for block acceleration in the downslope direction. As we consider the details of each term, block displacement will be non-dimensionalized by W, and velocities will be non-dimensionalized by the "natural" system velocity of $C_w = (gW \sin \delta)^{1/2}$, where the velocity length-scale is half the downslope vertical drop of the block.

Let us now consider the nature and properties of these forces. We have the fewest physical constraints on the interface dynamic friction force. We know it must drop below the static friction value for the slide to proceed. If the dynamic friction force drops to just slightly less than the static value, then the block slowly creeps down the slope — this will not generate any waves. At the other extreme, friction drops to zero and then the block accelerates as fast as it can, given the other resistive forces. We shall assume this extreme case for the initial phase of slider motion, i.e., $F_{\text{frict}}/F_{\text{grav}} = 0$. Eventually, the interface friction term must acquire a value of more than 1 to decelerate and finally arrest the motion of the slider block. Similar to the earthquake partitioning problem, time history of the interface friction force determines both the friction and seismic wave energies.

The hydrodynamic drag force basically depends on water density, slider frontal area, the non-dimensional drag coefficient D, and square of slider velocity as follows: $F_{\text{drag}} = D\rho\, bL\, V^2$. There is a long history of theoretical and experimental development in fluid dynamics over the details of hydrodynamic drag; object shape, surface roughness, and Reynolds number all influence the details of drag. Studies of hydrodynamic drag lead to the discoveries of turbulence and boundary layer separation, but all these details can be lumped into the non-dimensional D — which has a typical value of about 1, though it can be as low as 0.1 or 0.2 for streamlined bodies at moderately high Reynolds numbers. Given the large length scale of geologic blocks, the overall Reynolds number is very high as soon as the block moves. The dependence of D on V is ignored, but the model will be tested with values of D that cover the range from 0.1 to 1.

We now consider the tsunami wave force on the block. If a block moves through a water layer with a free surface, then any gravitational surface waves that are generated impart a wave resistance force back on the slider block. There is a long and rich history of work on the wave resistance problem, most of the famous solutions are for "ship waves" (e.g., LAMB, 1932; LIGHTHILL, 1978). LAMB (1932) and LIGHTHILL (1978) also give analytic solutions for the two-dimensional problem of objects moving along the bottom of a uniform depth water layer; or conversely, a water layer flowing over an obstacle. These classic solutions are for the steady-state problem where the object has moved with a constant velocity since $t = -\infty$. We focus on those cases where the dominant wavelength is much larger than the water depth. In these solutions, wave behavior is largely determined by the relationship between object velocity V, and wave velocity for long wavelengths, $C = (gH)^{1/2}$ where g is gravitational acceleration and H is water layer thickness. When the V/C ratio is much smaller than 1, then both steady-state wave amplitude and wave resistance force are small. As V/C approaches 1, both wave amplitude and resistive force grow and eventually we must abandon the premise of linear long-wave theory. If V/C is much larger than 1, the water layer just rushes over the object and returns to its original state without significant wave production. While these steady-state solutions are instructive, we need to solve the transient problem where the block starts at zero velocity, accelerates to some maximum velocity, and then eventually stops. The transient problem is more difficult than the steady-state problem, yet the classic treatments still provide the basis for developing a Green's function style solution appropriate for the transient problem. WARD (2000) shows how to integrate a particular source contribution for a 3-D tsunami wave generation problem. In comparison, we will use a 2-D Green's function that give waves propagating in the downslope and upslope directions. But for extending these results to a dynamic model, we need to use the current slider velocity to generate a wave, add it to the waves previously generated, and then calculate the wave resistance force that feeds back into the force balance equation that determines the change in slider velocity. There is the additional complication that wave velocity changes as a function of slider displacement, i.e., $C(d) = [g(H_o + d \sin \delta)]^{1/2}$, where H_o is initial water depth. Our experience in seismic source theory suggests that the first part of this problem — wave generation and propagation due to an unknown and variable velocity time history — can be written as the convolution between the "impulse" wave generation function of the triangular block and the velocity history function. For the second part of the problem, we simply use the average net wave pressure differences on the front and back faces of the block to calculate the wave resistance force. This entire cycle of wave generation and wave resistance can be written as an analytic convolution integral over a nonlinear function of velocity. The final wave resistance then depends on the sum of two convolution integrals for the downslope and upslope propagating waves. This non-dimensionalized form of the $F_{\text{tsunami}}/F_{\text{grav}}$ ratio can be written:

$$F_{\text{tsunami}}/F_{\text{grav}} = \rho/\Delta\rho \; b/W \sin \delta[A_> + A_<], \tag{3.4}$$

where $\rho/\Delta\rho$ is the non-dimensional density ratio of water to rock-water, $b/W \sin \delta$ is the non-dimensional "scale" for tsunami wave height, and $A_>$ and $A_<$ are the non-dimensional average wave-height force functions above the slider for down- and upslope waves. $A_>$ and $A_<$ are calculated from the wave-making functions, $M_>$ and $M_<$, which are given by:

$$M_> = [+\text{Box}(x; 0, W) - \text{Box}(x; W, 2W)] * \{1/[C/V - 1]\}(x) \tag{3.5a}$$

$$M_< = [-\text{Box}(x; 0, W) + \text{Box}(x; W, 2W)] * \{1/[C/V + 1]\}(x), \tag{3.5b}$$

where x measures the compressed/extended horizontal distance from the leading edge of the wave back to the slider as $\partial x/\partial d = C/V \pm 1$ for the downslope(−)/upslope(+) waves, normalized by W, Box(x; A, B) is the "boxcar" function that is 1 within the range from A to B, zero elsewhere, $\{1/[C/V \pm 1]\}(x)$ are the velocity kernel functions, and * represents the convolution integral over x from the furthest advance of the initial wave back to the slider.

Before proceeding, let me review the approximations and shortcomings of the above formulas. The slider block shape can be assumed to have a symmetric triangular cross section so that the impulse response for water surface vertical velocity is a positive "boxcar" function above the leading face and a negative "boxcar" function above the trailing face. Also, the velocity kernel function is evaluated at the block apex and water depth is assumed to be small enough that wave dispersion effects at the "boxcar" wavelength can be ignored. This dynamic solution for the free-falling block generates the largest waves in the transient regime, hence the classic steady-state solutions do not really emerge from the above-described solution. Finally, the over-arching assumption is that linear long-wave theory can be used for the generation and local propagation of the waves (see WATTS, 1998, for further discussion of linearity criteria).

3.3 Evaluation of Landslide Dynamics

Although an analytical formula has been derived for wave resistance, substitution of this full formulation into the force balance equation produces a highly nonlinear integro-differential equation for block velocity evolution. Here, the force balance of Equation (3.3) is written with full-normalization and substitution of explicit dependencies, except for the wave force term where the summary formula of Equation (3.4) is used:

$$\partial/\partial t(V/C_w) = \{\Delta\rho/\rho_a 1/(W/C_w))\}$$
$$\{1 - F_{\text{frict}}/F_{\text{grav}} - \Delta\rho/\rho D(V/C_w)^2 - \Delta\rho/\rho\ b/W \sin\delta[A_> + A_<]\}, \quad (3.6)$$

where all velocities are normalized by the natural system velocity $C_w = (gW\sin\delta)^{1/2}$, hence W/C_w is the system time scale, and ρ_a is the relevant slider density for inertia which includes the added mass coefficient (small for the low aspect slider).

The A functions remain much less than 1 for V/C much less than 1. If V/C is around 1, then $A_>$ grows as $\approx\Delta d/W$ for displacement Δd at $V/C \approx 1$. Thus, basic consideration of force magnitudes in Equation (3.6) demonstrates that the wave resistance term will be small if V/C stays much smaller than 1 during the landslide. If the wave resistance term can be ignored, then the primary force balance is between gravity, inertia, and hydrodynamic drag. The velocity initially increases linearly due to gravitational free-fall down the slope, but as the velocity approaches the limiting value for a particular choice of D, the rate decreases. This behavior is similar to that for a purely linear negative feedback which displays an exponential approach to the limiting velocity; there is just a quantitative difference for the (velocity)^2 feedback. If we make a reasonable choice for the $\Delta\rho/\rho$ density ratio to be 1, then the limiting non-dimensionalized (velocity)^2 is simply the inverse of the drag coefficient. For a choice of $D \approx 1$, this implies that slider velocity cannot exceed the natural system velocity of C_w.

To now consider the wave resistance force, the key physical parameter is the ratio of slider velocity to wave velocity V/C, where C depends on d as $C = (gH(d))^{1/2}$, where $H(d)$ is the variable water depth above the slider as it displaces toward deeper water. To make a large wave resistance force, the slider must move with a velocity close to the local wave velocity over a displacement that is a substantial fraction of W. If the wave velocity is faster than the limiting block velocity due to drag, then the wave amplitude and wave resistance will remain small. The geologic conclusion for this situation is quite simple: if $H_o/W \sin \delta \gg 1/D$, then the maximum tsunami wave height generated at the source region will always be a small fraction of b, the block height.

The above considerations indicate that larger tsunami waves are generated when the initial water depth above the slider H_o is much smaller than the characteristic vertical drop of the slider block, $W \sin \delta$ (WATTS, 1998). This basic fact is one of the most important conclusions of this paper; it also conforms to our intuition about wave generation by slides.

We now further investigate a particular case where the block can accelerate until the wave resistance term approaches 1 to balance the gravitational force. Choose the $\Delta\rho/\rho$ and $\Delta\rho/\rho_a$ density ratios to be 1, let the slider geometry ratio $b/W \sin \delta$ be 1, choose initial water depth to make the $H_o/W \sin \delta$ ratio be 1, and then choose a low drag coefficient of $D = 0.1$ so that the $C(d)$ line plots below the limiting drag velocity (see Fig. 4b). For the synthetic landslide generated by these parameter choices, the block displacement, velocity, and wave-making force are plotted as a function of time in Figure 4a. The physical connection between these variables is more easily seen in the "phase space" plot of *velocity squared* versus *displacement* (Fig. 4b). For the particular parameter choices made here, $V/C(0)$ reaches an initial peak value of about 1 at displacement $d/W = 1$, and the wave force term is about 1/2 which causes velocity to grow slowly until $d/W = 2$. This system path is plotted in Figure 4b. Eventually $V/C(0)$ begins to increase as the slider moves into deeper water, but note that the trajectory slope is less than the wave velocity line (bold line in Fig. 4b), hence the local $V/C(d)$ ratio will never approach 1 in deeper water. Note that somewhere to off the right edge of Figure 4, interface friction increases from zero such that slider velocity will decline back toward zero. The parameter values for the run in Figure 4 were chosen to maximize tsunami generation at a fixed ratio of $H_o/W \sin \delta = 1$. Thus, other parameter combinations will produce phase space paths that plot in the lower-right section of Figure 4b — and hence produce a smaller tsunami.

3.4 Some Details on Wave Amplitudes and Energy Partitioning

Let us now turn to some details of the generated tsunami wave heights and associated energy balance. One difference between the steady-state and transient solutions is that the transient case sends waves in both the downslope and upslope directions. In fact, for low V/C ratios, the wave heights are comparable, albeit quite small. As V/C approaches 1, wave heights become more asymmetric with higher waves in the direction of block motion. As seen above, even when choosing parameter values that maximize wave height, the $V/C(d)$ ratio can only

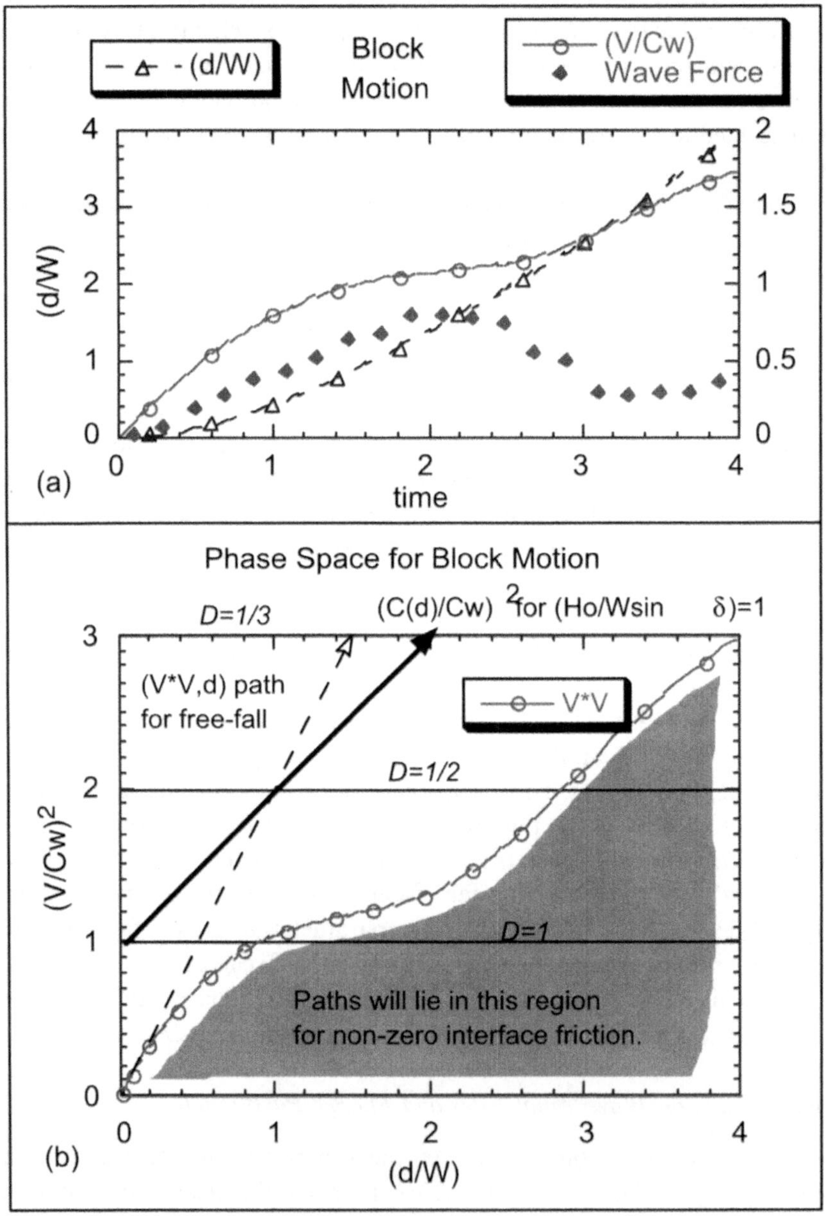

be close to 1 for a small sliding distance (compared to W). This initial transient wave is "up" in the downslope direction, "down" in the upslope direction (see waves in Figure 5 for the specific case discussed in section 3.3). If the dynamic interface friction can then jump to a very large value (i.e., $F_{frict}/F_{grav} > 5$) to quickly stop the slider while V/C is still close to 1, then large waves can be produced by this truncation transient — they can be nearly the same amplitude as the initial transients, though of opposite polarity. On the other hand, a gradual deceleration of the slider produces a very long wavelength ramp of very small wave amplitude.

Final energy conversion from gravitational into wave energy depends on the abruptness of motion termination and the total distance traveled beyond the initial $d = 2W$. To generate the largest waves for given values of D and $H_o/W \sin \delta$, let dynamic friction drop to zero and let the slider block displace to $d/W = 2$, then abruptly truncate the motion so that no energy is lost into friction but instead goes into the large truncation waves. For realistic choices of both D and $H_o/W \sin \delta$ equal to 1.0, then this abrupt truncation behavior can convert about 45% of the net gravitational energy into wave energy. For even more realistic cases where interface friction slows the block gradually so that $d/W \gg 2$, most of the additional gravitational energy release will be deposited into water drag and interface friction; thus, the wave conversion factor systematically declines with increasing d. Even if the initial transient waves have the maximum amplitude, the final value for energy conversion from gravitational into wave energy could be much less than 10% if the slide continues for great distances. Thus, maximum wave amplitude is a more robust system parameter than final energy conversion efficiency.

◀

Figure 4

(a) An example synthetic landslide is plotted. The non-dimensionalized block displacement d/W, velocity V/C_w, and wave force are graphed as a function of non-dimensionalized time (see text for definitions). The displacement function uses the left-hand axis, velocity and wave force use the right-hand axis. Parameters for this particular example are chosen to maximize tsunami wave generation; this is verified as the wave force approaches a value of 1 for a brief time. (b) "Phase space" to track the position and velocity of the block during a landslide. It is best to plot (velocity)^2, non-dimensionalized by C_w^2, *versus* displacement d, non-dimensionalized by W. Hydrodynamic drag limits the long-term slider velocity; horizontal lines show velocity bounds for drag coefficient D choices of 1/3, 1/2, and 1. Even more important is the $H_o/W \sin \delta$ ratio, initial water depth to slider vertical drop, since this ratio sets the vertical offset of wave velocity $C(d)^2$ lines (see the bold velocity line plotted for $(C(0)/C_w)^2 = H_o/W \sin \delta = 1$). The dashed line shows system trajectory for gravitational free-fall with no resistive forces; all system trajectories must fall below this line, though slider paths can initially follow this line for zero dynamic friction. To maximize tsunami wave generation, the hydrodynamic drag line should plot above the wave velocity line so that slider velocity can approach the wave velocity. The system path is plotted (the "$V*V$" function) for the landslide example shown in (a), where all system ratios are 1 except for $D = 0.1$ so that wave force is more important than hydrodynamic drag over the plotted displacement range.

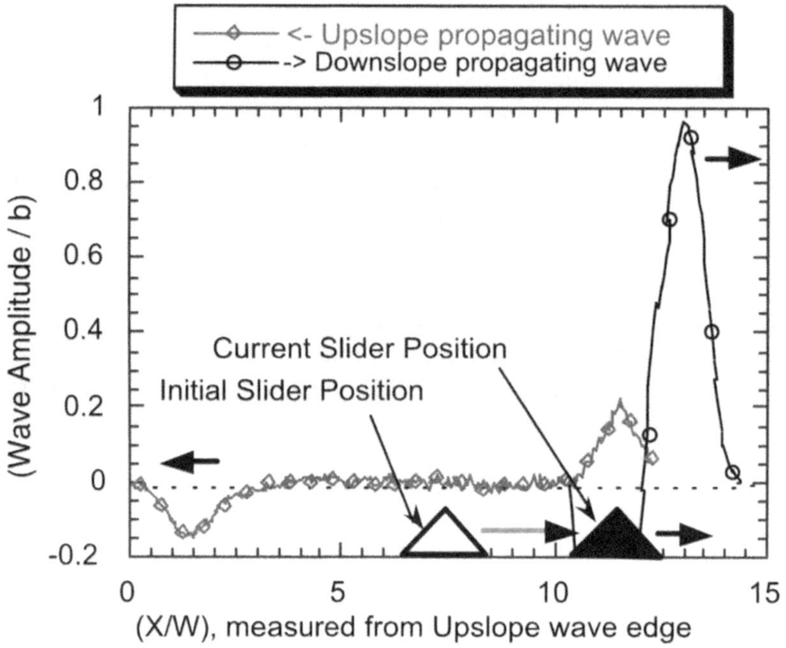

Figure 5
Plots of tsunami waves generated by the system path shown in Figure 4. Horizontal scale measures distance along sea surface, normalized by W, and amplitudes are normalized by b, slider block height. The choice of small D and $H_o/W \sin \delta$ produces nearly maximal wave heights, about $+1$ in the downslope direction and -0.15 in the upslope direction. Continuation of wave plots over the current position of slider block shows the truncation waves that would propagate away if the block were abruptly stopped. Note the large vertical exaggeration for the triangular block.

3.5 More Details on Wave Amplitude Systematics

To focus on the systematic quantitative aspects of these transient waves, the force feedback of waves upon block velocity results in a stable moderation of maximum tsunami wave height. If we specify both $H_o/W \sin \delta$ and D ratios to be small, which maximizes tsunami wave height, the tsunami wave amplitude in the downslope direction is never much larger than b. This natural buffering effect produces an excellent "rule of thumb": if $H_o/W \sin \delta < 1$, maximum tsunami wave height at the source region is the same as slider block height, to within 30% (see Fig. 6 for details). We also find a summary rule for downslope tsunami wave height when the $H_o/W \sin \delta$ ratio is larger. Set the system density, geometry, and drag ratios all to 1, then the ratio of maximum tsunami wave height to b, the slider block height, approximately depends on $H_o/W \sin \delta$ as follows:

$$(\text{max wave height})/b = [1 + H_o/W \sin \delta]^{-1}. \tag{3.7}$$

This approximate formula is valid for the $H_o/W \sin \delta$ ratio up to about 3 or so, though still valid at the 50% level for $H_o/W \sin \delta$ up to 10. Figure 6 schematically

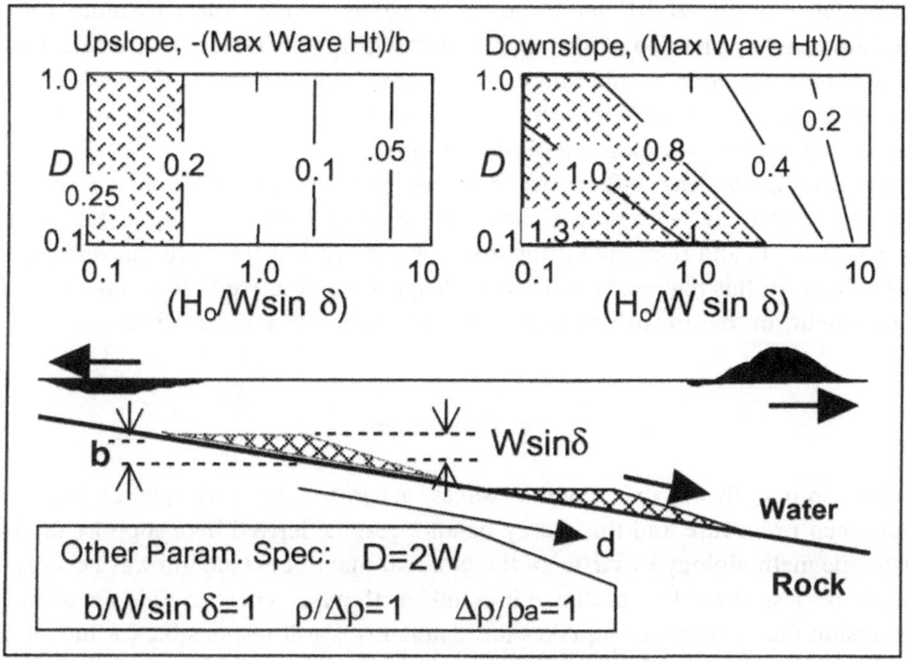

Figure 6
Summary dependence of tsunami wave heights on drag coefficient D and initial water depth $H_o/W \sin\delta$. The two graphs above up- and downslope propagating waves show approximate contours of normalized maximum wave heights (multiply by b for actual wave heights). The hachured regions show the realm of the nearly constant largest wave heights. For the downslope/upslope waves, the scale-length from leading edge of wave to peak wave amplitude changes from about $2W/1.5W$ at $H_o/W\sin\delta = 10$ to $1W$ in both hachured regions. This figure can be used to provide quick estimates of tsunami wave heights and pulse widths given the basic size, geometry, and water depth of candidate submarine landslides.

graphs the maximum wave amplitudes over variations in both $H_o/W \sin\delta$ and D. If we perturb the other system ratios, then the maximum wave height in the downslope direction for physically realizable parameter combinations is about twice the block height.

If the landslide occurs in a closed basin, then both downslope and upslope waves can be hazardous. But for many of the geologic settings of triggered landslides, the upslope propagating wave presents the greatest hazards to the nearby coasts. For both the 1992 Flores Island and 1998 PNG events, it was the immediate proximity of the coast to the upslope edge of the slide that caused terrible disasters. Thus, let us also look at the upslope wave amplitudes. The maximum amplitude in the upslope direction is always less than for the downslope direction. For values of $H_o/W \sin\delta$ more than 3, the upslope wave amplitude is about half the downslope amplitude. As $H_o/W \sin\delta$ becomes quite small, then the upslope amplitude is about one-fourth the downslope amplitude. Thus, we can extract another useful "rule of thumb": if the landslide initiates in very shallow water such that the $H_o/W \sin \delta$ ratio is less than 1, then the maximum tsunami wave amplitude generated in the source region is approximately $b/4$, where b is the block height. Of course, this initial source region height of the 2-D wave can be enhanced due to propagation effects near the shore. Figure 6 shows schematic graphs that summarize these amplitude results.

To summarize, the 2-D block dynamic problem produces a self-regulating feedback so that the V/C ratio achieves of value of 1 only for slip distances small compared to W, and then only if the initial water depth is less than the vertical drop of the slider. In this regime, tsunami wave heights are "buffered" to be about b, slider block height, in the downslope direction, and $b/4$ in the upslope direction.

4. *Conclusions*

The source theory of tsunami waves generated by earthquakes is a well-established procedure and the energy balance results derived here support the basic kinematic methodology by virtue of the fact that maximum tsunami wave energy will always be less than 1% of the minimum earthquake energy. This small energy conversion factor from earthquakes into tsunamis is still responsible for most of the destructive tsunamis in the world.

Marine surveys have now documented ancient submarine landslides that rival the greatest earthquakes in their total energy release. The expected annual occurrence of huge submarine landslides is still an unanswered question, but even small submarine landslides can cause large localized tsunamis, as evidenced by recent earthquake-triggered slides. My quantitative exploration of tsunami wave source theory from submarine landslides shows that gravitational to wave energy conversion can range from essentially zero to about 50% as a practical upper bound for reasonable geologic and hydrodynamic parameters. This wide range in energy conversion produces a highly variable efficiency in tsunami generation by submarine landslides. The key parameter is the ratio of initial water depth above the block-to-block vertical scale height; this ratio must be less than 1 for the landslide to generate large waves. Indeed, for future detailed numerical modeling, a glance at Figure 6 will already forecast whether significant tsunami waves can be generated for a particular physical setting. Turning to the practical key parameter of maximum tsunami wave height in the source area, the interaction of all the dynamic effects produces a remarkably stable and simple result: if the key ratio of water depth to block vertical scale height is small such that a significant tsunami will be generated, then the maximum tsunami height will be approximately equal to the block height — largely independent of the other parameters! The graphs in Figure 6 give more detailed information on maximum wave amplitudes for both the downslope and upslope propagating waves. Figure 6 provides a simple yet useful evaluation of tsunami hazards for submarine landslides.

Acknowledgments

Thanks to the organizers and participants in the "Workshop on the Prediction of Underwater Landslide and Slump Occurrence and Tsunami Hazards off of Southern California," held in Los Angeles, CA in March, 2000. This work was presented there, and I enjoyed the scientific discussions and feedback at this gathering of the world's tsunami community. I also appreciate the many discussions on earthquakes, tsunamis, and landslides over the years with my Michigan colleagues — especially Kenji Satake and Yuichiro Tanioka. Special thanks to Phil Watts for his encouragement and constructive comments. The research program on earthquakes, tsunamis, and related phenomenon over the years at the University of Michigan has been supported by the National Science Foundation (EAR-9725175 and OCE-9905503) and USGS (GR-02754).

References

ABE, K. (1979), *Size of Great Earthquakes of 1873–1974 Inferred from Tsunami Data*, J. Geophys. Res. *84*, 1561–1568.

DAHLEN, F. (1977), *The Balance of Energy in Earthquake Faulting*, Geophys. J. Roy. Astron. Soc. *48*, 239–261.

GRILLI, S. and WATTS, P. (1999), *Modeling of Waves Generated by a Moving Submerged Body: Applications to Underwater Landslides*, Engin. Analysis Boundary Elements *23*, 645–656.

HAMPTON, M. A., LEE, H., and LOCAT, J. (1996), *Submarine Landslides*, Rev. Geophys. *34*, 33–59.

HASEGAWA, H. and KANAMORI, H. (1987), *Source Mechanism of the Magnitude 7.2 Grand Banks Earthquake of November 1929: Double Couple or Submarine Landslide*? Bull. Seismol. Soc. Am. *77*, 1984–2004.

HASKELL, N. (1964), *Total Energy and Energy Spectral Density of Elastic Wave Radiation from Propagating Faults*, Bull. Seismol. Soc. Am. *54*, 1811–1842.

HIDAYAT, D., BARKER, J., and SATAKE, K. (1995), *Modeling the Seismic Source and Tsunami Generation of the December 12, 1992 Flores Island, Indonesia Earthquake*, Pure Appl. Geophys. *144*, 537–554.

IMAMURA, F., GICA, E., TAKAHASHI, T., and SUTO, N. (1995), *Numerical Simulation of the 1992 Flores Tsunami: Interpretation of Tsunami Phenomenon in Northeastern Flores Island and Damage at Babi Island*, Pure Appl. Geophys. *144*, 555–568.

IWASAKI, S., FURUMOTO, A., and HONZA, E. (1996), *Can a Submarine Landslide be Considered as a Tsunami Source*? Tsunami Hazards *14*(2), 89–100.

JOHNSON, C. and MADER, C. (1994), *Modeling of the 105 Ka Lanai Tsunami*, Tsunami Hazards *12*(1), 33–38.

KANAMORI, H. (1972), *Mechanisms of Tsunami Earthquakes*, Phys. Earth Planet. Int. *6*(5), 346–359.

KANAMORI, H. (1977), *The Energy Release in Great Earthquakes*, J. Geophys. Res. *82*, 2981–2987.

KAWATA, Y., BENSON, B., BORERRO, J., BORERRO, J. L., DAVIES, H., DELANGE, W., IMAMURA, F., LETZ, H., NOTT, J., and SYNOLAKIS, C. E. (1999), *Tsunami in Papua New Guinea was as Intense as First Thought*, *EOS*, Trans. AGU *80*, 101–105.

KIKUCHI, M. and FUKAO, Y. (1988), *Seismic Wave Energy Inferred from Long-period Body Wave Inversion*, Bull. Seismol. Soc. Am. *78*, 1707–1724.

LAMB, H. (1932), *Hydrodynamics* (Cambridge Univ. Press) 6th ed., 738.

LIGHTHILL, J. (1978), *Waves in Fluids* (Cambridge Univ. Press) 504.

MA, K., KANAMORI, H., and SATAKE, K. (1999), *Mechanism of the 1975 Kalapana Hawaii Earthquake Inferred from Tsunami Data*, J. Geophys. Res. *104*, 13,153–13,167.

MILLER, D. J. (1960), *The Alaska Earthquake of July 10, 1958: Giant Wave in Lituya Bay*, Bull. Seismol. Soc. Am. *50*, 252–266.

MOORE, J., NORMARK, W., and HOLCOMB, R. (1994), *Giant Hawaiian Landslides*, Annu. Rev. Earth Planet. Sci. *22*, 119–144.

MURTY, T. S. (2003), *Tsunami Generation from Submarine Landslides*, Pure Appl. Geophys. *160*, 2147–2153.

NOMANBHOY, N. and SATAKE, K. (1995), *Generation Mechanism of Tsunamis from the 1883 Krakatau Eruption*, Geophys. Res. Lett. *22*, 509–512.

OKAL, E. (1988), *Seismic Parameters Controlling Far-field Tsunami Amplitudes: A Review*, Natural Hazards *1*, 67–96.

PIPER, D., SHOR, A., FARRE, J., O'CONNELL, S., and JACOBI, R. (1985), *Sediment Slides and Turbidity Currents on the Laurentian Fan: Sidescan Sonar Investigations near the Epicenter of the 1929 Grand Banks Earthquake*, Geology *13*, 538–541.

RUFF, L. (1999), *Dynamic Stress Drop of Recent Earthquakes: Variations within Subduction Zones*, Pure Appl. Geophys. *154*, 409–431.

SATAKE, K. (1995), *Linear and Non-linear Computations of the 1992 Nicaragua Earthquake Tsunami*, Pure Appl. Geophys. *144*, 455–470.

SWEET, S. and SILVER, E. A. (2003), *Seismic Reflection Images of the Source Region of the 1998 Papua New Guinea Tsunami*, Pure Appl. Geophys. *160*, 1945–1968.

SYNOLAKIS, C. and TADEPALLI, S. (1993), *The Runup of Dipole Waves*, Proc. IUGG/IOC Inter. Tsunami Symp. 175–187.

TANIOKA, Y. (1999), *Analysis of the Far-field Tsunamis Generated by the 1998 Papua New Guinea Earthquake*, Geophys. Res. Lett. *26*, 3393–3396.

TAPPIN, D., WATTS, P., MCMURTY, G., LAFOY, Y., and MASUMOTO, T. (2003), *Prediction of Slump Generated Tsunamis: The July 17, 1998 Papua New Guinea Tsunami.*

TICHELAAR, B. and RUFF, L. (1993), *Depth of Seismic Coupling along Subduction Zones*, J. Geophys. Res. *98*, 2,017–2,037.

TITOV, V. and SYNOLAKIS, C. E. (1993), *A Numerical Study of Wave Runup of the September 2, 1992, Nicaraguan Tsunami*, Proc. IUGG/IOC Inter. Tsunami Symp., 627–635.

VASSILIOU, M. S. and KANAMORI, H. (1982), *The Energy Release in Earthquakes*, Bull. Seismol. Soc. Am. *71*(2), 371–387.

WARD, S. N., *Tsunamis*. In *Encyclopedia of Physical Science and Technology* (Academic Press, New York: 2000).

WATTS, P. (1998), *Wavemaker Curves for Tsunamis Generated by Underwater Landslides*, J. Waterway Port Coastal Ocean Eng. *124*(3), 127–137.

WATTS, P. and BORRERO, J. C. (2003), *Probabilistic Predictions of Underwater Landslides and Tsunami Amplitudes*

WINSLOW, N. W. and RUFF, L. J. (1999), *A Hybrid Method for Calculating the Radiated Wave Energy of Deep Earthquakes*, Phys. Earth Planet. Int. *115*, 181–190.

Pure appl. geophys. 160 (2003) 2177–2188
0033–4553/03/112177–12
DOI 10.1007/s00024-003-2425-x

Pure and Applied Geophysics

A Theoretical Comparison of Tsunamis from Dislocations and Landslides

EMILE A. OKAL[1] and COSTAS E. SYNOLAKIS[2]

Abstract—We use simple physical models to evaluate and compare the orders of magnitude of the energy generated into a tsunami wave by seismic dislocations and underwater slumps. We conclude that the two sources can generate tsunamis of comparable total energy. However, the slumping source is shown to be fundamentally dipolar in nature, which results in a low-frequency deficiency in the far-field. These simple conclusions corroborate the interpretation of the 1998 Papua New Guinea tsunami as being generated by an underwater slump.

Key words: Tsunamis, landslides, seismic sources.

1. Introduction

We present in this paper a number of theoretical discussions of the excitability of tsunami waves by both earthquakes sources (modeled as dislocations occurring in an elastic medium) and underwater landslides or slumps. We are motivated by the recent investigations of the catastrophic tsunami of 17 July 1998 in Papua New Guinea (hereafter, PNG), which suggest that its source was an underwater slump involving 4 km^3 of sedimentary material and occurring 13 mn after the mainshock (OKAL, 2002a; SYNOLAKIS *et al.*, 2002). Because they take place under water and are generally not witnessed directly, slumps remain very poorly known and only a few exceptional cases have been the subject of specific investigations, such as the 1929 Grand Banks slump (HASEGAWA and KANAMORI, 1987) or the Suva landslide of 1953 (HOUTZ, 1962). In both instances, the rupture of telegraphic cables provided the key evidence for slumping.

Our approach is to use very simple physical models of the deformation of the ocean floor to obtain the principal properties of the tsunamis generated by the two kinds of sources, with an aim at understanding the fundamental parameters controlling the energy dissipated into the tsunami wave. We must emphasize that the purpose of this paper is not to give exact theoretical solutions for actual case studies of seismic and slumping events, but rather to provide orders of magnitude of the

[1] Department of Geological Sciences, Northwestern University, Evanston, IL 60208, U.S.A.
[2] Department of Civil Engineering, University of Southern California, Los Angeles, CA 90089, U.S.A.

amount of energy which can be disseminated into the tsunami wave by the various sources.

2. The Dislocation Source

The generation of a tsunami by a dislocation source deforming the bottom of the ocean can be schematized by the model shown on Figure 1. In its simplest form, we consider a sudden uplift of a section of ocean floor of area S, moved vertically an amount δh. If this deformation is instantaneous, it results in the development of an identical hump on the ocean surface (Fig. 1a). The increase in the potential energy of the ocean is computed readily by displacing a volume of water $S \cdot \delta h$ from the bottom to the surface of the ocean:

$$\Delta W_1 = \rho_w g S \, \delta h \, H, \tag{1}$$

where ρ_w is the density of water, H the depth of the ocean column, and g the acceleration of gravity. This also represents the work of the pressure forces $S \rho_w g H$ displacing the ocean bottom a distance δh. Because the ocean is a non-viscous fluid, the hump is unstable and will flow back to the ocean surface, whose steady-state level will be unchanged on account of the very large lateral dimension of the ocean basin as compared to S (Fig. 1c). The center of mass of the displaced water, initially at height $\delta h/2$ above the ocean floor (solid dot on Fig. 1b) is transferred to the ocean surface at height H, so that the eventual change in potential energy is only

$$\Delta W_2 = \rho_w g S \, \delta h \, (H - \delta h/2). \tag{2}$$

The difference between (1) and (2) is the energy available to the gravitational oscillation induced by the collapse of the hump, which is the tsunami wave:

$$E_T = \Delta W_1 - \Delta W_2 = \frac{1}{2} \rho_w g S \, (\delta h)^2. \tag{3}$$

This very simple model is applicable because, in all practical cases, the source duration (i.e., the time over which the deformation of the ocean floor takes place) is short as compared to the time it takes for the tsunami wave to settle and eliminate the hump. The former is on the order of the dimension of the source, L, divided by the rupture velocity of the seismic source, v; the latter on the order of L/c, where c is the phase velocity of the tsunami. When the deformation takes place rapidly ($v \gg c$), hydraulic equilibrium is not reached at all times during the upwards motion of the ocean floor, the deformation is not reversible, more work is done than the eventual increase of potential energy in the new steady state, and the difference is funneled into the tsunami wave. Should the uplift take place very slowly ($v \ll c$), then equilibrium would be reached at all times during the deformation, which means that the hump on the surface would disappear faster

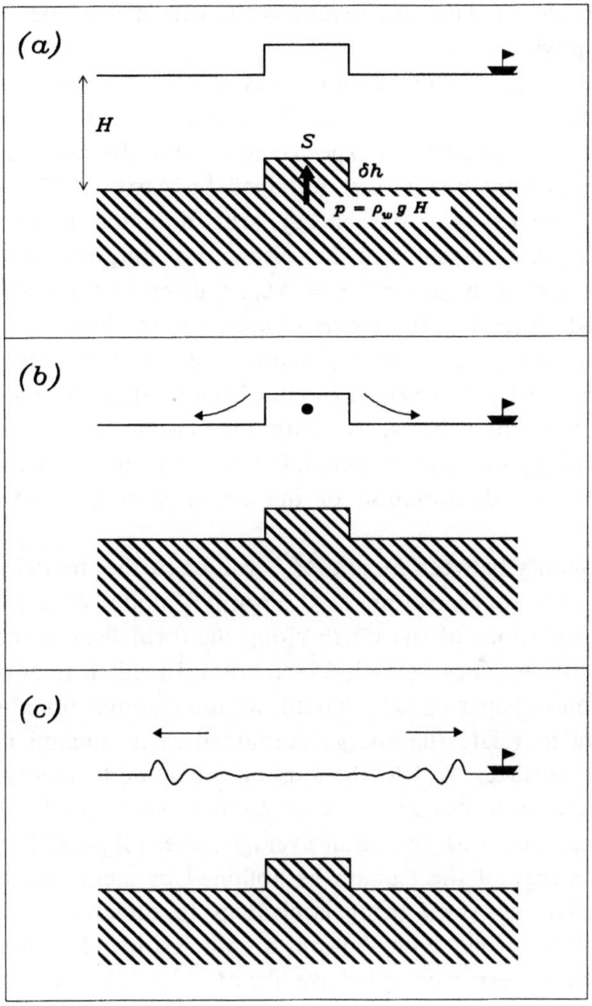

Figure 1
Excitation of a tsunami by a seismic dislocation. In this very simple model, a fraction of the ocean floor is suddenly uplifted, resulting in an immediate and identical hump on the ocean surface (*a*). Because the ocean is fluid, the hump is unstable and flows sideways (*b*), with the center of mass of the displaced material (solid dot) falling down by an amount $\delta h/2$. The resulting change in potential energy makes up the energy of the tsunami wave, which propagates away from the now defunct hump (*c*).

than it is being created; as a result, the pressure forces on the bottom would continuously adjust to the new height of the water column, the process would be thermodynamically reversible, and the work done to deform the ocean floor would be the exact difference in potential energies; there would be no energy available to the tsunami wave. In practice, for seismic sources, the rupture velocity v is on the order of 3 km/s, whereas the tsunami velocity $c = \sqrt{gH}$ is typically less than 250 m/s, even for the deepest ocean basins. Thus the regime is the irreversible one, and the model in Figure 1 is warranted.

When the size of the parent earthquake is increased, seismic scaling laws (e.g., GELLER, 1976) predict that δh scales as L, S as L^2, and the seismic moment M_0 as L^3. As a result, the energy E_T is expected to scale as M_0 raised to the power 4/3. This extremely important result, first obtained by AIDA (1977) and discussed by KAJIURA (1981), means that the energy sent into the tsunami wave by an earthquake source grows faster than the seismic moment, and hence than the total elastic energy released at the source, $E_e = \mathbf{M_0} : \boldsymbol{\varepsilon}$, given by the scalar product of the moment tensor $\mathbf{M_0}$ into the strain release tensor $\boldsymbol{\varepsilon}$, the latter expected to have a constant amplitude (in practice 10^{-4}), characteristic of the material properties of the rock. In other words, the fraction E_T/E_e of the total elastic energy release made available to the tsunami wave grows with the seismic moment like $M_0^{1/3}$. Note further that the energy E_T is, as expected, independent of the sign of δh, and hence of the polarity of the deformation of the ocean floor (i.e., reverse *vs.* normal faulting).

The proportionality between E_T and $M_0^{4/3}$ can be further modeled by considering a tsunami wave as a superposition of the normal modes of a specific branch of spheroidal free oscillations of the Earth along the formalism introduced by WARD (1980) and discussed by OKAL (1982, 1988, 1991). In this framework, discussed in detail in a companion paper (OKAL, 2003b), we use a conventional eigenfunction of the tsunami wave to relate the energy contained in a tsunami mode, E_l^m, to its amplitude at the surface, Y. We then use normal mode theory to express the amplitude Y excited by a seismic source of moment M_0 located immediately below the water-solid interface, with the result averaged over all possible focal mechanism geometries. The energy of the tsunami is obtained by summing up the individual modes, the summation over l being equivalent to integration over the angular frequency ω. High-frequency contributions are altered to reflect a corner frequency in the source spectrum, expected to behave like $M_0^{-1/3}$, while at the same time seismic scaling laws are used to account for the necessary extension of the width of the fault at depth (which affects the excitation of the tsunami mode) when the size of the earthquake, and hence M_0, grows. The final result is that the energy of the tsunami can be written

$$E_T = 0.22 \frac{\rho_w g}{\mu^{4/3}} \cdot \varepsilon_{\max}^{2/3} \cdot M_0^{4/3}, \tag{4}$$

where μ is the elastic rigidity of the solid Earth, and $\varepsilon_{\max}$ the characteristic strain release during the rupture. For $\mu = 7 \times 10^{11}$ dyn/cm^2, characteristic of the upper mantle, and $\varepsilon_{\max} = 10^{-4}$, this yields

$$E_T = 7 \times 10^{-17} M_0^{4/3} \tag{5}$$

where E_T is in ergs and M_0 in dyn-cm (OKAL, 2003b). Equation (4) is formally equivalent to the formula proposed by KAJIURA (1981), although the constants in (4) and (5) are greater than he proposes.

Equation (5) allows the computation of orders of magnitude of the energy of tsunamis generated by major earthquakes. For example, in the case of the 1964 Alaskan earthquake ($M_0 = 7.5 \times 10^{29}$ dyn-cm) (KANAMORI, 1970), it yields $E_T = 5 \times 10^{23}$ ergs. For the largest event ever recorded, the 1960 Chilean earthquake ($M_0 = 2 \times 10^{30}$ dyn-cm), the energy would reach $E_T = 2 \times 10^{24}$ ergs, which remains 1/100 of the total elastic energy released during the event. Incidentally, the Chilean earthquake also features the longest documented single episode of seismic rupture, with $L = 800$ km. Assuming that this number could be tripled to obtain a maximum conceivable size of subduction zone earthquake (although no such coherent fault system is, to our knowledge, documented on Earth), we would reach $L = 2500$ km, corresponding to $M_0 = 3 \times 10^{32}$ dyn-cm and $E_T/E_e = 4\%$; in this respect, the finite radius of the Earth, which scales the size of the tectonic plates, and hence to a large extent the maximum dimensions of earthquake sources, prevents the energy of the tsunami from catching up with the total available elastic energy.

3. The Slump Source

The case of an underwater slump can be modeled in extremely simplified fashion by considering (Fig. 2) that it consists of translating a mass of solid material along the sea floor, and hence, of combining a negative source of the type studied in Section 2 at the heel of the slump with a positive one at its toe. The slump source thus appears as dipolar in nature, with the total energy of the tsunami being the sum of the two contributions of each element of the dipole.

However, in this particular case, we can no longer assume that the displacement of the ocean floor takes place instantaneously as compared to the evolution of the tsunami wave. This is because, in the case of a gravitational slump, the velocity acquired by the slumping mass is controlled by the gravity field g. In practice, the maximum velocity reached during a slumping event on an inclined plane is given by $v = \sqrt{2gz}$, where z is the maximum vertical extent traveled by the slumping material. Then, v must be compared to the phase velocity c, their ratio being $\sqrt{2z/H}$. This number can never be large, since the slump has to be contained in the water column; in practice, a slump sliding 500 m in a 1500-m deep ocean would correspond to $v/c = 0.8$.

For such values of z and H, the regime would be neither fully irreversible (as in the case of the seismic dislocation), nor fully reversible (in which case the slump would be so slow as to generate no tsunami). Thus, our estimate of the energy radiated into the tsunami must be corrected by modeling quantitatively the

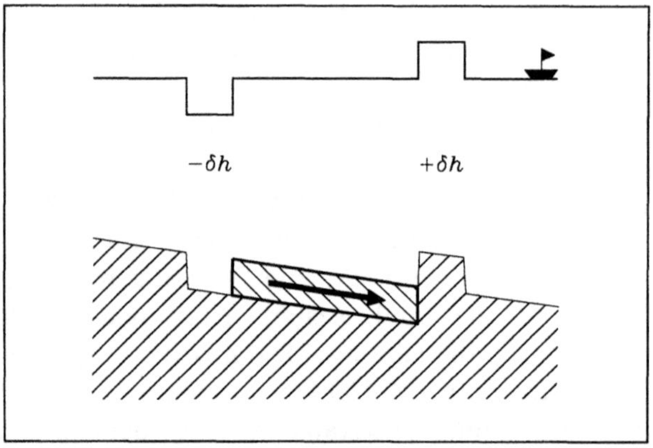

Figure 2
Generation of a tsunami wave by an underwater slide. In this highly simplified conceptual model, the moving slide or slump creates a bulge, and hence a positive hump on the surface, at its toe, while leaving a mass deficiency, and hence a trough at the surface, at its heel. As a result, the initial condition of the ocean surface takes a dipolar character.

development of the tsunami wave during the slumping. For this purpose, we use the schematic model of a block sliding on the bottom of an ocean of constant depth H, at a velocity v, starting from a standstill at $t = 0$ (Fig. 3a), and stopping abruptly at $t = T$. We assume that the block has the shape of a gaussian, so that the deformation can be given by:

$$\varepsilon(x,t) = A \exp[-k(x - vt)^2] \quad (0 \leq t \leq T). \tag{6}$$

The deformation of the surface of the ocean, $\eta(x,t)$, then satisfies

$$\frac{\partial^2 \eta}{\partial t^2} - c^2 \frac{\partial^2 \eta}{\partial x^2} = \frac{\partial^2 \varepsilon}{\partial t^2}. \tag{7}$$

The solution is obtained by combining ($\eta = \eta_1 + \eta_2$) the case (Fig. 3b) of a block moving indefinitely for all positive times ($t \geq 0$),

$$\eta_1(x,t) = AH(t)\left[\frac{v^2}{v^2 - c^2}\exp[-k(x - vt)^2] + \frac{v^2}{2c}\left(\frac{\exp[-k(x + ct)^2]}{c + v} + \frac{\exp[-k(x - ct)^2]}{c - v}\right)\right], \tag{8}$$

(where $H(t)$ is the Heaviside function), with that of a block of negative amplitude,$-A$, starting its motion at the time $t = T$ and at abscissa $x = L = v \cdot T$,

$$\eta_2(x,t) = -AH(\tau)\left[\frac{v^2}{v^2 - c^2}\exp[-k(\xi - v\tau)^2] + \frac{v^2}{2c}\left(\frac{\exp[-k(\xi + c\tau)^2]}{c + v} + \frac{\exp[-k(\xi - c\tau)^2]}{c - v}\right)\right], \tag{9}$$

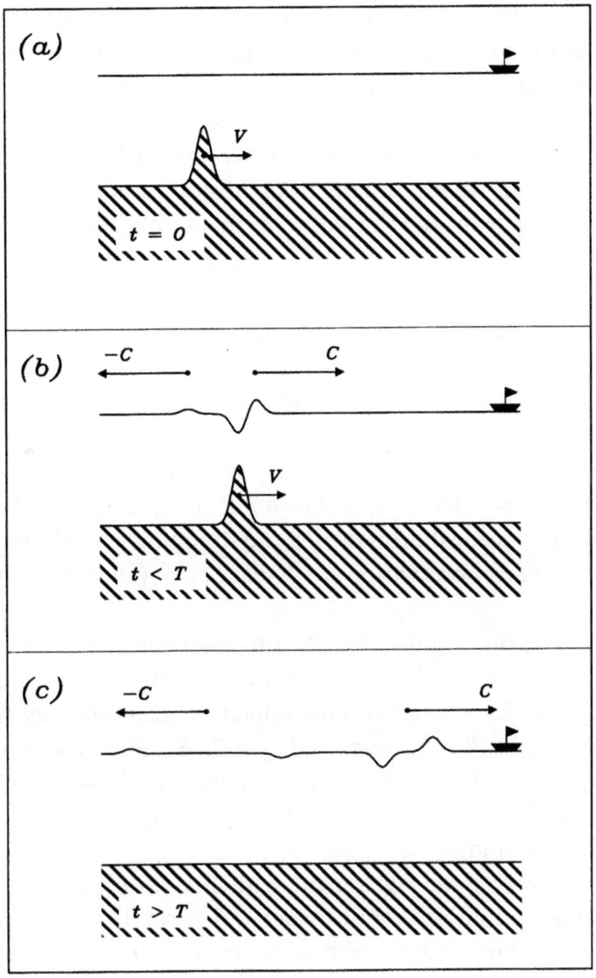

Figure 3
Tsunami wave generated by a bulge moving on the ocean floor at velocity V, illustrated in the subsonic case ($V < C$). The motion starts at $t = 0$ (a) and stops at $t = T$. For $t < T$, (b) represents the solution (8), in which the tsunami wave develops ahead of the deformation, while a smaller wave propagates in the opposite direction. After the motion has stopped (c), a stopping phase is added to the solution (10). The tsunami wave propagating outwards has the structure of a dipole, as does the wave propagating backwards, albeit with a reduced amplitude of the displacement, and an enhanced arm of the dipole. Note that the amplitude of the wave is scaled to the ocean deformation, but not to the thickness of the oceanic layer.

where $\xi = x - L$ and $\tau = t - T$. For $t \geq T$, the combined solution is then

$$\eta(x,t) = \frac{Av^2}{2c}\left[\frac{\exp[-k(x+ct)^2] - \exp[-k(\xi+c\tau)^2]}{c+v} + \frac{\exp[-k(x-ct)^2] - \exp[-k(\xi-c\tau)^2]}{c-v}\right]. \tag{10}$$

The first term in (10) represents the tsunami wave propagating to the left, i.e., upslope of the slump, while the second term propagates downslope.

This simple model suggests an order of magnitude of the effect of the finite duration of the slumping on the scaling of the energy available to the tsunami. With respect to the case of the fast, irreversible deformation generated by a seismic dislocation, the waves propagating upslope and downslope can thus be regarded as emanating from dipolar sources, the amplitude of the "charges" being modulated by

$$\alpha_{\rm up} = \frac{v^2}{2c\,(c+v)}; \quad \alpha_{\rm down} = \frac{v^2}{2c\,(c-v)} \tag{11}$$

or, on the average, $\alpha = v^2/2c^2$. Similar results were obtained by TINTI and BORTOLUCCI (2000).

In trying to apply this result quantitatively to the case of the PNG slump, we take appropriate values $H = 1500$ m ($c = 120$ m/s), an average thickness of the slide $h = 500$ m, and a total vertical drop during the slumping $z = 500$ m. These values are taken from the shipboard observations of TAPPIN *et al.* (1999) and SWEET and SILVER (2003).

We note, however, that, under the effect of buoyancy, the slumping sedimentary material of density ρ_s will be accelerated only by an effective gravity $\gamma = g(\rho_s - \rho_w)/\rho_s$, so that the maximum velocity reached during the slumping will be only $\sqrt{2\gamma z}$ or $v_{\rm max} = 70$ m/s, given $\rho_s/\rho_w \approx 2$. A more typical value of v during the slumping may be $v = 0.5\, v_{\rm max}$ or 35 m/s, leading to $v/c \approx 0.3$. The coefficient α is then about 0.04, leading to a dipole (hump-and-trough) of amplitude 20 m upon termination of the slumping.

This value is in very good agreement with that modeled by SYNOLAKIS *et al.* (2002), who obtained initial peak and trough values of 14 and 16 m respectively. If we further take $L = 5$ km, and a width of 4 km, the surface of the slump (20 km^2) then leads to a total energy $E_T \approx 8 \times 10^{20}$ ergs.

We emphasize once again that this value should not be considered more than an order of magnitude of the energy available to the tsunami, as generated by a crude model of the PNG slump. It is remarkable that it is only 40 times less than that computed above for the 1964 Alaskan earthquake, the second largest event ever recorded; conversely, and in the framework of Equation (5), the PNG energy would be equivalent to that of an earthquake of moment $M_0 = 4.5 \times 10^{28}$ dyn-cm, comparable to such large earthquakes as the 1922 Chilean ($M_0 = 4.2 \times 10^{28}$ dyn-cm), or the 1923 event in Kamchatka ($M_0 = 5.5 \times 10^{28}$ dyn-cm) (OKAL, 1992), both of which created substantial transpacific tsunamis. Why then did the PNG slump raise only negligible tsunamis at teleseismic distances? This question will be examined in the next section.

4. *Far-Field Behavior of Tsunamis Generated by Earthquakes and Slumps*

We recall very briefly in this section the principal results obtained by OKAL (1990), to which the reader is referred for full details. The far-field amplitude of tsunamis generated by either source can be most efficiently studied in the formalism of normal modes. The difference in far-field properties stems from the combination of two factors: the different mechanical nature of the source, and a fundamental difference in its time function.

4.1. Mechanical Nature

It has been now been known for about 45 years that earthquake dislocations can be described by a system of forces known as a double-couple (VVEDENSKAYA, 1956). On the other hand, landslides or slumps can be modeled as single forces, representing the reaction, on the solid Earth, of the acceleration of the sliding mass (KANAMORI *et al.*, 1984). Coefficients for the excitation of a normal mode of the Earth by either a single force or a double-couple can be readily derived theoretically (GILBERT, 1970). In the case of single forces, their expression was given by EISSLER and KANAMORI (1987), and their relation to the double-couple coefficients was discussed by OKAL (1990). In very simple terms, the excitation coefficient of a normal mode whose displacement eigenfunction is $\mathbf{u}$ by a single force $\mathbf{F}$ is proportional to the scalar product $\mathbf{F}\cdot\mathbf{u}$, while its excitation by a double-couple $\mathbf{M_0}$ is proportional to $\mathbf{M_0} : \boldsymbol{\varepsilon}$, where $\boldsymbol{\varepsilon}$ is the eigenstrain of the mode, and thus the ratio of the excitability of a given normal mode by a single force and a double-couple is proportional to its equivalent wavelength $\Lambda = 2\pi a/(l + 1/2) = 2\pi c/\omega$, where a is the radius of the Earth and l the mode's orbital number:

$$X_{SF/DC} = b\,\frac{F}{M_0}\Lambda, \tag{12}$$

where b is a coefficient varying from 1.25 in the case of Love waves, to 2.65 for Rayleigh waves and 6 for tsunamis. The physical meaning of Equation (12) is that a single force, being the spatial integral of a double-couple, should excite a given mode (or wave) more efficiently at lower (spatial or temporal) frequencies, and hence would be expected to be a better tsunami generator than a double-couple, as compared for example to their respective excitation of Rayleigh waves. However, this preliminary interpretation assumes comparable source time functions, and thus grossly overlooks the fundamentally different time histories of earthquakes and landslides.

4.2. Source Time Function

In the case of an earthquake dislocation, the time history of the seismic moment release, $M_0(t)$, coincides with that of the seismic displacement on the fault, $\Delta u(t)$. At long enough periods, it behaves fundamentally like a Heaviside function $H(t)$. By contrast, the time history of a single force $F(t)$ representative of a landslide must satisfy a condition of zero impulse, necessary to keep the closed Earth system globally unaccelerated. An appropriate source time function for $F(t)$ would be the full period of a sinusoid, expressing the acceleration of the slumping material, followed by its deceleration and eventual stop (HASEGAWA and KANAMORI, 1987). The source time function of the single force thus appears fundamentally as the second derivative of that of the double-couple, which will bring an additional factor of order ω^2 to the ratio (12). As a result, the ratio will behave like ω, rather than $1/\omega$, at low frequencies.

In conclusion, a landslide modeled as a single force with a realistic source time function will be a deficient generator of low-frequency energy in the far-field, and in particular of tsunamis, when compared with a dislocation modeled by the familiar step-function double-couple.

5. *Conclusion*

By using very simple models of dislocations and slumps on the ocean floor, as well as theoretical results from normal mode theory, we obtain two fundamental results.

First, the energy of the tsunami generated by an underwater slump of the dimensions of the PNG event is on the same order as that expected from a seismic dislocation of moment 4×10^{28} dyn-cm. That this is at all possible while the slump clearly involves a smaller volume of material illustrates the dependence of E_T on $(\delta h)^2$: the slump moves less material, but it moves it vertically 100 times as much as the comparable earthquake. Because the PNG slump was, at 4 km^3, of relatively modest size, these results confirm that large slumps, involving hundreds if not thousands of km^3, are potentially catastrophic tsunami generators, on a scale dwarfing the 1998 PNG disaster, itself the deadliest tsunami worldwide in 65 years.

Second, the slumping source is fundamentally of a dipolar nature, as compared to the monopolar seismic dislocation. This dipolar nature is evident from local modeling, but can be reconstructed in the formalism of sources represented by combinations of forces, once proper attention is given to the behavior of the relevant source time functions. As a result, i.e., at distances greater than a few times the arm of dipole, the tsunami radiated by a slumping event is expected to decay faster in the far-field than that of a dislocation source. This will occur as soon as the epicentral distance is one order of magnitude greater than the arm of the dipole, itself controlled by the total dimension L of the slump. In addition, this deficiency will be emphasized at low frequencies, and hence will be stronger than for regular seismic waves.

The combination of these two results—tsunami energy comparable to that of a major earthquake, but deficiency in the far-field—clearly requires a concentration of exceptionally large tsunami amplitudes in the near-field. This is exactly why the slumping model is required to account for the observed characteristics of the PNG event, namely exceptionally high runup on the local coast, and minimal if at all detectable amplitudes at transpacific distances.

Acknowledgements

The presentation of this research at the March 2000 workshop on tsunamis and landslides in Los Angeles was supported by the National Science Foundation.

References

AIDA, I. (1977), *Simulations of Large Tsunamis Occurring in the Past off the Coast of the Sanriku District*, Bull. Earthq. Res. Inst. Tokyo Univ. *52*, 71–101 (in Japanese).

EISSLER, H. K. and KANAMORI, H. (1987), *A Single-force Model for the 1975 Kalapana, Hawaii Earthquake*, J. Geophys. Res. *92*, 4827–4836.

GELLER, R. J. (1976), *Scaling Relation for Earthquake Source Parameters and Magnitudes*, Bull. Seismol. Soc. Am. *66*, 1501–1523.

GILBERT, J. F. (1970), *Excitation of the Normal Modes of the Earth by Earthquake Sources*, Geophys. J. Roy. Astr. Soc. *22*, 223–226.

HASEGAWA, H. S. and KANAMORI, H. (1987), *Source Mechanism of the Magnitude 7.2 Grand Banks Earthquake of November 18, 1929: Double-couple or Submarine Landslide?*, Bull. Seismol. Soc. Am. *77*, 1984–2004.

HOUTZ, R. E. (1962), *The 1953 Suva Earthquake and Tsunami*, Bull. Seismol. Soc. Am. *52*, 1–12.

KAJIURA, K. (1981), *Tsunami Energy in Relation to Parameters of the Earthquake Fault Model*, Bull. Earthq. Res. Inst. Tokyo Univ. *56*, 415–440.

KANAMORI, H. (1970), *The Alaska Earthquake of 1964: Radiation of Long-Period Surface Waves and Source Mechanism*, J. Geophys. Res. *75*, 5027–5040.

KANAMORI, H., GIVEN, J. W., and LAY, T. (1984), *Analysis of Seismic Waves Excited by the Mount St. Helens Eruption of May 18, 1980*, J. Geophys. Res. *89*, 1856–1866.

OKAL, E. A. (1982), *Mode-wave Equivalence and Other Asymptotic Problems in Tsunami Theory*, Phys. Earth Planet. Inter. *30*, 1–11.

OKAL, E. A. (1988), *Seismic Parameters Controlling Far-field Tsunami Amplitudes: A Review*, Natural Hazards, *1*, 67–96.

OKAL, E. A. (1990), *Single Forces and Double-couples: A Theoretical Review of their Relative Efficiency for the Excitation of Seismic and Tsunami Waves*, J. Phys. Earth *38*, 445–474.

OKAL, E. A. (1991), *Erratum [to Seismic Parameters Controlling Far-field Tsunami Amplitudes: A review]*, Natural Hazards *4*, 433.

OKAL, E. A. (1992), *Use of the Mantle Magnitude M_m for the Reassessment of the Seismic Moment of Historical Earthquakes. I: Shallow Events*, Pure Appl. Geophys. *139*, 17–57.

OKAL, E. A. (2003a), *T Waves from the Papua New Guinea Sequence and its Aftershocks: Timing the Tsunamigenic Slump*, Pure Appl. Geophys., *160*, 1843–1863.

OKAL, E. A. (2003b). *Normal Modes Energetics for Far-field Tsunamis Generated by Dislocations and Landslides*, Pure Appl. Geophys. *160*, 2189–2221.

SWEET, S. and SILVER, E. A. (2003), *Tectonics and Slumping in the Source Region of the 1998 PNG Tsunami from Seismic Reflection Images*, Pure Appl. Geophys. *160*, 1945–1968.

SYNOLAKIS, C. E., BARDET, J.-P., BORRERO, J. C., DAVIES, H. L., OKAL, E.A., SILVER, E. A., SWEET, S., and TAPPIN, D.R. (2002), *The Slump Origin of the 1998 Papua New Guinea Tsunami*, Proc. Roy. Soc. London, Ser. A *458*, 763–789.

TAPPIN, D. R. and 18 co-authors (1999), *Sediment Slump Likely Caused 1998 Papua New Guinea Tsunami*, EOS,Trans. Am. Geophys. Un. *80*, pp. 329, 334 and 340.

TINTI, S. and BORTOLUCCI, E. (2000), *Energy of Water Waves Induced by Submarine Landslides*, Pure Appl. Geophys. *157*, 281–318.

VVEDENSKAYA, A. V. (1956), *Determination of Displacements Fields by Means of Dislocation Theory*, Izv. Akad. Nauk SSSR, Ser. Geofiz. *6*, 227–284.

WARD, S. N. (1980), *Relation of Tsunami Generation and an Earthquake Source*, J. Phys. Earth *28*, 441–474.

Pure appl. geophys. 160 (2003) 2189–2221
0033–4553/03/112189–33
DOI 10.1007/s00024-003-2426-9

Pure and Applied Geophysics

Normal Mode Energetics for Far-field Tsunamis Generated by Dislocations and Landslides

EMILE A. OKAL[1]

Abstract—We apply the normal mode representation of tsunami waves, as introduced by WARD (1980) to the systematic study of the excitation of far-field tsunamis by both dislocation sources (represented by double-couples of moment M_0), and landslides (represented by single forces). Using asymptotic representations of the continuation of the tsunami eigenfunction into the solid Earth, we derive analytical expressions of the spectral amplitude generated by both systems. We show that the quadrupolar corrections defined by DAHLEN (1993) in the case of landslides can result in an increase of 1 to 2 orders of magnitude of the effective force. Even so, the spectrum of tsunami waves generated by landslides is found to be offset significantly to relatively high frequencies (10 mHz), where dispersion becomes important and eventually diminishes time-domain amplitudes. We proceed to calculate the total energy delivered into the tsunami modes by integrating the energy of multiplets for an average source geometry. In the case of dislocation sources, and taking into account the corner frequency of the source, we reproduce the scaling with $M_0^{4/3}$ which was derived from purely static arguments by KAJIURA (1981). We compare the directivity patterns of far-field tsunami waves by dislocations and landslides, and conclude that the latter cannot give rise to pronounced lobes of directivity for physically acceptable values of the velocity of the slide. Directivity thus constitutes a robust discriminant of the nature of the source which, when applied to the 1946 Aleutian tsunami in the far-field, requires generation by a dislocative source.

Key words: Tsumani theory, normal modes, landslides, directivity.

1. Introduction and Background

The catastrophic 1998 local tsunami in Papua New Guinea [PNG] featured characteristics which could not be modeled with a classical dislocation source, namely an excessive value of runup on the nearby coast, relative to the inferred slip on the fault plane, a peaked lateral distribution of the runup along the coast, and a significant delay of the wave at the shore. These properties suggested that the PNG tsunami was generated by an offshore underwater landslide, itself triggered with a 13-mn delay by the earthquake (e.g., SYNOLAKIS *et al.*, 2002), a model also supported by later mapping of a fresh slide using seismic reflection techniques (SWEET and SILVER, 2003).

The identification of a landslide as the source of the PNG event has led to a renewal of interest in the topic of tsunamigenic landslides, as expressed for example

[1] Department of Geological Sciences, Northwestern University, Evanston, IL 60208, U.S.A.

in this topical issue. Indeed, long before the physical representation of the earthquake source as a double-couple was first proposed by VVEDENSKAYA (1956), and the first seismic moment tensor computed from seismic waves by AKI (1966), a number of visionary scientists such as MILNE (1898), MONTESSUS DE BALLORE (1907) and GUTENBERG (1939) had proposed that large tsunamis could be, or even had to be, generated by underwater landslides. The successful modeling of the gigantic events of the 1960s and of their powerful tsunamis with double-couple sources (e.g., BEN-MENAHEM and ROSENMAN, 1972) temporarily eclipsed the importance of landslides as tsunami generators. Then, direct observation of a major aerial landslide during the 1980 explosion of Mount St. Helens led KANAMORI *et al.* (1984) to successfully model its seismic waves using single forces, and later HASEGAWA and KANAMORI (1987) and EISSLER and KANAMORI (1987) to apply this concept to the submarine cases of the 1929 Grand Banks and 1975 Hawaii earthquakes, which had generated locally destructive tsunamis. KANAMORI (1985) also speculated that the source of the 1946 Aleutian tsunami might have involved a landslide.

In this framework, it is important to assess theoretically the properties of tsunamis generated by both kinds of sources – seismic dislocations and landslides. In particular, it is crucial to understand which observable characteristics of tsunamis can be used as robust discriminants to identify the physical nature of the source.

In this study, we focus on far-field tsunamis, typically recorded at distances of several thousand km, and consider theoretically such properties as the frequency content of their amplitude spectrum, the total energy released into the tsunami wave, the azimuthal directivity of the far-field amplitude, and the scaling of such characteristics with source size. For this purpose, we use the normal mode formalism for tsunami waves, as described by WARD (1980, 1981, 1982), and further discussed by OKAL (1982a, 1988, 1990). It considers tsunamis as a special branch of spheroidal eigenmodes of a self-gravitating elastic Earth whose outermost shell is an ocean of depth h. The principal advantage of this approach is that it greatly simplifies the study of the coupling between the solid Earth and the water layer, i.e., the penetration of the tsunami eigenfunction into a substratum of finite rigidity, an approach originally found in POD"YAPOL'SKII (1970), and pursued by ALEKSEEV and GUSYAKOV (1976). It also allows the immediate computation of accurate estimates of the spectral amplitudes and energies excited into individual modes by any combination of forces, by using the classical formalism of normal mode excitation, introduced by GILBERT (1970). Finally, normal mode theory lends itself easily to the comparison of surface waves and tsunamis excited by the same seismic source.

This approach is also satisfying when considering that the alternative has traditionally involved two separate calculations utilizing inconsistent models of the ocean floor. The first step uses a homogeneous half-space of finite rigidity, but featuring no layering representative of any potential sedimentary structures and no overlying ocean, and computes the static surface displacement resulting from the earthquake dislocation, using such codes as MANSINHA and SMYLIE'S (1971) or OKADA's (1985); the second step then feeds the resulting field of vertical displacements as an initial condition into numerical codes solving the propagation of the tsunami of a water layer overlying an infinitely rigid substratum.

On the other hand, the normal mode formalism has several limitations, at least in the elementary form considered in the present paper. The most important one is that it assumes a laterally homogeneous structure, and as such, an ocean of constant depth. Similarly, it cannot handle the presence of continents, and in particular of a shoreline in the vicinity of the source. Finally, it uses the asymptotic expansion of Legendre polynomials, and as such is applicable only in the far field, i.e., at distances from the source of at least several wavelengths, or a few hundred km in the case of typical tsunamis. Nevertheless, within the framework of these limitations, it can give considerable insight into many properties of tsunami waves (WARD, 1980), and in particular it allows direct comparison with those of the familiar Rayleigh modes (OKAL, 1990), to which we will also refer regularly in the course of this study.

2. *Ingredients: Modes and Sources*

The Tsunami Eigenfunction in the Normal Mode Formalism

In the normal mode formalism, the tsunami wave is simply a special case of the general six-dimensional eigenfunction of a spheroidal free oscillation (of angular degree l), as described by SAITO (1967) whose notation we shall adopt. However, considerable insight can be obtained from the classical tsunami theory derived in the case of a flat-layered ocean overlying a perfectly rigid substratum; the mode-wave equivalence between the two approaches was explored by OKAL (1982a). In this particular scenario, the vertical displacement y_1 of the water varies linearly from 0 at the bottom to Y at the surface, $y_1 = Y(1 - z/h)$, the horizontal displacement eigenfunction, ly_3, is constant throughout the column, $ly_3 = Y/(hk)$, and the normal stress y_2 (the opposite of the overpressure) grows linearly from surface to bottom, $y_2 = -Y\rho_w gz/h$. In these formulae, h is the thickness of the water column, the depth coordinate z varies from 0 at the surface to h at the bottom, k is the horizontal wave number, ρ_w the density of water, and g the acceleration of gravity.

In the presence of a solid substratum of finite rigidity μ, OKAL (1988) derived the shallow structure of the eigenfunction in the solid Earth by showing that it behaves as a "pseudo-Rayleigh" wave, i.e., a combination of inhomogeneous P and S waves whose phase velocity C remains controlled by the water column and essentially unperturbed from the rigid case, while matching the overpressure $-y_2$ at the ocean floor, which is also found to be basically unchanged from the rigid case. The displacement y_1 at the ocean floor is then controlled by the impedance of the solid at the angular frequency ω,

$$Z = \frac{y_2}{y_1} = -\mu k \frac{(2-\kappa^2)^2 - 4\sqrt{(1-\kappa^2)(1-\kappa^2/3)}}{\kappa^2\sqrt{1-\kappa^2/3}} , \tag{1}$$

where $k = \omega/C = (l + 1/2)/a$ is the horizontal wavenumber, and $\kappa = C/\beta$ is always much smaller than 1; β is the shear velocity of the solid, whose Poisson ratio has been taken as 1/4. In the limit $\kappa \to 0$, one has $Z \to (4/3)\mu k$ (OKAL, 1988). The vertical displacement at the ocean floor, y_1, is then given by $-Y\rho_w g/Z$, which remains much smaller than the surface displacement Y at all periods less than several hours. (By contrast, in the case of a Rayleigh mode, the boundary condition $y_2 = 0$ at the surface imposes $Z = 0$, which in turn controls the phase velocity C through the classical result $\kappa = \sqrt{2 - 2/\sqrt{3}}$.)

In essence, the weak coupling between the water and solid layers dictates the value of C (or κ), and thus Z in (1), which in turns specifies the relative amplitude of the P and S components of the pseudo-Rayleigh wave, making it easy to compute its entire structure in the substratum, including any excitation coefficients by any combination of forces (OKAL, 1988, 1990).

The Energy Integrals I_1 and I_2

Several applications such as the evaluation of the excitation coefficients of the various modes require their normalization to unit energy through the computation of the energy integrals (KANAMORI and CIPAR, 1974).

$$I_1 = \int_0^a \rho r^2 y_1^2 dr \quad \text{and} \quad I_2 = \int_0^a \rho r^2 y_3^2 dr . \tag{2}$$

For Rayleigh waves, and in the limit of a wavelength much smaller than the Earth's radius ($l \gg 1$), one can approximate these integrals as:

$$I_1 = \int_0^\infty \rho_s a^2 u_z^2 dz \quad \text{and} \quad L^2 I_2 = \int_0^\infty \rho_s a^2 u_x^2 dz , \tag{3}$$

where a is the Earth's radius, and u_z and u_x are the vertical and horizontal components of the displacement field of the standard Rayleigh wave of a homogeneous half-space of density ρ_s. It is straightforward to derive from (3)

$$I_1 + L^2 I_2 = 2^{1/2} 3^{3/4} Y^2 \frac{C}{\omega} \rho_s a^2 , \tag{4}$$

where $L^2 = l(l+1)$. We have verified that (4) provides an acceptable approximation to the exact energy integrals; for example, for the mode ${}_0S_{100}(T = 97.25\,\text{s})$, it predicts ($2.75 \times 10^{25}$ g*cm^2 when using $\rho_s = 3.3$ g/cm^3, and for $Y = 1$ cm) as compared to 2.5×10^{25} g*cm^2 obtained from (2) using a full normal mode eigensolution derived for the stratified spherical PREM model (DZIEWONSKI and ANDERSON, 1981). Given the crude assumptions in our model (e.g., the absence of layering, and the extension of the integrals (3) to infinity), the agreement is excellent and justifies our methodology. *For tsunami modes*, on the other hand, the integrals in (2) can be limited in practice to the water layer, where the horizontal motion is dominant in the long-wavelength approximation ($a/lh \gg 1$), leading to

$$I_1 + L^2 I_2 \approx L^2 I_2 = \rho_w \frac{a^4}{L^2 h} Y^2 = \rho_w \frac{a^2 \cdot g}{\omega^2} Y^2 \ . \tag{5}$$

The difference in behavior with frequency of the energy in the two types of modes (ω^{-1} for Rayleigh waves, as opposed to ω^{-2} for tsunamis) is fundamental, and has important consequences on all aspects of spectral excitation by both dislocations and landslide sources.

Sources: Modeling of a Dislocation

Earthquake dislocations are modeled as double-couples of moment M_0, whose excitation of any Earth normal mode has been known since the seminal work of GILBERT (1970). We express it here for spheroidal modes in the formalism of KANAMORI and CIPAR (1974), which gives the component of vertical displacement at angular distance Δ as

$$u_r = y_1(a) M_0 \left[K_0 s_R P_l^0(\cos\Delta) - K_1 q_R P_l^1(\cos\Delta) + K_2 p_R P_l^2(\cos\Delta) \right] \cdot \cos(\omega t) \ , \tag{6}$$

which is equivalent to a spectral amplitude in the far field (KANAMORI and STEWART, 1976)

$$|u_r(\omega)| = M_0 \frac{y_1(a)}{\sqrt{\sin\Delta}} \sqrt{\frac{\pi}{2l} \frac{a}{U}} \left| s_R K_0 - p_R l^2 K_2 - i q_R l K_1 \right| \ . \tag{7}$$

In these formulae, a is the Earth's radius, ω its angular frequency, and U the equivalent group velocity ($U = a \cdot d\omega/dl$); P_l^m is the Legendre polynomial of degree l and order m. We refer to KANAMORI and CIPAR (1974) and KANAMORI and STEWART (1976) for the exact expressions of the excitation coefficients K_i (which depend only on frequency and source depth) and of the geometric coefficients p_R, q_R, s_R which depend only on the orientation of the dislocation relative to the Earth's surface and to the azimuth of the receiver.

As the size of the earthquake increases, source finiteness results in the saturation of spectral amplitude at any given frequency. This can be schematized across the frequency band by introducing a source spectrum characterized by a corner frequency ω_c below which interference effects are absent, and above which the source spectrum falls off as ω^{-3} (GELLER, 1976). An estimate of the scaling of ω_c with moment M_0 is given in Appendix 1 following GELLER (1976).

Numerical values. For a large earthquake, capable of generating a transpacific tsunami ($M_0 = 5 \times 10^{28}$ dyn-cm), Equations (A9) and (A11) predict corner periods of 49 s and 130 s, for Rayleigh waves and tsunamis, respectively. These values would grow to 166 and 445 s, respectively for a mega-thrust event of the size of the 1960 Chilean earthquake ($M_0 = 2 \times 10^{30}$ dyn-cm). Furthermore, it is interesting to note the influence of the rupture velocity V_R in Equation (A7); for an earthquake with slow rupture, such as the 1992 Nicaragua "tsunami earthquake," V_R could be as slow as 1 km/s (KANAMORI and KIKUCHI, 1993), which would increase the relevant corner periods by a factor of 1.5, to as much as 200 s for the tsunami of a large earthquake (5×10^{28} dyn-cm).

We will also consider an event of the size of the PNG earthquake ($M_0 = 3.7 \times 10^{26}$ dyn-cm), which generated a mediocre far-field tsunami, reaching only decimetric amplitudes at a few Japanese harbors, and unobserved at transpacific distances.

Sources: Modeling of a Landslide

The excitation of seismic waves in the Earth by the motion of a sliding mass at its surface was modeled by KANAMORI *et al.* (1984) in the case of the massive slide accompanying the 1980 eruption of Mount St. Helens, and their model was later applied to the 1929 Grand Banks and 1975 Kalapana underwater events (HASEGAWA and KANAMORI, 1987; EISSLER and KANAMORI, 1987). It considers that the Earth is subject to the reaction of the acceleration of the sliding mass, $-M\ddot{\mathbf{x}}$. Such sources can then be represented using single forces $\mathbf{F}$ as opposed to moment tensors $\mathbf{M_0}$. However, as discussed by OKAL (1990), a major difference arises, since the source must satisfy a condition of zero impulse, which brings an additional factor ω^2 into the source time function of the landslide at low frequencies. The high-frequency behavior of a landslide is difficult to describe, since it hinges on the particular development of the initial phases of sliding, and of the braking phase of the slide at its toe. OKAL (1990) has studied theoretically several models of source time function. We will consider here the function

$$F_0(t) = F_0 \cdot [H(t) - H(t - 2\tau_1)] \cdot \sin \frac{\pi t}{\tau_1} \tag{8}$$

proposed by KANAMORI *et al.* (1984) for the Mount St. Helens eruption. Here τ_1 characterizes the duration of the source, and $H(x)$ is the Heaviside function. Relative to a step function, (8) introduces an additional spectral factor

$$X_d(\omega) = \left|\frac{\dot{F}_0(\omega)}{F_0}\right| = \left|\frac{2\omega\omega_1}{\omega^2 - \omega_1^2} \cdot \sin\frac{\omega\tau_1}{2}\right| \quad (\omega_1 = \pi/\tau_1) \ , \tag{9}$$

which can be approximated by $X_d = \pi \frac{\omega^2}{\omega_1^2}$ (small ω) and $\frac{4}{\pi}\frac{\omega_1}{\omega}$ (large ω). This amounts to defining a *"corner frequency"* ω_c where the low- and high-frequency trends intersect:

$$\omega_c = \omega_1 \left(\frac{4}{\pi^2}\right)^{1/3} = 0.74\omega_1 \ . \tag{10}$$

Note that for other models of acceleration of the slump, the high-frequency spectrum may vary, but the low-frequency behavior will always be $\leq \omega^2$, in order to satisfy the zero-impulse condition.

In addition, DAHLEN (1993) has shown that the above description of a shallow landslide by a single force acting on an unperturbed elastic medium (which he termed the "Kanamori force, F_K") neglects the influence of the development of discontinuities at the ends of the sliding block, which are equivalent to a series of three quadrupolar terms. (Note a typographic error in DAHLEN's (1993) Figure 5: in the depiction of the surface quadrupoles, the first two should read " o • • o " and " $\begin{smallmatrix} \text{o} \\ \bullet \\ \bullet \\ \text{o} \end{smallmatrix}$ ", respectively.) In the formalism of OKAL (1982b), these terms contribute three second-order (three-index) moment components: $M_{\theta\theta\theta} = \frac{8}{3}\beta_0^2 MD$; $M_{\theta\phi\phi} = \beta_0^2 MD$; and $M_{\phi\theta\phi} = \frac{5}{3}\beta_0^2 MD$, with the radial axis $\hat{\mathbf{u}}_\theta$ oriented along the direction of sliding, $\hat{\mathbf{s}}$ in Dahlen's notation. Using the full set of three-index excitation coefficients derived in Equations (18) and (46) of OKAL (1982b), it is easy to verify that DAHLEN's (1993) simple correction to the Kanamori force F_K, namely the factor $f_D = \left(1 - \frac{8}{3}\frac{\beta_0^2}{C^2}\right)$, holds for tsunami modes in the limit $l \gg 1$. In this formalism, M is the mass of the sliding block, β_0 the shear velocity inside it, D the total displacement during the slide, and C the phase velocity of the relevant mode. Equation (8) can then be used to represent the landslide, with $F_0 = F_K \cdot f_D$.

A significant difference between tsunami and Rayleigh modes then arises from the difference in phase velocity C. DAHLEN (1993) was able to explain the general success of the Kanamori force in modeling such events as the Mount St. Helens landslide (for which there exists a precisely witnessed estimate of its kinematics), by noting that sliding structures are bound to be strongly brecciated, and therefore to feature low shear velocities for which the correction $\frac{8}{3}\frac{\beta_0^2}{C^2}$ becomes negligible. For example, if β_0 falls below 1 km/s, the Dahlen correction $(f_D - 1)$ becomes only -11% for a long-period Rayleigh wave with $C = 5$ km/s. However, for a tsunami mode with $C = 220$ m/s, the same material would entice $f_D = -55$, while the cohesive slumping of a crustal block ($\beta_0 = 3.5$ km/s) would lead to $f_D = -675$; note in particular that the negative values of f_D would result in a flip of phase. In real life, and for an underwater landslide, the situation is made more complex by the possible evolution of the slide into a turbidity current, in which case the value of β_0 could be regarded as evolving with time from a finite value to essentially zero.

Numerical estimates. Given the small number of well-studied cases of underwater landslides as sources of tsunamis, it is difficult to estimate typical values of the parameters F_K, τ_1, and f_D, let alone of their scaling with the overall size of the slide. For the 1929 Grand Banks slide, HASEGAWA and KANAMORI (1987) suggested $F_K = 1.4 \times 10^{20}$ dyn, but found little resolution of the duration τ_1 of the source, which they estimated at between 25 and 90 s. For the 1975 Kalapana event, ANDO (1979) and EISSLER and KANAMORI (1987) suggested a volume $V \approx 1000\,\text{km}^3$ and a characteristic time $\tau_1 \approx 90$ s. Despite its locally catastrophic tsunami, the 1998 PNG event was considerably smaller, with $V \approx 4\ \text{km}^3$ and $\tau_1 = 22$ s (SYNOLAKIS *et al.*, 2002; SWEET and SILVER, 2003).

In this framework, and for the purpose of comparing the order of magnitude of various characteristics of tsunamis generated by dislocations and landslides, we will consider a reference landslide with a volume $V = 500\ \text{km}^3$, $\tau_1 = 60$ s, and use a Dahlen correction $f_D = -50$ for the tsunami mode ($f_D = 0.9$ for a long-period Rayleigh mode). On an ocean floor sloping 10°, these values lead to $F_K = 2.2 \times 10^{20}$ dyn (note (DAHLEN, 1993) that the true density of the sliding material, *uncorrected for buoyancy*, should be used in the excitation calculations), and $F_0 = 1.1 \times 10^{22}$ dyn for the tsunami modes; these numbers may be applicable to the underwater landslide suggested as a source of the catastrophic near-field runup of the 1946 Aleutian tsunami (FRYER and WATTS, 2000; PLAFKER *et al.*, 2002). We also consider the slide discussed theoretically by OKAL and SYNOLAKIS (2003), which is inspired by the PNG event (albeit larger). It features a considerably smaller volume than our reference slide ($10\,\text{km}^3$; $F_K = 6.8 \times 10^{18}$ dyn), but we use a larger value of the correction, $f_D = -200$, the slide having been contained in the amphitheater, and thus presumably kept a greater level of cohesion. For that smaller, "PNG-type" slide, we use a shorter characteristic time, $\tau_1 = 22$ s.

3. *Average Excitation and Spectral Amplitudes*

Excitation by a Dislocation

The excitation of a normal mode (seismic or tsunami) by a shallow dislocation will in general depend on the geometry of the source, but an average value can be estimated as follows. The various coefficients K_i are expressed as combinations of the components of the mode's eigenfunction (KANAMORI and CIPAR, 1974). For example,

$$K_2 = \frac{2l+1}{4\pi\omega^2[I_1 + L^2 I_2]} \cdot \frac{y_3(r_s)}{r_s} , \tag{11}$$

where r_s is the distance from the seismic source to the Earth's center.

Rayleigh Waves

In the limit of a shallow source ($r_s \to a$), using (4) and the classical surface aspect ratio of the standard Rayleigh wave, $ly_3 = -0.68Y$, it is straightforward to obtain

$$K_2 \approx -0.034 \cdot \frac{Y^{-1}}{a^2 \rho_s} \cdot \frac{1}{C^2 l} \approx -\frac{0.00246}{C^2 l} , \tag{12}$$

if C is in km/s, K_2 in units of 10^{-27} dyn^{-1}, and for a standard normalization ($Y = 1$ cm). We have verified that (12) provides an acceptable estimate of K_2: sub-Moho values are computed at -2.43×10^{-6} and -1.46×10^{-6}, for ${}_0S_{50}$ and ${}_0S_{100}$, respectively, as compared to -3.04×10^{-6} and -1.50×10^{-6} in the PREM model (DZIEWONSKI and ANDERSON, 1981).

The pure dip-slip coefficient, K_1, which is proportional to the eigenstress component y_4 of the mode, vanishes at the surface, and hence can be taken as 0 in Equation (6). Finally, the coefficient K_0 behaves as $-5/3 \cdot L^2 K_2$ for the Rayleigh waves of a shallow source (OKAL, 1988). We can then assume an average value of 0.46 for the trigonometric function $|(5/3)s_R + p_R|$, take $C = 4.5$ km/s and $U = 3.8$ km/s on average, to obtain an asymptotic expression of the spectral amplitude

$$|u_r(\omega)| = 0.118 l^{1/2} \frac{M_0}{\sqrt{\sin \Delta}} , \tag{13}$$

where u_r is in cm*s and M_0 in 10^{27} dyn-cm. It predicts values of 1.74 cm*s at $T = 40$ s ($l = 222$) and 0.70 cm*s at $T = 250$ s ($l = 36$), for a moment of 10^{27} dyn-cm, observed at $\Delta = 90°$ in the absence of attenuation. These results are in general agreement with those obtained from the formalism of the mantle magnitude M_m (OKAL and TALANDIER, 1989) (2.04 cm*s and 0.60 cm*s, respectively), which indicates that our approximations provide adequate orders of magnitude of the scaling of spectral amplitudes with moment, despite the drastic simplifications used in the present study. Note in particular that we use constant values for C and U, in view of the fact that in the real Earth, they are slowly varying parameters in the range of frequencies considered (typically 2 to 10 mHz); we also neglect the influence that crustal layering would have on excitation coefficients.

Tsunami Modes

In the case of a tsunami mode, we use $Z = (4/3)\mu k$ and the structure of the pseudo-Rayleigh wave in the solid to infer $ly_3/y_1 \to -1/3$ at the top of the solid (OKAL, 1988 and *erratum*, 1991). Recalling (11) and (5), we obtain

$$K_2 \approx \frac{1}{8\pi\mu}\frac{Y^{-1}}{la^2}; \quad K_0 \approx L^2 K_2 \ . \tag{14}$$

We then consider a mantle source ($\mu = 7 \times 10^{11}$ dyn/cm^2), an ocean of depth $h = 5$ km ($C = U = 220$ m/s), and use an average value of 0.37 for the trigonometric coefficient $|s_R - p_R|$ to obtain

$$|u_r(\omega)| = 1.49 l^{1/2} \frac{M_0}{\sqrt{\sin \Delta}} \ , \tag{15}$$

which predicts spectral amplitudes of 28 cm*s and 15 cm*s at periods of 500 s ($l = 362$) and 1800 s ($l = 101$), respectively, for $M_0 = 10^{27}$ dyn-cm.

At this stage, an important conclusion is that the spectral amplitudes of Rayleigh waves and tsunamis excited by dislocations both grow with frequency as $\omega^{1/2}$ (under the assumption of a constant phase velocity). The difference in behavior of the energy integrals (4) and (5) is exactly compensated by the variable ratio, decaying like ω in the tsunami mode, between the horizontal motion in the solid, ly_3, and the surface amplitude of the wave, Y. Beyond the corner frequencies, the spectral amplitudes would decay as $\omega^{-5/2}$, as sketched on Figure 1. Note that this high-frequency behavior would be characteristic of a receiver feeling the full effect of saturation, i.e., lying *outside* the lobes of the directivity pattern; a slower fall-off with frequency beyond ω_c would be expected inside the lobes. Finally, we stress that, by using constant values of $C = U = \sqrt{gh}$, we neglect the effects of dispersion inherent at high frequencies in the exact expression, $C^2 = \frac{g}{k}\tanh(kh)$.

Excitation by a Landslide

As discussed for example by EISSLER and KANAMORI (1987) and OKAL (1990), the excitation of normal modes by single forces is expressed by an equation similar to (6)

$$u_r = y_1(a) F_0 \left[K_0^{SF} s_F P_l^0(\cos \Delta) + K_1^{SF} q_F P_l^1(\cos \Delta) \right] \cdot \cos(\omega t) \ , \tag{16}$$

where the coefficients K_i^{SF} are directly related to the K_j; for example, $K_1^{SF} = -r_s K_2$. For a sub-horizontal force, the first term in the bracket can be neglected, leading to

$$|u_r(\omega)| = F_0 \frac{y_1(a)}{\sqrt{\sin \Delta}} \sqrt{\frac{\pi}{2l}\frac{a}{U}} |q_F l K_1^{SF}| \tag{17}$$

or, on the average, given $\langle |q_F| \rangle = 2/\pi$,

$$|u_r(\omega)| = F_0 \frac{y_1(a)}{\sqrt{\sin \Delta}} \sqrt{\frac{2}{\pi l}\frac{a r_s}{U}} l |K_2| . \tag{18}$$

In the case of Rayleigh waves, we find

$$|u_r(\omega)| = 0.163 l^{-1/2} \frac{F_0 \cdot a}{\sqrt{\sin \Delta}} \ , \tag{19}$$

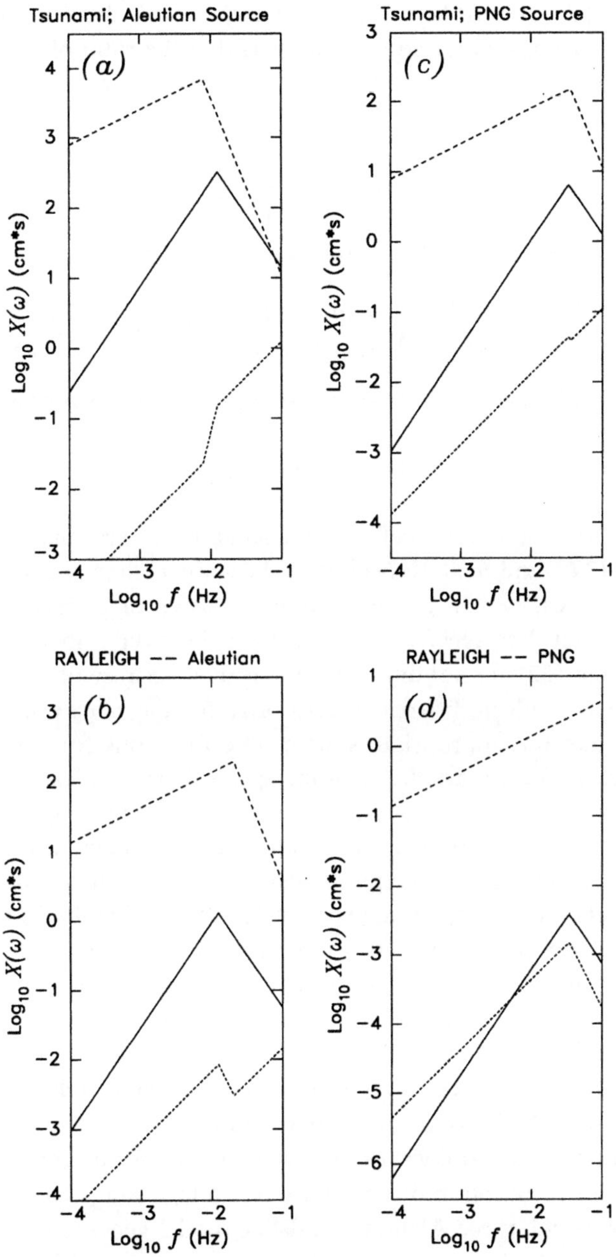

Figure 1

Comparison of the spectral amplitudes excited by simulated landslide sources (solid lines) and dislocations (dashed lines). The dotted lines are the ratios between the two. All scales are logarithmic. In the left frames, we consider the possible sources of the large 1946 Aleutian tsunami. The tsunami spectral amplitudes are shown in (*a*), the Rayleigh wave spectra in (*b*); In the right frames, we consider the PNG-type landslide source, and compare it to the dislocation source inverted by DZIEWONSKI *et al.* (1999) for the 1998 PNG earthquake. (*c*): tsunami waves; (*d*): Rayleigh waves.

where the product $F_0 \cdot a$ must be expressed in units of 10^{27} dyn*cm to obtain $u_r(\omega)$ in cm*s. The comparison of (13) and (19) suggests that the ratio of spectral amplitudes excited by a single force and a double-couple (at this stage with identical source time functions, arguably an unphysical situation) is

$$R_{SF/DC} = 1.38 \cdot \frac{1}{l} \cdot \frac{F_0 \cdot a}{M_0} \quad . \tag{20}$$

In the case of a tsunami mode, we would similarly obtain

$$|u_r(\omega)| = \frac{F_0}{\sqrt{\sin \Delta}} \sqrt{\frac{2}{\pi l}} \frac{ar_s}{U} l \frac{1}{8\pi \mu l a^2} = 3.22 l^{-1/2} \frac{F_0 \cdot a}{\sqrt{\sin \Delta}} \tag{21}$$

leading to a ratio of single-force to double-couple excitation for tsunami modes

$$T_{SF/DC} = 2.16 \cdot \frac{1}{l} \cdot \frac{F_0 \cdot a}{M_0} \quad . \tag{22}$$

In (20) and (22), the numerical coefficients 1.38 and 2.16 are smaller than computed in OKAL (1990) (2.65 and 6.00, respectively), because we have restricted ourselves to sub-horizontal forces, which are more representative of a mass movement on a shallow incline, but less efficient at exciting spheroidal modes. Note also that Equations (20) and (22) are not directly comparable, because of the different Dahlen corrections, which result in different values of F_0 for the same event.

We then invoke the appropriate source time functions for the dislocation and landslide sources, and compare the resulting spectral amplitudes on Figure 1. Frame (*a*) plots the spectral amplitudes of tsunami modes excited in the far field by the two kinds of sources proposed to model the 1946 Aleutian tsunami: a large dislocation ($M_0 = 5 \times 10^{28}$ dyn-cm) and a major landslide with a volume of 500 km^3, while Frame (*b*) presents the corresponding spectra for conventional seismic surface waves (Rayleigh). Several important results are immediately apparent:

First and foremost, the far-field tsunami amplitude of the reference landslide remains smaller than that of the dislocation by more than one order of magnitude; at frequencies typical of the observed tsunami (1 mHz), this deficiency reaches more than two orders of magnitude; the landslide would become a relatively more efficient tsunami generator only beyond the corner frequencies ($f \geq 10$ mHz), outside the realm of conventional tsunami waves and where many of our approximations break down, notably our assumption of undispersed propagation (C or U independent of frequency). In particular, the Aleutian landslide model generates a far-field tsunami whose spectral amplitudes are comparable to those of a dislocation of moment 2×10^{26} dyn-cm, a size of earthquake known not to excite detectable far-field tsunamis.

Second, in the case of the landslide, and precisely because the tsunami spectral amplitude is peaked around ω_c (i.e., around 10 mHz), and falls off fast (as $\omega^{3/2}$) at lower frequencies, it suffers significant dispersion when propagating large distances over a deep ocean. Indeed, the dispersion $|dU(\omega)/d\omega|$ is maximum around 150 s for a 5-km deep ocean where it reaches as much as 2 km/rd. It will affect slightly the spectral amplitudes (computed for constant C on Fig. 1), but strongly the spectral phases, resulting in the dispersion of tsunami energy, and a sharp reduction in time domain amplitude. By contrast, the tsunami generated by the dislocation source falls off slower with frequency (as $\omega^{1/2}$), and thus retains significant spectral amplitude at the typical frequencies (1 mHz) where dispersion is absent, thus allowing the retention of a significant time domain amplitude.

The bottom line from these calculations is that the combination of a lesser absolute level of excitation and a more pronounced shift of the spectral energy towards higher frequencies suffering strong dispersion, makes the landslide source a significantly deficient far-field tsunami generator, as compared to a classical dislocation.

In Frame (*b*), we similarly explore the excitation of Rayleigh waves by the two sources. The similarity between Equations (13) and (15), and (19) and (21) results in comparable behavior. However, the lesser value of the Dahlen correction f_D for Rayleigh waves results in an additional deficiency of the landslide source, reaching more than two orders of magnitude across the spectrum considered (typically 5 to 10 mHz).

In the case of the smaller event, inspired from the PNG source and simulated in Frames (*c*) and (*d*), the landslide source remains undetectable in the far field (note the translation of the vertical scale between the left and right frames). Also, for this model of a supposedly more cohesive slide, there is a greater scatter in the Dahlen correction ratios ($f_D = -200$ for the tsunami; 0.5 for Rayleigh waves), as compared to the Aleutian case ($f_D = -50$ for the tsunami; 0.9 for Rayleigh waves). As a result, the relative Tsunami-to-Rayleigh efficiency of the landslide is enhanced by an additional factor of ~ 7. In conclusion, a PNG-type landslide (even taken here as substantially larger than mapped by SWEET and SILVER (2003)) would remain undetectable in the far-field: its tsunami would be 20 times smaller than that of the earthquake; the ratio would reach 300 for its Rayleigh waves. Clearly, the absence of far-field signature cannot be an argument against a landslide model for the PNG local tsunami.

We wish to emphasize that formulae such as (13), (15), (19) and (21), or their plots on Figure 1 after convolution with the respective source time functions, should not be applied in the near field, and this for several reasons. First, the spectral amplitudes in those formulae are calculated using the asymptotic expansions of the Legendre polynomials, which break down as $\Delta \to 0$, in practice within one wavelength of the source. While Equations (6) and (16) remain in principle valid, (7) and (17) do not. Furthermore, propagation in the near field occurs entirely, from source to receiver, over variable bathymetry, a structure that our approach does not handle.

4. *The Energy of Tsunami Modes*

In this section, we address the question of the total energy excited into tsunami modes by dislocation and landslide sources, and of its scaling with the size of the source.

The Dislocation Source

The energy of an oscillating spheroidal singlet S_l^m of the Earth can be readily computed in its kinetic form as

$$E = \frac{1}{2} \int_{\text{Earth}} \rho \dot{\mathbf{u}}^2 \, dV = \frac{1}{2} \omega^2 (I_1 + L^2 I_2) \cdot A(m) \ , \tag{23}$$

where I_1 and I_2 were defined in (2) and the azimuthal integrals $A(m)$ are respectively,

$$A(0) = \frac{4\pi}{2l+1}; \quad A(\pm 1) = \frac{2\pi L^2}{2l+1}; \quad A(\pm 2) = \frac{2\pi L^4}{2l+1}. \tag{24}$$

To obtain the total energy excited into a multiplet, one simply adds that of the singlets (which are orthogonal on the sphere), weighted by the appropriate excitation coefficients.

• In the case of a *Rayleigh mode*, we have

$$E_{\text{multiplet}} = \frac{3^{3/4}}{2^{1/2}} \frac{\omega^2 \rho_s a^3}{l} Y^2 \cdot M_0^2 \left[A(0) c_0^2 K_0^2 + A(-1) c_{-1}^2 L^2 K_1^2 + A(1) c_1^2 L^2 K_1^2 + A(-2) c_{-2}^2 L^4 K_2^2 + A(2) c_2^2 L^4 K_2^2 \right] \ . \tag{25}$$

Here, the coefficients c_i are variations of s_R, q_R and p_R. They do not vary independently, but their average value for all focal mechanisms is on the order of 1/3. In first approximation, we can neglect the $m = \pm 1$ terms in (25), since K_1 vanishes asymptotically at the surface. Noting that $K_0 = -(5/3) L^2 K_2$ at the surface of a Poisson solid (OKAL, 1990), and substituting (12) and (24), we find the energy of a spheroidal multiplet as

$$E_{\text{multiplet}}^{\text{Rayleigh}} = N \frac{1}{a^3 C^2 \rho_s} \cdot L^2 M_0^2 \ , \tag{26}$$

where the non-dimensional constant N has the approximate value 4.8×10^{-3}. This energy of a single multiplet is, as expected (WARD, 1980), proportional to the *square* of the excitation M_0, and for a constant phase velocity C, grows as ω^2.

• In the case of a *Tsunami*, we need to replace the expression of the integrals I_1 and I_2 with (5), substitute $K_0 = L^2 K_2$, and use Equation (14) to approximate K_2 as a function of ω. The result is

$$E_{\text{multiplet}}^{\text{Tsunami}} = \frac{1}{288\pi} \cdot \frac{\rho_w g}{\mu a^2} \cdot l \cdot M_0^2 \ . \tag{27}$$

The comparison of (26) and (27) reaffirms the fundamental difference noted above, namely that the tsunami energy grows as ω, rather than ω^2. In other words, for the same dislocation source, the tsunami branch will be lower-frequency or more "red" than the Rayleigh wave†.

Integration over Frequency

There remains to integrate (26) and (27) over frequency to obtain the total energy carried by the Rayleigh and tsunami waves, replacing summation over l by integration over ω (weighted by a/U), and taking into account the source time function of the dislocation. The total energy for the branch becomes:

$$E = a \cdot \left[\int_0^{\omega_c} \frac{1}{U} E_{\text{multiplet}}(\omega) d\omega \quad + \quad \int_{\omega_c}^{\infty} \frac{1}{U} E_{\text{multiplet}}(\omega) \frac{\omega_c^6}{\omega^6} d\omega \right] \ , \tag{28}$$

where, as a further approximation, the integration bounds are extended to zero and infinite frequencies, on account of the characteristics of the source functions. Assuming that U and C can be taken as constant, we find

$$E^{\text{Rayleigh}} = 3.2 \times 10^{-3} \frac{1}{UC^4 \rho_s} \omega_c^3 M_0^2; \quad E^{\text{Tsunami}} = 8.3 \times 10^{-4} \frac{\rho_w g}{UC\mu^2} \omega_c^2 M_0^2 \ . \tag{29}$$

Note that the legitimacy of carrying the integral (28) to infinite frequency could be questioned, since our simplifying assumptions (e.g., constant C) would clearly break down as $\omega \to \infty$. However, the sharp fall off of the integrand with frequency ensures that the energy is almost entirely concentrated near ω_c, with $\omega < 2.15\omega_c$ contributing 90% of the second term of (28) (95% of the full energy integral), and thus rendering largely irrelevant any limitations on our assumptions at still higher frequencies.

After substituting the expression of the corner frequency (A8), we obtain the final result for Rayleigh waves:

$$E^{\text{Rayleigh}} = 0.055 \frac{1}{\kappa^3 \eta} \varepsilon_{\max} M_0 = 0.084 (\varepsilon_{\max} \cdot M_0) \ , \tag{30}$$

where η is the ratio of the group velocity U of the wave to the shear velocity, which we take as slightly lower than κ ($\eta = 0.85$), and $\varepsilon_{\max}$ the strain release during the rupture (see Appendix). Since ($\varepsilon_{\max} \cdot M_0$) represents the mechanical energy released at

† Note that there is no inconsistency between the fact that the energy in the two types of waves behaves differently with frequency while spectral amplitudes have the same dependence, for the latter is a local property of the eigenfunction at the surface of the Earth, while the former involves its integral over depth.

the source by the dislocation, we come to the simple conclusion that the energy channeled into the fundamental Rayleigh branch scales directly with M_0, and represents a constant fraction, about 8%, of the total seismic energy released by the source.

The situation is quite different in the case of the tsunami wave. Substituting (27), we obtain

$$E^{\mathrm{Tsunami}} = 5.5 \times 10^{-3} \frac{\rho_w}{\rho_s} \cdot \left(\frac{g\rho_s}{h^2 \mu^2}\right)^{1/3} \varepsilon_{\max}^{2/3} \cdot M_0^{4/3} \approx 1.04 \times 10^{-16} M_0^{4/3} \qquad (31)$$

if M_0 is in dyn-cm and E in ergs. For the reference Aleutian event, we find $E^{\mathrm{Rayleigh}} = 4.2 \times 10^{23}$ ergs, and $E^{\mathrm{tsunami}} = 1.9 \times 10^{22}$ ergs, or 0.4% of $(\varepsilon_{\max} \cdot M_0)$. For the smaller PNG event, these numbers would be 3.1×10^{21} ergs, 2.8×10^{19} ergs, and 0.08%, respectively. Note that the integrand in (28) falls off like ω^5 for $\omega \rightarrow \infty$, as opposed to ω^4 for Rayleigh waves. The question of the extension of (28) to infinite frequencies becomes even more irrelevant.

The energy of the tsunami is found to grow faster than M_0, which means that the fraction of released seismic energy which is channeled into the tsunami, itself grows like $M_0^{1/3}$, in other words scales like the linear dimension of the seismic source, L_f. This result agrees with OKAL and SYNOLAKIS' (2003), who used a simple model of the deformation of a rigid ocean floor to infer that the energy available to the tsunami would grow as the fourth power of the scaling dimension of a dislocation. Similarly, the scaling of E with $M_0^{4/3}$ was mentioned by AIDA (1977) and a full model derived by KAJIURA (1981), on the basis of the scaling of the static deformation of a homogeneous half-space by a seismic dislocation. Equation (31) can be directly compared to Kajiura's formula (12)

$$E_{\mathrm{Kaj}} = \frac{1}{2} \frac{\rho_w g}{\mu^2} \alpha^{2/3} M_0^{4/3} F_{\mathrm{Kaj}}.(\delta_f, \lambda, H, R) = \frac{1}{2^{4/3}} \frac{\rho_w g}{\mu^{4/3}} \varepsilon_{\max}^{2/3} M_0^{4/3} F_{\mathrm{Kaj}} \,, \qquad (32)$$

where the function $F_{\mathrm{Kaj}}.$ is a non-dimensional average of the square of the vertical deformation of the ocean bottom, scaled to the seismic slip on the fault, which depends on the geometry of faulting (dip angle δ_f and slip angle λ), on source depth H, and on the aspect ratio of the faulting area, $R = W/L_f$. Finally, in Equation (32), α is the ratio of M_0 to $S^{3/2}$, or $2^{-1/2}\mu\varepsilon_{\max}$ for a fault with $R = 1/2$ (see Appendix, Equation A4). The comparison of Equations (30) and (32) shows that our approach based on normal mode theory upholds KAJIURA's (1981) Equation (12), for an average value $F_{\mathrm{Kaj}}. = 0.57$ in the case of a 5-km deep ocean. However, this value is about 5 times larger than its maximum proposed by Kajiura (0.11). An additional shortcoming is that the expression of E^{Tsunami} in (32) is expected to depend on the ocean depth h (varying like $h^{-2/3}$), while KAJIURA's (1981) does not. We will show below that a minor refinement to our approach, namely taking into account the finite depth penetration of the source, resolves this latter discrepancy.

The growth of E^{Tsunami} as $M_0^{4/3}$ could appear as a paradox, especially since it would predict that for large enough earthquakes, the energy of the tsunami could become greater than the total elastic energy released during the earthquake, which grows only as M_0. However, it can be verified that this would occur for an earthquake of moment 9×10^{35} dyn-cm, which in turn would scale to a fault length $L_f = 37{,}000$ km; the latter, approaching the Earth circumference, would obviously violate the scaling laws upon which our discussion is based. Even the 1960 Chilean earthquake, at 2×10^{30} dyn-cm the largest seismic event ever measured, would channel only 1.3% of its released energy into its far-field tsunami. The physical origin of the paradox lies in the fact that a characteristic length in the problem, namely the ocean depth h, does not scale with the source. Finally, the scaling with $M_0^{4/3}$ is also close to the behavior suggested by WARD (1982) ($M_0^{1.5}$), based on an empirical fit of published parameters such as fault length L_f and rise time τ as a function of M_0.

The Landslide Source

To compute the energy of a normal mode excited by a landslide, we simply replace, in the second line of Equation (25), the average excitation of a double-couple with that of a single force:

$$F_0^2 \cdot \left[A(0)s_F^2 K_0^2 + A(-1)(c_{-1}^{SF})^2 L^2 (K_1^{SF})^2 + A(1)(c_1^{SF})^2 L^2 (K_1^{SF})^2\right] \quad . \tag{33}$$

For a sub-horizontal force, the first term in the bracket is negligible, the coefficients $c_{\pm 1}^{SF}$ are, on the average, $1/\pi$ (note that they differ from their double-couple counterparts in (25)), and $K_1^{SF} \approx -aK_2$. This leads to the energy of a multiplet:

$$E_{\text{multiplet}}^{\text{Rayleigh}} = 0.0012 \frac{F_0^2}{\rho_s a C^2}; \quad E_{\text{multiplet}}^{\text{Tsunami}} = 5 \times 10^{-4} \frac{\rho_w g}{\mu^2} \cdot \frac{1}{l} \cdot F_0^2 \quad , \tag{34}$$

which must then be integrated over frequency after weighting by the square of the appropriate source time function (10). In the case of Rayleigh waves,

$$E^{\text{Rayleigh}} = 0.0012 \frac{F_0^2}{\rho_s a C^2} \cdot \frac{a}{U} \left[\int_0^{\omega_c} \pi^2 \frac{\omega^4}{\omega_1^4} d\omega + \int_{\omega_c}^{\infty} \frac{16}{\pi^2} \frac{\omega_1^2}{\omega^2} d\omega\right] = 0.018 \cdot \frac{1}{\rho_s U C^2} \frac{(F_0^R)^2}{\tau_1} \tag{35}$$

and in the case of a tsunami,

$$E^{\text{Tsunami}} = 5 \times 10^{-4} \frac{\rho_w g}{\mu^2} \cdot F_0^2 \cdot \frac{a}{U} \left[\int_0^{\omega_c} \frac{\pi^2}{l} \frac{\omega^4}{\omega_1^4} d\omega + \int_{\omega_c}^{\infty} \frac{16}{\pi^2 l} \frac{\omega_1^2}{\omega^2} d\omega\right] \approx 0.0011 \frac{\rho_w g}{\mu^2} \cdot (F_0^T)^2 \quad , \tag{36}$$

where we have introduced superscripts ^{R} and ^{T} on F_0 to emphasize that its values in (35) and (36) differ, due to the Dahlen corrections. In the case of the Aleutian slide, we find energies of 5.7×10^{19} and 1.5×10^{21} ergs, for the Rayleigh and tsunami waves, respectively (we use a crustal value, 3×10^{11} dyn/cm^2, for μ). In the case of the smaller, PNG-type source, we find values of 4.7×10^{16} (Rayleigh) and 2.2×10^{19} ergs (tsunami).

Scaling with Slide Dimensions

We first note that, in the case of the smaller slide, the value computed above is significantly smaller than calculated ($E_{OS} = 8 \times 10^{20}$ ergs) from simple deformation arguments by OKAL and SYNOLAKIS (2003). The two figures can however be reconciled by noting that while (36) assumes a 5-km deep ocean, E_{OS} uses a shallower water column, $h = 1500$ m in the PNG epicentral area. Although neither Equation (36) nor OKAL and SYNOLAKIS' (2003) energy estimate expressly involve h, both do indirectly, through the Dahlen correction in the former and through the dynamic response of the water column, $\alpha = v^2/2C^2$ in the latter. When the deeper value $h = 5$ km is used, the estimate E_{OS} decreases to 7.2×10^{19} ergs, only three times the modal value. This discrepancy is not unexpected, since E_{OS} computes the infusion of energy into the initial condition of the gravitational wave, whereas the modal approach is expected to break down in the near field. Indeed, it is possible to explore theoretically the ratio between the full expressions of the two estimates of the energy generated by the landslide, assuming a similar average depth of the ocean:

$$\frac{E^{\text{Tsunami}}}{E_{OS}} = \frac{0.0011 \frac{\rho_w g}{\mu^2} \left[\rho_s S_s (\delta H) \cdot g \sin\theta \cdot \frac{8}{3} \frac{\beta_0^2}{C^2}\right]^2}{\rho_w g S_s (\delta H)^2 \cdot \frac{v^4}{4C^4}} = 0.03 \frac{g^2 \sin^2\theta}{v^4} S_s \approx \frac{1}{8} \frac{S_s}{D_a^2} \,, \tag{37}$$

where S_s is the surface area of the slide, θ its dip angle, and v the average velocity at which the material moves. The RHS of Equation (37) is obtained by taking v as half the maximum velocity $v_{\max}$ reached during the sliding (OKAL and SYNOLAKIS, 2003), which is approximately given by $v_{\max}^2 = 2gz$, where z is the total drop of the slide. In turn, z is approximately $D_a \tan\theta$, where D_a is the horizontal distance over which the material is accelerated. Note that, in general, D_a is only a fraction of the total length traveled, D, since acceleration stops when the slide encounters a flat bottom. Finally, we take $\cos\theta = 1$ for typical shallow-dipping slides. We conclude that a difference between E^{Tsunami} and E_{OS} is predictable from the aspect ratio of the area of the slide, and the number of slide lengths traveled during the motion: $E^{\text{Tsunami}}/E_{OS} = \frac{1}{8}(W/L_f) \cdot (L_f/D_a)^2$.

Finally, we consider the scaling of the tsunami energy (36) to the total change in potential energy ΔW resulting from the sliding. The latter drops a mass of ocean floor $\rho_s S_s(\delta h)$ a vertical distance z, and replaces it with water of density ρ_w. Hence

$\Delta W = (\rho_s - \rho_w) g z S_s(\delta h)$, and the tsunami efficiency of the slide, i.e., the fraction of the released potential energy channeled into the far-field tsunami is thus:

$$\frac{E^{\text{Tsunami}}}{\Delta W} = 7.8 \times 10^{-3} \cdot \frac{\rho_w}{\rho_s - \rho_w} \cdot \sin\theta \cdot \frac{S_s(\delta H)}{h^2 z} \quad . \tag{38}$$

In the case of the PNG slide, we take an average water depth of 1.5 km, a vertical drop of 1 km, and $\theta = 15°$, to obtain an efficiency of 0.9%; in the case of the Aleutian reference slide, with an average water depth of 3 km, and a vertical drop of 6 km (to the bottom of the Aleutian trench), along a 10° slope, we find a very similar value of 1.25% for the tsunami efficiency.

These numbers are significantly greater than their counterparts for dislocation sources, and thus landslides are found to be relatively more efficient than earthquakes at partitioning their total available energy into the far-field tsunami. This is not surprising, since the landslide occurs in the water in the first place, as opposed to the dislocation, which is released inside the elastic solid Earth. However, even large landslides release less total potential energy than do earthquakes, so that their absolute levels of far-field tsunami energy are no match for that of large dislocation sources.

5. *The Influence of Depth*

Until now, we have assumed in all our computations that the excitation coefficients K_i could be approximated by their asymptotic values as the centroid depth $a - r_s$ goes to zero. In particular, we have regarded as negligible the pure dip-slip coefficient K_1. However, when considering large earthquakes under our scaling model, the fault width W is expected to scale as $M_0^{1/3}$, and thus the vertical extent of the source can no longer be neglected. As the K_i's are themselves function of depth, their dependence will affect the growth of excitation with earthquake size, and this property must be analyzed carefully. Note, however, that this kind of correction is not warranted for landslides sources, which never penetrate the deep Earth, since their thickness is limited to at most a few km by the stratigraphic structure of the ocean floor.

Consider a source of finite depth, extending from the surface to a depth $H = W \sin\delta_f$, where δ_f is the dip of the fault. On the average, $H = 2W/\pi$, and thus H scales with source size as:

$$H = \frac{4^{1/3}}{\pi} \frac{M_0^{1/3}}{(\mu\varepsilon_{\max})^{1/3}} \quad . \tag{39}$$

All that is needed in principle is to replace the K_i's in Equation (25) by their values averaged over the vertical extent of the source: $\bar{K}_i = \frac{1}{H}\int_0^H K_i \cdot dz$.

Rayleigh Waves

We write the seismic displacement at any depth, $u_x = ly_3$, as the sum of its components in the inhomogeneous P and S waves making up the Rayleigh wave:

$$u_x = ly_3 = Ae^{-\omega\gamma z} + Be^{-\omega\delta z} \quad , \tag{40}$$

where the values $A = -1.6119\ Y$; $B = 0.9306\ Y$; $\gamma = 0.8475/C$; $\delta = 0.3933/C$ are obtained from elementary Rayleigh wave theory for a homogeneous half-space. It is straightforward to derive, for example, a correction Φ_2 to the coefficient K_2

$$\Phi_2 = \frac{1}{A+B}\bar{u}_x = \frac{1}{(A+B)H\omega} \cdot \left[\frac{A}{\gamma}(1 - e^{-\omega\gamma H}) + \frac{B}{\delta}(1 - e^{-\omega\delta H})\right] \tag{41}$$

which is used to replace K_2 in (25) with

$$\bar{K}_2 = K_2\Phi_2 = -0.0034\frac{Y^{-1}}{a_2\rho_s} \cdot \frac{1}{C^2 l}\Phi_2 \ . \tag{42}$$

A similar expression could be derived for the coefficient $\bar{K}_0$, and a finite value (going to zero for $H \to 0$) for $\bar{K}_1$. The resulting energy of the multiplet is then substituted into (28) for the integration over frequency. For the low-frequency part of the spectrum, the various integrals are merely exponential functions, and the computation is easy. For example, the term involving K_2 will be

$$E_{lf(2)} = (0.0034)^2 \frac{1}{(A+B)^2} \frac{2^{1/2}\pi}{3^{5/4}} \frac{M_0^2}{C^4 U H^2} \cdot \int_0^{\omega_c} \left[\frac{A}{\gamma}(1 - e^{-\omega\gamma H}) + \frac{B}{\delta}(1 - e^{-\omega\delta H})\right]^2 \cdot d\omega \ . \tag{43}$$

The details of the computation, available from the author on request, are of little interest; note however that since H scales with the seismic source, the product $H \cdot \omega_c/C = \xi_c \approx 1.38$ remains invariant with M_0. As such, the various exponentials in the coefficient Φ_2 (41) are always integrated to a *constant argument*, e.g., $-\omega_c\gamma H = -1.17$; $-2\omega_c\delta H = -1.09$; this property stems from the trivial observation that the corner frequency scales inversely to the linear dimension of the source. As a result, all integrals of the exponentials in (43) are proportional to Y^2/H, and $E_{lf(2)}$ scales as M_0^2/H^3, or M_0. As for the high-frequency part of the spectrum, it requires the computation of the exponential integrals

$$\mathbf{E}_\zeta(x) = \int_1^\infty t^{-\zeta} e^{-xt}\, dt \quad (\zeta = 6) \ , \tag{44}$$

(e.g., ABRAMOWITZ and STEGUN, 1965; p. 228), all arguments x being constants of the type $\omega_c\gamma H$, which are invariant under scaling of the source. The end result is that the corrected value of the energy excited into the Rayleigh wave can be written as:

$$E^{\text{Rayleigh}} = 0.0162 \frac{M_0}{\rho_s UC} (\mu \varepsilon_{\max}) = \frac{0.0162}{\kappa \eta} (M_0 \cdot \varepsilon_{\max}) = 0.0207 (M_0 \cdot \varepsilon_{\max}) \ . \tag{45}$$

We conclude that taking into account the finite depth of the source does not alter the scaling of energy with moment, but merely amounts to cutting the fraction of energy excited into the Rayleigh wave by a factor of 4. This number is not unreasonable, given that, at the corner frequency where spectral amplitude is maximum, the vertical extent of the fault is on the order of 0.5 to 1.2 skin depths of the shear and compressional potentials, respectively, so that the amplitude of the eigenfunctions is cut on the average by a factor exp(0.8) = 2.2; the energy should be cut by a factor of about 5, but some gain is achieved by allowing a contribution from the dip-slip coefficient K_1.

Tsunamis

The computation in the case of a tsunami proceeds in a conceptually similar fashion. For example, the structure of the pseudo-Rayleigh wave in the half-space leads to

$$u_x = l y_3 = \frac{\rho_w g}{\mu k} \cdot \left(A_3 e^{-\omega \gamma z} + B_3 e^{-\omega \delta z} \right) \ , \tag{46}$$

the coefficients A_3 and B_3 being now

$$\begin{aligned} A_3 &= -\frac{2 - \kappa^2}{(2 - \kappa^2)^2 - 4\sqrt{1 - \kappa^2}\sqrt{1 - \kappa^2/3}} \cdot Y; \\ B_3 &= \frac{2\sqrt{1 - \kappa^2}\sqrt{1 - \kappa^2/3}}{(2 - \kappa^2)^2 - 4\sqrt{1 - \kappa^2}\sqrt{1 - \kappa^2/3}} \cdot Y \end{aligned} \tag{47}$$

and the skin depth coefficients

$$\gamma = \frac{\sqrt{1 - \kappa^2/3}}{C}; \qquad \delta = \frac{\sqrt{1 - \kappa^2}}{C} \ . \tag{48}$$

An additional difficulty arises in the limit $\kappa \ll 1$ (typically $\kappa = 0.06$ for tsunamis) since both A_3 and B_3 are of order κ^{-2}, and hence a formulation such as (46) consists of subtracting two very large terms to find their finite difference, a notoriously unstable computational procedure. Nevertheless, the careful expansion of (46) in the vicinity of $\kappa \to 0$ can yield important results.

The deep point source double-couple. Before examining the influence of finite depth penetration on the generation of tsunami energy by a shallow dislocation source, we address the more general question of the variation of tsunami excitation with depth by a point source double-couple. This is an important matter, since it has often been assumed that only the shallowest sources, namely within a few km of the ocean floor, can excite significant tsunamis, including in the far field.

While the representation (46) is similar to the case of standard Rayleigh waves, we note that since $\kappa = C/\beta \ll 1$, both γ and δ are close to $1/C$, and both potentials (as well as all combinations thereof, including excitation coefficients) decay with depth essentially as $e^{-\omega z/C}$. This means that the skin depth of the pseudo-Rayleigh wave is comparable to the wavelength of the tsunami, typically hundreds of km. As a consequence, tsunami excitation in the far field is not significantly decreased in the first few tens of km inside the Earth. This result was obtained by OKAL (1988) on the basis of the computation of individual normal mode solutions (his Fig. 7), but was not previously rationalized theoretically. Although arguably not intuitive, it explains the generation of tsunamis detectable in the far field by intermediate depth earthquakes, such as the Tonga event of 22 June 1977 (TALANDIER and OKAL, 1979; LUNDGREN and OKAL, 1988).

In particular, the contribution of the K_2 terms to the energy of a multiplet excited by a point-source double-couple at depth z is then

$$E_{\text{multiplet}(2)} = \frac{1}{36\pi} \cdot \frac{M_0^2}{Y^2} \cdot \frac{\rho_w g}{a^2 \mu^2} \cdot l \cdot \left(A_3 e^{-\omega\gamma z} + B_3 e^{-\omega\delta z}\right)^2 \ . \tag{49}$$

This formula generalizes (27) to any focal depth z. (Note that if $z \to 0$, the bracket in (49) goes to $Y^2/16$, leading to a constant $1/(576\ \pi)$, which is half that in (27), since we have not accounted for the K_0 terms.)

The shallow source extending at depth. We proceed to average (49) (and similar terms emanating from K_0 and K_1) over depth (z ranging from 0 to H), and then to integrate over frequency, leading to

$$E_{lf(2)} = \frac{M_0^2}{36\pi} \cdot \frac{\rho_w g Y^{-2}}{\mu^2 UCH^2} \int_0^{\omega_c} \frac{1}{\omega} \left(\frac{A_3}{\gamma}\left(1 - e^{-\omega\gamma H}\right) + \frac{B_3}{\delta}\left(1 - e^{-\omega\delta H}\right) \right)^2 d\omega \tag{50}$$

as the low-frequency ($\omega < \omega_c$) part of the K_2 terms of the energy integral. When comparing this expression with (43), note the additional factor $1/\omega$ in the integrand. It leads to a more complex set of exponential integrals of the form

$$\mathbf{I}_{a,b}(\xi) = \int_0^{\xi} \frac{(1 - e^{-at})(1 - e^{-bt})}{t} dt \ . \tag{51}$$

Specifically, the integral in (50) is a sum of terms of the form $\frac{A'A''}{\gamma'\gamma''} \mathbf{I}_{C\gamma', C\gamma''}(\xi_c)$ where A' and A'' are either A_3 or B_3 (given in (47)), and γ' and γ'' either γ or δ (given by (48)), and the variable $\xi_c = \omega_c H/C$ is invariant upon scaling of the source (note that its value is larger in the case of a tsunami ($\xi_c \approx 9.8$) than for a Rayleigh wave). The leading terms in (50) can be evaluated by expanding A_3, B_3, γ, and δ in the vicinity of $\kappa = 0$. It is found that the integral in (50) is equivalent to

$$\frac{1}{16} \cdot C^2 Y^2 \cdot \left[\mathbf{I}_{1,1}(\xi_c) - 1\right] \; . \tag{52}$$

When the other terms (governed by K_0 and K_1) are included, the total low-frequency part of the integral adds to:

$$E_{lf} = \frac{M_0^2}{36\pi} \cdot \frac{\rho_w g C}{\mu^2 U H^2} \left[\frac{33}{8} \mathbf{I}_{1,1}(\xi_c) - \frac{23}{8}\right] \; . \tag{53}$$

As for the high-frequency part of the energy, it takes the form (again written for the K_2 terms only):

$$E_{hf(2)} = \frac{M_0^2}{36\pi} \cdot \frac{\rho_w g Y^{-2}}{\mu^2 U C H^2} \int_{\omega_c}^{\infty} \frac{1}{\omega} \frac{\omega_c^6}{\omega^6} \left(\frac{A_3}{\gamma} \left(1 - e^{-\omega\gamma H}\right) + \frac{B_3}{\delta} \left(1 - e^{-\omega\delta H}\right) \right)^2 d\omega \; . \tag{54}$$

In principle, this amounts to a combination of exponential integrals $\mathbf{E}_7(x)$ whose arguments x are combinations $\omega_c H \gamma'$ and $\omega_c H(\gamma' + \gamma'')$. Careful expansion near $\kappa \to 0$ shows that the leading term of the integral in (54) is

$$\frac{1}{6} \left(\frac{A_3}{\gamma} + \frac{B_3}{\delta} \right)^2 \underset{\kappa \to 0}{\approx} \frac{C^2 Y^2}{96} \; . \tag{55}$$

When the K_0 and K_1 terms are included, this expression becomes (33/48) $C^2 Y^2$, and the total energy of the tsunami wave (replacing Equation (31)):

$$E = \frac{M_0^2}{36\pi} \cdot \frac{\rho_w g}{\mu^2 H^2} \cdot \left[\frac{33}{8} \mathbf{I}_{1,1}(\xi_c) - \frac{35}{16}\right] \; . \tag{56}$$

It still formally depends on the ocean depth h through the variable $\xi_c = \frac{(28/5)^{1/3}}{(2\pi)^{1/6}} \kappa^{-2/3}$, which would vary from 9.8 for $h = 5$ km, to 14.5 for $h = 1500$ m. The corresponding values of $\mathbf{I}_{1,1}$, computed by expanding the integral into an alternating series, are 2.08 and 2.56, suggesting only a minor dependence. We then obtain for the 5-km ocean:

$$E = 6.328 \frac{M_0^2}{36\pi} \cdot \frac{\rho_w g}{\mu^2 H^2} = 0.219 \frac{\rho_w g}{\mu^{4/3}} \cdot \varepsilon_{\max}^{2/3} \cdot M_0^{4/3} \approx 7.4 \times 10^{-17} M_0^{4/3} \tag{57}$$

if M_0 is in dyn-cm and E in ergs. With respect to Equation (31), depth averaging results in a correction of only 30%, which is explained by the generally slower decay of the eigenpotentials with depth, as compared to Rayleigh waves. Note in particular that E no longer formally depends on the depth h of the oceanic column, because of the slow variation of $\mathbf{I}_{1,1}$ in the range of values considered. We can regard (57) as the final expression of the energy of a far-field tsunami generated by a dislocation of moment M_0, as predicted by normal mode theory.

It is rather difficult to assess the accuracy of this formula, because there does not exist, to our knowledge, a universal experimental method for measuring the energy of large transoceanic tsunamis. As summarized by IIDA (1963), various methods have been used in this respect, which fall basically into two categories: (i) attempts at integrating backwards to the source the kinetic energy flux estimated from tidal gauge records at distant stations; and (ii) using some form of scaling law relating tsunami energy to earthquake magnitude. Obviously, only (i) can be considered an independent physical measurement of the energy of the tsunami wave. On Figure 2, we compare the values predicted by (57) to those reported by IIDA (1963) as having been obtained following (i). The large scatter reflects the poor quality of those measurements, which likely included near-field observations, and could have been affected by site response at far-field tidal gauge stations.

A more remarkable result is the total similarity of the corrected expression (57) with KAJIURA's (1981) formula (12) (see Equation (32)), which he obtained from the computation of a purely static deformation field. In this respect, the present derivation represents an interesting example of the power of the normal mode formalism to directly handle the coupling between liquid ocean and solid Earth. On the other hand, the corresponding value of F_{Kaj}. equivalent to (57) remains too high.

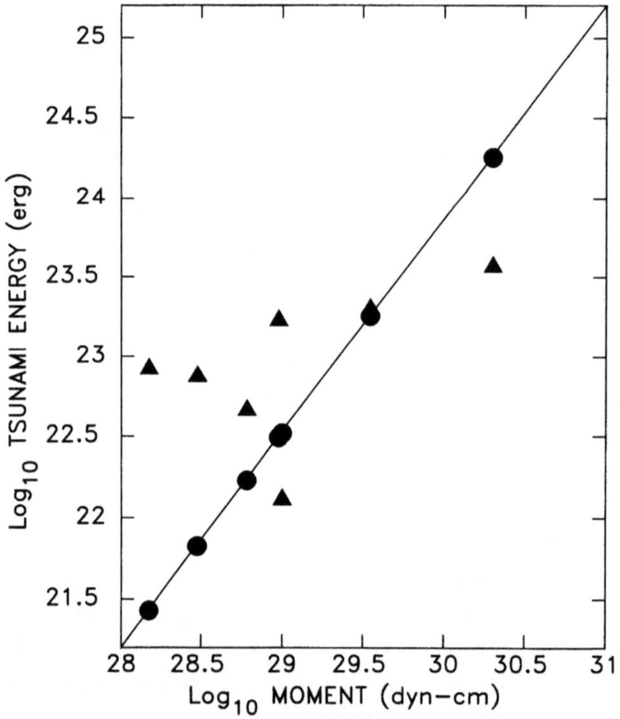

Figure 2
Comparison of values of tsunami energy obtained from (57) with estimates identified by IIDA (1963) as obtained from energy flux computations. Only those events for which a seismic moment is available in the literature (KANAMORI, 1972b; OKAL, 1992) are used. The energy values resulting from (57) are plotted as solid dots; values reported by IIDA (1963) are shown as triangles.

6. *Directivity*

We examine here in very general terms the directivity patterns of tsunamis generated by both dislocations and landslides. We recall that the azimuthal dependence of the amplitude radiated by a seismic source – dislocation or landslide – is affected both by the orientation of the source (resulting in the preferential excitation of the various azimuthal orders m in (6)), and by the spacio-temporal evolution of the position of the source. The former, called *radiation pattern*, is a point-source property; the latter, called *directivity* is rooted in the finite dimension of the source. We refer to BEN-MENAHEM (1961) for the definition of the directivity function DIR generated by a source of length L_f, moving at a velocity V_R ("rupture velocity") in the azimuth ϕ from the direction of observation in the far field:

$$DIR(\phi) = \frac{\sin X}{X}; \qquad X = \frac{\omega L_f}{2C} \cdot \left(\frac{C}{V_R} - \cos\phi \right) . \tag{58}$$

The parameter DIR is to be multiplied into the spectral amplitude predicted in the azimuth ϕ, as given by (7) or (17). Its expression results from the simple decomposition of the moving source into an integral of elementary sub-sources along its path in space and time. Note that $DIR = 1$ over an appreciable range of frequencies requires $X = 0$.

In order to develop a significant directivity pattern, two conditions are necessary in (58). First, the ratio $\omega L_f / V_R$ must be large, in other words the duration of the source, L_f / V_R must be at least comparable to the period of the wave; otherwise the source can be regarded as instantaneous, and all values of X are small, regardless of ϕ. Second, the values of X must change significantly with the azimuth ϕ, which excludes the case $C \gg V_R$. These simple remarks lead to fundamentally different regimes, depending on the ratio V_R/C, which have fundamental application to tsunamigenic events.

* *If the rupture is "sonic,"* i.e., V_R comparable to C, then (58) is maximum for $\cos\phi \approx 1$ or $\phi \approx 0$: in the direction (and sense) of rupture, the interference between all elements of the source is constructive and the amplitude maximum. This is the classical case of seismic surface waves from large earthquakes featuring conventional rupture, for which $C \approx V_R \approx 3.5$ to 4 km/s (Figure 3a).

* *If the rupture is hypersonic,* i.e., $V_R \gg C$, then $X = 0$ has the exact solution $\cos\phi = C/V_R \ll 1$, essentially at right angles from the rupture. This is the case of tsunamis generated by large seismic dislocations (typically, $V_R/C = 15$), as first exemplified by BEN-MENAHEM and ROSENMAN (1972) in the case of the 1964 Alaskan earthquake. OKAL and TALANDIER (1991) later discussed the width of the directivity lobes and concluded that tsunamis had generally much narrower lobes than the mantle waves generated by the same dislocations, as illustrated on Figure 3b.

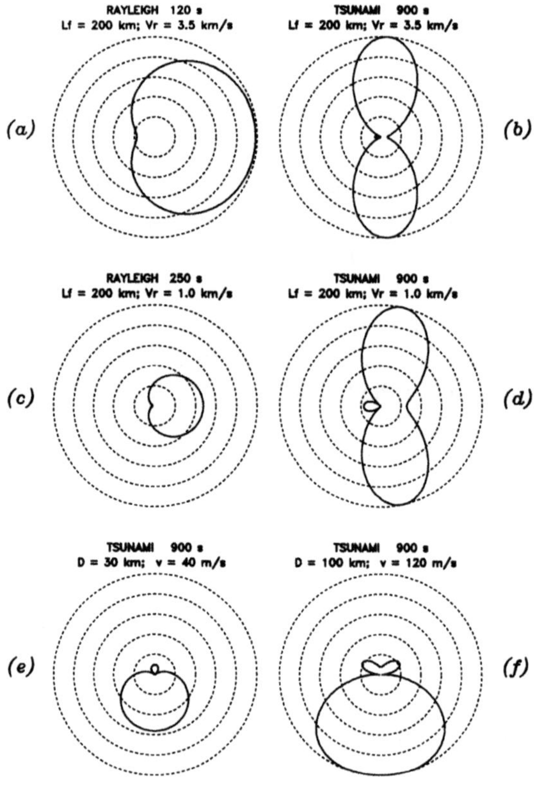

Figure 3

Theoretical directivity patterns computed for a number of tsunamigenic sources. All diagrams are azimuthal plots of the function $DIR(\phi)$ defined in (58). The dashed circles are drawn from $DIR = 0.2$ to $DIR = 1$, by steps of 0.2 units. In the top diagrams, a conventional dislocation source of length $L_f = 200$ km is rupturing to the right. (*a*): Rayleigh waves at 250 s; (*b*): Tsunami waves at 900 s. Note the narrower lobe, oriented about 90° from the rupture. In the central diagrams, we consider a slower rupture ("tsunami earthquake"); note the destructive interference of mantle Rayleigh waves (*c*), and the essentially unperturbed digram for tsunamis (*d*). In the bottom diagrams, we consider the tsunami wavefield of a landslide source, which moves towards the bottom of the figure; we consider a shallower ocean with $C = 120$ m/s. Note in (*e*) the destructive interference in all azimuths. (*f*): Only for unrealistic values of the velocity v does a constructive interference become possible; it does not, however, feature a narrow lobe comparable to those in (*b*) and (*d*).

* *If the rupture is subsonic*, i.e., $V_R < C$, then there is no solution to $\cos\phi = C/V_R$, and interference is destructive in all azimuths (Fig. 3e). Only at the longest periods, much greater than the duration of the source, can it become constructive, but then $DIR \to 1$ for all ϕ; azimuthal directivity is never present.

In this very general and classical framework, we discuss two special cases of relevance to tsunami genesis.

• We first focus on the so-called "tsunami earthquakes," generating tsunamis of disproportionate amplitude relative to their conventional seismic waves. Such events, which include the 1946 Aleutian and 1992 Nicaraguan earthquakes (KANAMORI, 1972a; KANAMORI and KIKUCHI, 1993), have been attributed to an exceptionally slow rupture along the fault plane, due either to propagation in a sedimentary wedge (FUKAO, 1979), or to a jerky mode of rupture in the presence of asperities on the fault (TANIOKA *et al.*, 1997), with proposed velocities V_R as low as 1 km/s (POLET and KANAMORI, 2000). For such values, the rupture becomes subsonic for Rayleigh waves, with the smallest value of the parameter $C/V_R - \cos\phi$ being 2.5. For $L_f = 200$ km, the function *DIR* remains less than 0.5 in all azimuths for $T \leq 250$ s, resulting in destructive interference across the conventional seismic spectrum (Fig. 3c). However, the rupture remains hypersonic for tsunami waves ($V_R/C = 5$), which keep a very strong directivity pattern (Fig. 3d), the lobes being slightly displaced (about 10 to 15°) away from the direction perpendicular to rupture. The conclusion of this remark is that all large dislocation sources, however slow their rupture may be, generate tsunamis featuring strong directivity.

• The situation is quite different in the case of *landslides*. The evolution of an underwater landslide is obviously a more complex and less well understood process than that of a seismic dislocation, since the texture and cohesion of the material may change with time, for example during the development of a turbidity current. Nevertheless, there is ample evidence that large underwater landslides may occur over several tens of km (e.g., LEE, 1989), and possibly hundreds of seconds, occasionally thousands if a turbidity current is involved, as documented for example by the relative timing of cable breaks (HEEZEN and EWING, 1952, 1955). In this framework, and as a simplified model, Equation (58) can be used with L_f being the total displacement D of the slide, as defined in Section 2, p. 10, and V_R the velocity of the moving block, called v in (37).

In Figure 3e, we present a directivity pattern computed for $D = 30$ km and $v = 40$ m/s. These figures are taken as possibly representative of the landslide proposed for the 1946 Aleutian event. However, the value of v must be regarded as an upper bound on physically acceptable material velocities during underwater landslides. For example, in the case of the 1929 Grand Banks earthquake, PIPER *et al.* (1985, 1989) have estimated velocities of 10 to 20 m/s, both inside the valleys channeling the initial slump, and later in the turbide phase on the abyssal plain. On Frames (e) and (f), the source is moving to the bottom of the picture, to ease the comparison with the dislocation cases (earthquake rupture will propagate along the coastline, and landslides along the steepest gradient, expected to at right angles to the coast). We conclude that the parameters *DIR* remain small, expressing destructive interference in all azimuths. In (f), we investigate theoretically the unrealistic case of a sonic landslide in a shallow source region ($v = C = 120$ m/s). The interference becomes constructive in the direction of the slide, but the lobe is broad; it would take a totally unphysical value of the velocity, $v = 250$ m/s, to develop a large, narrow lobe in the far field, which, incidentally, would then be at right angles from the direction of sliding.

The conclusion of this section is that azimuthal directivity can be a robust discriminant between dislocations and landslides as generators of tsunamis in the far field. The presence of narrow lobes of strong low-frequency amplitude at transoceanic distances requires generation by a conventional dislocation.

7. Conclusion and Perspective: The 1946 Aleutian Tsunami

We have obtained theoretical results based on normal mode theory allowing the comparison of the efficiency of landslides and dislocations as generators of far-field tsunamis. We show that the modeling of tsunamis generated by landslides will be strongly affected by the quadrupolar correction derived by DAHLEN (1993), which can add a factor of 50 to 200 to the monopole force used to model the landslide. We find that landslides are always deficient generators in the far field; the bulk of their spectral amplitude is generated in a range of frequencies where significant dispersion acts to lessen the resulting amplitude of the wave. We obtain general formulae scaling the total energy carried in the far field by a tsunami wave with the size of the source; in the case of dislocations, we are able to reproduce exactly (except for a numerical factor) the expression proposed by KAJIURA (1981) using an approach based exclusively on static deformation of the ocean floor, and in particular the scaling of energy with $M_0^{4/3}$. Finally, we show that, within acceptable ranges of their kinematic parameters, landslides cannot generate strongly lobed directivity patterns in the far field. Their observation can thus be regarded as a robust discriminant of the nature of the tsunami source.

The conclusions obtained in this paper provide a framework for discussing the possible sources of the great 1946 Aleutian tsunami. We recall that this event was catastrophic both in the near field, where it eradicated the Scotch Cap lighthouse, and in the far field, where it destroyed waterfronts in Hawaii, at a cost of 159 human lives, and wrought destruction in the Marquesas Islands and possibly as far away as Antarctica. Recent field work based on the interview of elderly witnesses and on the surveying of preserved watermarks, as well as ongoing seismological investigations have quantified the following parameters:

(a) In the near field, runup reaches 42 m at Scotch Cap on Unimak Island and such values are concentrated along approximately 35 km of coastline (PLAFKER *et al.*, 2002);

(b) In the far field, runup reaches 16 m in Hawaii (LANDER and LOCKRIDGE, 1989), and between 6 and 13 m in the Marquesas and Easter Island (OKAL *et al.*, 2002), these values being measured overland, i.e., outside river beds;

(c) Runup is much lower (2.7 m) at Juan Fernández Island, suggesting a large azimuthal variation between this location and Easter (OKAL *et al.*, 2002);

(d) The seismic signature of the event is exceedingly slow, making the precise determination of its source parameters difficult. Mantle surface waves suggest a seismic moment of at least 5×10^{28} dyn-cm (measured around 200 s), but the absence of multiple passages on Wiechert records at Uppsala rules out a further growth of this figure beyond 10^{29} dyn-cm (OKAL, 1992).

(e) The relocation of all aftershocks of the 1946 event suggests a fault length $L_f = 200$ km (OKAL and LÓPEZ, 2002).

As detailed by OKAL *et al.* (2002), the small tsunami amplitude at Juan Fernández, as compared to Easter and the Marquesas, requires a strong directivity effect focusing the energy of the tsunami in the azimuths $150 \pm 15°$, which as discussed in Section 6, is incompatible with generation by a landslide (note that our examples were computed at 900 s, the dominant period observed on far-field maregrams and reported by witnesses). Also, and as discussed in Section 3, the amplitudes of the tsunami in the far field are much too large to be explained by a landslide of an acceptable volume. Thus, the far-field characteristics of the 1946 Aleutian tsunami rule out its generation by a simple landslide. On the other hand, the extreme amplitude of runup in the near field is more than 6 times larger than any acceptable value of the slip Δu on the fault, and its distribution along the near-field coast is incompatible with generation by a dislocation (HOFFMAN *et al.*, 2002). We present as an inescapable conclusion that both phenomena – a large slow earthquake, and a major landslide – must have occurred concurrently during the 1946 Aleutian event.

Appendix 1: Corner Frequency as a Function of Moment

In this section, we derive an estimate of the scaling of the corner frequency ω_c with moment M_0, in the framework of GELLER (1976). Starting with his Equation (6), we study the evolution with frequency of the finiteness function

$$\Psi = \left| \frac{\sin \omega\chi_\tau}{\omega\chi_\tau} \cdot \frac{\sin \omega\chi_{L_f}}{\omega\chi_{L_f}} \cdot \frac{\sin \omega\chi_W}{\omega\chi_W} \right| , \tag{A1}$$

where τ is rise time, L_f source length and W source width, $\chi_\tau = \tau/2$, $\chi_{L_f} = \frac{L_f}{2V_R}$ (obtained by averaging the classical expression of directivity over azimuth), and

$$\chi_W = \frac{W}{2C} \cdot \langle |\cos\phi \sin\theta| \rangle = \frac{W}{\pi C} = \frac{L_f}{2\pi C} \tag{A2}$$

(obtained from GELLER (1976) after noticing that one averages the trigonometric function over the sphere, first over azimuth ϕ $(2/\pi)$, then over colatitude θ $(1/2)$. In these formulae, C is phase velocity, V_R rupture velocity, L_f fault length, and W fault width $(S = L_f W)$; we further assume a fault area of constant aspect ratio $W/L_f = 1/2$.

At long periods ($\omega \to 0$), $\Psi \to 1$. At short periods ($\omega \to \infty$), Ψ is at most

$$\frac{1}{\omega\chi_\tau \cdot \omega\chi_{L_f} \cdot \omega\chi_W \cdot} = \frac{8\pi V_R C}{\omega^3 \cdot \tau L_f^2} \quad . \qquad \text{(A3)}$$

In scaling the source, we assume that the strain release, ε, has a constant value corresponding to the threshold at which rupture takes place in the rock, $\varepsilon_{\max}$, which we take as 10^{-4}. Then the displacement Δu along the fault plane is approximately $\varepsilon_{\max} W$, or $\Delta u = \varepsilon_{\max} \frac{L_f}{2}$, and the seismic moment

$$M_0 = \mu S \Delta u = \frac{1}{\sqrt{2}} \mu \varepsilon_{\max} S^{3/2} = \frac{1}{4} \mu L_f^3 \varepsilon_{\max} \quad , \qquad \text{(A4)}$$

the constant 1/4 changing only slightly according to the geometry of rupture.

The rise time τ can be related to fault size through

$$\tau = \frac{8\sqrt{2}}{7\pi^{3/2}} \frac{L_f}{\beta} \quad , \qquad \text{(A5)}$$

which follows from GELLER's (1976) Equation (5) by setting $L_f = 2W$. Hence,

$$\Psi \underset{\omega\to\infty}{\approx} 7(\pi/2)^{5/2} \frac{V_R C \beta \mu \varepsilon_{\max}}{\omega^3 M_0} = \left(\frac{\omega_c}{\omega}\right)^3 \qquad \text{(A6)}$$

if we define

$$\omega_c = 7^{1/3} (\pi/2)^{5/6} \left(\frac{V_R}{\beta} \kappa\right)^{1/3} \cdot \beta \cdot \left(\frac{\mu \varepsilon_{\max}}{M_0}\right)^{1/3} \quad . \qquad \text{(A7)}$$

If we further assume a constant ratio between rupture and shear velocities, $V_R/\beta = 0.8$, we obtain finally

$$\omega_c = (7/5)^{1/3} \frac{\pi^{5/6}}{2^{1/6}} \cdot \kappa^{1/3} \cdot \beta \cdot \left(\frac{\mu \varepsilon_{\max}}{M_0}\right)^{1/3} \quad . \qquad \text{(A8)}$$

- In the case of Rayleigh waves in a Poisson solid, $\kappa = \sqrt{2 - 2/\sqrt{3}} = 0.9194$, and thus for a typical event in the upper mantle ($\mu = 7 \times 10^{11}$ dyn/cm^2; $\beta = 4.6$ km/s),

$$\omega_c = 4.77 \times 10^8 \, M_0^{-1/3} \qquad \text{or} \qquad T_c = 1.32 \times 10^{-8} \, M_0^{1/3} \qquad \text{(A9)}$$

where ω_c is in rd/s (T_c in s), and M_0 in dyn-cm.

- In the case of a tsunami wave, $\kappa = C/\beta = \left(\frac{\rho_s g h}{\mu}\right)^{1/2}$, so that

$$\omega_c = (7/5)^{1/3} \frac{\pi^{5/6}}{2^{1/6}} \cdot \beta \cdot \left(\frac{\sqrt{\mu \rho_s g h} \varepsilon_{\max}}{M_0}\right)^{1/3} \qquad \text{(A10)}$$

For an ocean of depth 5 km, and with the same parameters as previously used,

$$\omega_c = 1.78 \times 10^8 M_0^{-1/3} \qquad \text{or} \qquad T_c = 3.53 \times 10^{-8} M_0^{1/3} \tag{A11}$$

The corner frequency for tsunami waves is about 0.4 times that for Rayleigh waves.

Acknowledgments

I thank the conveners who invited me to the Los Angeles Workshop on landslides and tsunamis in March 2000. The comments of Bob Geller on an early version of these computations are appreciated, as well as recent discussions with Steve Kirby, Homa Lee and Bill Normark. The paper was improved through the comments of an anonymous reviewer. I am grateful to my field companions, principally Costas Synolakis and George Plafker, for helping me become acquainted first hand with the awesome scale of the 1946 Aleutian tsunami, both in the near and far fields. Parts of this work were supported by the National Science Foundation.

References

ABRAMOWITZ, M. and STEGUN I. A., *Handbook of Mathematical Functions* (Dover, New York, 1965), 1046 pp.

AIDA, I. (1977), *Simulations of Large Tsunamis Occurring in the Past off the Sanriku District*, Bull. Earthq. Res. Inst. Tokyo Univ. *52*, 71–101 (in Japanese).

AKI, K. (1966), *Generation and Propagation of G-waves from the Niigata Earthquake of June 16, 1964, Part 2. Estimation of Earthquake Moment, Released Energy, and Stress-strain Drop from the G-wave Spectrum*, Bull. Earthq. Res. Inst. Tokyo Univ. *44*, 73–88.

ALEKSEEV, A. S. and GUSYAKOV, V. K., *Numerical modeling of tsunami and seismic surface wave generation by a submarine earthquake*, Proc. Tsunami Res. Symp. (eds. R. A. Heath and M. M. Creswell) Roy. Soc. New Zealand, Wellington, (1976) pp. 243–252.

ANDO, M. (1979), *The Hawaii Earthquake of November 29, 1975: Low Dip Angle Faulting Due to Forceful Injection of Magma*, J. Geophys. Res. *84*, 7616–7626.

BEN-MENAHEM, A. (1961), *Radiation of Seismic Surface Waves from Finite Moving Sources*, Bull. Seismol. Soc. Am. *51*, 401–435.

BEN-MENAHEM, A. and ROSENMAN, M. (1972), *Amplitude Patterns of Tsunami Waves from Submarine Earthquakes*, J. Geophys. Res. *77*, 3097–3128.

DAHLEN, F. A. (1993), *Single-force Representation of Shallow Landslide Sources*, Bull. Seismol. Soc. Am. *83*, 130–143.

DZIEWONSKI, A. M. and ANDERSON, D. L. (1981), *Preliminary Reference Earth Model*, Phys. Earth Planet. Inter. *25*, 297–356.

DZIEWONSKI, A. M., EKSTRÖM, G. and MATERNOVSKAYA, N. (1999), *Centroid-moment Tensor Solutions for July–September 1998*, Phys. Earth Planet. Inter. *114*, 99–107.

EISSLER, H. K. and KANAMORI, H. (1987), *A Single-force Model for the 1975 Kalapana, Hawaii Earthquake*, J. Geophys. Res. *92*, 4827–4836.

FRYER, G. J. and WATTS, P. (2000), *The 1946 Unimak Tsunami: Near-source Modeling Confirms a Landslide*, EOS, Trans. Am. Geophys. Union *81*(48), F748–F749 (abstract).

FUKAO, Y. (1979), *Tsunami Earthquakes and Subduction Processes Near Deep-sea Trenches*, J. Geophys. Res. *84*, 2303–2314.

GELLER, R. J. (1976), *Scaling Relations for Earthquake Source Parameters and Magnitudes*, Bull. Seismol. Soc. Am. *66*, 1501–1523.

GILBERT, J. F. (1970), *Excitation of the Normal Modes of the Earth by Earthquake Sources*, Geophys. J. Roy. astr. Soc. *22*, 223–226.

GUTENBERG, B. (1939), *Tsunamis and Earthquakes*, Bull. Seismol. Soc. Am. *29*, 517–526.

HASEGAWA, H. S. and KANAMORI, H. (1987), *Source Mechanism of the Magnitude 7.2 Grand Banks Earthquake of November 18, 1929: Double-couple or Submarine Landslide?*, Bull. Seismol. Soc. Am. *77*, 1984–2004.

HEEZEN, B. C. and EWING, W. M. (1952), *Turbidity Currents and Submarine Slumps, and the 1929 Grand Banks (Newfoundland) Earthquake*, Am. J. Sci. *250*, 849–873.

HEEZEN, B. C. and EWING, W. M. (1955), *Orléansville Earthquake and Turbidity Currents*, Bull. Am. Assoc. Petrol. Geol. *39*, 505–514.

HOFFMAN, I., SYNOLAKIS, C. E. and OKAL, E. A. (2002), *Systematics of the Distribution of Tsunami Run-up along Coastlines in the Near-field for Dislocation Sources with Variable Parameters*, EOS, Trans. Am. Geophys. Union *83*(22), WP54 (abstract).

IIDA, K. (1963), *A Relation of Earthquake Energy to Tsunami Energy and the Estimation of the Vertical Displacement in a Tsunami Source*, J. Earth Sci. Nagoya Univ. *11*, 49–67.

KAJIURA, K. (1981), *Tsunami Energy in Relation to Parameters of the Earthquake Fault Model*, Bull. Earthq. Res. Inst. Tokyo Univ. *56*, 415–440.

KANAMORI, H. (1972a), *Mechanisms of Tsunami Earthquakes*, Phys. Earth Planet. Inter. *6*, 346–359.

KANAMORI, H. (1972b), *Tectonic Implications of the 1944 Tonankai and the 1946 Nankaido Earthquakes*, Phys. Earth. Planet. Inter. *5*, 129–139.

KANAMORI, H. (1985), *Non-double-couple seismic source*, Proc. XXIIIrd Gen. Assemb. Intl. Assoc Seismol. Phys. Earth Inter., Tokyo, p. 425, (abstract).

KANAMORI, H. and CIPAR, J.J. (1974), *Focal Process of the Great Chilean Earthquake, May 22, 1960*, Phys. Earth Planet. Inter. *9*, 128–136.

KANAMORI, H. and KIKUCHI, M. (1993), *The 1992 Nicaragua Earthquake; A Slow Tsunami Earthquake Associated with Subducted Sediments*, Nature *361*, 714–716.

KANAMORI, H. and STEWART, G.S. (1976), *Mode of the Strain Release along the Gibbs Fracture Zone, Mid-Atlantic Ridge*, Phys. Earth Planet. Inter. *11*, 312–332.

KANAMORI, H., GIVEN, J. W. and LAY, T. (1984), *Analysis of Seismic Body Waves Excited by the Mount St. Helens Eruption, of May 18, 1980*, J. Geophys. Res. *89*, 1856–1866.

LANDER, J. F. and LOCKRIDGE, P. A. (1989), *United States Tsunamis (including United States Possessions)*, 1690–1988, Natl. Geophys. Data Center, Natl. Atmos. Ocean. Admin. Publ. 41–2, Boulder.

LEE, H.J., *Undersea landslides; extent and significance in the Pacific Ocean*. In Proc 28th International Geological Congress Symposium on Landslides; Extent and Economic Significance (eds. E. E. Brabb and B. L. Harrod) (Balkema, Rotterdam, 1989) pp. 367–379.

LUNDGREN, P. R. and OKAL, E. A. (1988), *Slab Decoupling in the Tonga Arc: The June 22, 1977 Earthquake*, J. Geophys. Res. *93*, 13,355–13,366.

MANSINHA, L. and SMYLIE, D. E. (1971), *The Displacement Field of Inclined Faults*, Bull. Seismol. Soc. Am. *61*, 1433–1440.

MILNE, J., *Earthquakes and Other Earth Movements*, 376 pp. (Paul, Trench, Trübner & Co., London, 1898).

MONTESSUS DE BALLORE, F., *La Science Séismologique* 579 pp. (A. Colin, Paris, 1907).

OKADA, Y. (1985), *Surface Deformation Due to Shear and Tensile Faults in a Half-space*, Bull. Seismol. Soc. Am. *75*, 1135–1154.

OKAL, E. A. (1982a), *Mode-wave Equivalence and Other Asymptotic Problems in Tsunami Theory*, Phys. Earth Planet. Inter. *30*, 1–11.

OKAL, E. A. (1982b), *Higher Moment Excitation of Normal Modes and Surface Waves*, J. Phys. Earth *30*, 1–31.

OKAL, E. A. (1988), *Seismic Parameters Controlling Far-field Tsunami Amplitudes: A Review*, Natural Hazards *1*, 67–96.

OKAL, E. A. (1990), *Single Forces and Double-couples: A Theoretical Review of their Relative Efficiency for the Excitation of Seismic and Tsunami Waves*, J. Phys. Earth *38*, 445–474.

OKAL, E. A. (1991), *Erratum (to "Seismic Parameters Controlling Far-field Tsunami Amplitudes: A Review")*, Natural Hazards *4*, 433.

OKAL, E. A. (1992), *Use of the Mantle Magnitude M_m for the Reassessment of the Seismic Moment of Historical Earthquakes. I: Shallow Events*, Pure Appl. Geophys. *139*, 17–57.

OKAL, E A. and LÓPEZ, A.M. (2002), *New Seismological Results on the 1946 Aleutian Earthquake*, EOS, Trans. Am. Geophys. Union *83*(22), WP54, (abstract).

OKAL, E. A. and SYNOLAKIS, C. E. (2003), *Theoretical Comparison of Tsunamis from Dislocations and Landslides*, Pure Appl. Geophys. *160*, 2177–2188.

OKAL, E. A. and TALANDIER, J. (1989), *M_m: A Variable Period Mantle Magnitude*, J. Geophys. Res. *94*, 4169–4193.

OKAL, E. A. and TALANDIER, J. (1991), *Single-station Estimates of the Seismic Moment of the 1960 Chilean and 1964 Alaskan Earthquakes, Using the Mantle Magnitude M_m*, Pure Appl. Geophys. *136*, 103–126.

OKAL, E. A., SYNOLAKIS, C. E., FRYER, G .J., HEINRICH, P., BORRERO, J. C., RUSCHER, C., ARCAS, D., GUILLE, G. and ROUSSEAU, D. (2002), *A Field Survey of the 1946 Aleutian Tsunami in the Far Field*, Seismol. Res. Letts. *73*, 490–503.

PIPER, D. J. W., SHOR, A. N., FARRE, J. A., O'CONNELL, S. and JACOBI, R. (1985), *Sediment Slides and Turbidity Currents on the Laurentian Fan: Sidescan Sonar Investigations near the Epicenter of the 1929 Grand Banks Earthquake*, Geology *13*, 538–541.

PIPER, D. J. W., SHOR, A. N. and HUGHES CLARKE, J. E. (1988), *The 1929 "Grand Banks" Earthquake, Slump and Turbidity Current*, Geol. Soc. Am. Sp. Paper *229*, 77–92.

PLAFKER, G., OKAL, E. A. and SYNOLAKIS, C. E. (2002), *A New Survey of the 1946 Aleutian Tsunami in the Near Field: Evidence for a Large Underwater Landslide at Davidson Bank*, Seismol. Soc. Am. *73*, 259 (abstract).

POD"YAPOL'SKII, G. S., *Generation of the Tsunami Wave by the Earthquake*. In *Tsunamis in the Pacific Ocean* (ed. W. M. Adams) (East-West Center Press, Honolulu, 1970) pp. 19–32.

POLET, Y. and KANAMORI, H. (2000), *Shallow Subduction Zone Earthquakes and their Tsunamigenic Potential*, Geophys. J. Intl. *142*, 684–702.

SAITO, M. (1967), *Excitation of Free Oscillations and Surface Waves by a Point Source in a Vertically Heterogeneous Earth*, J. Geophys. Res. *72*, 3689–3699.

SWEET, S. and SILVER, E. A. (2003), *Tectonics and Slumping in the Source Region of the 1998 Papua New Guinea Tsunami from Seismic Reflection Images*, Pure Appl. Geophys. *160*, 1945–1968.

SYNOLAKIS, C. E., BARDET, J. -P., BORRERO, J.C., DAVIES, H.L., OKAL, E.A., SILVER, E.A., SWEET, S. and TAPPIN, D.R. (2002), *The Slump Origin of the 1998 Papua New Guinea Tsunami*, Proc. Roy. Soc. (London), Ser. A *458*, 763–789.

TALANDIER, J. and OKAL, E. A. (1979), *Human Perception of T Waves: The June 22, 1977 Tonga Earthquake Felt on Tahiti*, Bull. Seismol. Soc. Am. *69*, 1475–1486.

TANIOKA, Y., RUFF, L. J. and SATAKE, K. (1997), *What Controls the Lateral Variation of Large Earthquake Occurrence along the Japan Trench?*, The Island Arc *6*, 261–266.

VVEDENSKAYA (1956), *Opredelenie polej smeshchenij pri zemletryasenniyakh s pomoshch'yu teorii dislokatsij*, Izv. Akad. Nauk SSSR, Ser. Geofiz. *6*, 277–284 (in Russian).

WARD, S. N. (1980), *Relationship of Tsunami Generation and an Earthquake Source*, J. Phys. Earth *28*, 441–474.

WARD, S. N. (1981), *On Tsunami Nucleation: I. A Point Source*, J. Geophys. Res. *86*, 7895–7900.

WARD, S. N. (1982), *On Tsunami Nucleation: II. An Instantaneous Modulated Line Source*, Phys. Earth Planet. Inter. *27*, 273–285.

GPSR Compliance

The European Union's (EU) General Product Safety Regulation (GPSR) is a set of rules that requires consumer products to be safe and our obligations to ensure this.

If you have any concerns about our products, you can contact us on ProductSafety@springernature.com

In case Publisher is established outside the EU, the EU authorized representative is:

Springer Nature Customer Service Center GmbH
Europaplatz 3
69115 Heidelberg, Germany

Batch number: 10726322

Printed by Printforce, the Netherlands